Developmental Biology

A COMPREHENSIVE SYNTHESIS

Volume 4

Manipulation of Mammalian Development

Developmental Biology
A COMPREHENSIVE SYNTHESIS

Volume 1 OOGENESIS
Edited by Leon W. Browder

Volume 2 THE CELLULAR BASIS OF MORPHOGENESIS
Edited by Leon W. Browder

Volume 3 THE CELL SURFACE IN DEVELOPMENT AND CANCER
Edited by Malcolm S. Steinberg

Volume 4 MANIPULATION OF MAMMALIAN DEVELOPMENT
Edited by Ralph B. L. Gwatkin

Developmental Biology

A COMPREHENSIVE SYNTHESIS

Volume 4

Manipulation of Mammalian Development

Edited by
RALPH B. L. GWATKIN
The Cleveland Clinic Foundation
Cleveland, Ohio

PLENUM PRESS • NEW YORK AND LONDON

Library of Congress Cataloging in Publication Data

(Revised for vols. 3 & 4)

Developmental biology.

Includes bibliographies and index.
Contents: v. 1. Oogenesis— —v. 3. The cell surface in development and cancer—v. 4. Manipulation of mammalian development.
1. Developmental biology—Collected works. I. Browder, Leon W.
QH491.D426 1985 574.3 85-3406
ISBN-13:978-1-4612-9265-4 e-ISBN-13:978-1-4613-2143-9
DOI: 10.1007/978-1-4613-2143-9

Cover illustration: Scanning electron micrograph of the rearranged blastomeres from a 16-cell mouse embryo after separation with calcium-free medium. Cell fate is determined by asymmetry of the new cell contacts, not by the original position of the blastomeres in the embryo. (From Chapter 9 by Martin H. Johnson.)

Softcover reprint of the hardcover 1st edition 1986

A Division of Plenum Publishing Corporation
233 Spring Street, New York, N.Y. 10013

To my wife, Selma,
and to my children, Sharon, Nadine, and David

Contributors

Carole L. Banka Department of Reproductive Medicine, University of California—San Diego, La Jolla, California 92093

Barry D. Bavister Wisconsin Regional Primate Research Center and Department of Veterinary Science, University of Wisconsin—Madison, Madison, Wisconsin 53715

Patricia G. Calarco Department of Anatomy, University of California, San Francisco, California 94143

Katherine Gordon Integrated Genetics, Framingham, Massachusetts 01701

Bela J. Gulyas Pregnancy Research Branch, National Institute of Child Health and Human Development, National Institutes of Health, Bethesda, Maryland 20205

W. C. D. Hare Agriculture Canada, Animal Diseases Research Institute, Nepean, Ontario, Canada K2H 8P9

Susan Heyner Philadelphia College of Pharmacy and Science, Philadelphia, Pennsylvania 19104. *Present address:* Department of Obstetrics and Gynecology, Albert Einstein Medical Center, Northern Division, Philadelphia, Pennsylvania 19141

Martin H. Johnson Department of Anatomy, University of Cambridge, Cambridge CB2 3DY, England

Edward J. Kollar Department of Oral Biology, School of Dental Medicine, The University of Connecticut Health Center, Farmington, Connecticut 06032

Klaus Kratochwil Institute of Molecular Biology, Austrian Academy of Sciences, A-5020 Salzburg, Austria

Frank B. Kuzan Animal Reproduction Laboratory, Colorado State University, Fort Collins, Colorado 80523. *Present address:* Department of Obstetrics and Gynecology, University of Washington, Seattle, Washington 98195

James McGrath The Wistar Institute, Philadelphia, Pennsylvania 19104

Frank H. Ruddle Department of Biology, Yale University, New Haven, Connecticut 06511

George E. Seidel, Jr. Animal Reproduction Laboratory, Colorado State University, Fort Collins, Colorado 80523

Davor Solter The Wistar Institute, Philadelphia, Pennsylvania 19104

Alan Trounson Centre for Early Human Development, Monash University and Queen Victoria Medical Centre, Melbourne, Australia 3000

Ian Wilmut AFRC Animal Breeding Research Organization, Dryden Laboratory, Roslin, Midlothian EH25 9PS, United Kingdom

Preface

Developmental biology has been transformed recently by discoveries in the fields of molecular biology, cell biology, and immunology. New ways of manipulating mammalian development are uncovering control mechanisms and enabling us to apply them in solving practical problems in animal production and human health. This book outlines some of these new manipulations and how they have contributed to the present state of developmental biology.

Chapter 1 describes gene transfer by microinjection of cloned recombinant DNA into zygotes. Although the factors that affect transformation frequencies and integration sites are still unknown, such techniques offer a number of exciting prospects. Research models for human disease could be artificially created and desirable characteristics in agricultural animals could be enhanced.

The theme of cell-to-cell transfer is continued in Chapters 2 and 3. Chapter 2 describes pronuclear transplantation by Sendai virus-induced fusion of the karyoplast with the enucleated embryo. Using this procedure, it has been demonstrated that both male and female genomes are essential for normal development, although the reason for this is not yet understood. Chapter 3 describes studies on the fusion of whole oocytes.

Perhaps the greatest practical achievement of developmental biology to date is *in vitro* fertilization–embryo transfer (IVF–ET). Chapter 4 describes IVF–ET of laboratory and of farm animals. Although satisfactory culture conditions for cleavage and blastulation have not yet been achieved for most animals and complete development from fertilization to blastocyst has not been accomplished, the results have been impressive. Recently cynomolgus monkey oocytes were fertilized *in vitro* and then transferred to a rhesus foster mother. Birth of the former species from the latter illustrates the possibility of preserving endangered species by IVF-ET.

Turning to human IVF-ET (Chapter 5), there has been rapid progress, although community and legal restrictions severely limit experimentation. Fortunately the requirements for human IVF–ET have turned out to be less stringent than those for many animals. Nevertheless, 70–80% of patients with transferred embryos do not become pregnant, and we do not know the reason.

Ways of selecting fully competent gametes and more optimal culture conditions will probably overcome these problems eventually. It should be possible to sort out these factors with nonhuman primate or even rodent models.

It looks as if we are on the threshold of a revolution in animal breeding with the development of techniques for sex selection. To date these have involved the removal of a few cells from the blastocyst for chromosome analysis. However, as this procedure is inefficient and expensive, commercial prospects are dim. It should give way shortly to sexing of sperm, for which methods are being developed in laboratory animals. Both pre- and postfertilization sexing are reviewed in Chapter 6.

Future developments in sexing and in IVF–ET require successful gamete and embryo freezing. The latter is summarized in Chapter 7. The techniques described permit the intercontinental shipping of embryos in place of adult animals. This approach reduces the chances of spreading disease, since the transferred embryos are protected by passively acquired immunity from the adoptive mother. Furthermore, early embryos lack the specific tissues needed for many pathogens to multiply. Frozen stocks of the embryos of mutant laboratory animals also provide an efficient and inexpensive means of maintaining strains without the genetic drift that would occur with serial matings. The transfer of bovine embryos from prize cows, each worth thousands of dollars, has now become a significant industry; techniques have also been developed for splitting the embryos to further increase reproductive capacity (Chapter 8).

In addition to these practical applications, our ability to dissociate embryos enzymatically and then to reassemble their component cells and tissues is being utilized to analyze the basic processes of development. Formation of the blastocyst has been shown to involve the polarization of cells by their physical localization on the outside of the early embryo. These cells form tight junctions with one another that seal off the inner cells so that they remain unpolarized and develop into the inner cell mass. This in turn triggers secretion of the blastocoel fluid by the outer cells, now called the trophectoderm (Chapter 9). At a later stage of development, the inductive action of the mesenchyme on the epidermis can be analyzed by the now classic techniques of separation and recombination of these tissues both *in vivo* and *in vitro*. Such experiments are described in Chapters 10 and 11.

The final two chapters cover the use of antibodies, both polyclonal and monoclonal, to track the appearance and disappearance of specific antigens during development and to block developmental processes, such as compaction, so that the antigens responsible for these processes can be identified.

The history of science records that progress is critically dependent on the development of new techniques by which the laws governing phenomena can be elucidated. This volume shows how true this is of developmental biology. The manipulation of development by these techniques has only just begun. Much remains to be learned, and the future looks promising.

Ralph B. L. Gwatkin

Contents

Chapter 11 • Tissue Combination and Organ Culture Studies in the Development of the Embryonic Mammary Gland

Klaus Kratochwil

Chapter 12 • Immunogenetic Approaches to the Analysis of Mammalian Development

Susan Heyner

Chapter 13 • The Immunological Approach to the Study of Preimplantation Mammalian Development

Carole L. Banka and Patricia G. Calarco

Chapter 1

Gene Transfer into Mouse Embryos

KATHERINE GORDON and FRANK H. RUDDLE

1. Introduction

Recombinant DNA technology has made it possible to isolate and clone many of the genes of higher eukaryotes, permitting analysis of the structure of individual genes at the nucleotide level of resolution. Using somatic cell genetic methodologies, the linkage relationships of cloned genes can be determined at both intermediate and gross levels of resolution (Ruddle, 1981). While these methodologies provide a means for analyzing structure in great detail, they are not amenable to an analysis of function—nor do they by themselves provide a way to work out structure–function relationships. This can only be done by gene transfer systems in which the cloned genes of interest are introduced by transfection into somatic cell populations *in vitro*. Although such transfection systems have provided a wealth of information regarding gene structure–function relationships in the regulation of gene expression, it becomes necessary to extend gene transfer methodologies to the intact embryo in order to analyze the epigenetic control of gene function in the context of normal development.

Using the system of pronuclear injection, it has been demonstrated by many groups that transfected genes can be incorporated into the genome and can be transmitted sexually through numerous generations. It has also been shown that transfected genes are expressed and in many instances, exhibit tissue-specific gene regulation. Within just the last five years, mammalian embryo gene transfer has emerged as a spectacular technology capable of generating embryonic lethal mutations, delineating *cis*-acting regulatory sequences in the context of a developing interactive system and targeting oncogenesis to particular tissues.

This chapter summarizes the transgenic mouse system in terms of methodologies and principal research findings. It also outlines the possible future use of embryo gene transfer as a system for the study of mammalian development and as a means of establishing animal analogues of human genetic diseases.

KATHERINE GORDON • Integrated Genetics, Framingham, Massachusetts 01701. FRANK H. RUDDLE • Department of Biology, Yale University, New Haven, Connecticut 06511.

2. Gene Transfer in Somatic Cells

Within the past decade, molecular cloning in conjunction with gene transfer techniques in somatic cells has provided the experimental tools with which to probe gene regulation at the molecular level. Procedures for the introduction of cloned genes into cells fall into several categories, including (1) gene transfer by calcium phosphate precipitation (Wigler *et al.*, 1977), (2) liposome-mediated gene transfer (Fraley and Papahadjopoulos, 1982), (3) protoplast fusion (Schaffner, 1980; Rassoulzadegan *et al.*, 1982), (4) virus-mediated gene transfer (Hamer and Leder, 1979; Mulligan *et al.*, 1979), (5) receptor-mediated endocytosis (Stavridis and Psallidopoulos, 1982), and (6) glass needle microinjection (Anderson *et al.*, 1980; Capecchi, 1980). Of these procedures, calcium phosphate precipitation and microinjection have been the most actively exploited systems for gene delivery into somatic cells (reviewed in detail by Scangos and Ruddle, 1981). The former technique relies on the capability of cells to engulf DNA delivered as a very fine calcium phosphate precipitate and to incorporate it into their genomes. (Gene transfer by this method works most efficiently in the presence of carrier DNA.) Two types of selection procedure are commonly used to isolate rare transformants from mass population of cells: (1) selectable genes, such as thymidine kinase, [selected with hypoxanthine/aminopterin/thymidine (HAT)], to rescue mutant phenotypes; and (2) dominant selectable markers, such as *eco gpt* and the bacterial neomycin gene, to isolate transformants directly from nonmutant cell lines. The ability of unlinked DNAs to be cotransferred along with the selectable gene has enabled virtually any cloned gene or even random portions of entire eukaryotic genomes to be easily and efficiently recovered among transformed cells.

Problems frequently arise in the interpretation of studies utilizing gene transfer in somatic cells as a model system for the study of gene regulation *in vivo*. DNA rearrangements and mutations frequently occur during the course of gene transfers (Calos *et al.*, 1983; Razzaque *et al.*, 1983), particularly those using carrier DNA. A second problem is that many differentiated cell types essential for the study of expression of particular genes are difficult to maintain in culture or are poor recipients for gene transfer. Foremost, however, is the drawback that gene transfer into a single cell type trivializes much of the subtle complexities of gene regulation in the organism as a whole.

Interactions among different cell types and processes leading to differentiation and morphogenesis can only be understood within the context of an animal rather than somatic cell model system. For this reason, many investigators have attempted to use gene transfer systems that lead to retention of exogenous sequences in all the cells of the adult organism by gene transfer into embryos. In this way, a piece of DNA can be simultaneously introduced into differentiated cells of many types at a single chromosomal locus. It then becomes possible for the expression of altered genes to be studied in the context of a normal cellular and developmental environment. Such genetically manipulated organisms are referred to as "transgenic". In the following pages, several techniques are discussed for the transformation of animals by introducing DNA into embryos. The potential usefulness of the mouse embryo transformation system in the production of animal models for human diseases, and in particular, the factors that effect gene expression are emphasized.

3. Mouse Chimeras and Teratocarcinomas

Mammalian transformation was accomplished using embryonic chimeras several years before the availability of recombinant DNA techniques. Experiments conducted more than 20 years ago demonstrated that embryonic cleavage cells of distinct genotypes could be aggregated, implanted into uteri of pseudomothers, and could contribute to the genomes of mouse hybrids containing cells derived from both donors (see review by Mintz, 1974). In later experiments it was demonstrated that teratocarcinoma or embryonal carcinoma (EC) cells maintained in tissue culture or passaged *in vivo* could be used in the formation of chimeras (Brinster, 1974; Mintz, 1974, 1979; Mintz and Illmensee, 1975; Papaioannou *et al.*, 1975; Papaioannou *et al.*, 1978; Stewart and Mintz, 1981; Stewart, 1982). The teratocarcinoma cells could be introduced by direct microinjection into the blastocoel cavity of the recipient embryo or by aggregation with preimplantation embryos, in some cases leading to transformation of the germ line (Mintz and Illmensee, 1975; Illmensee and Mintz, 1976; Cronmiller and Mintz, 1978; Stewart and Mintz, 1981; for review see Mintz, 1983). Hybrid cells derived from the fusion of human fibrosarcoma cells to mouse thymidine kinase (−) teratocarcinoma cells and hybrids between rat hepatoma hypoxanthine phosphoribosyltransferase HPRT (−) cells and mouse teratocarcinoma cells were also shown to be capable of contributing to the genetic constitution of chimeric mice (Illmensee *et al.*, 1978; Illmensee and Croce, 1979). Interestingly, while both types of hybrid and teratocarcinoma cells alone form tumors in nude mice, their malignant potential is not realized either in the developing embryo or in adult mice derived from fusions involving these cells.

Genetic selection of mutant teratocarcinoma cell lines has been accomplished by two procedures and, in one case, the contribution of these cells in viable chimeras has been demonstrated. One approach, used by Dewey *et al.* (1977), demonstrated that mutant EC cells deficient in HPRT could be directly selected with 6-thioguanine and used in the formation of mouse chimeras. A second approach used to produce mutant teratocarcinoma cell lines is DNA-mediated gene transfer (DMGT). Several groups have shown that teratocarcinoma cells are effective recipients in DMGT experiments (Dewey *et al.*, 1977; Watanabe *et al.*, 1978; Pellicer *et al.*, 1980; Wagner and Mintz, 1982), yet no laboratory to date has reported the generation of chimeric mice from teratocarcinoma cells modified by DNA gene transfer.

The exciting potential of this system lies in the ability to create genetic alterations at single loci in the mouse genome experimentally and to study the expression of these mutations *in vivo*. It should be possible to engineer mutant teratocarcinoma cell lines containing integrated copies of recombinant DNAs added by DMGT and to produce chimeric mice containing the gene of interest.

Major obstacles in the general applicability of these techniques are the low frequencies with which teratocarcinoma cells contribute to the formation of viable germ cells and frequent karyotype abnormalities arising during the cell culture of teratocarcinoma cells. Moreover, this procedure involves the "transfer" of a whole genome in addition to the gene of interest, making the results more difficult to interpret.

4. Viral Transformation of Mouse Embryos

Mouse embryos can be infected by several types of virus, including SV40 and Moloney murine leukemia virus (Mo-MuLV), yet neither of these viruses is able to replicate in the early embryos (Jaenisch *et al.*, 1975; Jaenisch and Berns, 1977; Harbers *et al.*, 1981). Since the presence of neither virus in embryos interferes to an appreciable extent with subsequent embryonic development, murine viruses have been introduced into embryos and viral gene expression studied in adults which developed from these embryos. When SV40 DNA was introduced into the blastocoel cavity of mouse embryos by microinjection and the embryos nurtured to term in pseudomothers, sequences homologous to SV40 DNA were discovered in several tissues of mice born from infected embryos (Jaenisch and Mintz, 1974). In one set of experiments, germ line transmission of SV40 sequences was not found (Jaenisch and Mintz, 1974), leading to the conclusion that the SV40 DNA had not integrated, although the patterns of integration were not directly analyzed by these investigators. In other experiments, most SV40 sequences derived from infected embryos were unintegrated DNA copies (see Kelly and Condamine, 1982; Willison *et al.*, 1983).

By contrast, transformation of mouse embryos by Mo-MuLV typically leads to integration into the host genome and transmission of the viral sequences to subsequent generations. Jaenisch's group has shown that Mo-MuLV can be introduced into adult mice by at least three routes: (1) direct infection of preimplantation embryos, (2) cocultivation of preimplantation embryos with virus-producing mouse cells, and (3) by injection of a virus-producing cell into the blastocyst cavity. Mice born from Mo-MuLV-infected preimplantation embryos tended to be somatic and germ cells mosaics, often containing multiple integration sites (Jähner and Jaenisch, 1980). After being mated and segregated in the second generation, lines of transformed true-breeding mice were established containing single copies of the viral DNA integrated at unique loci. The lines were called Mov-1–Mov-12, each representing a substrain of mice with single copies of Mo-MuLV at different chromosome sites (Jähner and Jaenisch, 1980; Jaenisch *et al.*, 1981). These mouse lines differ phenotypically from one another in the extent and timing of viral production. Among the mice that became viremic, most eventually developed acute thymus-dependent leukemia between 3 weeks and 2 months of age depending on the strain.

A strain of mouse termed Mov-13 was produced from injection of Mo-MuLV into mid-gestation embryos *in vivo* (Jaenisch, 1980). Transformation of mid-gestation embryos differs significantly from infection of preimplantation embryos in two respects: (1) the frequency of subsequent germ line transmission is very low, and (2) viral RNA is produced in extremely high amounts in virtually all tissues examined. By contrast, mice derived from infection of preimplantation embryos produce viral RNA only in tissues of the lymphatic lineage (spleen and thymus). The differences in organ trophism and overt levels of gene expression may be related to the preferential integration of viral DNA at distinct chromosomal sites during infection at different periods of embryogenesis. It is also possible that gene activity is repressed at preimplantation stages, as suggested by methylation studies (see Section 5.8).

Stuhlmann *et al* (1984) have used a defective retroviral vector for gene transfer into post-implantation mouse embryos. They injected cells from a defective virus-producing cell line which expressed murine sarcoma virus (MSV) and a selectable gene, *gpt*. The cells were coinjected with helper virus in order to facilitate spread of the virus. Mice were analyzed for expression of the injected sequences either at day-19 of gestation or as adults. Recombinant MSV-*gpt* virus was produced in several tissues of animals analyzed at either stage, including brain, kidney, lung, spleen and thymus, in about one-quarter of the animals that expressed viremia.

Experiments dealing with the transformation of preimplantation mouse embryos by Mo-MuLV suggest that expression of the introduced DNA sequences in adult mice is highly dependent on site of integration in the genome. Integration at certain sites, such as those present in Mov-4–Mov-3 mice, is not consistent with expression, whereas integration at other sites leads to the expression of viremia. Studies of DNA-mediated gene transfer in tissue culture cells have shown that integration can occur at many loci (Robins *et al.*, 1981). Levels of expression of the introduced sequences may be dependent on the sites of integration. Cloned recombinant DNA introduced into mouse embryos by microinjection can also integrate at many different sites in the genome. Possible correlations between integration sites and levels of gene expression in transgenic mice are discussed in detail in a later section.

Two substrains of mice generated by infection of embryos with Mo-MuLV which failed to become viremic showed evidence of gene rearrangement at or near the site of integration. Gene rearrangements also have been documented in transgenic mice and tissue cells modified by gene transfer as discussed earlier. These and other factors likely to affect the levels of gene expression in embryo transformation experiments are discussed in Section 5.8.

5. Transformation of Mouse Embryos with Recombinant DNA

The fact that foreign DNA sequences can be introduced into the mammalian genome by gene transfer into embryos raises the intriguing possibility of being able to design and engineer alterations in the normal genetic constitution of a mouse. Particularly exciting is the availability of cloned probes for microinjection that code for genes that have been associated with genetic diseases. The β-globin gene is one example in which mutations which alter gene expression are associated with genetic diseases such as sickle cell anemia and β-thalassemia. Techniques for the introduction of exogenous DNA sequences into mouse zygotes were pioneered by Jaenisch and Mintz (1974), who microinjected whole viral genomes into the blastocoel cavities of early embryos. Procedures have recently been developed for gene transfer into *Xenopus* (Rusconi and Schaffner, 1981; Etkin and Roberts, 1983) and *Drosophila* embryos (Rubin and Spradling, 1982; Spradling and Rubin, 1982) but unfortunately cannot be discussed here in detail.

5.1. Methods

The experimental procedures by which cloned DNA molecules are microinjected into fertilized mouse eggs, the eggs implanted into foster mothers for the gestation period, and the resultant progeny analyzed for the retention of the introduced sequences were recently reviewed in detail by Gordon and Ruddle (1984). Only the salient features of the methodology and particular variations introduced by different laboratory groups are discussed here.

The procedure followed by our laboratory is outlined below. Donor mice are superovulated with injections of pregnant mare serum (PMS) and human chorionic gonadotropin (hCG) in order to enhance the production of large numbers of eggs for microinjection. On the day before microinjection, the superovulated donor mice are mated to fertile males and randomly cycling adult female mice are mated to vasectomized males. The pseudopregnant mice become hormonally receptive as foster mothers for embryo transfer. One-cell fertilized eggs are recovered from the donor mice, cumulus cells are removed by incubation in hyaluronidase, and the embryos are extensively washed in embryo-culturing medium (Gordon and Ruddle, 1984). The eggs are placed into a drop of culture medium under oil on a tissue culture dish and are held in place with gentle suction from a holding pipette. The microinjection needles are filled with DNA solution, then with Fluorinert, which is used as a hydraulic fluid for operation of both the holding pipette and injection needle. Both are attached to syringe micrometers. By gently adjusting micromanipulators, the needle is inserted into one of the pronuclei and an advance on one of the micrometers expels DNA solution into the pronucleus. A slight swelling is evident, indicating that approximately 1 picoliter (pl) of solution has been delivered to the egg. The eggs surviving the microinjection procedure are incubated for about 1 hr and then surgically implanted into the oviducts of foster mothers. After a normal gestation period, the offspring can be analyzed for integration of the introduced sequences.

We use randomly bred CD-1 females (Charles River Laboratories) as foster mothers, and the donor eggs are obtained from CD-1 females × B6D2F1 males (Jackson Laboratory). Other groups use different strains of mice as donors for microinjection: C57BL/6 × CBA/H (Costantini and Lacy, 1981); C57 × SJL (Brinster *et al.*, 1981); C3H × C57BL6 (E. F. Wagner *et al.*, 1981); C3H × DBA/2 (T. A. Stewart *et al.*, 1982); and C57BL/6j × LT/SV (T. E. Wagner *et al.*, 1981).

The possibility that the recipient hybrid strain of mouse used for microinjection has any bearing on the frequency of transformation or expression of injected sequences has not been systematically studied. Some investigators have used inbred strains of mouse as recipients (Bürki and Ullrich, 1982; Brinster *et al.*, 1985). The latter authors have compared the efficiency of gene transfer into inbred C57 eggs with transfer into C57 × SJL embryos. On many different levels, they found the inbred mice to be less efficient as recipients. Less eggs were obtained per donor, injected eggs had a lower viability, and the frequency of integration among survivors was markedly reduced. Yet positive mice did express the exogenous sequences. These authors estimate that there is approximately an overall decrease in efficiency of 10-fold using inbred versus hybrid strains. One group has used a genetic mutant (lit/lit) as the recipient for gene transfer with the normal growth hormone gene (Hammer *et al.*, 1984) and has succeeded in compensating for the genetic defect. Most groups of investigators implant the microinjected embryos after a brief period of incubation, but at least two laboratories (Bürki and Ullrich, 1982; T. E. Wagner *et al.*, 1981) culture the embryos for several days to the morula/blastocyst stage prior to implantation into the uteri of foster mothers.

Brinster *et al.* (1985) have carefully optimized several of the parameters critical in microinjection and integration of foreign genes. The buffer composition does appear to have an effect in that high concentrations of magnesium or EDTA minimize survival frequencies. Integration frequencies were highest in 1.0mM EDTA (58%) but the overall efficiency was reduced when considering the fraction of eggs surviving the manipulations (9.4%). Total overall efficiencies were highest in the presence of 0.1 mM EDTA. There was a slightly better frequency of integration when sequences were introduced into the male (17%) versus female (13%) pronucleus. In addition, these investigators attempted to inject both nuclei of 2-cell embryos with low success rates (14%). Injection of DNAs into the cytoplasm gave virtually no integrants. They also compared different molecular forms and concentrations of DNA used for injection, and these are discussed in following sections.

5.2. Gene Systems

Stable transformation achieved by microinjection of cloned DNA into fertilized mouse eggs and subsequent development of the eggs *in utero* has been successfully demonstrated with many different genes (summarized in Table I). Several groups have reported the microinjection and transformation of mice with the thymidine kinase (Tk) gene from Herpes Simplex Virus (HSV). In one case, the HSV-Tk gene was cloned into a recombinant plasmid containing a portion of the viral genome from SV40 (Gordon *et al.*, 1980). A second group injected a construction in which the Tk gene was cloned into a plasmid containing the human β-globin gene (E. F. Wagner *et al.*, 1981), while a third group microinjected plasmid molecules in which the coding portion of the HSV-Tk gene had been fused to the pro-

Table I. Mouse Embryo Transformation Using Microinjected Cloned DNA

Genes	Source(s)	Copies injected[a]	Transformation frequency[b]	References
Alpha-fetoprotein	Mouse	260	5/19 (26%)	Krumlauf *et al.* (1985)
Elastase-1	Rat	200	7/37 (19%)	Swift *et al.* (1984)
β-galactosidase/ globin promoter	*E. coli* Mouse	N.S.[h]	2/12 (17%)	Humphries *et al.* (1985)
β-globin	Rabbit	20,000	N.S.	T. E. Wagner *et al.* (1981)
β-globin	Rabbit	50–20000	9/24 (38%)	Costantini and Lacy (1981)
α+β-globin	Mouse/human	N.S.	N.S.	Chada *et al.* (1985)
β-globin	Human	200	1/6	Humphries *et al.* (1985)
GH[c]	Human	N.S.	N.S.	Kuschner and Martin (1983)
GH	Rat/human	2000	N.S. (15%)	Selden *et al.* (1983)
GH/elastase-1 promoter	Human/rat	N.S.	N.S.	Ornitz *et al.* (1985)
GH/MT[d] promoter	Rat/mouse	600	7/21 (33%)	Palmiter *et al.* (1982*a*)
GH/MT promoter	Human/mouse	1000	33/101 (33%)	Palmiter *et al.* (1983)
GH/MT promoter	Rat/mouse	1000	7/41 (6%)	Hammer *et al.* (1984)
GH	Human	6000	6/20 (30%)	E. F. Wagner *et al.* (1983)
GH releasing factor/MT promoter	Human/mouse	1600	14/59 (24%)	Hammer *et al.* (1985*a*)
Immunoglobulin (κ light chain)	Mouse	440	6/11 (54%)	Brinster *et al.* (1983)
Immunoglobulin (μ heavy chain)	Mouse	50	5/13 (38%)	Grosschedl *et al.* (1984)
Immunoglobulin (μ and κ chains)	Mouse	N.S.	1/5 (20%)	Rusconi and Köhler (1985)
Insulin	Human	30000[i]–50000	2/60 (3%)	Bürki and Ullrich (1982)
Interferon	Human	N.S.	N.S.	Gordon and Ruddle (1983)
Interferon cDNA	Human	1000–35000	1/10 (10%)	Gordon and Ruddle (1981)
MHC[e]-1 gene	Porcine	500	1/83	Frels *et al.* (1985)
MHC-2 gene	Mouse	N.S.	2/37 (5%)	LeMeur *et al.* (1985)
MHC-2 gene	Mouse	200	3/30 (10%)	Yamamura *et al.* (1985)
Myc/MMTV[f] promoter	Mouse/mouse	500	N.S.	T. A. Stewart *et al.* (1984)
Myosin light-chain	Rat	500	4/26 (15%)	Shani (1985)
Somatostatin/MT promoter	Rat	Several hundred	11/43 (26%)	Low *et al.* (1985)

Table I. *(Continued)*

Genes	Source(s)	Copies injected[a]	Transformation frequency[b]	References
SV-40/insulin 2 promoter	SV40/rat	N.S.	N.S.	Hanahan (1985)
SV-40-Tk/MT promoter	SV40-HSV/ mouse	240	15/64 (23%)	Brinster *et al.* (1984)
SV-40-GH/MT promoter	SV40-human/ mouse	830	10/31 (32%)	Brinster *et al.* (1984)
SV-40 coinjected with myc	SV40/MC29 virus	15000	1/13 (7%)	Small *et al.* (1985)
Tk[g]/β-globin	HSV/human	N.S.	5/33 (15%)	E. F. Wagner *et al.* (1981)
Tk/β-globin	HSV/human	2500	2/62 (3%)	T. A. Stewart *et al.* (1982)
Tk/MT promoter	HSV/mouse	200	7/41 (20%)	Brinster *et al.* (1981)
Tk/SV-40 origin	HSV/SV-40	1000–30000	3/78 (4%)	Gordon *et al.* (1980) Gordon and Ruddle (1982)
Transferrin	Chicken	140–500	11/44 (25%)	McKnight *et al.* (1983)

[a]Number of plasmid copies injected per nucleus.
[b]Number of positive mice/number of mice analyzed.
[c]GH, growth hormone.
[d]MT, metallothionine.
[e]MHC, major histocompatibility complex.
[f]MMTV, mouse mammary tumor virus.
[g]Tk, thymidine kinase.
[h]N.S., not stated or not studied.
[i]Injections were nuclear and cytoplasmic.

moter of the metallothionine-1 gene from mouse (Brinster *et al.*, 1981; Palmiter *et al.*, 1982*b*). The latter recombinant construct directed the synthesis of biologically active HSV in response to induction with cadmium. Genetic transformation of mice has also been achieved with the growth hormone genes from rat (Palmiter *et al.*, 1982*a*), human (Kuschner and Martin, 1983; E. F. Wagner *et al.*, 1983), and human/rat fusion recombinant genes (Selden *et al.*, 1983). The growth hormone structural gene has been introduced into embryos in constructs preceeded by the promoter from the mouse metallothionine gene (Palmiter *et al.*, 1982b; Hammer *et al.*, 1984) or the rat elastase-1 promoter (Ornitz *et al.*, 1985). In addition, Hammer *et al.* (1985a) have introduced fusion genes into mice which contain the gene for growth hormone releasing factor under the control of the metallothionine promoter.

Complementary DNA (cDNA) coding for human α-interferon (Gordon and Ruddle, 1981), an α-interferon gene (Gordon and Ruddle, 1983), a human insulin gene (Bürki and Ullrich, 1982), a chicken transferrin gene (McKnight *et al.*, 1983), an elastase-1 gene (Swift *et al.*, 1984), an alpha-fetoprotein gene (Krumlauf *et al.*, 1985), the myosin light-chain gene (Shani, 1985) and the somatostatin gene (Low *et al.*, 1985) have also been used for mammalian transformation experiments.

Two groups have introduced the rabbit β-globin gene into mouse embryos (Costantini and Lacy, 1981; T. E. Wagner *et al.*, 1981); one group has introduced a mouse-human fusion β-globin gene (Chada *et al.*, 1985), and another a human globin gene cluster (Humphries *et al.*, 1985). Three groups have introduced and studied expression of immunoglobulin genes (Brinster *et al.*, 1983; Grosschedl *et al.*, 1984; Rusconi and Köhler, 1985). Several groups have been studying transgenic mice with introduced major histocompatibility genes of both class 1 (Frels *et al.*, 1985) and class 2 (LeMeur *et al.*, 1985; Yamamura *et al.*, 1985). In addition, several groups have recently introduced oncogenic genes into mouse embryos (Brinster *et al.*, 1984; T. A. Stewart *et al.*, 1984; Hanahan, 1985; Small *et al.*, 1985). The use of tissue-specific promoters in these constructs, such as MMTV (T. A. Stewart *et al.*, 1984) or the insulin promoter (Hanahan, 1985), has targeted gene expression and oncogenesis to particular tissues and promises to be an invaluable research tool in studies of the genesis of particular neoplasms.

The majority of these introduced sequences have led to germ line transmission to successive generations and expression of the newly introduced DNA. The frequencies of transformation vary considerably in these reports—from 1% to a high of 55%. The factors which may influence this broad range of transformation frequency are discussed in the following section.

5.3. Transformation Frequencies

Table I shows that high transformation frequencies have now been achieved with a variety of genes and by many different groups. The highest frequencies of gene transfer into embryos obtained by microinjection are 55% (6 positive mice of 11 analyzed), reported by Brinster *et al.* (1983) and 38% (9 positive mice of 24 analyzed), achieved by Costantini and Lacy (1981) and Grosschedl *et al.* (1984) (see Table I). For comparison, the frequency of transformation by microinjection in tissue culture cells ranges from about 5–20% (Anderson *et al.*, 1980; Capecchi, 1980; Yamaizumi *et al.*, 1983).

In several studies high transformation frequencies were obtained using restriction fragments with noncohesive ends (for example Swift *et al.*, 1984; Hammer *et al.*, 1985*a*; Low *et al.*, 1985). Other investigators have achieved significant frequencies using either intact plasmids (Shani, 1985) or linearized molecules (Brinster *et al.*, 1984; Grosschedl *et al.*, 1984). In a study by Brinster *et al.* (1985) transformation frequencies were compared using a plasmid introduced in different molecular forms. Supercoiled and blunt-ended linear molecules gave rise to transformation frequencies of 7.7% and 8.5% respectively. Linears with similar ends produced transgenic mice at 24% frequency and linear molecules with different ends at frequencies of 31%. This large discrepancy between supercoiled and linear molecules (excluding blunt-ended molecules) was reproduced with many plasmid constructs and appears to have no relationship to the size of the molecules introduced. On the whole, linear molecules appear to integrate about fivefold more efficiently than circular molecules. Interestingly, the presence of head-to-tail concatomeric sequences in the DNA of most mice transformed by linear DNA from a variety of labs indicates that the linear fragments may recircularize and amplify prior to integration or may have been involved in gene amplification following integration at an early stage of embryogenesis (see Palmiter *et al.*, 1982*b*; and Brinster *et al.*, 1985). Brinster *et al.* (1985) have analyzed the state of injected DNAs 24 hours after introduction into pronuclei. When either blunt ended linears or linears with heterologous ends were injected, a large fraction of the molecules migrated at the position of supercoiled molecules after 24 hours. In view of this, the fact that linear molecules integrate with a greater efficiency than supercoils is an enigma. In some as yet poorly understood way, the free ends act to stimulate recombination, perhaps by activating ligation processes. In any event, the egg does appear to possess and utilize ligation machinery.

Molecular integrity may also be important in promoting a high rate of recombination after the introduction of restriction fragments, plasmids or bacteriophage into embryos. A possible mechanism may involve the formation of large stable intracellular concatomers. By analogy, exogenous DNAs introduced into *Xenopus* embryos (Rusconi and Schaffner, 1981) and tissue culture cells (Pellicer *et al.*, 1978) become degraded following gene transfer and tend to be stabilized by the formation of molecular aggregates (Willecke *et al.*, 1979).

As can be seen from Table I, the number of copies of DNA injected per nucleus has been varied in different experiments from 50 (Grosschedl *et al.*, 1984) to tens of thousands (Gordon and Ruddle, 1981; T. E. Wagner *et al.*, 1981; Small *et al.*, 1985). Brinster *et al.* (1985) have systematically examined the integration efficiency and overall frequency of gene transfer (including embryo viability) as functions of copy number injected. The number of copies of DNA introduced per pronucleus was varied between 5 and 24,000. Though the latter concentration gave the highest efficiency of integration (31%), the best overall efficiency was achieved with 530 molecules of DNA injected per pronucleus.

Another factor to consider in the analysis of transformation frequencies is the coding portion of the gene itself. There is no reason to suspect *a priori* that all genes are capable of recombination with the embryonic genome to equal extents. Genes may differ from one another in secondary conformation, in ability to associate with endogenous proteins, and in inherent stability, any of which may affect the ability of a particular gene to recombine with the embryonic genome. Brinster *et al.* (1982) have shown that the pMK plasmid is capable of undergoing transient expression in mouse embryos after microinjection; most likely other plasmids are similarly expressed in mouse eggs before integration. Since transient expression of the introduced sequences must be compatible with embryonic development, the number of viable transformants obtained using certain recombinant constructs may under-represent the actual number of recombination events in the microinjected mouse embryos. In some cases, selection for viable transformants may inadvertently select for integration at inactive chromosomal sites or constructs that fail to express transiently. Microinjection of genes whose products are normally expressed during embryogenesis (e.g., metallothionine and transferrin) may lead to a greater frequency of stable integration than genes that are never normally expressed during early development.

The number of copies of the transforming DNA sequences identified in the genomes of transgenic mice varies considerably from one copy (or less than one in the cases of mosaics) to several hundred gene copies. The presence of single copies of cloned DNA in transgenic mice often occurs in the same experiments which generated mice with tens or hundreds of gene copies. Typically the newly introduced DNA is arranged in multiple copies of head-to-tail concatomers. Multiple copies of microinjected sequences in tandem arrays may have been derived from recombination events prior to integration into the embryo genome or from gene duplications following integration. These possibilities have not been experimentally distinguished. Interestingly, all mice produced by infecting embryos with Mo-MuLV, rather than microinjecting recombinant clones, contained integrated proviral DNA at single copy levels.

5.4. Mosaicism

Most transgenic mice studied to date appear to be derived from single integration events at single loci which presumably represent integrations taking place at the one-cell stage. Yet there is ample evidence for mosaicism most likely due to rearrangements or integration events taking place later in development. Presumably, at least a fraction of the introduced plasmid sequences can be retained and stabilized without integration in the embryonic nucleus. There are several experimental indications of mosaicism, as discussed below; the number of mosaic transgenic mice represent approximately 15% of the total, but this may be an underestimate since many investigators have not reported frequencies of mosaicism among their transgenic mice. Experimental indications of mosaicism possibly arising from delayed integration events are (1) reduced frequency of genetic transmission to successive generations below Mendelian levels (Gordon and Ruddle, 1982; T. A. Stewart *et al.*, 1982; Kuschner and Martin, 1983; McKnight *et al.*, 1983; Grosschedl *et al.*, 1984; Palmiter *et al.*, 1984; Swift *et al.*, 1984; Hanahan, 1985; Low *et al.*, 1985); (2) different Southern blot patterns of hybridization in fetal and placental tissues (E. F. Wagner *et al.*, 1981; Bürki and Ullrich, 1982); (3) enhanced copy numbers of transgenic sequences in progeny mice compared with the transgenic parent (Costantini and Lacy, 1981; Palmiter *et al.*, 1984; Swift *et al.*, 1984; Hanahan, 1985). Foreign injected species may be present in only a fraction of the cells of the tissue being examined in the founder animals. This is apparent as less than one copy of the foreign sequence per cell (Wagner *et al.*, 1981; McKnight *et al.*, 1983; Krumlauf *et al.*, 1985; Rusconi and Köhler, 1985; Shani, 1985), but may lead to an increase in copy number in subsequent generations; (4) enhancement in the levels of expression of newly introduced genes in progeny mice (Lacy *et al.*, 1983; McKnight *et al.*, 1983; Ornitz *et al.*, 1985; Rusconi and Köhler, 1985);

and (5) quantitatively or qualitatively variable band patterns on Southern blot hybridizations of different tissues from the same mouse (T. A. Stewart *et al.*, 1982; Kuschner and Martin, 1983).

The failure of transforming DNA sequences to be transmitted through the germ line can also be explained by a lack of integration into the host genome. At least two groups of investigators have reported transgenic mice that retained extrachromosomal nonintegrated copies of injected DNA sequences (Gordon *et al.*, 1980; Gordon and Ruddle, 1982*b*; Selden *et al.*, 1983). The possibility that sequences are replicated and maintained as free elements during embryonic development is unlikely due to the experiments of Brinster *et al.* (1985) who showed that the amount of injected DNA does not change following a 24-hr incubation. Another explanation for the lack of integration is that the DNA molecules microinjected into the egg were preserved in an undegraded form and partitioned into cells of the developing mouse. A similar situation was described earlier in which infection of mouse embryos by SV40 led to the presence of extrachromosomal copies of viral DNA in some tissues of adult mice. Since mouse embryos are nonpermissive for SV40 viral DNA replication, an analogous explanation must be proposed to explain the retention of free DNA in the cells of adult mice.

5.5. Chromosomal Integration

Most transgenic mice contain multiple gene copies integrated at a single chromosomal loci. Yet, the integration of recombinant sequences at multiple sites has also been suggested from the data of several laboratories (Brinster *et al.*, 1981; Bürki and Ullrich, 1982; E. F. Wagner *et al.*, 1981, 1983; Brinster *et al.*, 1983; Lacy *et al.*, 1983; Stewart *et al.*, 1984; Low *et al.*, 1985; Swift *et al.*, 1985). In some cases, multiple integrations have been used to produce additional segregating lines of mouse upon breeding.

With the exceptions noted in Section 5.4, there are multiple compelling lines of evidence that recombinant DNA in transgenic mice becomes integrated into high molecular weight genomic DNA. The types of evidence for integration are as follows:

1. One approach is to look for the acquisition of novel restriction fragments compared to the sequence that was injected. Specifically, mouse DNA/vector DNA junction fragments on genomic Southern blots indicate integration. Ideally, two junction fragments should be evident per integration.
2. Germ line transmission of cloned DNA and inheritance of Southern blotting patterns also provide evidence for the integration of exogenous DNA into genomes of transgenic mice. The inheritance of expressing phenotypes among progeny of transgenic mice is a common observation which would also indicate stable germ line integration.

3. Genetic mapping is also conclusive evidence for integration (Lacy *et al.*, 1983; Woychik *et al.*, 1985).
4. Molecular cloning of junction fragments is a procedure that could be used to demonstrate integration into the chromosomes of transgenic mice.

Integration junction fragments have been cloned from transformants in tissue culture experiments and mice derived from Mo-MuLV infected embryos. In experiments with infection of mouse embryos by Mo-MuLV (see above), Harbers *et al.* (1981) cloned the Mov-3 locus into a bacterial plasmid and introduced the recombinant retroviral sequences back into recipient mouse embryos by microinjection. In one animal that was extensively analyzed, flanking DNA and vector sequences were lost as though a cDNA copy of the cloned Mov-3 locus had been inserted. The DNA base sequence information of this and other similar transgenic mouse junction sites may yield valuable information on the presence of any preferred sites for integration of DNA sequences and may provide data on the mechanism of recombination. This type of information may also eventually be essential for designing vector systems capable of targeting exogenous DNA sequences into specific chromosomal loci.

5.6. Integration Sites

All data collected to date suggest that integration of foreign DNA in tissue culture cells, Mo-MuLV in mouse embryos, and cloned DNA in transgenic mice, can occur at many different loci, possibly at random. The question as to whether direct homology between cloned exogenous DNA and genomes of the injected mouse embryos plays a role in the process of recombination, thereby governing sites of integration, has been addressed by several experiments. First, Southern blots on the DNA from pMK mice show no evidence of integration into the endogenous metallothionine locus. Second, C. Stewart *et al.* (1983) failed to find evidence of homologous recombination in experiments in which cloned proviral DNA from Mov-3 mice, including flanking sequences, was injected into mouse embryos. Third is the observation made by many laboratories that no two independent transgenic mice in a given experiment have been reported to share identical junction restriction fragments which would have been predicted by homologous integrations. Fourth, Lacy *et al.* (1983) analyzed the sites of integration of the foreign DNA in five mice which had been independently derived from the microinjection of a rabbit β-globin gene which is approximately 81% homologous to the β-globin gene of the mouse (van Ooyen *et al.*, 1979). *In situ* hybridization to prebanded metaphase chromosome spreads was performed on the material from these mice. Intense hybridization to single sites on chromosomes 1, 3, 12, and 14 were apparent in four of the mice, and a fifth mouse had two sites of hybridization—one on chromosome 3 and the other on chromosome 12 (distinct from subchromosomal regions of hybridization seen in the other mice). The endogenous β-globin locus maps to chromosome 7, and thus the patterns of integration are inconsistent with recombination strictly mediated by homology-dependent mechanisms.

Whether recombination events are totally random in these experiments or the sites of integration are established by other factors, such as local chromosomal topology, remains to be determined. It will be informative to examine in detail integration junction fragments by molecular cloning and sequencing in order to determine whether there are preferred sites for integration.

5.7. Rearrangements and Mutations

Several laboratories have described internal DNA deletions and rearrangements within the foreign DNA in transgenic mice (Gordon *et al.*, 1980; Gordon and Ruddle, 1981; E. F. Wagner *et al.*, 1981; T. A. Stewart *et al.*, 1982). The expression of rearranged sequences may be repressed in these mice due to deletions of structural or controlling DNA.

Another class of DNA rearrangements observed in transgenic mice involves the rearrangements of the host chromosome. A chromosomal translocation was identified involving the locus of integration of an interferon sequence introduced into the mouse egg by microinjection (J. Gordon *et al.*, manuscript submitted). In this case, a reciprocal translocation involving chromsomes 2 and 12 was seen, and the translocation was correlated with male sterility. The specific defect involves abnormal spermatogenesis.

Insertional mutations may be created which lead to the disruption of structural or regulatory DNA sequences as a consequence of gene integration. One estimate holds that as many as 10–20% of transgenic mice contain such recessive mutations (Palmiter and Brinster, 1985). Sequences may integrate at sites which are critical for embryonic development and the mutations only appear evident upon attempts to breed the mice to homozygosity (Jaenisch *et al.*, 1983; E. F. Wagner *et al.*, 1983). Litters obtained from appropriate heterozygote crosses appeared small in these experiments and excluded mice with homozygous genotypes due to embryo mortality. In experiments by R. Jaenisch's group, the integration of Mo-MuLV at the Mov-3 locus interrupted the structural gene coding for α1(I) collagen (Schnieke *et al.*, 1983). This recessive mutation was only lethal in homozygous form, causing embryonic death between days 12 and 13 of gestation (Jaenisch *et al.*, 1983).

In two other groups of experiments the detection of insertional mutations depended upon fortuitous integration of introduced DNAs into genes whose phenotypes were nonlethal and detectable. In experiments by Palmiter *et al.* (1984), attempts to breed a mouse line to homozygosity led to the observation of transmission distortion in which the line could only become transmitted by females. Though the male mice are fertile and their sperm contained the integrated DNA, these mice did not pass on the transgenic sequences to progeny.

These observations suggest that the integration interrupted the expression of a gene active during spermatogenesis.

In another series of experiments, P. Leder's group has produced mice which contain a limb deformity caused by insertional mutagenesis and again was revealed in crosses designed to breed homozygotes (Woychik *et al.*, 1985). This defect causes fusion of the bones of the lower limbs and feet. Interestingly, the insertion maps to a chromosomal locus close to the "limb deformity" locus which was described many years ago and produces mice with similar phenotypes. Insertional mutations will undoubtedly prove useful for the cloning of such genes and for the generation of mutations and identification of genes which are important in gametogenesis and embryogenesis.

5.8. Gene Expression

It is now abundantly clear from the data of many laboratories that DNA sequences introduced into transgenic mice by microinjection are most often expressed in appropriate target tissues. In most cases the specificity is dictated by the promoter sequences, but in other examples the product or phenotype is manifest in tissues apparently specified by the coding moiety of the introduced DNA. Since the chromosomal integration sites for transgenic sequences appear to be random in location, position effects add some variability to the levels and reproducibility of expression among different mice. Nonetheless, expression of exogenous sequences in the transgenic mouse system promises to be invaluable in the dissection of many complex biological problems. Some specific applications which will be exploited in future years are: (1) analysis of genetic diseases for which no mouse models are currently available, (2) analysis of biological phenomena by the introduction of random mutations, (3) study of sequences important in embryonic development, (4) molecular dissection of DNA sequences conferring *cis*-acting regulatory specificity, and (5) study of oncogenesis. The following pages will discuss these applications and experimental parameters in the transgenic mouse system which appear to be important in the regulation of introduced sequences.

5.8.1. Tissue-Specific Gene Expression

The metallothionine promoter has been used to direct transcription of several genes in transgenic mice, including herpes thymidine kinase (Brinster *et al.*, 1981), rat and human growth hormone (Palmiter *et al.*, 1982*a*, 1983), human growth hormone-releasing factor (Hammer *et al.*, 1985*a*), somatostatin (Low *et al.*, 1985), and constructs containing the SV40 large T antigen (Brinster *et al.*, 1984). These data are summarized in Table II. Fusion gene constructs directed by the metallothionine promoter are primarily expressed in liver tissue, the major site of endogenous metallothionine expression, with some exceptions as noted in the following sections. Some mice have also been shown

Table II. Expression of Introduced DNAs in Transgenic Mice

Gene	Predominant cell type	Predominant phenotype	Reference
Alpha-fetoprotein	yolk sac, fetal liver, fetal brain	N[f]	Krumlauf *et al.* (1985)
Elastase-1	pancreas	N	Swift *et al.* (1984)
β-globin	bone marrow, blood	N	Chada *et al.* (1985)
β-globin	testes, skeletal muscle	N	Lacy *et al.* (1983)
GH[a]/elastase promoter	pancreas	N	Ornitz *et al.* (1985)
GH/MT[b] promoter	liver	increased growth	Palmiter *et al.* (1983)
Rat growth hormone/MT promoter	liver	reduced fertility, increased growth	Palmiter *et al.* (1982*a*)
GH releasing factor/MT promoter	kidney, gut, pancreas	increased growth rate, pituitary hyperplasia	Hammer *et al.* (1985*a*)
Immunoglobulin (κ)	spleen	N	Brinster *et al.* (1983)
Immunoglobulin (μ)	spleen, lymph node, thymus, B and T cells	N	Grosschedl *et al.* (1984)
Immunoglobulin (μ and κ)	B cells	N	Rusconi and Köhler (1985)
MHC-1[c]	peripheral blood lymphocytes	N	Frels *et al.* (1985)
MHC-2	B cells, monocytes	N	LeMeur *et al.* (1985); Yamamura *et al.* (1985)
Myc/MMTV[d] promoter	salivary gland, intestine (small), mammary gland	breast adenocarcinoma	T. A. Stewart *et al.* (1984)
Myosin light-chain	skeletal muscle	N	Shani (1985)
Somatostatin/MT promoter	anterior pituitary	N	Low *et al.* (1985)
SV-40/insulin promoter	pancreas	disordered islets of Langerhans, islet hyperplasia, islet tumors	Hanahan (1985)
SV-40-Tk[e]/MT promoter	brain, thymus, kidney	tumor of choroid plexus, thymic hypertrophy, abnormal kidney pathology	Brinster *et al.* (1984)
SV-40-Tk/GH promoter	brain, thymus, kidney	tumor of choroid plexus, thymic hypertrophy, abnormal kidney pathology	Brinster *et al.* (1984)
Tk/MT promoter	liver, kidney	N	Brinster *et al.* (1981, 1982)
Transferrin	liver	N	McKnight *et al.* (1983)

[a]GH, growth hormone.
[b]MT, metallothionine.
[c]MHC, major histocompatibility complex.
[d]MMTV, mouse mammary tumor virus.
[e]Tk, thymidine kinase.
[f]normal or not stated.

to express the fusion genes to various extents in other tissues, including kidney, intestine, heart, brain, spleen and testes. Possible factors controlling the primary sites of expression are discussed below.

Brinster *et al.* (1981) provided the first piece of evidence that transgenic mice can express foreign sequences in a tissue-specific manner and that expression could be regulated by the promoter sequences. They showed that herpes-specific thymidine kinase is expressed in some tissues of mice containing integrated copies of the metallothionine Tk fusion plasmid (*pMK*). In their experiments, expression is induced *in vivo* by intraperitoneal injection of cadmium. Two mice out of four analyzed show strong HSV-Tk expression in liver tissue on the basis of several criteria: (1) levels of HSV-Tk-specific mRNA were high relative to the negative controls, (2) total thymidine kinase activity was significantly inhibited in one of the mice by HSV-Tk antibody, and (3) one of the mice showed no significant inhibition in the ^{125}IdC assay with H_4Udr. Both mice that expressed HSV-Tk synthesized significantly more viral Tk mRNA in liver than kidney although for the most part the total thymidine kinase activity was approximately equal in the two tissues. The tissue-specific pattern of HSV-Tk RNA accumulation after induction accurately reflects the distribution of metallothionine-1 RNA in the tissues. This indicates that the introduced MT promoter is regulated with fidelity in these transgenic mice and that the MT promoter sequence can impose tissue specificity on a proximal gene which would normally be expressed in a different distribution of tissues.

Palmiter *et al.* (1982*b*) quantitated the levels of HSV-Tk induction by cadmium and dexamethasone in several lines of transgenic mice. Mice were injected with cadmium or dexamethasone and partial hepatectomies were performed after 4 or 18 hr. Assays were separated by intervals of 2–4 weeks. Levels of induction with cadmium ranged from ninefold to 46-fold (over the noninduced controls), yet dexamethasone failed to induce any HSV-Tk activity. The most likely explanation for the lack of induction with dexamethasone is that the hormone-responsive sequences were not present on the fusion recombinant plasmids which were microinjected.

In another series of experiments, Palmiter *et al.* (1982*a*) microinjected a fusion gene containing the metallothionine promoter and the rat growth hormone gene into mouse eggs and adult mice were analyzed for gene expression. There was a reasonably good correlation between the number of integrated copies of *pMGH*, the number of *pMGH* mRNA molecules produced per cell, the levels of growth hormone produced, and the physiological growth of the animals. The animals were provided with supplementary zinc in their water at the time of weaning, but their growth rates were accelerated even before the induction. Presumably, expression of the fusion genes was induced by endogenous signal(s). All positive mice expressed greater levels of rat growth hormone to some extent than negative control values, induction values over controls ranging approximately twofold to more than 800-fold. The level of accumulation of *pMGH* mRNA molecules was much higher than the induced levels of HSV-Tk RNA reported previously (Brinster *et al.*, 1981; Palmiter *et al.*, 1982*b*). These workers attribute the difference to the greater stability of *MGH* RNA relative to Tk RNA. Other explanations are also possible. Clearly, high levels of rat growth hormone are not detrimental to the developing mouse. The significance of this study in the construction of animal disease models is discussed in Section 6.

Human growth hormone-releasing factor (GRF) was also injected into mice in constructs with the metallothionine promoter (Hammer *et al.*, 1985*a*). The fusion gene was primarily expressed in the liver in 11 of 14 mice which showed expression of the transgenic DNA. As in previous experiments, the animals were maintained on diets with zinc supplementation to enhance expression. There was poor correlation between gene copy number and either GRF RNA levels, plasma GRF protein levels or relative growth levels of the mice. Yet, many of the mice grew significantly larger than the control littermates, as much as two times larger. The growth rates are significantly slower than the growth rates of transgenic mice previously described containing fusion genes which express human or rat growth hormone. The authors attribute this difference to the presumed localized release of growth hormone in the anterior pituitary cells of transgenic mice which express GRF compared to the high circulating levels of growth hormone in the mice previously described.

The metallothionine promoter fused to the somatostatin gene has been introduced into transgenic mice and is expressed primarily in the anterior pituitary as determined by hormone concentration (per mg tissue) and liver, kidney, spleen, lung, heart, testes, and muscle to much lesser extents (Low *et al.*, 1985). The reasons why in this experiment the primary site of synthesis of protein product are neither determines by the normal site of metallothionine biosynthesis (liver) nor the normal site of somatostatin production (posterior pituitary) remain unclear. It is unlikely that position effects are the sole cause since the integration sites would be expected to be random and the unusual cell-type expression among many mice in these experiments is reproducible.

High levels of somatostatin were produced by the anterior pituitary cells of these transgenic mice (94 ng/mg protein) and the product was immunoreactive. The precursor somatostatin produced by these cells was accurately post-translationally processed to 14 and 28 amino acid peptides, the normal sizes of the products, whereas the liver and kidney synthesized a much smaller version of the final product. Somatostatin normally inhibits the release of growth hormone, yet the transgenic mice with high levels of expression grew normally. The reasons for this are unclear but probably relate to inappropriate recognition of signals possibly due to the localized site of expression.

McKnight *et al.* (1983) reported that chicken transferrin RNA was preferentially transcribed in liver tissue in five of seven transgenic mice analyzed. Transferrin, like metallothionine, is normally transcribed in the liver, and these foreign genes therefore appear to be undergoing an accurate form of expression. McKnight *et al.* (1983) also found that the chicken transferrin was secreted into the serum of transgenic mice. The levels of transferrin in the serum did not directly reflect the amount of transferrin RNA accumulated in these animals. Although the distribution of transcripts observed by McKnight *et al.* (1983)

accurately reflects the native tissue-specificity of the introduced genes, the overall levels of expression of these genes are low in comparison to the endogenous counterparts. Possible reasons for the sluggish transcription rates are discussed in a following section.

Several groups have been using the transgenic mouse system to study physiological regulation of gene expression in the immune system. Brinster *et al.* (1983) introduced a plasmid into embryos containing a rearranged functional κ-chain gene from a myeloma line. Both RNA and protein were produced from the transgenic DNA in all mice examined but levels did not correlate with gene copy number. Levels of accumulation of RNA were very high, approaching the levels of production of κ RNA in normal spleen cells, and the pattern of gene expression was accurately tissue-specific. The usefulness of the transgenic mouse system lies in the ability to introduce rearranged genes into all tissues, whereas normally gene rearrangement would occur in only a single cell type during developmental ontogeny. RNAs were produced in abundance in spleen tissue, but not at all in liver, and expression appeared to be restricted to cells of the B lineage (Storb *et al.*, 1984). High levels of expression of this gene may be due to the presence of endogenous *cis* "enhancer" sequences within the immunoglobulin gene (Banerji *et al.*, 1983; Gillies *et al.*, 1983; Queen and Baltimore, 1983). An important implication of these findings is that the rearrangement of immunoglobulin κ-chain genes during normal development of cells of the B cell lymphoid lineage is not the sole requirement for transcriptional activation nor is it solely responsible for the tissue trophism of gene expression.

Grosschedl *et al.* (1984) introduced a rearranged μ heavy chain gene into mice. Four of five mice examined expressed the foreign gene specifically in spleen, lymph node, and thymus tissue. The foreign heavy chains were capable of combining with endogenous light chains to form function antibodies which bind the hapten NP. In agreement with the data presented above, these experiments demonstrate that gene rearrangement is not absolutely required in order to determine the tissue specificity of gene expression. Yet interestingly, the foreign gene was expressed in B as well as T cells, suggesting that rearrangement may be a requirement for restricting the specificity of expression within the lymphoid lineage.

Ritchie *et al.* (1984) have also used this system to study the mechanism of allelic exclusion and to distinguish between a so-called "stochastic" model in which statistical improbability argues against two functional rearrangements occuring in the same cell and a regulated system in which the first functional rearrangement feeds back regulatory signals and prevents further rearrangements. They produced hybridomas from transgenic mice and showed evidence of a regulated pathway for allelic exclusion. They showed that expression of the rearranged transgenic κ-chain and association with a functional heavy chain is sufficient to prevent rearrangement of the endogenous κ gene. As a control, when the transgenic κ locus was inactive, the endogenous genes could rearrange. Yet with no functional heavy chain, expression of the transgenic κ locus was insufficient to prevent further rearrangement of the endogenous locus.

Frels *et al.* (1985) introduced a class 1 porcine major histocompatibility gene (MHC) into mouse embryos. A positive mouse and resultant second generation progeny were accessed for cell surface expression using a monoclonal antibody which differentiated porcine from murine MHC. All positive mice which were examined expressed immunoreactive antigen on the surfaces of peripheral blood lymphocytes. Interestingly, though, the "foreign" porcine antigen appeared to be tolerated by the transgenic mice. It functioned as a bonafide transplantation antigen and induced graft rejection when tissue was transplanted to a control mouse.

The elastase gene has also been shown to be expressed in a tissue-specific manner in the pancreas of transgenic mice (Swift *et al.*, 1984). Five of five mice analyzed show elevated levels of RNA in the pancreas relative to other tissues with four of those mice exhibiting levels greater than endogenous levels in rats (10,000 molecules/cell). One of the mice expressed as high as 120,000 molecules per cell. Minor amounts of elastase RNA were evident in other tissues of some of the mice. Expression was not absolutely correlated to gene copy number, though the mouse with the highest number of gene copies (100) had the greatest level of expression. The construct used to generate these transgenic mice was a 23 kb piece of genomic DNA containing 7 kb upstream and 5 kb downstream from the coding portion of the gene.

Ornitz *et al.* (1985) constructed fusions between the elastase promoter and the human growth hormone gene and were able to achieve accurate expression of the growth hormone gene in the pancreas. High levels of transcription were observed in a total of 13 of 18 mice, reaching as high as 39,400 molecules per cell in one transgenic mouse. In no case was expression in other tissues detected at levels over 10 molecules per cell. Three versions of the fusion gene were injected with different amounts of 5′ flanking elastase sequence. Even the construct with as little as 213 bp of promoter directed accurate tissue-specific expression. The pancreatic cells expressing the fusion constructs were shown to be the ascinar cells by immunofluorescence. There seemed to be no correlation between gene copy number and level of gene expression.

Among other genes which appear to exhibit tissue-specific regulation in transgenic mice is the gene for rat myosin light-chain 2, which is expressed exclusively in skeletal muscle in all expressing transgenic mice (2 of 3 mice examined; Shani, 1985). In one of the mouse lines the level of RNA transcribed was 5–10% greater than in a control differentiated myogenic cell line.

Chada *et al.* (1985) demonstrated accurately regulated tissue-specific expression of hybrid mouse/human β-globin genes in transgenic mice. Four of seven positive mice analyzed in which a coding restriction fragment had been injected showed expression in the blood, bone marrow and spleen, whereas the other three mice showed no detectable expression at all. The maximum level of expression in the blood of one of their mice was 2% of the total β-globin RNA. No expression was detected in three mice in which the entire plasmid was injected. This fusion gene is expressed correctly in a developmental context despite the overall low level of expression (Magram *et al.*, 1985; see Section 5.8.2). In experiments with rabbit β-globin genes, this group had previously been unable to detect expression of these sequences in appropriate tissues (Lacy *et al.*, 1983). Possible reasons for the difference in expression are discussed in a Section 5.8.2.

Alpha-fetoprotein genes have also been shown to be capable of tissue-specific gene regulation (Krumlauf *et al.*, 1985). Minigene constructs containing either 14 or 7 kb of 5′ flanking DNA were used to produce 3 or 12 lines of transgenic mice, respectively. All mice expressed the AFP minigene in a tissue-specific manner, in the yolk sac, fetal liver and fetal gut, but not the kidney, heart or brain. The maximum level of expression in one line of mouse was 25% of the endogenous AFP level. Variability in expression of the AFP minigenes could not be correlated to either copy number of the gene sequences or amount of 5′ flanking DNA in the gene construct used for injection.

Constructs described above which contained metallothionine promoter sequences are examples of inducible gene systems since it has been shown that the levels of gene expression can be enhanced in many cases with stimulation by cadmium or zinc. Another example of a fusion gene introduced into transgenic mice with an inducible promoter is the mouse mammary tumor virus (MMTV) promoter which was fused to the *myc* oncogene (T. A. Stewart *et al.*, 1984). The goal was to produce strains of mice in which the levels and timing of expression of the *myc* gene could be regulated by supplies of exogenous glucocorticoids and use these mice to study the role of this gene in oncogenesis. Thirteen transgenic mice were obtained of which 11 show expression in at least one tissue. Two of three mice analyzed show expression in the intestine and mammary gland, whereas most mice show expression in the salivary gland. One mouse showed expression in all tissues analyzed. Interestingly, two of the original founder females developed mammary adenocarcinomas at the second or third pregnancies, presumably related to hormonal signals acting upon the MMTV promoter. The propensity to develop such tumors was inheritable in subsequent generations. Activation of the *myc* gene alone, however, does not appear to be sufficient to cause the development of carcinomas since not all mice which express these genes develop tumors.

Brinster *et al.* (1984) have introduced plasmids containing the SV40 early region into mice. The plasmids contained the metallothionine promoter fused to the thymidine kinase gene or the metallothionine promoter fused to the growth hormone gene, as these constructs had been shown to be transcriptionally active and non-delitereous to transgenic mice. Many of the mice in these experiments develop heritable brain tumors of the choroid plexus epithelium. Many mice also exhibit thymic hypertrophy and abnormal kidney pathology. Since normal numbers of mice were obtained from these experiments it appears that the SV40 sequences are not detrimental to mouse development. These authors suggest that the SV40 sequences may have undergone an inactivation during early development, drawing an analogy to the inactivation which occurs when retroviruses are used to infect preimplantation embryos (Jaenisch, 1976). One explanation for the genesis of such specific brain tumors in the high percentage of transgenic mice is that the SV40 sequences are specifically activated in that tissue at a well defined developmental stage. Consistent with this hypothesis is the finding that the transgenic sequences become amplified in five out of six cell lines derived from tumor tissue. Also consistent is the increased levels of expression of SV40 T antigen and SV40 RNA in the tumor cells and cell lines derived from the tumors relative to low levels from other tissues. Small *et al.* (1985) also found a transgenic mouse which developed papilloma of the choroid plexus. They had coinjected SV40 sequences and the v-*myc* oncogene.

Hanahan (1985) has specifically targeted expression of the SV40 T antigen in transgenic mice to the β-cells of the pancreas by putting the SV40 DNA under control of the insulin-2 promoter. Mice which express the SV40 sequences do so only in the pancreas and expression is correlated to the genesis of genetically heritable β-cell tumors. The phenotype is first manifested as densely packed cells and hyperplasia. In addition, the organization of α- and δ-cells becomes highly disrupted. Yet, the tumors arise from only a few islets, suggesting that a complementary molecular event is required for transformation. The ability to induce specific types of tumor in transgenic mice provides an exceptional experimental approach toward analyzing auxillary molecular events which are required for expression of oncogenesis.

To summarize the experiments on tissue-specific gene expression, most groups find no absolute correlation between gene copy number and either tissue trophism or overall levels of gene expression, suggesting an influence on gene expression from the neighboring chromosomal environment. Some genes can be expressed at or above the endogenous levels whereas other genes are only expressed to a fraction of the endogenous levels. In the experiments presented here, genes with native promoters tend to be expressed in appropriate target tissues. Constructs in which the metallothionine promoter was fused to thymidine kinase, growth hormone and growth hormone-releasing factor were expressed primarily in the liver, as dictated by the specificity of the promoter region, whereas MT constructs with somatostatin or SV40 T antigen were expressed in the anterior pituitary or choroid plexus, respectively. Presumably, in these latter two systems, gene-specific signals which can target expression to particular tissues are dominant over controls emanating from the promoter. Clearly, *cis*-acting sequences can be sufficient to target gene expression to particular tissues at specific times but alone may not reflect all the interactions that are needed for proper regulation.

5.8.2. Developmentally Regulated Gene Expression

Transgenic mice can be used in at least two ways to study gene expression in development. Already described are experiments delineating the generation and potential usefulness of insertional mutations, many of which are lethal mutations, in dissecting developmental phenomena. Another powerful use for the transgenic mouse system in studying developmental processes is in the delineation of *cis*- and *trans*-acting sequences which regulate developmental programs. In the same way that the sequences upstream sufficient for gene expression in adult tissues were mapped by Ornitz *et al.* (1985), it will be possible to determine which *cis*-acting sequences are important in the activation of genes during normal development.

Two groups have looked at regulation of transgene expression in specific tissues during development. In both cases, the genes that were analyzed were expressed in the appropriate temporal period and in correct target tissues. Magram *et al.* (1985) analyzed the developmental pattern of expression in lines of transgenic mice containing copies of a mouse/human β-globin hybrid gene (Chada *et al.*, 1985). In this construct the mouse portion contributed the 5′ end of the gene plus 1.2 kb of 5′ flank and the human portion contained the 3′ end with 2 kb of 3′ flank. Beginning at 12 days of gestation, normal expression of the adult β-globin gene occurs in the circulating erythrocytes derived from the fetal yolk sac and in the fetal liver, and starting at day 16 of gestation, in the spleen and bone marrow. These investigators found that the fusion transgene was inactive in fetal blood prior to day-12 but later was expressed in fetal liver tissue and bone marrow. This is the pattern predicted by the normal expression of the gene. However, though the relative levels of endogenous RNA are approximately equal in fetal liver and bone marrow, the expression ratios of the transgenic DNAs differ. In two mice, higher levels were found in the fetal liver, whereas in another two mouse lines, higher levels were found in adult bone marrow. Influences of chromosomal position or heterologous DNA in the construct may have been important.

Krumlauf *et al.* (1985) also found accurately regulated developmental expression of transgenic sequences. AFP is normally produced in the visceral endoderm of the yolk sac, fetal liver and fetal gut. As discussed previously, expression of the AFP minigene was appropriately regulated, in that all mice which expressed the minigene did so only in the correct target tissues. In normal development transcription of the AFP gene is turned off at birth and in the transgenic mice studied by these authors, expression is turned off in temporal concert with the endogenous gene. In these experiments, the presence of as little as 7 kb of 5′ flanking DNA was consistent with accurate tissue and developmental regulation of the minigene AFP construct. Further experiments along these lines will be informative in the functional mapping of specific *cis*-acting upstream sequences in the control of gene activity in development.

If foreign genes are ever expressed in inappropriate tissues or at inappropriate periods in prenatal development these expression events may lead to embryo lethality in a high proportion of cases. Probably, in most cases, expression of a foreign gene would not interfere with expression of the endogenous genes in the developing embryo based on what has been found in transgenic adults. However, in a positive sense, the presence of an active aberrant gene product in an embryonic tissue may interfere with the normal processes of differentiation by, for example, confusing normal cell–cell recognition signals. Thus, even if modified gene constructs could lead to inappropriate developmental regulation, mice exhibiting aberrant patterns of expression may not be found due to developmental selection.

5.8.3. Complementation of Genetic Defects

Recently the transgenic mouse system has been used to compensate for some types of genetic defects by complementation. In one case, embryo gene therapy has been shown to be capable of supplementing the growth of dwarf mice and in another example, enabling immunodeficient mice to mount full antibody responses. As mentioned earlier, mouse embryos of the dwarf genotype, lit/lit, were used as recipients for gene transfer with the rat growth hormone gene directed by the MT promoter (Hammer *et al.*, 1984). As had been the case for normal strains of mice receiving this gene (Palmiter *et al.*, 1982*a*; Palmiter *et al.*, 1983), these mice grew to larger sizes than control siblings. The serum growth hormone levels in the transgenic animals reached as high as 20 times the wild type control levels and nearly 400 times the level of homozygous mutant controls. Yet, there was no absolute correlation between levels of growth at 16 weeks, serum growth hormone levels and gene copy number. The transgenic mutants grow to a size even greater than non-mutant control animals, by 90 days reaching a size 1.5 times larger than such controls.

Two groups have used a strain of mouse with a genetic defect in the major histocompatability complex as recipients for gene therapy (LeMeur *et al.*, 1985; Yamamura *et al.* 1985). The class 2 antigens, A and E, each consist of 2 subunits, α and β, which associate to form dimers on the cell surface of B lymphocytes and monocytes. C57BL/6 mice of the *b* haplotype contain a 5′ deleted E gene and hence fail to produce α subunits. E β is produced but remains in the cytoplasm. These groups introduced the E α gene into mice and in the transgenic mice expressing this gene heterodimers could be found on the cell surfaces of appropriate cells. Whereas the immunodeficient mutant mice fail to mount an antibody response to some antigens, including poly (L-glutamic acid-L-lysine-L-phenylalanine), the ability to respond antigenically to this agent was restored in the transgenic animals.

Genetic complementation experiments such as these promise to be extremely useful, not so much in gene therapy *per se*, but in the delineation of sequences necessary and sufficient for the elicitation of biological phenomena. Since the complementation of a genetic defect by definition requires functional gene activity, these types of experiments not only score for gene expression, but biologically active gene expression.

5.8.4. Factors Involved in the Variability of Gene Expression

As described in the preceding pages, many genes which have been introduced into mice are expressed to high levels in the tissues which the "normal" gene counterparts are expressed. In some cases the levels of expression have approached or exceeded the levels obtained with corresponding endogenous genes (Grosschedl *et al.*, 1984; Swift *et al.*, 1984; Shani, 1985), whereas in other groups of experiments the expression levels obtained were low in comparison to endogenous levels (McKnight *et al.*, 1983, Chada *et al.*, 1985; Krumlauf *et al.*, 1985). Many investigators have also noted variability from mouse to mouse, both in terms of the number of gene copies which become integrated and in the levels of gene expression. Yet it seems from all the systems examined thus far that high gene copy number does not necessarily correlate to high levels or accurate forms of gene expression (see for example, Swift *et al.*, 1984; Hammer *et al.*, 1985*a*; Krumlauf *et al.*, 1985; Ornitz *et al.*, 1985). Possible factors giving rise to the variability in levels of gene expression among different gene systems and among individual mice in given experiments are discussed in the following paragraphs.

Several parameters may affect the levels of gene expression and tissue trophism observed in transgenic mice:

1. *Gene construct:* Appropriate regulatory sequences may be absent from some of the recombinant constructs introduced into mouse embryos, particularly since many of the microinjected genes are cloned from heterologous species. In some cases, expression may require integrations at chromosomal sites adjacent to appropriate promoters. Lacy *et al.* (1983) found evidence for expression of the rabbit β-globin gene in only two of nine transgenic mice analyzed and in contrast to most of the experiments presented above, the genes appear to be expressed in inappropriate tissues. Several tissues from seven of the mice were analyzed in detail for RNA expression (erythropoietic tissue, brain, liver, kidney, lung, skeletal muscle, and testes), using a sensitive S1 protection assay. Only two mice, numbers 4 and 23, showed any evidence of expression. In mouse 4, rabbit β-globin RNA was detected primarily in skeletal muscle, and in mouse line 23 the protected RNA was demonstrated only in testes tissue. The aberrant tissue specific pattern of gene expression was inherited in successive generations from both expressing mice. Conceivably, the construct which was introduced into the transgenic mice lacked appropriate *cis*-acting regulatory sequences necessary for the expression in correct target tissues. Possibly, the positive mice which expressed β-globin in inappropriate tissues had the foreign sequences integrated next to random endogenous promoters. In contrast, genes which are under the control of the metallothionine promoter typically give rise to high proportions of mice which express the introduced gene in the correct target tissues (see Table II). Since the metallothionine promoter was originally derived from mouse DNA, it may more efficiently direct transcription within the mouse genome by being able to accurately respond to murine inductive stimuli. Yet, several other genes have been successfully expressed in correct tissues in mice, either with their own promoters or heterologous promoters and so it is impossible to predict *a priori* which sequences will be expressed accurately based on the construct alone. Another possibility is that certain vectors contain "poison" sequences which are incompatible with appropriate gene expression. Chada *et al.* (1985) and Krumlauf *et al.* (1985) suggest that vector sequences may be involved in the variability of gene expression. While it is clear that high levels of genetic transformation can be achieved with plasmids containing vector sequences, it

might be necessary to remove such sequences in order to assure consistently high levels of gene expression in correct target tissues.

2. *Methylation:* Methylation of the introduced recombinant sequences may play a role in the modulation of gene expression in transgenic mice. In one example, Palmiter *et al.* (1982*b*) were able to correlate the pattern of methylation of the MT/HSV-Tk gene with expression, although a similar correlation was impossible to make in the analysis of expression in another lineage of animals. Lacy *et al.* (1983) state that some DNA sequences analyzed in transgenic mice were highly methylated. In this regard, R. Jaenisch's group has found that Mo-MuLV DNA tends to become highly methylated upon integration in early mouse embryos and embryonal carcinoma cells (Jähner *et al.*, 1982; C. L. Stewart *et al.*, 1982; Simon *et al.*, 1983). DNA extracted from such mice fails to induce foci in NIH/3T3 cells and is inactive in the XC plaque assay. The most likely explanation is that the methylation of Mo-MuLV DNA upon integration inhibits expression of the viral genes. Upon molecular cloning of DNA from a Mov-3 mouse, a process that removes methyl groups, the sequences became highly infectious in infectivity assays in tissue culture. It is possible that molecularly cloned sequences may also undergo extensive *de novo* methylation upon integration that temporarily or permanently inactivates these sequences. The degree of gene methylation may be related to the chromosomal site of integration. In another series of experiments, no expression was found of the *E. coli* galactokinase gene in transgenic mice, presumably due to methylation (Humphries *et al.*, 1985). When 5-azacytidine was administered to these mice, these sequences became partially demethylated and expression of gal-K became detectable. However, the evidence that methylation is solely responsible for variability in levels of gene expression is conflicting, even among mice from a single experiment (see above). More experimental work needs to be done in order to clarify whether *de novo* methylation plays a critical part in the regulation of gene activity in transgenic mice.
3. *Position effects:* Variability in the levels of gene expression among mice in a given experiment and the inability to draw correlations between these levels and gene copy number is usually attributed to "position" effects. Literally, this refers to expression levels being more a function of the chromosomal site of integration than the gene and control sequences themselves. As discussed in a previous section, it appears to be the case that genes integrate into the host genome in a presumably random fashion, and thus may be present at any chromosomal site. It is possible that integrated genes become transcriptionally repressed due to the general inactivation of certain chromosomal domains. Genes may sometimes integrate into subchromosomal regions of heterochromatin or into other regions of the genome that are never transcriptionally activated. In this regard, it is interesting that tissue-specific expression of several genes in *Drosophila* transformants is largely independent of

position effects (Goldberg *et al.*, 1983; Scholnick *et al.*, 1983; Spradling and Rubin, 1983). Yet, in order for gene expression in transgenic mice to provide accurate information on the normal processes of gene regulation in mammalian organisms, *in vivo* activity of the introduced genes should reflect the restrictions which the local chromosomal topology place upon the integrated foreign gene. Correlations between chromosomal localization and phenotypic expression have been made by a few laboratories (Lacy *et al.*, 1983; C. Stewart *et al.*, 1983). More studies are needed in which integrated DNA sequences in transgenic mice are mapped to specific chromosomal loci in order to draw conclusions about position effects and gene expression.

4. *Gene rearrangements:* Gene rearrangements are another possible explanation for the lack of correlation between numbers of gene copies and levels of gene expression. Among mice with multiple gene copies, for example, it is unclear whether all copies or just a few are transcriptionally active. Gene and chromosomal rearrangements may be produced in the genesis of transgenic mice, one class of which are the insertional mutations created by the integration of new DNA discussed in detail in a previous section. Small, undetectable local DNA alterations might also lead to the inactivation of transgenic sequences. In order to locate and characterize DNA rearrangements, it may be necessary to sequence large portions of integrated DNA and adjacent mouse DNA of endogenous origin. It may also be informative in the search for functionally significant rearrangements to analyze in tissue culture the expression of cell lines derived from transgenic mice.
5. *Harmful expression of some genes:* Direct phenotypic selection may take place in transgenic mice to inhibit the expression of genes with deleterious effects on the organism. Expression may be inhibited by integrations at nonpermissive chromosomal sites (Lacy *et al.*, 1983), or be repressed due to *de novo* methylation (Jähner *et al.*, 1982). *In vivo* selection may operate to underrepresent expressing genes whose products are harmful to the developing embryo or adult mouse. Transient or stable expression of harmful or aberrantly expressed gene products may disrupt development or make such embryos less viable than controls. High incidences of embryo lethality may be indicative of such phenomenon (Humphries *et al.*, 1985). Gene transfer experiments with mouse embryos that have yielded high frequencies of transformation probably used gene systems and vectors in which expression was harmless to the developing mouse.

6. Summary, Conclusions, and Perspectives

Gene transfer into the murine genome was accomplished nearly a decade ago by use of chimeras and teratocarcinomas; however, the low frequencies of transfer into the germ line and other difficulties stemming from mosaicism and karyotypic abnormalities in chimeric mice have limited the general usefulness of this procedure in achieving transformation in mammalian embryos. The introduction of cloned genes into teratocarcinoma cells, selection for a mutant phenotype, and transfer of those cells into mouse embryos holds some promise as a technique to employ mouse chimeras for gene transfer into mice.

Infection with animal viruses and retroviral vectors provides another way to introduce exogenous DNA into mouse embryos. Infection with Mo-MuLV has been utilized to characterize the relationship between sites of integration and gene function in developing and adult mice. Gene transfer by microinjection of cloned recombinant DNA has been used by many laboratories for the transfer of DNAs into mouse embryos. The factors affecting transformation frequencies and sites of integration are unknown at present, although it seems that integration is not strictly mediated by homology-dependent events. Many genes have been introduced into mouse embryos by these procedures and many of these are expressed at high levels in appropriate tissues.

No realistic possibility exists at the present time for the utilization of embryo gene transfer in the medical field for the correction of genetic defects for several reasons. First, in order to effectively provide "gene therapy" it would be necessary to determine the genotype of each recipient egg, a technical impossibility. The genetic diseases that would be amenable to germ line intervention are recessive diseases and there would be only a 25% chance of any one embryo derived from heterozygous parents being a homozygous recessive. Moreover, it would be impossible to distinguish the normal from abnormal embryos. Second, the frequencies of transformation are so low as to exclude work on human beings on ethical grounds. Third, the parameters effecting chromosomal integration sites and gene expression have not been fully characterized. Until it becomes experimentally possible to target the newly introduced DNA into expressable chromosomal sites and actively replace or supplement defective genes, the possibility of gene therapy through manipulation of embryos is remote. Yet, efforts to provide gene therapy in somatic tissues have been promising, leading to expression of a modified phenotype (Anderson, 1984). In contrast to embryo gene therapy, gene therapy in somatic tissues would not lead to germ line propagation of the manipulated genotype.

The agricultural field may benefit in the years ahead by being able to transfer genes into the embryos of farm animals. Success in achieving genetic transformation of rabbits, sheep and pigs using microinjection has recently been achieved (Hammer *et al.*, 1985*b*). Genes which improve the ability of cattle to produce milk or of animals to grow to maturity in shortened lengths of time, and eliminate or help to combat diseases are good candidates for germ line modification.

The usefulness of embryo gene transfer in the medical field lies in the creation of animal models for human diseases. Such models are an important part of testing drug regimens and studying the etiology of human disease. Few such models are currently available (e.g., Fleischman and Mintz, 1979; Martinell *et al.*, 1981; see Green, 1975). Gene transfer into embryos offers the hope of establishing models for a wider range of human diseases. An example is provided by the experiments of Palmiter *et al.* (1982*a*) in which mice expressing human growth hormone grew at vastly accelerated rates, relative to control siblings. These mice provide a model for giganticism, and it will now be possible to study the factors important in the genesis and progression of this disease. These mice express human growth hormone at high levels and the presence of an intact endogenous growth hormone gene may not be a complicating factor in the interpretation of their data. In addition, the ability to induce tumors of precise lineages using transgenic mouse technology will be invaluable in the field of oncology.

Embryo gene transfer may provide a useful practical system in the future to study gene expression during embryonic development and may provide an appropriate model system for the study of birth defects. The exquisite interaction of hundreds of thousands of gene products leads to the morphogenesis of a complex metazoan organism from a single egg, a process that has remained elusive to developmental biologists despite more than a century of research. By disrupting genes that are critical in development, it may now be possible to investigate the physiological function of specific gene products in embryogenesis. Moreover, transformation of embryos leading to insertional mutation events and the aberrant or enhanced expression of DNAs during embryogenesis may aid in our understanding of the processes by which differential gene activation lead to development.

Note Added in Proof

The number of laboratories skilled in the technical implementation of embryo microinjection technology has been increasing at a rapid rate within the last few years. In general the very recently published reports support information presented in this review: (1) Coding information specified in fusion constructions is often expressed in a tissue-specific manner in transgenic mice as dictated by the specifity of the promoter (Adams *et al.*, 1985; Babinet *et al.*, 1985; Overbeek *et al.*, 1985; Ross and Solter, 1985; Townes *et al.*, 1985; Khillan *et al.*, 1986; Overbeek *et al.*, 1986). (2) Other sequences may also be important in directing the tissues for expression of transgenes, in some cases overriding the specificity of the promoters (Chisari *et al.*, 1985; Palmiter *et al.*, 1985; Stout *et al.*, 1985; Swanson *et al.*, 1985; Low *et al.*, 1986). (3) Many recent reports (including some mentioned above in other contexts) have used transgenic mice as disease models and to study various aspects of gene regulation (Messing *et al.*, 1985; Storb *et al.*, 1985; Weaver *et al.*, 1985; Bieberich *et al.*, 1986; Chada *et al.*, 1986; Gordon, 1986; Hammer *et al.*, 1986; Krumlauf *et al.*, 1986; Leder *et al.*, 1986; Selden *et al.*, 1986; Weaver *et al.*, 1986).

ACKNOWLEDGMENTS. We would like to extend special thanks to D. Brown and H. Lieberman for their critical comments on this manuscript and the many scientists who contributed information before publication. K. G. was supported by a fellowship from the Damon-Runyon Foundation.

References

Adams, J. M., Harris, A. W., Pinkert, C. A., Corcoran, L. M., Alexander, W. S., Cory, S., Palmiter, R. D., and Brinster, R. L., 1985, A c-*myc* oncogene driven by immunoglobulin enhancers induces lymphoid malignancy in transgenic mice, *Nature* **318**:533–538.

Anderson, W. F., 1984, Prospects for Human Gene Therapy, *Science* **226**:401–409.

Anderson, W. F., Killos, L., Sanders-Haigh, L., Kretschmer, P. J., and Diacumakos, E. G., 1980, Replication and expression of thymidine kinase and human globin genes microinjected into mouse fibroblasts, *Proc. Natl. Acad. Sci. USA* **77**:5399–5403.

Babinet, C., Farza, H., Morello, D., Hadchouel, M., and Pourcel., C., 1985, Specific expression of hepatitis B surface antigen (HBsAg) in transgenic mice, *Science* **230**:1160–1163.

Banerji, J., Olson, L., and Schaffner, W., 1983, A lymphocyte-specific cellular enhancer is located downstream of the joining region in immunoglobulin heavy chain genes, *Cell* **33**:729–740.

Bieberich, C., Scangos, G., Tanaka, K., and Jay, G., 1986, Regulated expression of a murine class I gene in transgenic mice, *Mol. Cell. Biol.* **6**:1339–1342.

Brinster, R. L., 1974, The effect of cells transferred into the mouse blastocyst on subsequent development, *J. Exp. Med.* **140**:1049–1056.

Brinster, R. L., Chen, H. Y., Trumbauer, M., Senear, A. W., Warren, R., and Palmiter, R. D., 1981, Somatic expression of Herpes thymidine kinase in mice following injection of a fusion gene into eggs, *Cell* **27**:223–231.

Brinster, R. L., Chen, H. Y., Warren, R., Sarthy, A., and Palmiter, R. D., 1982, Regulation of metallothionein-thymidine kinase fusion plasmids injected into mouse eggs. *Nature (Lond.)* **296**:39–42.

Brinster, R. L., Ritchie, K. A., Hammer, R. E., O'Brien, R. L., Arp, B., and Storb, U., 1983, Expression of a microinjected immunoglobulin gene in the spleen of transgenic mice, *Nature (Lond.)* **306**:332–336.

Brinster, R. L., Chen, H. Y., Messing, A., van Dyke, T., Levine, A. J., and Palmiter, R. D., 1984, Transgenic mice harboring SV40 T-antigen genes develop characteristic brain tumors, *Cell* **37**:367–379.

Brinster, R. L., Chen, H. Y., Trumbauer, M. E., Yagle, M. K., and Palmiter, R. D., 1985, Factors affecting the efficiency of introducing foreign DNA into mice by microinjecting eggs, *Proc. Natl. Acad. Sci. USA* **82**:4438–4442.

Bürki, K., and Ullrich, A., 1982, Transplantation of the human insulin gene into fertilized mouse eggs, *EMBO J.* **1**:127–131.

Calos, M. P., Lebkowski, J. S., and Botchan, M. R., 1983, High mutation frequency in DNA transfected into mammalian cells, *Proc. Natl. Acad. Sci. USA* **80**:3015–3019.

Capecchi, M. R., 1980, High efficiency transformation by direct microinjection of DNA into cultured mammalian cells, *Cell* **22**:479–488.

Chada, K., Magram, J., Raphael, K., Radice, G., Lacy, E., and Costantini, F., 1985, Specific expression of a foreign β-globin gene in erythroid cells of transgenic mice, *Nature* **314**:377–380.

Chada, K., Magram, J., and Costantini, F., 1986, An embryonic pattern of expression of a human fetal globin gene in transgenic mice. *Nature* **319**:685–689.

Chisari, F. V., Pinkert, C. A., Milich, D. R., Filippi, P., McLachlan, A., Palmiter, R. D., and Brinster, R. L., 1985, A transgenic mouse model of the chronic hepatitis B surface antigen carrier state, *Science* **230**:1157–1160.

Costantini, F., and Lacy, E., 1981, Introduction of a rabbit β-globin gene into the mouse germ line, *Nature (Lond.)* **294**:92–94.

Cronmiller, C., and Mintz, B., 1978, Karyotypic normalcy and quasi-normalcy of developmentally totipotent mouse teratocarcinoma cells, *Dev. Biol.* **67**:465–477.

Dewey, M. J., Martin, D. W., Martin, G. R., and Mintz, B., 1977, Mosaic mice with teratocarcinoma-derived mutant cells deficient in hypoxanthine phosphoribosyltransferase, *Proc. Natl. Acad. Sci. USA* **74**:5564–5568.

Etkin, L. D., and Roberts, M., 1983, Transmission of integrated sea urchin histone genes by nuclear transplantation in *Xenopus laevis*, *Science* **221**:67–69.

Fleischman, R. A., and Mintz, B., 1979, Prevention of genetic anemias in mice by microinjection of normal hematopoietic stem cells into the fetal placenta, *Proc. Natl. Acad. Sci. USA* **76**:5736–5740.

Fraley, R. and Papahadjopolos, D., 1982, Liposomes: The development of a new carrier system for introducing nucleic acids into plant and animal cells, *Curr. Topics Microbiol. Immunol.* **96**:171–191.

Frels, W. I., Bluestone, J. A., Hodes, R. J., Capecchi, M. R., and Singer, D. S., 1985, Expression of a microinjected porcine class 1 major histocompatiblity complex gene in trangenic mice, *Science* **228**:577–580.

Gillies, S. D., Morrison, S. L., Oi, V. T., and Tonegawa, S., 1983, A tissue-specific transcription enhancer element is located in the major intron of a rearranged immunoglobulin heavy chain gene, *Cell* **33**:717–728.

Goldberg, D. A., Posakony, J. W., and Maniatis, T., 1983, Correct developmental expression of a cloned alcohol dehydrogenase gene transduced into the *Drosophila* germ line, *Cell* **34**:59–73.

Gordon, J. W., 1983, Studies of foreign genes transmitted through the germ line of transgenic mice, *J. Exp. Zool.* **228**:313–324.

Gordon, J. W., 1986, A foreign dihydrofolate reductase gene in transgenic mice acts as a dominant mutation, *Mol. Cell. Biol.* **6**:2158–2167.

Gordon, J. W., and Ruddle, F. H., 1981, Integration and stable germ line transmission of genes injected into mouse pronuclei, *Science* **214**:1244–1246.

Gordon, J. W., and Ruddle, F., H., 1982, Germ line transmission in transgenic mice, in: *Embryonic Development, Part B: Cellular Aspects*, pp. 111–124, Alan Liss, New York.

Gordon, J. W., and Ruddle, F. H., 1983, The production of transgenic mice, in: *Eukaryotic Gene Expression* (A. Kumar, ed.), George Washington University Biochemistry Symposium, pp. 109–116, Plenum Press, New York.

Gordon, J. W., and Ruddle, F. H., 1984, Gene transfer into mouse embryos: Production of transgenic mice by pronuclear injection, in: *Methods in Enzymology 101* (R. Wu, L. Grossman, and K. Moldave, eds.), pp. 411–433, Academic Press, New York.

Gordon, J. W., and Ruddle, F. H., 1985, DNA-mediated genetic transformation of mouse embryos and bone marrow—a review, *Gene* **33**:121–136.

Gordon, J. W., Scangos, G. A., Plotkin, D. J., Barbosa, J. A., and Ruddle, F. H., 1980, Genetic transformation of mouse embryos by microinjection of purified DNA, *Proc. Natl. Acad. Sci. USA* **77**:7380–7384.

Green, E. L. (ed.), 1975, *Biology of the Laboratory Mouse*, Dover Publications, New York.

Grosschedl, R., Weaver, D., Baltimore, D., and Costantini, F., 1984, Introduction of a μ immunoglobulingene into the mouse germ line: Specific expression in lymphoid cells and synthesis of functional antibody, *Cell* **38**:647–658.

Hamer, D. H., and Leder, P., 1979, Expression of the chromosomal mouse B^{maj}-globin gene cloned in SV40, *Nature (Lond.)* **281**:35–40.

Hammer, R. E., Palmiter, R. D., and Brinster, R. L., 1984, Partial correction of murine hereditary growth disorder by germ-line incorporation of a new gene, *Nature (Lond.)* **311**:65–67.

Hammer, R. E., Brinster, R. L., Rosenfeld, M. G., Evans, R. M., and Mayo, K. E., 1985*a*, Expression of human growth hormone-releasing factor in transgenic mice results in increased somatic growth, *Nature (Lond.)* **315**:413–416.

Hammer, R. E., Pursel, V. G., Rexroad, C. E., Wall, R. J., Bolt, D. J., Ebert, K. M., Palmiter, R. D., and Brinster, R. L., 1985*b*, Production of transgenic rabbits, sheep and pigs by microinjection, *Nature (Lond.)* **315**:680–683.

Hammer, R. E., Idzerda, R. L., Brinster, R. L., and McKnight, G. S., 1986, Estrogen regulation of the avian transferrin gene in transgenic mice, *Mol. Cell. Biol.* **6**:1010–1014.

Hanahan, D., 1985, Heritable formation of pancreatic β-cell tumours in transgenic mice expressing recombinant insulin/simian virus 40 oncogenes, *Nature (Lond.)*, **315**:115–122.

Harbers, K., Jähner, D., and Jaenisch, R., 1981, Microinjection of cloned retroviral genomes into mouse zygotes: Integration and expression in the animal, *Nature (Lond.)* **293**:540–542.

Humphries, R. K., Berg, P., DiPietro, J., Bernstein, S., Baur, A., Nienhuis, A. W., and Anderson, W. F., 1985, Transfer of human and murine globin-gene sequences into transgenic mice, *Am. J. Hum. Genet.* **37**:295–310.

Illmensee, K., and Croce, C. M., 1979, Xenogeneic gene expression in chimeric mice derived from rat-mouse hybrid cells, *Proc. Natl. Acad. Sci. USA* **76**:879–883.

Illmensee, K., and Mintz, B., 1976, Totipotency and normal differentiation of single teratocarcinoma cells cloned by injection into blastocysts. *Proc. Natl. Acad. Sci. USA* **73**:549–553.
Illmensee, K., Hoppe, P. C., and Croce, C. M., 1978, Chimeric mice derived from human-mouse hybrid cells, *Proc. Natl. Acad. Sci. USA* **75**:1914–1918.
Jaenisch, R., 1976, Germ line integration and Mendelian transmission of the exogenous Moloney leukemia virus, *Proc. Natl. Acad. Sci. USA* **73**:1260–1264.
Jaenisch, R., 1980, Retroviruses and embryogenesis: Microinjection of Moloney leukemia virus into midgestation mouse embryos, *Cell* **19**:181–188.
Jaenisch, R., and Berns, A., 1977, Tumor virus expression during mammalian embryogenesis, in: *Concepts in Mammalian Embryogenesis* (M. I. Sherman, ed.), pp. 267–303 MIT Press, MA.
Jaenisch, R., and Mintz, B., 1974, Simian virus 40 DNA sequences in DNA of healthy adult mice derived from preimplantation blastocysts injected with viral DNA, *Proc. Natl. Acad. Sci. USA* **71**:1250–1254.
Jaenisch, R., Fan, H., and Croker, B., 1975, Infection of preimplantation mouse embryos and of newborn mice with leukemia virus: Tissue distribution of viral DNA and RNA and leukemogenesis in the adult animal, *Proc. Natl. Acad. Sci. USA* **72**:4008–4012.
Jaenisch, R., Jähner, D., Nobis, P., Simon, I., Löhler, J., Harbers, K., and Grotkopp, D., 1981, Chromosomal position and activation of retroviral genomes inserted into the germ line of mice, *Cell* **24**:519–529.
Jaenisch, R., Harbers, K., Schnieke, A., Löhler, J., Chumakov, I., Jähner, D., Grotkopp, D., and Hoffmann, E., 1983, Germline integration of Moloney murine leukemia virus at the *Mov13* locus leads to recessive lethal mutation and early embryonic death, *Cell* **32**:209–216.
Jähner, D., and Jaenisch, R., 1980, Integration of Moloney leukaemia virus into the germ line of mice: correlation between site of integration and virus activation, *Nature (Lond.)* **287**:456–458.
Jähner, D., Stulmann, H., Stewart, C. L., Harbers, K., Lohler, J., Simon, I., and Jaenisch, R., 1982, *De novo* methylation and expression of retroviral genomes during mouse embryogenesis, *Nature (Lond.)* **298**:623–628.
Kelly, F., and Condamine, H., 1982, Tumor viruses and early mouse embryos, *Biochim. Biophys. Acta* **651**:105–141.
Khillan, J. S., Schmidt, A., Overbeek, P. A., de Crombrugghe, B., and Westphal, H., 1986, Developmental and tissue-specific expression directed by the α_2 type I collagen promoter in transgenic mice, *Proc. Natl. Acad. Sci. USA* **83**:725–729.
Krumlauf, R., Hammer, R. E., Tilghman, S. M., Brinster, R. L., 1985, Developmental regulation of α-fetoprotein genes in transgenic mice, *Mol. Cell. Biol.* **5**:1639–1648.
Krumlauf, R., Chapman, V. M., Hammer, R. E., Brinster, R., and Tilghman, S. M., 1986, Differential expression of α-fetoprotein genes on the inactive X chromosome in extraembryonic and somatic tissues of a transgenic mouse line, *Nature* **319**:224–226.
Kurtz, D. T., 1981, Hormone inducibility of rate α 2u globulin genes in transfected mouse cells, *Nature* **291**:629–631.
Kuschner, P. J., and Martin, D. W., 1983, Transfer of the human growth hormone gene to mice, *J. Cell. Biochem.* [Suppl.] 7A:144.
Lacy, E., Roberts, S., Evans, E. P., Burtenshaw, M. D., and Costantini, F. D., 1983, A foreign β-globin gene in transgenic mice: Integration at abnormal positions and expression in inappropriate tissues, *Cell* **34**:343–358.
Leder, A., Pattengale, P. K., Kuo, A., Stewart, T. A., and Leder, P., 1986, Consequences of widespread deregulation of the c-*myc* gene in transgenic mice. Multiple neoplasms and normal development, *Cell* **45**:485–495.
LeMeur, M., Gerlinger, P., Benoist, C., and Mathis, D., 1985, Correcting an immune-response deficiency by creating E_α gene transgenic mice, *Nature (Lond.)* **316**:38–42.
Low, M. J., Hammer, R. E., Goodman, R. H., Habener, J. F., Palmiter, R. D., and Brinster, R. L., 1985, Tissue-specific posttranslational processing of pre-prosomatostatin encoded by a metallothionein–somatostatin fusion gene in transgenic mice, *Cell* **41**:211–219.
Low, M. J., Lechan, R. M., Hammer, R. E., Brinster, R. L., Habener, J. F., Mandel, G., and Goodman, R. H., 1986, Gonadotroph-specific expression of metallothionein fusion genes in pituitaries of transgenic mice, *Science* **231**:1002–1004.
Magram, J. Chada, K., and Costantini, F., 1985, Developmental regulation of a cloned adult β-globin gene in transgenic mice, *Nature (Lond.)* **315**:338–340.

Martinell, J., Whitney, J. B., Popp, R. A., Russell, L. B., and Anderson W. F., 1981, Three mouse models of human thalassemia, *Proc. Natl. Acad. Sci. USA* **78**:5056–5060.

McKnight, G. S., Hammer, R. E., Kuenzel, E. A., and Brinster, R. L., 1983, Expression of the chicken transferrin gene in trangenic mice, *Cell* **34**:335–341.

Messing, A., Chen, H. Y., Palmiter, R. D., and Brinster, R. L., 1985, Peripheral neuropathies, hepatocellular carcinomas and islet cell adenomas in transgenic mice, *Nature* **316**:461–463.

Mintz, B., 1974, Gene control of mammalian differentiation, *Annu. Rev. Genet.* **8**:411–470.

Mintz, B., 1979, Teratocarcinoma cells as vehicles for introducing mutant genes into mice, *Differentiation* **13**:25–27.

Mintz, B., 1983, Manipulating the genotype of developing mice, in: *Gene Structure and Regulation in Development* (S. Subtelny and F. C. Kafatos, eds.), pp. 113–133, Alan Liss, New York.

Mintz, B., and Illmensee, K., 1975, Normal genetically mosaic mice produced from malignant teratocarcinoma cells, *Proc. Natl. Acad. Sci. USA* **72**:3585–3589.

Mulligan, R. C., Howard, B. H., and Berg, P., 1979, Synthesis of rabbit β-globin in cultured monkey kidney cells following infection with a SV40 β-globin recombinant genone, *Nature (Lond.)* **227**:108–114.

Norstedt, G., and Palmiter, R., 1984, Secretory rhythm of growth hormone regulates sexual differentiation of mouse liver, *Cell* **36**:805–812.

Ornitz, D. M., Palmiter, R. D., Hammer, R. E., Brinster, R. L., Swift, G. H., and MacDonald, R. J., 1985, Specific expression of an elastase-human growth hormone fusion gene in pancreatic acinar cells of transgenic mice, *Nature* **313**:600–602.

Overbeek, P. A., Chepelinsky, A. B., Khillan, J. S., Piatigorsky, J., Westphal, H., 1985, Lens-specific expression and developmental regulation of the bacterial chloramphenicol acetyltransferase gene driven by the murine α A-crystalline promoter in transgenic mice, *Proc. Natl. Acad. Sci. USA* **82**:7815–7819.

Overbeek, P. A., Lai, S. P., Van Quill, K. R., and Westphal, H., 1986, Tissue-specific expression in transgenic mice of a fused gene containing RSV terminal sequences, *Science* **231**:1574–1577.

Palmiter, R. D., and Brinster, R. L., 1985, Transgenic mice, *Cell* **41**:343–345.

Palmiter, R. D., Brinster, R. L., Hammer, R. E., Trumbauer, M. E., Rosenfeld, M. G., Birnberg, N. C., and Evans, R. M., 1982*a*, Dramatic growth of mice that develop from eggs microinjected with metallothione in growth hormone fusion genes, *Nature (Lond.)* **300**:611–615.

Palmiter, R. D., Chen, H. Y., and Brinster, R. L., 1982*b*, Differential regulation of metallothionein-thymidine kinase fusion genes in transgenic mice and their offspring, *Cell* **29**:701–710.

Palmiter, R. D., Norstedt, G., Gelinas, R. E., Hammer, R. E., and Brinster, R. L., 1983, Metallothionein-human GH fusion genes stimulate growth of mice, *Science* **222**:809–814.

Palmiter, R. D., Wilkie, T. M., Chen, H. Y., and Brinster, R. L., 1984, Transmission distortion and mosaicism in an unusual transgenic mouse pedigree, *Cell* **36**:869–877.

Palmiter, R. D., Chen, H. Y., Messing, A., and Brinster, R. L., 1985, SV40 enhancer and large-T antigen are instrumental in development of choroid plexus tumours in transgenic mice, *Nature* **316**:457–460.

Papaioannou, V. E., McBurney, M. W., Gardner, R. L., and Evans, M. J., 1975, Fate of teratocarcinoma cells injected into early mouse embryos, *Nature* **258**:70–72.

Papaioannou, V. E., Gardner, R. L., McBurney, M. W., Babinet, C., and Evans, M. J., 1978, Participation of cultured teratocarcinoma cells in mouse embryogenesis. *J. Embryol. Exp. Morph.* **44**:93–104.

Pellicer, A., Wigler, M., Axel, R., and Silverstein, S., 1978, The transfer and stable integration of the HSV thymidine kinase gene into mouse cells, *Cell* **14**:133–141.

Pellicer, A., Wagner, E. F., Kareh, A. E., Dewey, M. J., Reuser, A. J., Silverstein, S., Axel, R., and Mintz, B., 1980, Introduction of a viral thymidine kinase gene and the human β-globin gene into developmentally multipotential mouse teratocarcinoma cells, *Proc. Natl. Acad. Sci. USA* **77**:2098–2102.

Queen, C., and Baltimore, D., 1983, Immunoglobulin gene transcription is activated by downstream sequence elements, *Cell* **33**:741–748.

Ragg, H., and Weissmann, C., 1983, Not more than 117 base pairs of 5′-flanking sequence are required for inducible expression of a human IFN-α gene, *Nature (Lond.)* **303**:439–442.

Rassoulzadegan, M., Binetruy, B., and Cuzin, F., 1982, High frequency of gene transfer after fusion between bacteria and eukaryotic cells, *Nature (Lond.)* **295**:257–259.

Razzaque, A., Mizusawa, H., and Seidman, M. M., 1983, Rearrangement and mutagenesis of a shuttle vector plasmid after passage in mammalian cells, *Proc. Natl. Acad. Sci. USA* **80**:3010–3014.

Ritchie, K. A., Brinster, R. L., and Storb, U. 1984, Allelic exclusion and control of endogenous immunoglobulin gene rearrangement in K transgenic mice, *Nature* **312**:517–520.

Robins, D. M., Ripley, S., Henderson, A. S., and Axel, R., 1981, Transforming DNA integrates into the host chromosome, *Cell* **23**:29–39.

Ross, S. R., and Solter, D., 1985, Glucocorticoid regulation of mouse mammary tumor virus sequences in transgenic mice, *Proc. Natl. Acad. Sci. USA* **82**:5880–5884.

Rubin, G. M., and Spradling, A. C., 1982, Genetic transformation of *Drosophila* with transposable element vectors, *Science* **218**:348–353.

Ruddle, F. H., 1981, A new era in mammalian gene mapping: Somatic cell genetics and recombinant DNA methodologies. *Nature (Lond.)* **294**:115–120.

Rusconi, S., and Köhler, G., 1985, Transmission and expression of a specific pair of rearranged immunoglobulin μ and κ genes in a transgenic mouse line, *Nature* **314**:330–334.

Rusconi, S., and Schaffner, W., 1981, Transformation of frog embryos with a rabbit β-globin gene *Proc. Natl. Acad. Sci. USA* **78**:5051–5055.

Scangos, G., and Ruddle, F. H., 1981, Mechanisms and application of DNA-mediated gene transfer in mammalian cells—A review, *Gene* **14**:1–10.

Schaffner, W., 1980, Direct transfer of cloned genes from bacteria to mammalian cells, *Proc. Natl. Acad. Sci. USA* **77**:2163–2167.

Schnieke, A., Harbers, K., and Jaenisch, R., 1983, Embryonic lethal mutation in mice induced by retrovirus insertion into the α1(I) collagen gene, *Nature (Lond.)* **304**:315–320.

Scholnick, S. B., Morgan, B. A., and Hirsh, J., 1983, The cloned dopa decarboxylase gene is developmentally regulated when reintegrated into the *Drosophila* genome, *Cell* **34**:37–45.

Selden, R. F., Wagner, T. E., Tischer, E., and Goodman, H. M. 1983, Microinjection of a human/rat growth hormone gene into mouse embryos, *J. Cell. Biochem.* [Suppl.] **7A**:143.

Selden, R. F., Skóskiewicz, M. J., Howie, K. B., Russell, P. S., and Goodman, H. M., 1986, Regulation of human insulin gene expression in transgenic mice, *Nature* **321**:525–528.

Shani, M., 1985, Tissue-specific expression of rat myosin light-chain 2 gene in transgenic mice, *Nature (Lond.)* **314**:283–286.

Simon, D., Stuhlmann, H., Jähner, D., Wagner, H., Werner, E., and Jaenisch, R., 1983, Retrovirus genomes methylated by mammalian but not bacterial methylases are non-infectious, *Nature (Lond.)* **304**:275–277.

Small, J. A., Blair, D. G., Showalter, S. D., and Scangos, G. A., 1985, Analysis of a transgenic mouse containing simian virus 40 and v-*myc* sequences, *Mol. Cell. Biol.* **5**:642–648.

Spradling, A. C., and Rubin, G. M., 1982, Transposition of cloned P elements into *Drosophila* germ line chromosomes, *Science* **218**:341–347.

Spradling, A. C., and Rubin, G. M., 1983, The effect of chromosomal position on the expression of the *Drosophila* xanthine dehydrogenase gene, *Cell* **34**:47–57.

Stavridis, J. C., and Psallidopoulos, M. 1982, Use of transferrin as a gene-carrier to the erythroid cells of the marrow, *Cell. Mol. Biol.* **28**:15–18.

Stewart, C., Harbers, K., Jähner, D., and Jaenisch, R., 1983, X chromosome-linked transmission and expression of retroviral genomes microinjected into mouse zygotes, *Science* **221**:760–762.

Stewart, C. L., 1982, Formation of viable chimaeras by aggregation between teratocarcinomas and preimplantation mouse embryos, *J. Embryol. Exp. Morph.* **67**:167–179.

Stewart, C. L., Stuhlmann, H., Jähner, D., and Jaenisch, R., 1982, *De novo* methylation, expression, and infectivity of retroviral genomes introduced into embryonal carcinoma cells, *Proc. Natl. Acad. Sci. USA* **79**:4098–4102.

Stewart, T. A., and Mintz, B., 1981, Successive generations of mice produced from an established culture line of euploid teratocarcinoma cells, *Proc. Natl. Acad. Sci USA* **78**:6314–6318.

Stewart, T. A., Wagner, E. F., and Mintz, B., 1982, Human β-globin gene sequences injected into mouse eggs, retained in adults, and transmitted to progeny, *Science* **217**:1046–1048.

Stewart, T. A., Pattengale, P. K., and Leder, P., 1984, Spontaneous mammary adenocarcinomas in transgenic mice that carry and express MTV/myc fusion genes, *Cell* **38**:627–637.

Storb, U., O'Brien, R. L., McMullen, M. D., Gollahon, K. A., and Brinster, R. L., 1984, High expression of cloned immunoglobulin κ gene in transgenic mice is restricted to B lymphocytes, *Nature* **310:**238–241.

Storb, U., Denis, K. A., Brinster, R. L., and Witte, O. N., 1985, Pre-B cells in κ-transgenic mice, *Nature* **316:**356–358.

Stout, J. T., Cen, H. Y., Brennand, J., Caskey, C. T., and Brinster, R. L., 1985, Expression of human HPRT in the central nervous system of transgenic mice, *Nature* **317:**250–252.

Stuhlmann, H., Cone, R., Mulligan, R. C., and Jaenisch, R., 1984, Introduction of a selectable gene into different animal tissue by a retrovirus recombinant vector, *Proc. Natl. Acad. Sci. USA* **81:**7151–7155.

Swanson, L. W., Simmons, D. M., Arriza, F., Hammer, R., Brinster, R., Rosenfeld, M. G., and Evans, R. M., 1985, Novel developmental specificity in the nervous system of transgenic animals expressing growth hormone fusion genes, *Nature* **317:**363–366.

Swift, G. H., Hammer, R. E., MacDonald, R. J., and Brinster, R. L., 1984, Tissue-specific expression of the rat pancreatic elastase 1 gene in transgenic mice, *Cell* **38:**639–646.

Townes, T. M., Chen, H. Y., Lingrel, J. B., Palmiter, R. D., and Brinster, R. L., 1985, Expression of human β-globin genes in transgenic mice: Effects of a flanking metallothionein human growth hormone fusion gene, *Mol. Cell. Biol.* **5:**1977–1983.

van Ooyen, A., van den Berg, J., Mantei, N., and Weissman, C., 1979, Comparison of total sequence of a cloned rabbit β-globin gene and its flanking regions with a homologous mouse sequence, *Science* **206:**337–344.

Wagner, E. F., and Mintz, B., 1982, Transfer of nonselectable genes into mouse teratocarcinoma cells and transcription of the transferred human β-globin gene, *Mol. Cell. Biol.* **2:** 190–198.

Wagner, E. F., Stewart, T. A., and Mintz, B., 1981, The human β-globin gene and a functional viral thymidine kinase gene developing mice. *Proc. Natl. Acad. Sci. USA* **78:**5016–5020.

Wagner, E. F., Covarrubias, L., Stewart, T. A., and Mintz, B., 1983, Prenatal lethalities in mice homozygous for human growth hormone gene sequences integrated in the germ line, *Cell* **35:**647–655.

Wagner, T. E., Hoppe, P. C., Jollick, J. D., Scholl, D. R., Hodinka, R. L., and Gault, J. B., 1981, Microinjection of a rabbit β-globin gene into zygotes and its subsequent expression in adult mice and their offspring, *Proc. Natl. Acad. Sci. USA* **78:**6376–6380.

Watanabe, T., Dewey, M. J., and Mintz, B., 1978, Teratocarcinoma cells as vehicles for introducing specific mutant mitochondrial genes into mice, *Proc. Natl. Acad. Sci. USA* **75:**5113–5117.

Weaver, D., Costantini, F., Imanishi-Kari, T., and Baltimore, D., 1985, A transgenic immunoglobulin mu gene prevents rearrangement of endogenous genes, *Cell* **42:**117–127.

Weaver, D., Reis, M. H., Albanese, C., Costantini, F., Baltimore, D., and Imanishi-Kari, T., 1986, Altered repertoire of endogenous immunoglobulin gene expression in transgenic mice containing a rearranged mu heavy chain gene, *Cell* **45:**247–259.

Wigler, M., Silverstein, S., Lee, L.-S., Pellicer, A., Cheng, Y.-c., and Axel, R., 1977, Transfer of purified Herpes virus/thymidine kinase gene to cultured mouse cells, *Cell* **11:**223–232.

Willecke, K., Klomfab, M., Mierau, R., and Dohmer, J., 1979, Intraspecies transfer via total cellular DNA of the gene for hypoxanthine phosphoribosyltransferase into cultured mouse cells, *Mol. Gen. Genet.* **170**:179–185.

Willison, K., Babinet, C., Boccara, M., Kelly, F., 1983, Infection of preimplantation mouse embryos with simian virus 40, in: *Teratocarcinoma Stem Cells*, Cold Spring Harbor Conference on Cell Proliferation, **10:**307.

Woychik, R. P., Stewart, T. A., Davis, L. G., and Leder, P., 1985, An inherited limb deformity created by insertional mutagenesis in a transgenic mouse, *Nature* **318**:36–40.

Yamaizumi, M., Horwich, A. L., and Ruddle, F. H., 1983, Expression and stabilization of microinjected plasmids containing the herpes simplex virus thymidine kinase gene and polyoma virus DNA in mouse cell, *Mol. Cell. Biol.* **3:**511–522.

Yamamura, K., Kikutani, H., Folsom, V., Clayton, L. K., Kimoto, M., Akira, S., Kashiwamura, S., Tonegawa, S., and Kishimoto, T., 1985, Functional expression of a microinjected E_{α}^{d} gene in C57BL/6 transgenic mice, *Nature (Lond.)* **316:**67–69.

Chapter 2

Nuclear and Cytoplasmic Transfer in Mammalian Embryos

JAMES McGRATH and DAVOR SOLTER

1. Introduction

Manipulation of embryos (surgical and/or chemical) is an important tool in the experimental embryology of mammals. By altering the number of cells in the embryo, the relationship of cells to each other, combinations of cells from different embryos, and so forth, much has been learned about mechanisms which control normal mammalian development (for reviews, see Mintz, 1974; Gardner, 1978; Johnson, 1981). This chapter is concerned, however, only with procedures that alter the nuclear–cytoplasmic composition of the early mammalian embryo. These alterations include the introduction of foreign nuclei or cytoplasm into intact or enucleated early mammalian ova. The response of the transplanted nucleus to its new cytoplasmic environment may permit insight into the nature of nuclear–cytoplasmic communication. In addition, the ability of the transplanted nucleus to assume the role of the embryo nucleus should permit a functional assessment of nuclear potential. Similarly, the transfer of foreign cytoplasm will expose the host nucleus to cytoplasmic signals not normally encountered and thus serve to define cytoplasmic controls in development. Since experimental manipulations of this type require the technical capability to transfer nuclei or cytoplasm, a description of a nuclear–cytoplasmic transfer method devised by the authors is presented. The ability to refashion the nuclear and cytoplasmic components of the early mammalian embryo should enhance our understanding of how these two fundamental cellular compartments interact in normal development.

2. Nuclear Transfer in Amphibians versus Mammals

A basic developmental question, addressed by the method of nuclear transfer, relates to the uniqueness of the embryonic pronuclei. Can one replace the

JAMES McGRATH and DAVOR SOLTER • The Wistar Institute, Philadelphia, Pennsylvania 19104.

genomic material of the zygote with the genomic material of the other cells without disrupting normal development? If not, why? This question has been addressed in nuclear transplantation analyses of amphibian embryo development and the results show that nuclei from early embryonic stages can support normal development (Briggs and King, 1952; King and Briggs, 1955; Gurdon, 1962*a*). Nuclei isolated from later-staged embryonic or adult cells, however, exhibit a diminished capacity to support normal development (King and Briggs, 1956; Subtelney, 1965; DiBerardino and King, 1967; Gurdon, 1962*b*). Although these experiments clearly show that early embryonic nuclei are developmentally totipotent it is unclear whether nuclei from later stages of development share a similar potential (for a review, see Danielli and DiBerardino, 1979).

Early amphibian and mammalian development differ, however, in several important respects and one cannot assume that results obtained in amphibian species will apply to mammals. The amphibian egg possesses a large store of maternal mRNA and is able to undergo substantial development in the presence of transcriptional inhibitors (for a review, see Davidson, 1976). In the presence of similar inhibitors, however, mouse embryo development is completely blocked (for a review, see Johnson, 1981). These results indicate that immediate and presumably correct gene activity must take place if mammalian development is to proceed. Therefore in mammals, in distinction to amphibians, experimentally transferred nuclei must be reprogrammable and functional from the beginning. In addition, any experimental design that presumes to introduce or exchange nuclei in the early mammalian embryo must recognize and accommodate the relative small size and physical fragility of the latter.

3. Nuclear Transfer in Nonenucleated Mammalian Embryos

Investigators have attempted to introduce nuclei into nonenucleated early mammalian embryos using either microsurgical injection or viral-mediated fusion. While microsurgical injection permits more direct control over nuclear introduction, this approach is hampered by the relatively limited ability of the mammalian embryo to survive micropipette penetration. Alternatively viral-mediated fusion, which permits a high survival of the manipulated embryos, frequently results in only a minority of the embryos fusing with the donor cell and incorporating the donor nucleus. Despite these difficulties both methods have been successfully employed and a summary of the relevant studies is presented in Table I.

3.1. Nuclear Transfer *via* Viral-Mediated Cell Fusion

Inactivated Sendai virus has been employed to fuse together various combinations of unfertilized ova, fertilized embryos, later-staged embryonic cells and adult somatic cells. Since Sendai virus-mediated fusion of oocytes is ad-

Table I. Introduction of a Foreign Nucleus into Nonenucleated Mammalian Ova and Embryos

Donor	Recipient	Method of introduction	Reference
Spleen cells Bone marrow cells	Fertilized and unfertilized ova 2-cell embryos	Sendai virus	Graham (1969)
Embryonic fibroblasts Adult somatic cells Lymphocytes	Unfertilized ova	Sendai virus	Baranska and Koprowski (1970)
Embryonic fibroblasts Bone marrow cells	8–16-cell embryos	Sendai virus	Graham (1972)
Lymph node cells Bone marrow cells	2- or 8-cell embryos	Sendai virus	Lin *et al.* (1973)
Adult somatic cells	2- or 4-cell embryos	Sendai virus	Bernstein and Mukherjee (1972, 1973)
Rabbit morula nuclei	Rabbit unfertilized ova	Microinjection or Sendai virus	Bromhall (1975)
12–18-cell mouse embryo nuclei	Fertilized 1-cell embryo	Microinjection	Modlinski (1978)
Follicle cells	Unfertilized ova	Sendai virus	Tarkowski and Balakier (1980)
Polar body nuclei	Fertilized 1-cell embryo	Microinjection	Modlinski and McLaren (1980)
Inner cell mass nuclei Trophectoderm nuclei	Fertilized 1-cell embryo	Microinjection	Modlinski (1981)

dressed in Chapter 3 of this volume, only those investigations attempting to fuse later-staged embryonic or adult cells and early embryo blastomeres are considered here.

In the first experiments to introduce a foreign nucleus into a mammalian embryo, Graham (1969, 1972) fused spleen cells, bone marrow cells, or embryonic fibroblasts with unfertilized ova or early cleavage stage mouse embryos. Histological analysis subsequently revealed the presence of donor nuclei in the cytoplasm of some embryos. Continued development, however, was limited to a single cleavage division. Subsequent attempts to introduce foreign nuclei from either embryonic or adult cells using inactivated Sendai virus (Baranska and Koprowski, 1970; Lin *et al.*, 1973; Bromhall, 1975) met with similar results. Namely, donor nuclei were sometimes observed in some embryos, however continued development was limited and the presence of donor nuclei in developing blastomeres could not be definitively ascertained. Embryo cytoplasm was, however, shown to induce both nuclear swelling (Graham, 1969; Tarkowski and Balakier, 1980) and altered RNA synthesis (Berstein and Murkherjee, 1972, 1973) in the newly introduced nuclei.

3.2. Nuclear Transfer *via* Microsurgical Injection

The introduction of a foreign nucleus into a mammalian embryo *via* microinjection was first achieved when Bromhall (1975) successfully microinjected morula stage nuclei into unfertilized rabbit ova. Although a small number of these embryos developed to the blastocyst stage, evidence demonstrating the continued presence of the introduced nucleus was lacking. Modlinski (1978), however, through the use of cytological markers and chromosomal analysis demonstrated the successful microinjection of 8–16-cell stage mouse embryo nuclei into fertilized mouse embryos and the contribution of the injected nuclei to the formation of tetraploid blastocysts. In a subsequent study, inner cell mass cell and trophectoderm cell nuclei were also introduced into fertilized mouse embryos (Modlinski, 1981). The results again showed that injected nuclei could participate in normal preimplantation stage development, since tetraploid nuclei possessing the inner cell mass cell marker chromosomes were present in manipulated blastocysts. The introduction of trophectoderm nuclei was not, however, compatible with successful preimplantation development.

In all the nuclear transfer studies described above, the recipient embryos were not enucleated. Donor nucleus introduction therefore resulted in the production of tetraploid embryos. Tetraploidy, *per se*, however, would not explain the poor development observed in many of these studies since tetraploid embryos can develop at a high proportion through the preimplantation stages (Snow, 1973) and can occasionally develop to term (Snow, 1975). The results of Modlinski (1978, 1981) conclusively show that certain nuclei (12–18-cell and inner cell mass cell) can combine with the host embryo nucleus and normal preimplantation development can occur. At the same time, Modlinski's results show that the presence of trophectoderm nuclei will prevent development. Thus the possibility exists that the presence of the nuclei introduced by other investigators (e.g., adult somatic cell nuclei) may have similarly interfered with subsequent development.

In addition to the transfer of interphase nuclei, microinjection has also been used to introduce spermatozoan nuclei into the cytoplasm of fertilized and unfertilized ova (Uehara and Yanagimachi, 1976, 1977; Thadani, 1980; Markert, 1982). These studies report that subsequent to injection and activation of the ovum cytoplasm, the spermatozoan nucleus is transformed into a male pronucleus irrespective of the species origin of the ovum (Uehara and Yanagimachi, 1976; Thadani, 1980). Of obvious interest, but yet unknown, is the extent to which these interspecies hybrids will undergo subsequent development.

4. Nuclear Transfer in Enucleated Mammalian Embryos

An assessment of the full development potential of nuclei newly introduced into the cytoplasm of fertilized embryos requires the removal of the host male and female pronuclei. Pronuclear removal had initially been accomplished in investigations of haploid and uniparental embryos (Modlinski, 1975,

1980; Hoppe and Illmensee, 1977; Markert and Petters, 1977). In all cases, a micropipette was used to penetrate the plasma membrane of the ovum, a single pronucleus was completely or partially drawn into the pipette, and the pipette withdrawn. Subsequently, the removal of both pronuclei was combined with the microinjection of a foreign nucleus into the newly fertilized mouse embryo (Illmensee and Hoppe, 1981; Hoppe and Illmensee, 1982). These experiments, which produced the first complete nuclear transplant mammalian embryos, used donor nuclei derived from inner cell mass cells or trophectoderm cells of normal or parthenogenetic mouse embryos. In both sets of experiments, the nuclei derived from inner cell masses, but not from trophectoderm, supported development to term and a total of seven adult nuclear-transplant mice have been produced. The phenotype of these mice corresponded to the genotype of the donor nucleus. Despite several attempts, however, these results have not been repeated by other investigators. Our recent results demonstrate that nuclei isolated from cleavage stage embryos and transferred to enucleated zygotes do not support development (McGrath and Solter, 1984*c*).

The method described by Illmensee and Hoppe requires extreme caution in its performance, since the embryo is very sensitive to micropipette penetration and under the best of circumstances, the majority of embryos do not survive this procedure. In addition, this method does not permit the introduction of early preimplantation stage nuclei, since the latter are prohibitively large for mechanical introduction. We have therefore, developed a nuclear transplantation method (McGrath and Solter, 1983*a*) that combines microsurgery with cell fusion and that does not require micropipette penetration of egg plasma membrane. In addition, this method does permit the introduction of the large preimplantation stage nuclei. This procedure is well tolerated by the mouse embryo; to date, more than 100 adult nuclear-transplant mice have been produced (McGrath and Solter, 1983*a*, 1984*a*,*b*).

5. Microsurgical/Cell Fusion Nuclear Transplantation Procedure

The following description of our nuclear transplantation procedure should be viewed as a general outline. Individual experiments may require modifications so that, for example, the dimensions of the enucleation/injection pipette, the type of glass from which it is fashioned, and so forth will depend upon the size of the transferred nucleus and the specific aims of the experiments. In addition, readers desiring a more complete description of relevant microsurgical equipment and of its specific application to the early mammalian embryo are referred to an excellent review by Gardner (1978).

5.1. Preparation of Microinstruments

Throughout the microsurgical manipulations, the embryo is fixed in place with a pipette held in place while nuclei are either removed or introduced *via* an enucleation/injection pipette. Capillary tubing (Pyrex or Soft Flint; Drummond Scientific, Broomall, Pa.) 1.00 mm O.D. and 0.65 mm I.D. is used for the manufacture of both holding pipettes and enucleation/injection pipettes.

5.2. Enucleation/Injection Pipette

The enucleation/injection pipette is pulled on vertical pipette puller (David Kopf Instruments, DKI Model 700C) and the tip of the pipette is broken at the desired outside diameter by fusing the pipette to a glass anvil on a De Fonbrune microforge as diagrammed in Figure 1. The pipette tip is subsequently beveled on a grinding wheel coated with a thin layer of diamond paste (Astro-Met diamond compound; GCA/Precision Scientific, Chicago, Ill.). The outer surface of the beveled pipette is subsequently rinsed with dilute hydrofluoric acid (1 : 10 v/v for Soft Flint glass; 1 : 4 for Pyrex glass) to remove any residual diamond paste and to reduce the outside diameter of the pipette. The tip of the pipette is subsequently sharpened on a microforge using a 0.1-mm platinum wire as diagrammed in Figure 2. Subsequent to sharpening, the pipette tip is given five to six rinses in Nonidet P40 (NP40), followed by 10 rinses in distilled water. Excess detergent is removed from the outside of the pipette with paper toweling. The NP40 treatment is used in place of a silicladding agent. The enucleation/injection pipette is controlled by an oil-filled Beadouin syringe.

5.3. Holding Pipette

Capillary tubing, initially pulled by hand over a microburner, is broken on a glass anvil at an outside diameter of 75 μm in the same manner described for

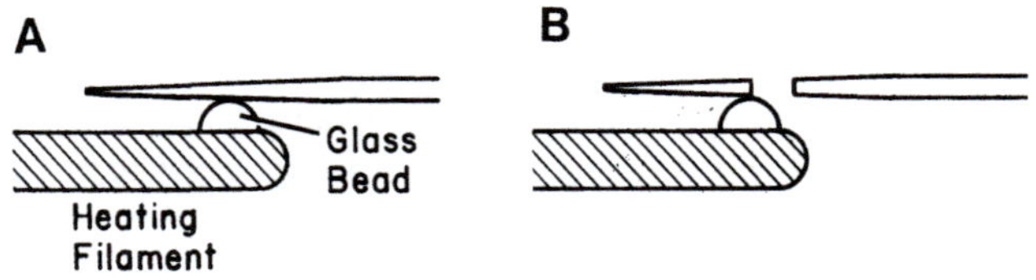

Figure 1. In order to manufacture enucleation/injection pipettes, capillary tubing, pulled on a vertical pipette puller, is placed on a De Fonbrune microforge equipped with a 0.1-mm (36-gauge) platinum filament. Under maximum airflow, the glass anvil, located at the tip of the heated filament, is brought into contact with the horizontal pipette (A). After the pipette has fused with the anvil, the filament is switched to the "off" position, resulting in retraction of the filament and fracture of the micropipette (B). Holding pipettes are similarly broken at the desired outside diameter, except that a 0.3-mm (28-gauge) platinum filament is employed. The holding pipette is polished by discontinuing airflow and placing the pipette tip in close proximity to the filament tip.

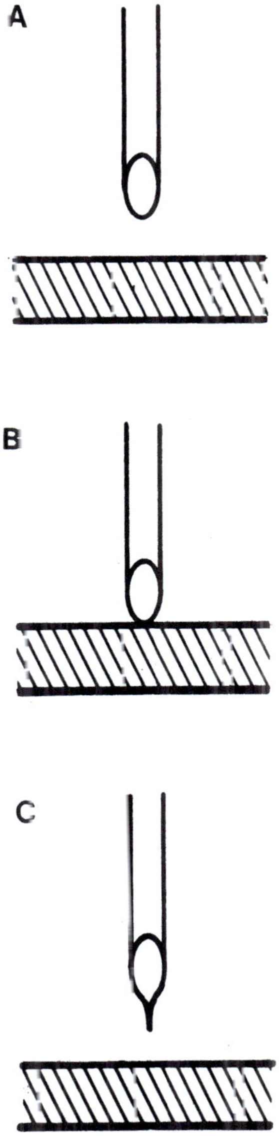

Figure 2. The beveled tip of the enucleation/injection pipette is sharpened by vertically positioning the pipette above a heated 0.1-mm platinum filament (A). Under maximum airflow, the tip is brought into contact with the filament (B). Subsequent withdrawal of the pipette results in the sharpened tip shown (C).

the enucleation/injection pipette. The pipette tip is polished by bringing it in close proximity to a heated 0.3-mm platinum wire on the microforge.

5.4. Sendai Virus

Sendai virus is prepared as described by Harris *et al.* (1966) and Giles and Ruddle (1973). Briefly, day 9 embryonated chicken eggs are candled and surface-sterilized with 70% ethanol. An active seed stock of the virus is diluted

$(1:10^{-5})$ in Hank's balanced salt solution (HBSS) containing 3 mg/ml bovine serum albumin (HBSS + BSA); 0.2 ml of the diluent is injected into the allantoic cavity of the eggs. The site of injection is covered with melted paraffin wax. Eggs are incubated for 72 hr in a humidified atmosphere at 37°C. Incubation is terminated by placing the eggs at −20°C for 30 min. Virus is harvested on ice by removing the egg shell and the outer and inner shell membrane and aspirating the allantoic fluid with an 18-gauge needle connected to a 10-ml syringe. Virus that is to be inactivated is pooled and centrifuged at 2000 rpm for 20 min to remove cell debris. The supernatant is subsequently centrifuged at 16,000 g for 1 hr. Following the second centrifugation, the supernatant is discarded, and HBSS + BSA is added to the pellet (at a volume approximately equal to 1/10th of the original volume). The pellets are softened by overnight storage at 4°C, followed by resuspension of the virus by vortexing. Additional HBSS + BSA may be added to the virus suspension at this time to achieve the desired hemagglutination titer. Hemagglutination titers are performed in 96-well round-bottomed microtiter plates (Flow Labs) using a 0.5% suspension of chicken red blood cells (Giles and Ruddle, 1973).

Virus is inactivated (Neff and Enders, 1968) by adding 1% of β-propiolactone (BPL) (diluted in cold phosphate-buffered saline) to the virus suspension, on ice, to achieve a final concentration of 0.025% BPL. Virus–BPL mixture is stored for 24 hr at 4°C, after which it is placed in a 37°C water bath for 20 min. Inactivated virus is subsequently aliquoted (100 μl), fast-frozen in an acetone-dry ice bath, and stored at −70°C.

5.5. Embryo Isolation and Microsurgery

One-cell stage embryos are isolated from the ampullary region of oviducts from mated females on the day of vaginal plug detection. Cumulus cells are removed with media containing hyaluronidase (Sigma, 500 NF units/ml). Embryos are washed through 4 drops of Whitten's medium (Whitten, 1971) containing 100 μM Na_2 EDTA (Abramczuk *et al.*, 1977) and cultured under silicone oil in an atmosphere of 5% O_2, 5% CO_2, and 90% N_2 at 37°C. Fifteen to 30 min before microsurgery, embryos are placed in bicarbonate-buffered Whitten's medium containing cytochalasin B (Sigma, 5 μg/ml) and colcemid (Sigma, 0.1 μg/ml) and returned to the incubator. Nuclear transplantation is performed on a Leitz Laborlux II fixed-stage microscope with bright-field optics using Leitz micromanipulators. All manipulations are performed at room temperature in hanging drops in a Leitz oil chamber (Fig. 3).

5.5.1. Enucleation

Previous methods for enucleating newly fertilized mouse embryos have succeeded in removing one or both pronuclei (Modlinski, 1975, 1980; Hoppe and Illmensee, 1977; Markert and Petters, 1977; Illmensee and Hoppe, 1981;

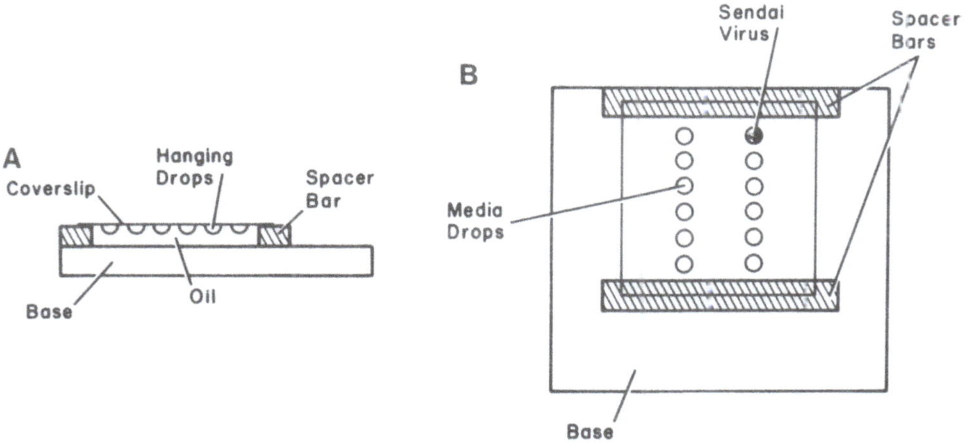

Figure 3. The Leitz oil chamber, measuring 47 × 40 mm with a 4-mm base, has two spacer bars 3 mm in height that serve as supports for a coverslip previously treated with Prosil-28 (Arthur H. Thomas, Philadelphia, Pa.). A thin layer of vacuum grease applied to the spacer bars secures the coverslip. Small drops of HEPES-buffered Whitten's medium containing cytoskeletal inhibitors are applied to the undersurface of the coverslip and the chamber is filled with oil (silicone oil or paraffin oil). The depth of the microdrops is increased by supplying additional medium. A single microdrop may be reserved for inactivated Sendai virus. (A). Side view; (B) overhead view.

Hoppe and Illmensee, 1982). Enucleation, however, oftentimes results in significant embryo death due to micropipette penetration of the embryo plasma membrane. This loss of embryos may be partially ameliorated by prior incubation of the embryos in cytochalasin B (Hoppe and Illmensee, 1977; Modlinski, 1980). To further reduce embryo mortality, we devised an enucleation procedure, shown diagrammatically in Figure 4, which permits the removal of both pronuclei in a membrane-bound pronuclear karyoplast. Embryo survival which typically exceeds 95% is critically dependent on the presence of the cytoskeletal inhibitors cytochalasin B (Hoppe and Illmensee, 1977) and colcemid (McGrath and Solter, 1983*a*). In the absence of these inhibitors, the integrity of the plasma membrane is not maintained during enucleation, and significant embryo mortality results. Enucleation is performed as described in the legend to Figure 4.

5.5.2. Nuclear Introduction

A foreign nucleus may be introduced into the enucleated embryo either by microinjection or by chemical/biological-mediated fusion. Embryo survival following microinjection in our hands typically averaged only 30%, while Sendai virus-mediated fusion of early embryo karyoplasts yielded survival rates in excess of 90%. A diagrammatic representation of a pronuclear transfer is shown in Figure 5.

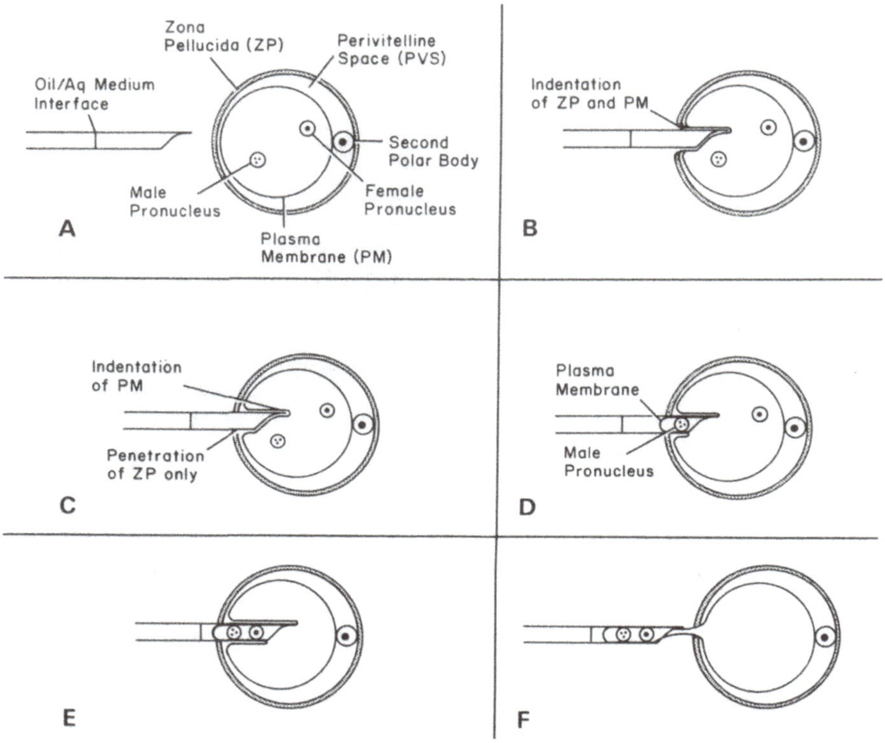

Figure 4. Diagrammatic representation of the removal of the pronuclei from the one-cell stage mouse embryo. The embryo, held in place by a suction pipette (not shown), is oriented with the second polar body opposite to the approach of the enucleation pipette (A). This orientation facilitates the relocation of the original site of pipette penetration when a second nucleus is introduced following enucleation. The zona pellucida (ZP) and embryo plasma membrane are indented by the advancing pipette (B) until the ZP is penetrated (C). Considerable deformation of the embryo may be necessary. The pipette tip is subsequently moved to overlie a pronucleus and the latter, preceded by a portion of the embryo plasma membrane, is aspirated into the pipette (D). The second pronucleus is similarly aspirated (E) and the pipette is withdrawn. Upon continued withdrawal (F) and in the presence of cytoskeletal inhibitors as described in the text, the pronuclear karyoplast will separate from the embryo cytoplasm.

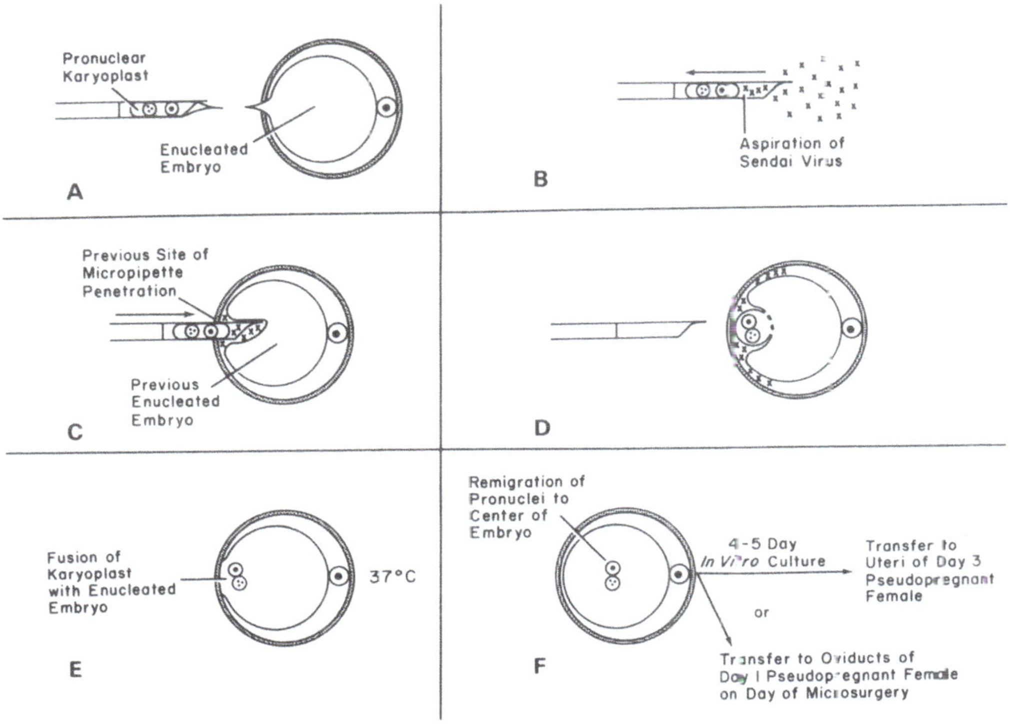

Figure 5. Diagrammatic representation of the introduction of pronuclei into a previously enucleated 1-cell stage mouse embryo. The pipette containing the pronuclear karyoplast is placed in a drop containing inactivated Sendai virus (approximately 3000 hemagglutinating units/ml) and a small volume of the virus suspension is aspirated into the pipette (B). The pipette, after placement in a separate drop containing a previously enucleated 1-cell stage embryo, is advanced through the previous site of enucleation (C) and the contents of the pipette are injected into the perivitelline space (D). After transfer to a 37°C incubator, the pronuclear karyoplast fuses with the enucleated embryo (E). The number of embryos which incorporated the donor nuclei can be determined on a dissecting microscope. The newly introduced pronuclei subsequently migrate to the center of the embryo (F). Nuclear transplant embryos may be cultured *in vitro* or transferred directly to the oviducts of pseudopregnant females.

6. Uses of the Nuclear Transfer Procedure

Several experiments using the described nuclear transfer procedure have been performed. These experiments have included analysis of the persistence of cytoplasmically inherited products in the embryo, maternally inherited lethal mutations, and gynogenetic and androgenetic embryo development.

6.1. Cytoplasmic Inheritance

In an initial series of experiments, the possible cytoplasmic inheritance of stage-specific embryonic antigen-3 (SSEA-3) (Shevinsky *et al.*, 1982) was considered. SSEA-3 is present on unfertilized ova and preimplantation-stage embryos. Embryos of several mouse strains (C3H/He, DBA/2), however, do not express SSEA-3. F1 embryos from reciprocal crosses between SSEA-3 positive and negative stains (e.g., C57BL/6J × C3H/He) are antigen-positive if derived from an antigen-positive female but antigen-negative if derived from an antigen-negative female. Embryos isolated from F1 adult females obtained from reciprocal crosses of antigen-positive and -negative strains (C57BL/6J × C3H/He), however, are always antigen-positive. These results suggested that the expression of SSEA-3 is maternally inherited in the embryo, but not in the adult. We determined whether SSEA-3 antigen expression observed on F1 embryos is derived from the unfertilized egg or from preferential expression of maternal genome by performing reciprocal pronuclear transfers between homozygous C57BL/6J (antigen-positive) and C3H/He (antigen-negative) one-cell-stage embryos (McGrath and Solter, 1983*b*). Enucleated SSEA-3-positive C57BL/6J zygotes containing pronuclei from SSEA-3-negative C3H/He zygotes remained SSEA-3-positive throughout preimplantation development, while reciprocal transfer embryos (C3H/He enucleated zygotes with C57BL/6J nuclei) remained negative. These results show conclusively that SSEA-3 antigen expression on preimplantation stage embryos is inherited solely from the unfertilized ovum. This experiment demonstrates the usefulness of nuclear transplantation in differentiating cytoplasmic inheritance from postfertilization activation of the maternal genome and its potential for analyzing the timing of activation of the maternal and paternal genome.

6.2. Maternally Inherited Mutations

In another series of experiments (McGrath and Solter, 1984*a*), we addressed the mechanism of action of the maternally inherited embryonic lethal mutation T hairpin (T^{hp}). This mutation, a member of the *T/t* complex, is a deletion of a proximal portion of chromosome 17 (Bennett *et al.*, 1975; Silver *et al.*, 1979), which results in a characteristic short-tailed phenotype when heterozygous. $T^{hp}/+$ embryos that inherit the maternal T^{hp} chromosome die during the second half of gestation, while those that inherit the paternal T^{hp} develop into viable short-tailed adults (Johnson, 1974, 1975). These results can be explained in either of two ways. One hypothesis is that during oocyte maturation, some essential product(s) are not made, thus resulting in a cytoplasmic deficiency in T^{hp} oocytes. Alternatively a portion of paternal chromosome 17 may be normally inactive during embryogenesis, thus requiring a functional maternal chromosome 17 for normal development (see McLaren, 1979). To differentiate these possibilities, we performed reciprocal pronuclear transfers between $T^{hp}/+$ and $+/+$ zygotes. When $+/+$ pronuclei from albino matings were introduced into enucleated one-cell stage embryos from pigmented short-tailed $T^{hp}/+$ females, 23% of the nuclear transplanted embryos developed to term. All progeny, as expected, were albino and normal tailed. When the reciprocal pronuclear transplantation was performed, however, only 9% of the embryos developed to term and all viable progeny were normal tailed. The placement of maternal $T^{hp}/+$ pronuclei into $+/+$ cytoplasm did not, therefore, rescue the maternal lethal effect of T^{hp}; thus no viable short-tailed progeny were observed. These experimental findings are consistent with the proposals that (1) unlike maternal chromosome 17, a portion of paternal chromosome 17 is inactive during some period of development, or (2) maternal chromosome 17 is required to activate paternal chromosome 17. Thus pairing of the normally inactive paternal chromosome 17 with the maternally derived deleted T^{hp} chromosome 17 results in embryo lethality. These experiments represent an example in which nuclear transfer can be used to analyze the difference between maternally and paternally inherited genes.

6.3. Maternal Versus Paternal Genome and Parthenogenesis

The proposed differential expression of the inherited maternal and paternal genomes during mammalian development was more extensively addressed in a third series of experiments (McGrath and Solter, 1984*b*). Diploid mouse parthenotes, despite their normal chromosome complement, undergo very limited postimplantation development and never develop to term (Graham, 1974). Proposed mechanisms to explain the developmental failure of parthenotes have included the absence of an extrachromosomal contribution by the fertilizing spermatozoon, the improper activation of the ovum, or the exposure of otherwise hidden recessive lethal mutations.

Several experimental approaches have been used to distinguish between these possibilities. The production of homozygous diploid embryos, in which one of the pronuclei was microsurgically removed and the remaining pronucleus was diploidized by an overnight incubation in cytochalasin B, resulted in the birth of gynogenetic and androgenetic homozygous diploid mice (Hoppe and Illmensee, 1977). These results would indicate that the developmental failure of parthenogenones is due to the absence of an extragenomic contribution by the fertilizing spermatozoon. Attempts by other investigators to produce homozygous diploid mice have resulted in the production of homozygous diploid embryos (Modlinksi, 1980; Markert, 1982). These embryos have not, however, developed to term, and Markert (1982) reports that homozygous diploid embryos, like parthenogenones, die during the first half of gestation. A somewhat different approach to the question of parthenogenetic death was recently described by Surani and Barton (1983). The latter investigators produced triploid fertilized eggs by suppressing the extrusion of the second polar body. Such triploid eggs were then made diploid by either removing the male pronucleus or one of the female pronuclei. Diploid eggs containing both female and male pronuclei developed normally, while those containing two female pronuclei (one being the nucleus of second polar body) died at various times during the first half of gestation. Since those embryos behaved very similarly to parthenotes, despite the fact that they were derived from normally fertilized eggs, Surani and Barton suggested that extensive homozygosity is the cause of parthenogenetic failure. A similar suggestion was also made by Markert (1982). The experiments by Surani and Barton were carefully controlled and would seem at odds with the results obtained by Hoppe and Illmensee (1977). However, excessive homozygosity was not entirely proven as the cause of developmental failure of parthenotes by the experiments of Surani and Barton (1983). One could, for example, argue that the female pronucleus derived from the suppression of the second polar body is not quite identical to the normal female pronucleus and, therefore, the resulting diploid gynogenone is not chromosomally normal. In addition, the ability of the male pronucleus to support development when diploidized as reported by Hoppe and Illmensee (1977) was not addressed. In an attempt to provide some definite answers to these questions, we performed a series of pronuclear transfers that resulted in the production of embryos containing two female pronuclei (biparental gynogenones) or two male pronuclei (biparental androgenones). The development of these embryos was compared with the development of control embryos in which the male or the female pronucleus was replaced with an isoparental pronucleus from another embryo. A diagrammatic representation of this experiment is shown in Figure 6.

The results of an extensive number of transfers in which appropriate genet-

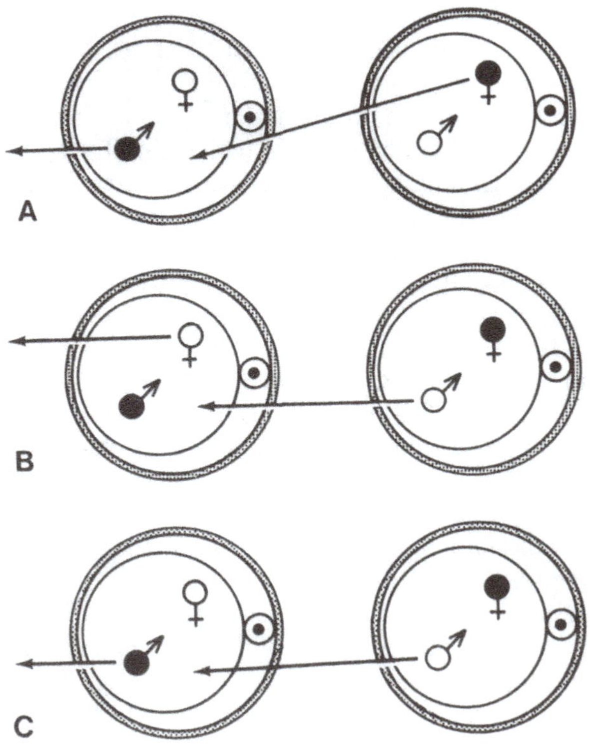

Figure 6. Diagrammatic representation of the production of diploid gynogenetic, androgenetic and control nuclear-transplant embryos (McGrath and Solter, 1984*b*). In the example shown, 1-cell stage embryos obtained from reciprocal matings of BALB/c (albino) and C57Bl/6J (black) mice are isolated, and a single pronucleus is microsurgically removed and replaced with a pronucleus from a second embryo. Gynogenones are produced by removing the C57Bl/6J male pronucleus and introducing a C57Bl/6J female pronucleus from the adjacent embryo (A). Similarly, androgenones are produced by removing the BALB/c female pronucleus and introducing a BALB/c male pronucleus (B). Control nuclear transplants are those embryos in which a pronucleus (in this case the male pronucleus) is removed and a second pronucleus (also male) is introduced (C). Gynogenetic and androgenetic embryos will therefore be agouti, while control nuclear-transplant embryos will be albino.

ic markers were used to verify the genotype of the manipulated embryos showed that in no case did biparental androgenones or gynogenones develop to term, while control nuclear transplant embryos did (McGrath and Solter, 1984*b*). In addition, the death of these embryos, like parthenotes, occurred during the early postimplantation period. Identical findings have been reported by other investigators (Barton *et al.*, 1984; Surani *et al.*, 1984). The logical conclusion from these results is that both maternal and paternal genomes are essential for normal development. Therefore, parthenotes do not die because of an absence of an extragenomic spermatozoon contribution or from lethal homozygosity, but die simply because the paternal genome is absent.

The implication of our experiments is that the maternal and paternal contributions to the embryonic genome are functionally different. It is possible that each of them expresses only a subset of genes necessary for development or alternatively that the maternal and paternal genomes must interact in a specific way in order for normal development to proceed. These results are consistent with the previously presented data concerning the maternal lethal T^{hp} deletion. In the latter study it was concluded that a portion of maternal and paternal chromosome 17 is differentially expressed during development. All these data suggest that at some time, very likely during gametogenesis, both maternal and paternal genomes undergo reversible functional modification so that the male and female pronuclei are not equivalent. Such differences can obviously persist for some time, since the paternally inherited *X* chromosome is recognized as distinct and preferentially inactivated in extraembryonic membranes (Takagi and Sasaki, 1975; West *et al.*, 1977). However, reversibility must be an essential feature of all these modifications since the paternal genome in one generation can and does become the maternal genome in the following generation and *vice-versa*. It is hoped that future experimentation will demonstrate how the maternal and paternal contributions to the embryonic genome are programmed to function differently during embryogenesis, how these differences are maintained, and for how long.

7. Cytoplasmic Transfer

The fertilized ovum contains a large store of cytoplasmic constituents the role and precise fate of which are largely unknown. Almost certainly these constituents are responsible for directing those developmental events that occur before the onset of nuclear gene activity. The transfer of cytoplasm into the early mammalian embryo either by microinjection or by fusing a membrane-bound cytoplast as described above should be a useful tool in analyzing the role of these cytoplasmic components in directing development. An example of such a study was recently reported (Muggleton-Harris *et al.*, 1982).

A high proportion of one-cell-stage embryos from certain "developmentally competent" mouse strains successfully develop *in vitro* to the blastocyst stage. One-cell stage embryos from other strains (developmentally incompetent) undergo complete developmental arrest *in vitro* at the 2-cell stage ("2-cell block"). This developmental incompetence is maternally inherited, since genetically identical F1 embryos obtained from matings between competent and incompetent strains will develop if derived from a competent strain female but will be subject to the 2-cell block if derived from an incompetent strain female (Goddard and Pratt, 1983). These data would imply that *in vitro* culture conditions show a "deficiency" in the metabolic apparatus of embryos of some strains that prevents normal development. In order to determine the nature of this deficiency, Muggleton-Harris *et al.* (1982) microinjected a small amount of cytoplasm (approx 8 pl or approx 4% of the egg volume) from developmentally competent 1- or 2-cell-stage embryos into developmentally incompetent 1- or 2-cell stage embryos. Injected embryos were then able to develop *in vitro* to the blastocyst stage. Cytoplasm derived from 2-cell stage embryos was also injected and observed to support an even greater proportion of developing embryos. These data indicate that the phenomenon of 2-cell block can be corrected by the addition of a small amount of cytoplasm. The question of whether this block is caused by the absence of the substance(s) made during oogenesis or postfertilization is not quite resolved. The fact that cytoplasm from 2-cell stage embryos had a greater ability to support development might suggest postfertilization genomic activity. Alternatively, fertilization may cause the eventual release of previously stored and inactive substances. Further experiments would be necessary to resolve these questions; however, it is apparent from this investigation that cytoplasmic transfer represents yet another manipulation that will prove useful in elucidating the control mechanisms of early development.

8. Conclusions

Using a combination of chemical agents, microsurgical manipulations and virus-mediated membrane fusion, we can dissect and recombine the various components of the mammalian embryo. Thus, the respective roles of the paternal and maternal genomes, egg plasma membrane and egg cytoplasm in regulating developmental events can be assessed. Several described experiments indicate that these techniques will open new experimental possibilities, which will greatly expand and probably change our understanding of control mechanisms necessary to ensure normal development.

ACKNOWLEDGMENTS. This work was supported in part by U.S. Public Health Service Grants HD-12487 and HD-17720 from the National Institute of Clinical Health and Human Development, and by Grants CA-10815 and CA-25875 from the National Cancer Institute.

References

Abramczuk, J., Solter, D., and Koprowski, H., 1977, The beneficial effect of EDTA on development of mouse one-cell embryos in chemically defined medium, *Dev. Biol.* **61**:378–383.

Baranska, W., and Koprowski, H., 1970, Fusion of unfertilized mouse eggs with somatic cells, *J. Exp. Zool.* **174:**1–14.

Barton, S. C., Surani, M. A. H., and Norris, M. L., 1984, Role of paternal and maternal genomes in mouse development, *Nature (Lond.)* **311:**374–376.

Bennett, D., Dunn, L. C., Spiegelman, M., Artzt, K., Cookingham, J., and Schermerhorn, E., 1975, Observations on a set of radiation-induced dominant T-like mutations in the mouse, *Genet. Res.* **26:**95–108.

Bernstein, R. M., and Mukherjee, B., 1972, Control of nuclear RNA synthesis in 2-cell and 4-cell mouse embryos, *Nature (Lond.)* **238:**457–459.

Bernstein, R. M., and Mukherjee, B. B., 1973, Cytoplasmic control of nuclear activity in preimplantation mouse embryos, *Dev. Biol.* **34:**47–65.

Briggs, R., and King, T. J., 1982, Transplantation of living nuclei from blastula cells into enucleated frogs' eggs, *Proc. Natl. Acad. Sci. USA* **38:**455–463.

Bromhall, J. D., 1975, Nuclear transplantation in the rabbit egg, *Nature (Lond.)* **258:**719–722.

Danielli, J. F., and DiBerardino, M. A., (eds.), 1979, Nuclear Transplantation, *International Review of Cytology* (Supplement 9), Academic Press, New York.

Davidson, E. H., 1976, *Gene Activity in Early Development*, Academic Press, New York.

DiBerardino, M. A., and King, T. J., 1967, Development and cellular differentiation of neural nuclear-transplants of known karyotype, *Dev. Biol.* **15:**102–128.

Gardner, R. L., 1978, Production of chimeras by injecting cells or tissue into the blastocyst, in: *Methods in Mammalian Reproduction* (J. C. Daniel, Jr., ed.), pp. 137–165, Academic Press, New York.

Giles, R. E., and Ruddle, F. H., 1973, Production of sendai virus for cell fusion, *In Vitro* **9:**103–107.

Goddard, M. J., and Pratt, H. P. M., 1983, Control of events during early cleavage of the mouse embryo. An analysis of the "2-cell block," *J. Embryol. Exp. Morphol.* **73:**111–133.

Graham, C. F., 1969, The fusion of cells with one- and two-cell mouse embryos, in: The Wistar Institute Symposium Monograph No. 9: *Heterospecific Genome Interaction*, pp. 19–35, The Wistar Institute Press, Philadelphia.

Graham, C. F., 1972, Genetic manipulation of mouse embryos, *Adv. Biosci.* **8:**263–273.

Graham, C. F., 1974, The production of parthenogenetic mammalian embryos and their use in biological research, *Biol. Rev.* **49:**399–422.

Gurdon, J. B., 1962*a*, Adult frogs derived from the nuclei of single somatic cells, *Dev. Biol.* **4:**256–273.

Gurdon, J. B., 1962*b*, The developmental capacity of nuclei taken from intestinal epithelium cells of feeding tadpoles, *J. Embryol. Exp.* **10:**622–640.

Harris, H., Watkins, J. F., Ford, C. E., and Schoefl, G. I., 1966, Artificial heterokaryons of animal cells from different species, *J. Cell Sci.* **1:**1–30.

Hoppe, P. C., and Illmensee, K., 1977, Microsurgically produced homozygous–diploid uniparental mice, *Proc. Natl. Acad. Sci. USA* **74:**5657–5661.

Hoppe, P. C., and Illmensee, K., 1982, Full-term development after transplantation of parthenogenetic embryonic nuclei into fertilized mouse eggs, *Proc. Natl. Acad. Sci. USA* **79:**1912–1916.

Illmensee, K., and Hoppe, P. C., 1981, Nuclear transplantation in Mus musculus: Developmental potential of nuclei from preimplantation embryos, *Cell* **23:**9–18.

Johnson, D. R., 1974, Hairpin tail: A case of post-reductional gene action in the mouse egg?, *Genetics* **76:**795–805.

Johnson, D. R., 1975, Further observations on the hairpin-tail (T^{hp}) mutation in the mouse, *Genet. Res.* **24:**207–213.

Johnson, M. H., 1981, The molecular and cellular basis of preimplantation mouse development, *Biol. Rev.* **56:**463–498.

King, T. J., and Briggs, R., 1955, Changes in the nuclei of differentiating gastrula cells, as demonstrated by nuclear transplantation, *Proc. Natl. Acad. Sci. USA* **41:**321–325.

King, T. J., and Briggs, R., 1956, Serial transplantation of embryonic nuclei, *Cold Spring Harbor Symp. Quant. Biol.* **21:**271–290.

Lin, T. P., Florence, J., and Oh, J. O., 1973, Cell fusion induced by a virus within the zona pellucida of mouse eggs, *Nature (Lond.)* **242:**47–49.

Markert, C. L., 1982, Parthenogenesis, homozygosity, and cloning in mammals, *J. Hered.* **73:**390–397.

Markert, C. L., and Petters, R. M., 1977, Homozygous mouse embryos produced by microsurgery, *J. Exp. Zool.* **201:**295–302.

McGrath, J., and Solter, D., 1983*a*, Nuclear transplantation in the mouse embryo by microsurgery and cell fusion, *Science* **220:**1300–1302.

McGrath, J., and Solter, D., 1983*b*, Nuclear transplantation in mouse embryos, *J. Exp. Zool.* **228:**355–362.

McGrath, J., and Solter, D., 1984*a*, Maternal T^{hp} lethality in the mouse is a nuclear, not cytoplasmic, defect, *Nature (Lond.)* **308:**550–551.

McGrath, J., and Solter, D., 1984*b*, Mouse embryogenesis requires a maternal and paternal genome, *Cell* **37:**179–183.

McGrath, J., and Solter, D., 1984*c*, Inability of mouse blastomere nuclei transferred to enucleated zygotes to support development *in vitro*, *Science* **226:**1317–1319.

McLaren, A., 1979, The impact of pre-fertilization events on postfertilization development in mammals, in: *Maternal Effects in Development* (D. R. Newth and M. Balls, eds.), pp. 287–320, Cambridge University Press, New York.

Mintz, B., 1974, Gene control of mammalian differentiation, *Annu. Rev. Genet.* **8:**411–470.

Modlinski, J. A., 1975, Haploid mouse embryos obtained by microsurgical removal of one pronucleus, *J. Embryol. Exp. Morphol.* **33:**897–905.

Modlinski, J. A., 1978, Transfer of embryonic nuclei to fertilized mouse eggs and development of tetraploid blastocysts, *Nature (Lond.)* **273:**466–467.

Modlinski, J. A., 1980, Preimplantation development of microsurgically obtained haploid and homozygous diploid mouse embryos and effects of pretreatment with cytochalasin B on enucleated eggs, *J. Embryol. Exp. Morphol.* **60:**153–161.

Modlinski, J. A., 1981, The fate of inner cell mass and trophectoderm nuclei transplanted to fertilized mouse eggs, *Nature (Lond.)* **292:**342–343.

Modlinski, J., and McLaren, A., 1980, A method for visualizing the chromosomes of the second polar body of the mouse egg, *J. Embryol. Exp. Morphol.* **60:**93–97.

Muggleton-Harris, A., Whittingham, D. G., and Wilson, L., 1982, Cytoplasmic control of preimplantation development *in vitro* in the mouse, *Nature (Lond.)* **299:**460–462.

Neff, J. M., and Enders, J. F., 1968, Poliovirus replication and cytopathogenicity in monolayer hamster cell cultures fused with beta propiolactone-inactivated sendai virus, *Proc. Soc. Exp. Biol. Med.* **127:**260–267.

Shevinsky, L. H., Knowles, B. B., Damjanov, I., and Solter, D., 1982, Monclonal antibody to murine embryos defines a stage-specific embryonic antigen expressed on mouse embryos and human teratocarcinoma cells, *Cell* **30:**697–705.

Silver, L. M., Artzt, K., and Bennett, D., 1979, A major testicular cell protein specified by a mouse T/t complex gene, *Cell* **17:**275–284.

Snow, M. H. L., 1973, Tetraploid mouse embryos produced by cytochalasin B during cleavage, *Nature (Lond.)* **244:**513–515.

Snow, M. H. L., 1975, Embryonic development of tetraploid mice during the second half of gestation, *J. Embryol. Exp. Morphol.* **34:**707–721.

Subtelny, S. J., 1965, On the nature of the restricted differentiation-promoting ability of transplanted Rana pipiens nuclei from differentiating endoderm cells, *J. Exp. Zool.* **159:**59–92.

Surani, M. A. H., and Barton, S. C., 1983, Development of gynogenetic eggs in the mouse: Implications for parthenogenetic embryos, *Science* **222:**1034–1036.

Surani, M. A. H., Barton, S. C., and Norris, M. L., 1984, Development of reconstituted mouse eggs suggests imprinting of the genoma during gametogenesis, *Nature (Lond.)* **308:**548–550.

Takagi, N., and Sasaki, M., 1975, Preferential inactivation of the paternally derived X chromosome in the extraembryonic membranes of the mouse, *Nature (Lond.)* **256:**640–642.

Tarkowski, A. K., and Balakier, H., 1980, Nucleo-cytoplasmic interactions in cell hybrids between mouse oocytes, blastomeres and somatic cells, *J. Embryol. Exp. Morphol.* **55:**319–330.

Thadani, V. J., 1980, A study of hetero-specific sperm–egg interactions in the rat, mouse, and deer mouse using in vitro fertilization and sperm injection, *J. Exp. Zool.* **212:**435–453.

Uehara, T., and Yanagimachi, R., 1976, Microsurgical injection of spermatozoa into hamster eggs with subsequent transformation of sperm nuclei into male pronuclei, *Biol. Reprod.* **15**:467–470.

Uehara, T., and Yanagimachi, R., 1977, Behavior of nuclei of testicular, caput and cauda epididymal spermatozoa injected into hamster eggs, *Biol. Reprod.* **16**:315–321.

West, J. D., Frels, W. I., Chapman, V. M., and Papaioannou, V. E., 1977 Preferential expression of the maternally derived X chromosome in the mouse yolk sac, *Cell* **12**:873–882.

Whitten, W. K., 1971, Nutrient requirements for the culture of preimplantation embryos in vitro, *Adv. Biosci.* **6**:129–138.

Chapter 3

Oocyte Fusion

BELA J. GULYAS

1. Introduction

Membrane fusion is paramount to many cellular and subcellular functions of eukaryotic cells. It occurs continuously between intracellular organelles. For example, endocytotic vesicles are formed by the fusion of smaller vesicles, which in turn coalesce with components of the Golgi or lysosomes (see review, Meldolesi *et al.*, 1979). Membrane fusion also plays a prominent role in endocytosis of extracellular substances and exocytotic discharge of cellular products or by-products. Partial or complete membrane fusion can occur also between cells. A partial fusion of plasma membranes of two opposing cells takes place during the formation of zona occludens, which are essential in cell–cell communications and in the maintenance of spacial polarity of cells with respect to each other (see reviews, Sheridan, 1976; Papahadjopoulos *et al.*, 1979). Spontaneous fusion of cells without the addition of exogenous chemical fusing agents or viruses occurs as a natural phenomenon during the formation of polykaryote cells. Spontaneous cell-to-cell fusion represents a very specific event; that is, under normal conditions the extent of fusion is regulated by the participant cells. Common to all spontaneous cell fusion is that the nature and the cause of the fusion are not well understood.

One of the best recognized spontaneous cell fusions occurs during fertilization (see review, Gwatkin, 1976). The various aspects of fertilization, including gamete fusion, have been considered in great detail in previous reviews (see review, Bedford and Cooper, 1979), yet the complexities of membrane interaction of this spontaneous cell-to-cell fusion remain unresolved. It is fair to assume that normal fusion of homologous gametes at fertilization requires specialized characteristics of their respective surfaces. However the mechanism of membrane fusion and the factors that regulate this phenomenon are unknown.

This chapter reviews aspects of oocyte fusion, emphasizing an examination of whether experimentally fusing two metaphase II oocytes could mimic

BELA J. GULYAS • Pregnancy Research Branch, National Institute of Child Health and Human Development, National Institutes of Health, Bethesda, Maryland 20205.

fertilization and initiate embryonic development without the participation of sperm.

2. Cell Fusion

A host of agents have been tested for their cell-fusing potential. Also, the molecular mechanisms involved in membrane fusion, be it membrane models or live cells, have been discussed in detail by numerous authorities. For the purposes of oocyte fusion, we shall be concerned only with viral and polyethylene glycol- (PEG) induced fusion and give some considerations to a new approach that, in the opinion of this reviewer, will revolutionize cell fusion technology.

2.1. Fusing Agents

Under pathological conditions, virus-induced cell fusion was reported more than 100 years ago (see reviews, Papahadjopoulos *et al.*, 1979; Poste and Pasternak, 1979), although cell fusion in a tissue culture system utilizing various viruses was accomplished only 30 years ago. Among the many viruses tested, Sendai virus (Harris and Watkins, 1965) is the most useful and the most often employed. This is because Sendai virus is easy to prepare, relatively stable, and considered nonpathogenic for humans. Viral fusion is done by inactivated virus particles without replication of the virus.

During the past decade, however, cell hybridization has been attained by numerous membrane-active agents. Most chemical fusogenic agents are unsaturated, or medium-chain length saturated, carboxylic acids, or esters of fatty acids (Lucy, 1975, 1978). Cell fusion induced by PEG has become almost a standard procedure in hybrid cell formation during recent years. PEG offers the further advantages over viral fusion in that it permits fusion of a variety of cell types with fairly high efficiency; also, it is not pathogenic. Membrane-active properties of a wide range of PEG compounds and other agents were recently evaluated (Lucy, 1978; Klebe and Mancuso, 1981). It is therefore of no surprise that the work on fusing mammalian oocytes, either in our laboratory or elsewhere, utilized either Sendai virus or PEG.

In spite of the efficiency of either Sendai virus or PEG, when looking into the future, it is fair to predict that cell hybridization techniques will be revolutionized by the electric field-induced fusion system. Whereas fusion with virus or PEG is achieved by membrane perturbations that often lead to unphysiological conditions, the field-induced fusion technique avoids all these disadvantages (see review, Zimmermann and Vienken, 1982). To date only sea urchin eggs have been subjected to field-induced fusion with good success (Richter *et al.*, 1981), but there is no reason to believe that this system could not be applied successfully to mammalian gametes as well.

2.2. Mechanisms of Fusion

The molecular mechanism(s) responsible for fusion of cell membranes of two or more apposing cells, as well as the theoretical aspects involved, received considerable attention in previous reviews (Lucy, 1978; Papahadjopoulos *et al.*, 1979; Poste and Pasternak, 1979). Furthermore, it is not in the purview of this chapter to consider these aspects of membrane fusion in great detail. Here, only a brief and general account will be given on the steps involved using either inactivated Sendai virus or PEG, as the fusing agent.

In general, the process of cell fusion can be broken down to three distinct stages (Fig. 1). In the first stage, plasma membranes of adjacent cells are brought into close proximity during agglutination (Knutton and Pasternak, 1979). A close membrane contact is an obligatory prerequisite for cell fusion. In the case of viral fusion, sialic acid molecules of adjacent cell surfaces are crosslinked by virus particles and cause agglutination. During fusion with PEG, cells either agglutinate instantaneously or they have to be agglutinated with lectins, such as phytohemagglutinin (PHA). During PEG treatment, the cells shrink due to removal of water by the hydrophilic PEG molecules. In the second step, fusion occurs at small loci where close contacts were previously established. In the third step, the small loci of fusion proceed to expand, and eventually a single spherical hybrid cell is formed. This last step is usually accompanied with increased permeability resulting in swelling. In case of viral fusion, K^+ leaks out Na^+ leaks in accompanied by an influx of water. The permeability changes caused by PEG are not understood. Although the molecular mechanisms of

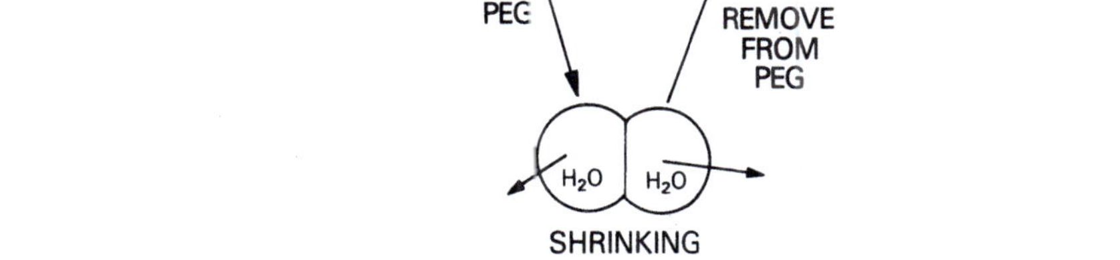

Figure 1. Diagrammatic presentation of general aspects of cell fusion. In the first stage, plasma membranes of cells are agglutinated. In the second stage, fusion occurs at small loci; when fusing with PEG, cells lose water and shrink. Fusion starts after removal from the PEG solution. In the third stage, the sites of fusion expand and the cell swells and rounds up. Swelling may be accompanied by efflux of K^+ and influx of NA^+. (Adapted from Knutton and Pasternak, 1979.)

viral-induced versus PEG-induced membrane fusion differ somewhat, the three steps described above are applicable in both instances.

3. Oocyte Hybrids

The capability to produce cell hybrids, more specifically hybrids between homologous or heterologous oocytes, and between oocytes and embryonic or somatic cells, opened up a new avenue to explore questions in gametogenesis and developmental biology that could not be asked before. Here we shall examine some of the most pertinent questions.

3.1. Fusion of Immature Oocytes

In mammalian oocytes, meiotic maturation is believed to be governed by the ooplasm. Although indirect observations indicate the presence of a maturation factor (MF), also called maturation-promoting factor and maturation-inducing factor, it has not yet been isolated and characterized. Oocyte fusion offers an effective way to analyze the influence of MF on germinal vesicle breakdown (GVBD) and chromatin condensation. However, caution must be exercised in interpretation of the data, since it has not yet been demonstrated that the factor responsible for oocyte maturation is the same one that induces premature chromatin condensation.

Fully grown oocytes from adult mice can undergo spontaneous progression from dictyate to metaphase II during *in vitro* culture. However, small oocytes, or oocytes from sexually immature mice, are not capable of resuming meiosis *in vitro*, presumably because they lack the MF. Consequently, when two immature porcine oocytes were fused with inactivated Sendai virus, the fusion products contained two intact germinal vesicles (GV) that failed to break down after 3–6 hr incubation after fusion (Fulka, 1983). During further incubation for 26–30 hr, the breakdown of GV in these fusion products occurred at a similar rate as in untreated control immature oocytes. However, when oocytes that first matured *in vitro* for 26–30 hr were fused with immature oocytes, GVBD occurred in 91% of the immature oocytes within 3–6 hr (Fulka, 1983). This finding clearly demonstrates that a cytoplasmic factor appears in those maturing oocytes that progress to late diakinesis or metaphase I; this factor in turn can induce chromatin condensation and GVBD in the immature oocyte after oocyte fusion.

The results were more dramatic demonstrating the effects of a MF when GV-stage mouse oocytes from immature mice were fused with metaphase II oocytes using Sendai virus (Balakier, 1978). In all fusion products of this type, GVBD of the immature oocyte occurred immediately after fusion. This was followed by chromosome condensation shortly thereafter. Results from utilizing heterologous oocytes suggest that the MF is not species specific. For example, fusion of a metaphase II rabbit oocyte and a GV-stage mouse oocyte resulted in rapid GVBD and condensation of the mouse oocyte chromosomes (Fulka, 1983).

It is contended that a factor similar to MF is also present in dividing embryonic and somatic cells. This is supported by several experimental fusions between oocytes and cleavage-stage blastomeres, as well as oocytes or blastomeres and somatic cells (Graham, 1969; Baranska and Koprowski, 1970; Bromhall, 1975; Tarkowski and Balakier, 1980; Szollosi *et al.*, 1980). In general, cell fusion was found to be a good way to introduce nuclei into 1- and 2-cell embryos. The response of the donor nucleus to the embyonic cell environment depended a great deal on the donor cell in question. Synchronization of the cell cycle of donor and recipient cells enhanced normal division. Most of the fusion of oocytes and blastomeres, or oocytes and blastomeres with somatic cells, resulted in no, or a limited number of, cell divisions. In some instances the problem derives from irregular or scattered chromosomal arrangements, as revealed by electron microscopy (Szollosi *et al.*, 1980). Although much remains to be clarified, these studies showed that cell-fusion techniques can be advantageously utilized to investigate and manipulate nucleocytoplasmic interactions during oocyte maturation.

3.2. Fusion of Two Metaphase II Oocytes

Equally challenging questions can be asked concerning fertilization and regulation of cleavage stage development through fusion of two metaphase II oocytes or fusion of two parthenogenetically activated haploid eggs (Graham, 1971*a*). Parthenogenetically activated mouse eggs can develop, at best, to the forelimb bud stage, when transferred to pseudopregnant foster mothers (Kaufman *et al.*, 1977). It has been suggested that the fertilizing spermatozoon contributes some extragenetic factor essential for early development (Graham, 1974; Whittingham, 1980; Markert and Seidel, 1981), and in the absence of this factor the parthenogenones cannot fully develop. Whether or not this is the case, experimentally fusing two oocytes one can mimic fertilization and initiate embryonic development in the absence of any sperm component. Initial work on this topic was performed at the MRC Experimental Embryology and Teratology Unit, utilizing oocytes from two different F1 (CBA/NIMR × C57BL/NIMR and (C57BL/6J/Lac × CBA/Ca/Lac) adult female mice (Gulyas *et al.*, 1984). Recently, this approach was repeated, modified, extended, and improved, utilizing oocytes from the F1 C57BL/6N × CBA/N (B6CB/N) mice.

3.2.1. Obtaining Oocytes

Metaphase II oocytes were obtained from B6CB/N females. The letter N stands for National Institutes of Health Genetic Resources. The C57BL/6N mice were derived from C57BL/6J, and the CBA/N were derived from CBA/Ca H line. Unless indicated otherwise, heretofore unpublished data in this chapter pertain to the B6CB/N F1 oocytes. Adult females were superovulated with injections of 5 IU of pregnant mares serum (PMS; Calbiochem) at 1600–1700 hr and 5 IU of chorionic gonadotropin (hCG; Sigma) 48 hr later. Oocytes in cumulus mass were recovered from the ampullary region of the oviducts 14–16 hr after hCG injection. The cumulus masses were released directly into hyaluronidase, 115 U/ml (Sigma). The freed oocytes were washed in phosphate-buffered saline (PBS) or culture medium M2 (Fulton and Whittingham, 1981) without bovine serum albumin (BSA). Zonae pellucidae were removed with 0.04% α-chymotrypsin (Sigma) in PBS (Cuthbertson *et al.*, 1981; Cuthbertson, 1983). The oocytes were washed in several changes of PBS, and the zona-free oocytes were kept in M2 under oil in a Falcon (#1007) dish. Zona-free oocytes even in the presence of protein are extremely sticky and it is hard to remove them from the surface of the dish without incurring damage, except when using this particular type of dish.

3.2.2. Agglutination

For oocyte fusion, a close apposition of membrane surfaces to be fused is obligatory. For achieving close membrane contact, zona-free oocytes, without their first polar bodies, were agglutinated into pairs with PHA (Sigma, type V). A concentration of 400 μg/ml (Mintz *et al.*, 1973) in M2 medium, supplemented with 2.5% BSA, was found to be optional. The tip of each glass pipette was fire-polished to eliminate membrane damage from sharp edges and burs. A new pipette was used for each step. Approximately 50 μl of PHA was placed in a Falcon (#1007) dish under oil along with three or four rows of 4 drops of M2, about 50 μl each. Ten to 20 oocytes were discharged into the drop of PHA one at a time. Initial contact between two oocytes was established as they were released into the PHA. To enhance agglutination and to increase the area of contact between two oocytes, oocyte pairs were first gently pushed together with a tip of a fire-polished glass pipette, then drawn through the small orifice of the pipette, thereby compressing the two oocytes (Fig. 2A). Having obtained the desired degree of agglutination (60–90 sec) the oocyte pairs were transferred through several changes of M2 (50–100 μl) under oil in the same dish, each time paying attention to keep oocyte pairs out of each other's contact. Agglutinated pairs were allowed to stand in M2 for 15–20 min before further manipulation.

3.2.3. Activation of Development

Both oocytes in an agglutinated pair were in metaphase II. Since the aim was to obtain diploid oocyte fusion products (OFP), it was essential that

Figure 2. Production and developmental progress of OFP. (A) Two zona-free metaphase II oocytes agglutinated by PHA; (B) OFP from two oocytes; (C) two second polar bodies being released during fusion process; (D) two second polar bodies released after completion of fusion; (E) OFP with two second polar bodies and two pronuclei; (F) two-cell stage from OFP, with polar bodies still attached; (G) four-cell stage from OFP, blastomeres arranged in a crosswise manner; (H) different configurations of 4-cell stages from OFP (note the number of cell-to-cell contacts; dotted lines connect sister blastomeres); (I) morulae from OFP; (J) blastocyst from OFP.

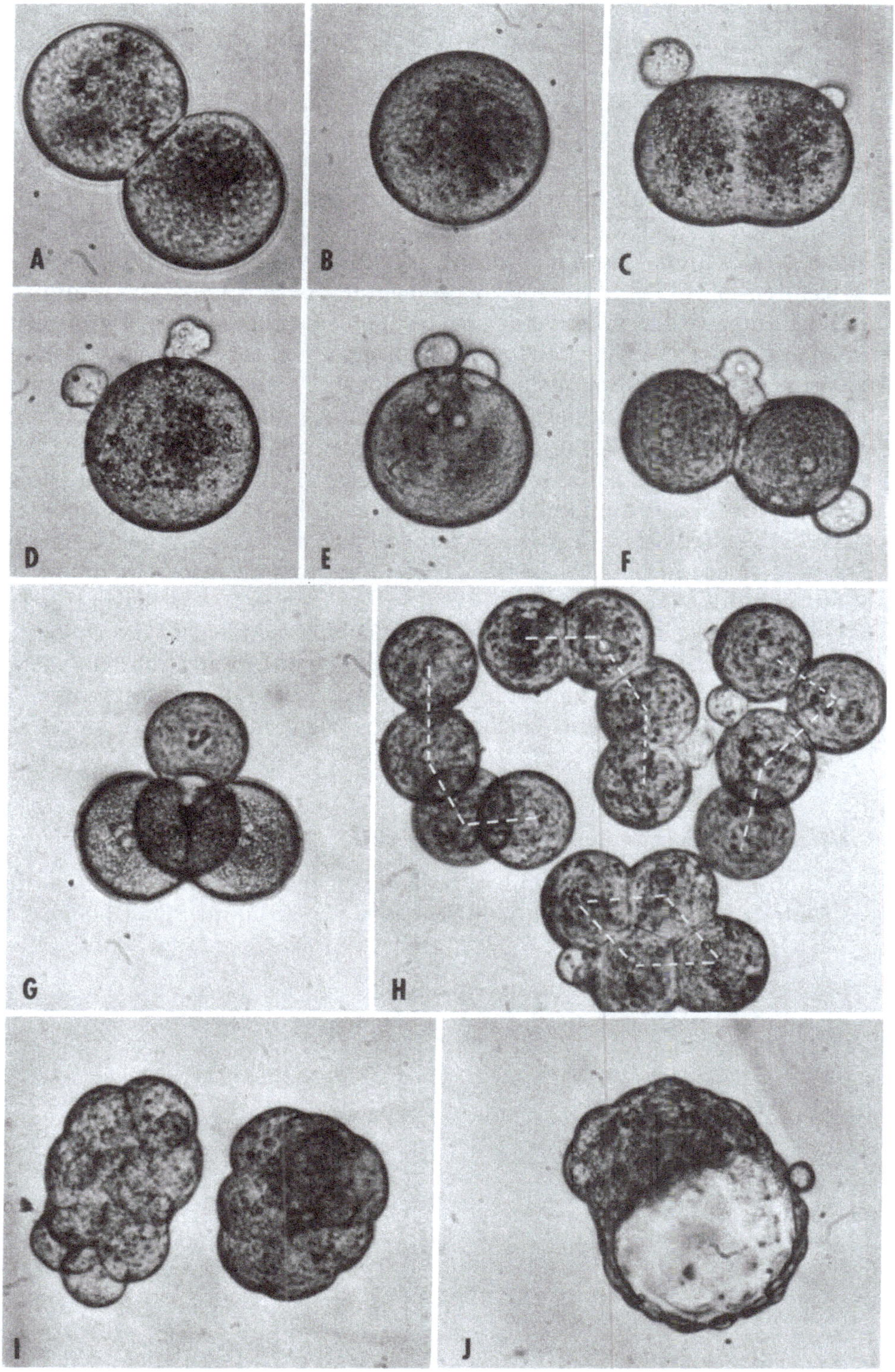
A
B
C
D
E
F
G
H
I
J

meiosis be completed in both oocytes. Soupart (1980, 1982) reported that mouse oocytes complete second meiotic division after fusion with inactivated Sendai virus. Our early experience with oocytes from the F1 (CBA/NIMR × C57BL/NIMR and C57BL/6JLac × CBA/Ca/Lac) mice indicated that fusion of agglutinated oocytes with PEG induced second meiotic division only in a limited number of oocytes (Gulyas *et al.*, 1984). Further studies on oocytes of B6CB/N mice confirmed these observations (see Fig. 3). Working with somatic cells Schneiderman *et al.* (1979) greatly reduced the toxic effects of PEG by deleting Ca^{2+} from the fusion medium. Oocytes behaved differently from somatic cells under these conditions, as shown in Figure 3. With increased concentration of PEG in the Ca^{2+}-free medium, oocyte activation diminished and the number of rapidly lysing oocytes increased (not shown). Furthermore, when agglutinated oocyte pairs were exposed to PEG cortical reaction failed to occur (B. J. Gulyas, unpublished observations). Therefore, in order to ensure oocyte activation, agglutinated oocyte pairs were treated with 7–8% alcohol (Cuthbertson *et al.*, 1981) in medium 16 (Whittingham, 1971) for 4 min. After activation, the oocyte pairs were washed in several changes of M2 and allowed to stand for 15–20 min in M2 under oil at 37°C. Within 15–30 min after activation, usually shortly after exposure to PEG, each original oocyte in a pair extruded a second polar body (Fig. 2B,C), one from each meiotic spindle. In addition to alcohol being instrumental in the completion of second meiotic division, we recently demonstrated that alcohol treatment of mouse oocytes also elicits cortical reaction in a manner similar to fertilization (Gulyas and Yuan, 1985). This is in contrast to Sendai virus fusion of oocytes, where both completion of meiosis and cortical reaction were induced in a single step (Soupart, 1982).

3.2.4. Fusion of Activated Oocytes with PEG

A 50% (v/w) PEG solution was prepared as follows. To 5 ml of two times concentrated M2 medium without BSA, added 5 g liquid PEG (Sigma, 1000

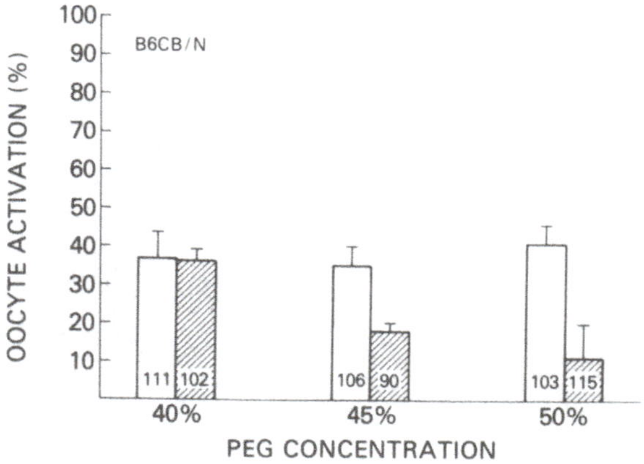

Figure 3. Activation (formation of second polar body) of oocytes after exposure to increasing concentrations of PEG with or without the presence of Ca^{2+}. Each bar represents four experiments (mean ± SE). Numbers in bars are the total number of oocytes examined. [(□): with Ca^{2+}; (▨): without Ca^{2+}.]

M_r). After mixing thoroughly, it was allowed to cool to ambient temperature; the pH was then adjusted to 7.55. Filtered 50% solution was stored at room temperature for not longer than one week. A 45% solution was prepared immediately before use from the 50% stock, using single-strength M2 medium without BSA for dilution.

A 45% solution of PEG gave optimal fusion in the testing of 2-cell stages (Section 4.2); this concentration was adapted to fuse oocytes. Concentrations lower than 45% resulted in fewer or no cell fusion, and a higher concentration resulted in a high frequency of cell lysis. The optimal time of exposure of the oocytes to PEG was 90–120 sec. For fusion, 15–20 agglutinated pairs of oocytes were placed in 200 μl of 45% PEG under a stereomicroscope. The amount of fluid transferred with the oocytes was kept minimal (5–10 μl) to avoid further dilution of the PEG solution. Twenty to 30 sec after transfer to PEG, the oocytes momentarily became invisible, then gained a glassy appearance. They shrank considerably in the PEG but regained their original size once transferred to M2 after PEG treatment. At the end of the treatment, oocyte pairs were collected in a minimum amount of PEG solution and were transferred through four changes of 50–100 μl drops of M2. The oocyte pairs were allowed to rest for 30–45 sec in each drop of M2. During this time, the oocytes gradually regained their original size and shape. Care was taken to transfer a minimum amount (10–20 μl) of fluid from one M2 drop to the next. In this manner, the PEG was gradually diluted out of the oocytes. PEG-exposed oocyte pairs were kept in M2 medium under oil at 37°C on a warming plate and were observed periodically. Fusion occurred a maximum of 30 min after exposure to PEG (Fig. 2B). The fusion efficiency (Table I) in this study was slightly better (80%) than in our earlier studies (71%) (Gulyas *et al.*, 1984). The entire procedure, preceding and including PEG treatment, is summarized diagrammatically in Figure 4.

3.3. Developmental Potential of OFP

OFP from two metaphase II oocytes have diploid number of chromosomes after completion of second meiotic division. Completion of meiosis occurred either during (Fig. 2C) or shortly after fusion (Fig. 2D). Each set of chromosomes is equal to what an oocyte would contribute toward the genetic complement of the zygote. An OFP can also be considered as a special form of parthenogenone because activation of development and establishment of diploidy was achieved without any contribution from a sperm. Two hours after fusion two female pronuclei appeared (Fig. 2E). Electron microscopic observations have shown the pronuclei to be single entities with nucleoli similar to those one observes in fertilized eggs (Fig. 5). Cortical granules were sparse; when they were present, they were usually in the deeper cortical regions. In those OFP that were incubated in medium 16, first (Fig. 2F) and second cleavages (Fig. 2G) occurred 16–20 and 40–44 hr after fusion, respectively. In the absence of zona containment, the blastomeres formed various two-dimensional arrangements (Fig. 2H) many

Table I. *In Vitro* Development of OFP[a]

F_1 Hybrid	No. of pairs	No. fused	No. incubated	Developmental stage reached by OFP				
				2-cell	4-cell	8-cell	Morula	Blastocyst
CBA/NIMR × C57BL/NIMR[b]	73	52 (71)	52	45 (86)	26 (50)	18 (34)	18 (34)	7 (13)
C57BL/6J/Lac × CBA/Ca/Lac[b]	14	8 (57)	8	6 (75)	3 (37)	3 (37)	2 (25)	1 (12)
C57BL/6N × CBA/N	195	157 (80)	134	112 (71)	86 (55)	71 (45)	48 (31)	25 (19)

[a]Numbers in parentheses represent percentages.
[b]From Gulyas *et al.* (1984).

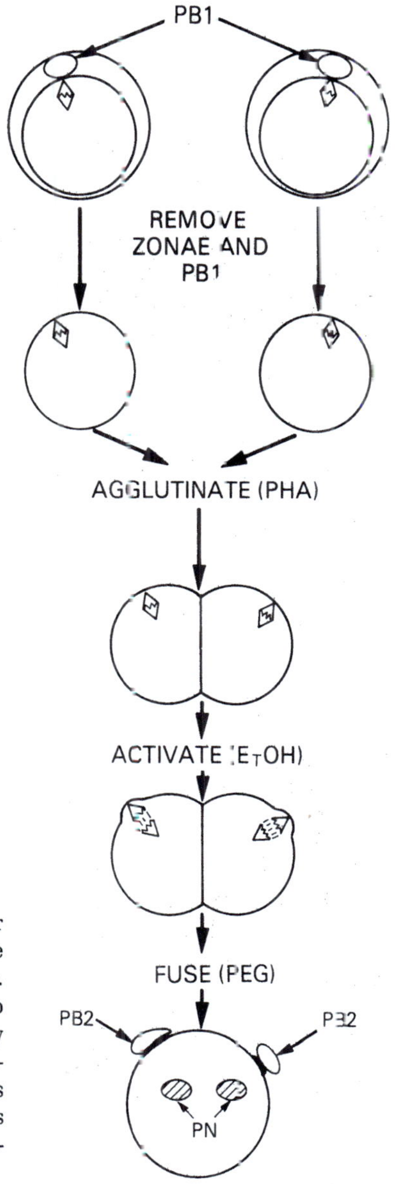

Figure 4. A diagrammatic summary of the procedure for fusing metaphase II oocytes. First, the zonae pellucidae are removed, at which time the first polar bodies (PB1) are lost. In the next step, zona-free oocytes are agglutinated into pairs with PHA, then activated in alcohol (EtOH), whereby second meiotic division is completed by each meiotic spindle. In the last step, agglutinated and activated oocyte pairs are exposed to PEG. After removal from PEG the oocytes fuse and later form two pronuclei. The final product contains two pronuclei and two second polar bodies.

with low numbers of cell-to-cell contacts. Progression of *in vitro* development of embryos from OFT was slower than those of untreated fertilized eggs. The developmental rate of OFP lagged approximately 18–24 hr behind *in vitro* development of *in vivo* fertilized eggs (Fig. 6). Both first and second cleavage divisions lasted approximately 24 hr, after which cleavage occurred at a faster rate. Exposure of the oocytes to PHA at ambient temperatures might have contributed to the slowed development (Mintz *et al.*, 1973). Those cleavage stages from OFP that managed to maintain a close cell-to-cell contact and attain compaction at the 8–16-cell stage usually continued to develop into morulae

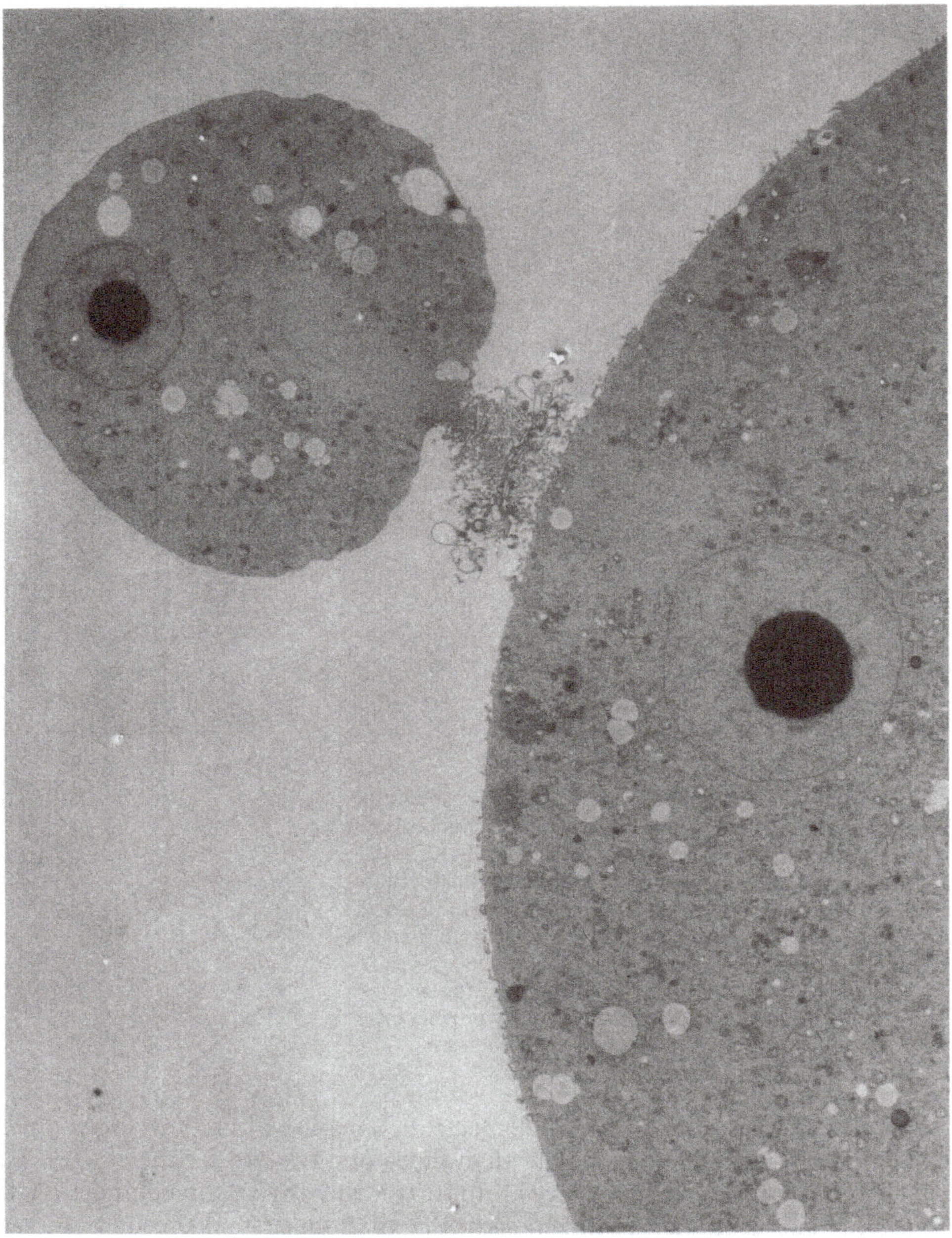

Figure 5. A portion of an OFP with pronucleus and one polar body. Note the absence of cortical granules beneath the plasmalemma.

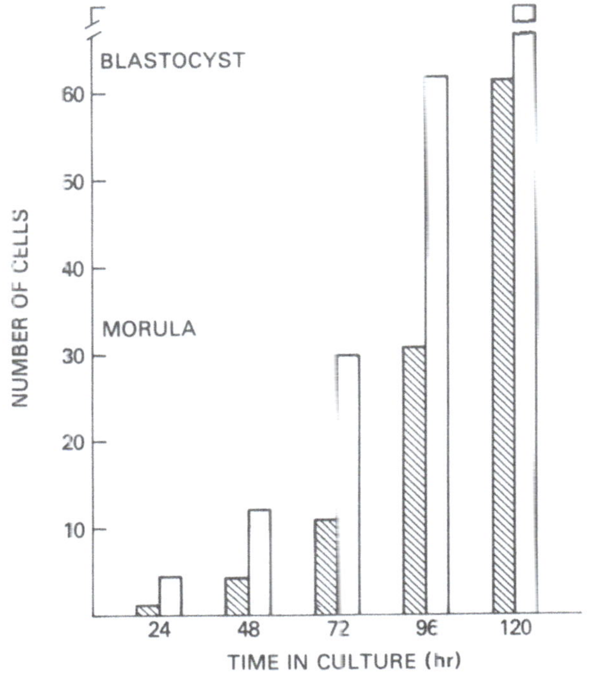

Figure 6. *In vitro* development of OFP as compared with *in vivo* fertilized eggs (B6CB/N)). Developmental rate of OFP lags 24 hr behind the fertilized eggs. [(▧): OFP (B6CB/N); (□): fertilized (B6CB/N).]

(Fig. 2I) and blastocysts (Fig. 2J). As shown in Table I, in our previous study (Gulyas *et al.*, 1984) about 30% of the fusion products reached morula, and 12–13% reached blastocyst stage. In more recent studies, development of OFP from B6CB/N mice into blastocysts was improved (19%), but it was still considerably lower than development of fertilized zygotes (see Table II, group 1). In either case, development of OFP using PEG was severalfold greater than those obtained by viral fusion (Soupart, 1980, 1982).

Aside from doubling the mass and surface area, the absence of the zona pellucida and exposure to hyaluronidase, pronase, PHA, alcohol, and PEG could have easily contributed either singly, or in synergy, to the reduced rate of development of OFP. To explore which one of these factors might have influenced development of OFP, as controls, intact and zona-free fertilized eggs were treated in a similar manner to oocytes undergoing fusion and then incubated *in vitro*. Judging from these results, perhaps the greatest contributor to the reduced development rate of OFP was the low numbers of cell-to-cell contacts at the 4-cell stage and beyond, in many of the developing embryos in the absence of the zona pellucida. This was well documented in the control experiments presented in Table II. Whereas 86% of nontreated zygotes developed to blastocysts (group 1), only 49% of zona-free zygotes developed to blastocysts (group 6). The contribution of other factors (hyaluronidase, pronase, PHA, and PEG) to the reduced development either singly (groups 2, 3, and 4 and groups 6, 7 and 8) or in combination (groups 5 and 9), in the presence or

Table II. *In Vitro* Development of Intact and Zona-Free Fertilized Mouse Eggs (B6CB/N) Following PHA, PEG, or PHA and PEG Treatment[a]

Group no.	Treatments				No. of eggs	Developmental stage reached in culture				
	Hyaluronidase	Zona removed	PHA	PEG		2-cell	4-cell	8-cell	Morula	Blastocyst
1	−	−	−	−	70	66 (94)	62 (89)	62 (89)	60 (86)	60 (86)
2	+	−	−	−	71	71 (100)	68 (96)	65 (92)	64 (90)	52 (73)
3	+	−	+	−	77	77 (100)	77 (100)	71 (92)	71 (92)	52 (68)
4	+	−	−	+	77	77 (100)	70 (91)	67 (87)	67 (87)	46 (60)
5	+	−	+	+	78	77 (99)	75 (96)	75 (96)	69 (88)	51 (65)
6	+	+	−	−	84	80 (95)	70 (83)	53 (63)	53 (63)	41 (49)
7	+	+	+	−	90	84 (93)	77 (86)	62 (69)	62 (70)	47 (52)
8	+	+	−	+	88	86 (98)	75 (85)	57 (65)	54 (61)	42 (48)
9	+	+	+	+	76	73 (96)	62 (82)	50 (66)	49 (64)	38 (50)

[a]Numbers in parentheses represent percentages.

in the absence of zona pellucida (groups 2–5 vs. 6–9), was considerably less compared with the effects of the lack of zona pellucida. In view of these observations on control embryos, a 19% yield of blastocysts from OFP is considerably different from the 48–52% yield (Table II) from the zona-free zygotes. Transfer of OFP into empty zonae could conceivably increase the development rate of OFP (Soupart, 1982; B. J. Gulyas, unpublished observations).

Alcohol treatment activates completion of the second meiotic division and triggers cortical reaction in a manner much like that at fertilization (Gulyas and Yuan, 1985). However, 6-min exposure of mouse oocytes to 8% alcohol produced chromosomal nondisjunction in nearly 19% of oocytes (Kaufman, 1982) whereas a 3-min exposure was adequate for activation and nondisjunction was negligible. Thus, decreasing the duration of alcohol treatment in the present study to 4 min might have contributed to the increase in *in vitro* development of OFP to blastocysts (Table I), as compared with our earlier studies (Gulyas *et al.*, 1984).

Electron microscopic observations on blastocysts obtained by virus-induced oocyte fusion have shown cellular differentiation into well-developed inner cell mass and trophoblast cells (Soupart, 1982). Recently we made similar observations on blastocysts obtained by PEG fusion (B. J. Gulyas, unpublished data). Except for the absence of the zona pellucida and the presence of some larger than normal cells, the fine structure of OFP-derived blastocyst is identical with that of fertilization-derived blastocysts.

In our previous study (Gulyas *et al.*, 1984), cleavage-stage embryos or blastocysts from OFP were transferred to the oviduct or the uterus of day 1 or day 3 pseudopregnant foster mothers, respectively. Of the seven F1 embryos from OFP transfers, six gave a decidual cell response and one developed to the 14 somite stage. This is in comparison with 96% decidual cell response and 87% embryonic tissue formation from control embryo transfers. More recently 121 morulae and blastocysts (B6CB/N) were transferred to 3- or 4-day pseudopregnant recepients and allowed to develop. No viable young resulted from these embryo transfers.

The cause for the developmental arrest of these, and any other type of mammalian parthenogenones after transfer into foster mothers remains obscure. Several factors may be considered: (1) doubling of the nuclear–cytoplasmic ratio; (2) the possibility that both X chromosomes may remain active; (3) and the possible role of recessive lethal mutations. Some of these ideas have met with serious objections (see reviews, Graham, 1971*a*, 1974; Kaufman, 1975). Furthermore, Kaufman *et al.* (1978) demonstrated that one of the X chromosomes of diploid mouse parthenogenones was inactivated on schedule as in normal embryos. Lastly, totally homozygous uniparental female mice had been reported to develop to live offspring following nuclear-transfer experiments (Hoppe and Illmensee, 1977). This led to the postulation that the fertilizing spermatozoon not only initiates completion of meiosis and triggers cortical reaction, but it might provide some yet unidentified extragenetic factor essential for early development as well (Graham, 1974; Whittingham, 1980;

Markert and Seidel, 1981). More recent observations, however, demonstrated that death of parthenogenones is not due to a lack of cytoplasmic factors from the sperm (Mann and Lovell-Badge, 1984); instead, it was suggested that death of mammalian parthenotes is due to lack of paternal contribution to the embryonic genome (McGrath and Solter, 1984). For further detail, see Chapter 2.

3.4. Overcoming Developmental "Block" through Oocyte Fusion

The metabolic requirements and, therefore, the developmental potential of mouse zygotes from various strains (inbred or outbred) and from hybrids differ considerably. Whereas zygotes from some inbred strains, particularly from hybrids between them, can develop into normal blastocysts when cultured from the 1-cell stage, zygotes from outbred strain generally cannot develop. In these embryos, development is arrested at the 2-cell stage (hereafter referred to as "2-cell block").

The inability of zygotes from certain strains to develop *in vitro* is now believed to be an artifact of the culture system. Early cleavage in the mouse embryo is under a degree of maternal control, and it is the egg alone that determines whether the embryo "blocks" at 2-cell or continues cleavage (Goddard and Pratt, 1983). When F1 metaphase II oocytes (CBA/NIMR × C57BL/NIMR or C57BL/6J/Lac × CBA/Ca/Lac) were fused with MF1 oocytes (outbred strain demonstrating "2-cell block"), the fusion products surmounted the "2-cell block," and 10% of them developed to blastocysts *in vitro* (Gulyas *et al.*, 1984). These results confirmed previous observations by Muggleton-Harris *et al.* (1982), showing that the injection of cytoplasm from nonblocking embryos into zygotes of a "blocking" strain to enhance development of these embryos beyond the 2-cell stage. Furthermore, these observations confirmed the cytoplasmic nature of the defect. Both cytoplasmic injection and oocyte fusion are potential future useful approaches in identifying the cytoplasmic factor involved in the "2-cell block."

4. Fusion of Fertilized Eggs and Cleavage-Stage Cells

Fusion techniques have been effectively employed to examine the mechanism of cellular regulation and the effects of polyploidy on development. Of particular interest to mammalian developmental biologists has been the developmental potential of tetraploid embryos. Tetraploidy can be achieved by fusing either two fertilized eggs or the blastomeres of early cleavage stages. Other means of producing tetraploid mammalian embryos are also available, but are not relevant to the present discussion. We also found that the fusion efficiency of a PEG preparation can be reliably and conveniently tested using late 2-cell-stage embryos.

4.1. Fusion of Fertilized Eggs

Several attempts were made to produce tetraploid embryos by fusing 1-cell fertilized eggs (Tarkowski, 1961; Graham, 1971*b*). Using inactivated Sendai virus, 66% of 1-cell eggs fused, 19% of which developed into blastocysts (Graham, 1971*b*). When fused cells were combined into pairs, 100% of this type of fusion product developed into blastocysts.

4.2. Testing Fusing Efficiency of PEG Preparations: Fusion of Blastomeres

In order to perform fusion of metaphase II oocytes at a high rate of efficiency a reliable method was needed to determine the optimal conditions for fusion and to monitor the quality of the PEG preparations. It was discovered early on that fusion efficiency of PEG preparations could be monitored quantitatively using late 2-cell-stage mouse embryos.

The parameters that influence oocyte or blastomere fusion and postfusion viability of the fusion products are (1) size (molecular weight) of the PEG, (2) concentration, (3) pH, (4) duration of exposure, (5) divalent ions (Ca^{2+}) and (6) source (purity) of the PEG.

Previous experience in our laboratory (Gulyas *et al.*, 1984) and by others (Eglitis, 1980; Spindle, 1981) has shown that a 1000-M_r PEG preparation is most suitable for fusing mouse blastomeres. Therefore all testing and fusion was done with 1000-M_r PEG solution. Furthermore, lysis of cells after exposure to PEG was found to be minimal when the PEG solution was kept at pH 7.5–7.6, preferably at pH 7.55. These two parameters were kept constant in all our testing. PEG (1000) from two sources (Baker and Sigma) was tested for optimal concentration, duration of exposure, and presence of Ca^{2+} in an effort to determine optimal conditions for blastomere fusion.

Two-cell stage embryos were obtained between 1300 and 1400 hr of day 2 of pregnancy (day of plug counted as day 1). A small number of the embryos were already at the 3- or 4-cell stage. The zona was left intact in all these experiments in order to avoid introducing yet another variable. After exposure to 40, 45, or 50% PEG, the embryos were washed in several changes of M2 and kept in M2 under oil at 37°C for 1 hr. Progress of fusion (Fig. 7A–C) was checked periodically. Fusion products, PEG-exposed nonfused 2-cell stages, and control 2-cell stages were cultured *in vitro* and their progress of development was recorded daily.

The results of fusing blastomeres are summarized in Figure 8. In the presence of Ca^{2+}, 50% Sigma PEG for 90-sec exposure, 45% Sigma PEG for 120 sec, 45% Baker PEG for 90 sec, and 40% Baker PEG for 120-sec exposure gave equally good fusion efficiency at 76%, 82%, 72%, and 82%, respectively. A similar fusion efficiency was attained by Spindle (1981) fusing the blastomeres of zona-enclosed 2-cell stages. However, the duration of exposure to PEG to

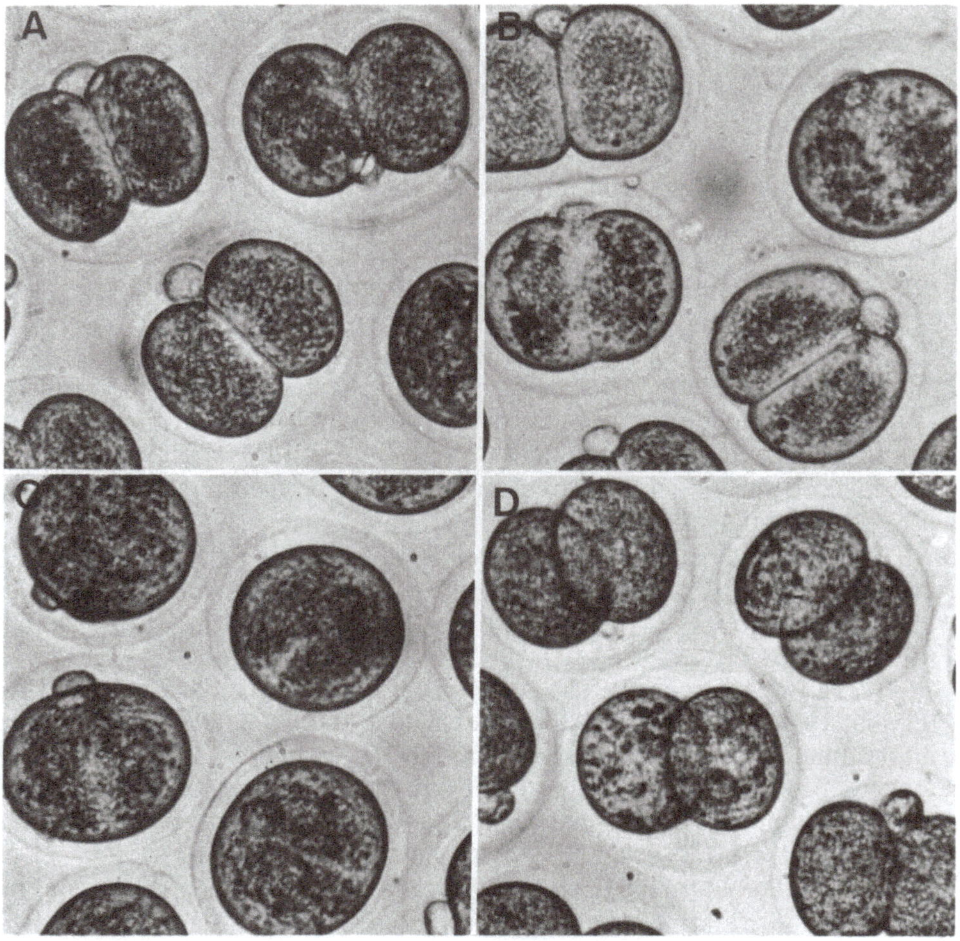

Figure 7. Fusion of blastomeres of 2-cell stage embryos. (A) Untreated 2-cell-stage embryos; (B) 2-cell stages during early phase of fusion; (C) fusion products from 2-cell embryos; (D) first cleavage of fusion product.

obtain optimum fusion in B6CB/N eggs was 1½ to 2 times that reported by Spindle (1981). Dissociated blastomeres of 2- or 4-cell stages can also be fused with either virus (Graham, 1971*b*) or PEG (Eglitis, 1980; Spindle, 1981), although with lower efficiency. It should also be noted that whereas for fusion of somatic cells a 50–55% concentration of PEG is optimum (see review, Mercer and Baserga, 1982), for fusion of mouse oocytes and blastomeres the optimum concentration is 45%. These differences are likely to reflect differences in the kind and number of cell surface components within the cell membranes.

Some cell lysis occurred in the presence of Ca^{2+} at higher PEG concentrations, particularly after long exposures (Eglitis, 1980; Spindle, 1981; present work). Using a Ca^{2+}-free fusion protocol, Schneiderman *et al.* (1979) were able to greatly reduce the toxic effects associated with PEG-induced fusion of somatic cells. This protective effect of Ca^{2+} ion exclusion was particularly more apparent with Baker PEG. The precise nature of protection from cell lysis by Ca^{2+} ion exclusion is not clear (Mercer and Baserga, 1982). Our results fusing blastomeres of 2-cell-stage mouse embryos did not confirm these observations. With one exception, an increase in PEG concentration and the exposure time to PEG, fusion efficiency decreased (Fig. 8). The one exception was the 40–50% Sigma PEG after 120-sec exposure. Cell lysis rapidly increased with increase in PEG concentration and duration of exposure to PEG (unpublished observations). Equally important in our consideration was the postfusion viability and ability of the fusion product to develop to blastocysts after fusion, a topic discussed in the next section.

4.3. Postfusion Development of Tetraploid Embryos

A number of studies dealt with the development of tetraploid embryos Here we shall be concerned only with those studies in which tetraploidy was produced through virus or PEG-induced cell fusion. In our experiments the developmental potential of both tetraploid embryos from blastomere fusion and nonfused PEG-exposed diploid embryos was a function of PEG concentra-

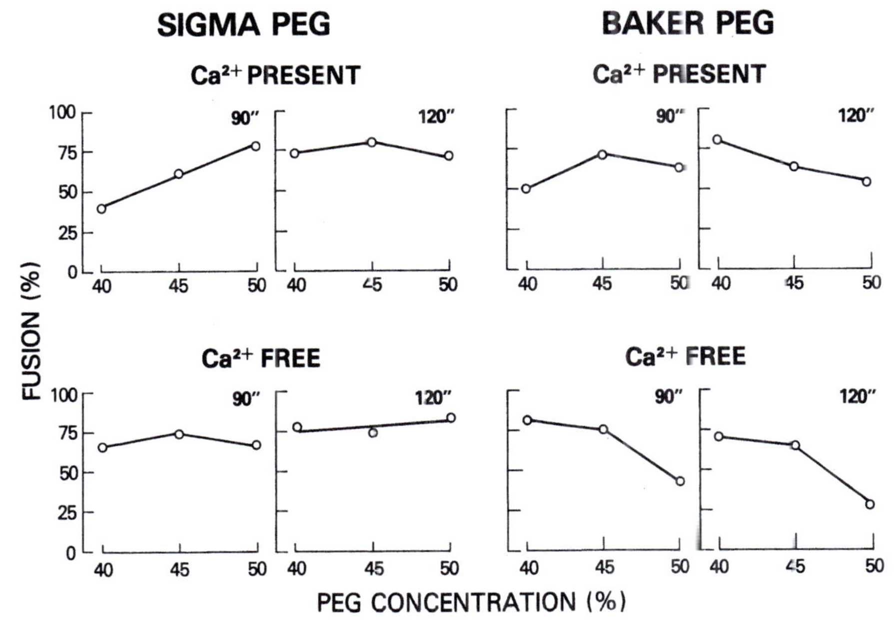

Figure 8. Summary of an extensive study testing fusion efficiency of three concentrations (40, 45, and 50%) of PEG from two different sources (Sigma and Baker). Two-cell stage embryos were exposed to PEG for either 90 or 120 sec. The fusion was done either in the presence or in the absence of Ca^{2+}. Each point of every graph represents four experiments on 50 2-cell-stage embryos.

tion (Fig. 9). Regardless of the source of PEG, the presence of Ca^{2+}, and the duration of PEG exposure, higher PEG concentration was inversely related to the development of tetraploid embryos into blastocysts. A possible exception were those embryos that were fused either in 40 or 45% Sigma PEG for 120 sec in Ca^{2+}-free fusion medium. The decline in embryonic development with exposure to increased concentrations of PEG indicated cytotoxicity of the fusing agent. This was also supported by the similar regression demonstrated in the development of nonfused PEG-exposed diploid embryos. By contrast, exposure to Sendai virus did not prevent subsequent development to blastocysts (Graham, 1971*b*). Instead, cytotoxicity was prominent at the time of viral fusion.

Embryos obtained either by fusion of blastomeres at the 2-cell stage or at the 4-cell stage were found to be uniformly tetraploid (Eglitis, 1980; Spindle, 1981). Embryos constructed from four tetraploid blastomeres developed significantly better than did those constituted of two or of a single tetraploid blastomere. Blastocysts developed from fused blastomeres (2-cell) with PEG (Fig. 10A–C) had significantly fewer cells in the inner cell mass (5.9 ± 1, $N = 12$) than normal diploid control (22.7 ± 2, $N=13$) blastocysts (Spindle, 1981). In general, embryos developing *in vitro* (experimental or control) in the presence or in the absence of zona pellucida had fewer cells at the blastocyst stage (Graham, 1971*b*) than did those controls developing *in vitro*. The rate of blastulation could be enhanced by increasing the starting number of cells (Eglitis, 1980). It is generally thought that the slowed-down development and

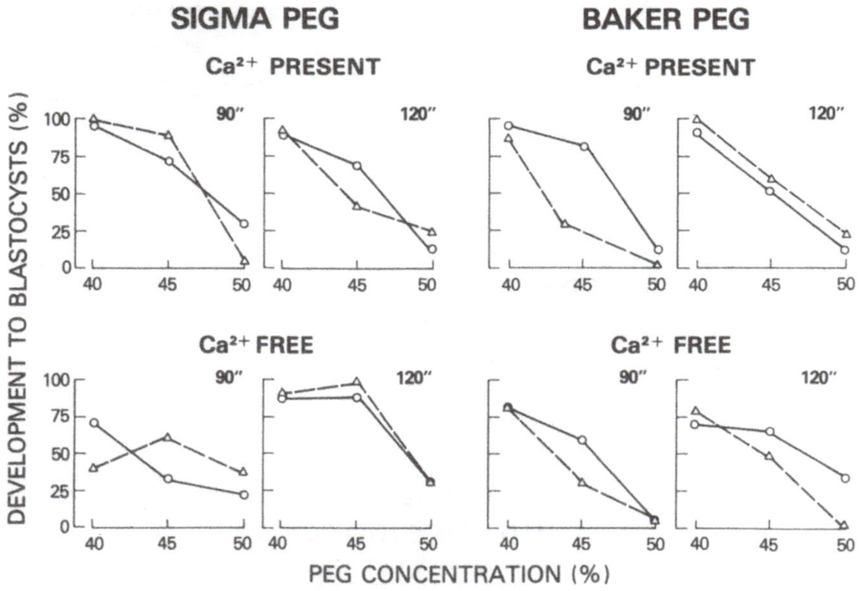

Figure 9. Development of fusion products (—) and PEG-exposed, but nonfused, 2-cell stages (- - -) from Fig. 8 into blastocysts. For nomenclature, see Figure 8 legend.

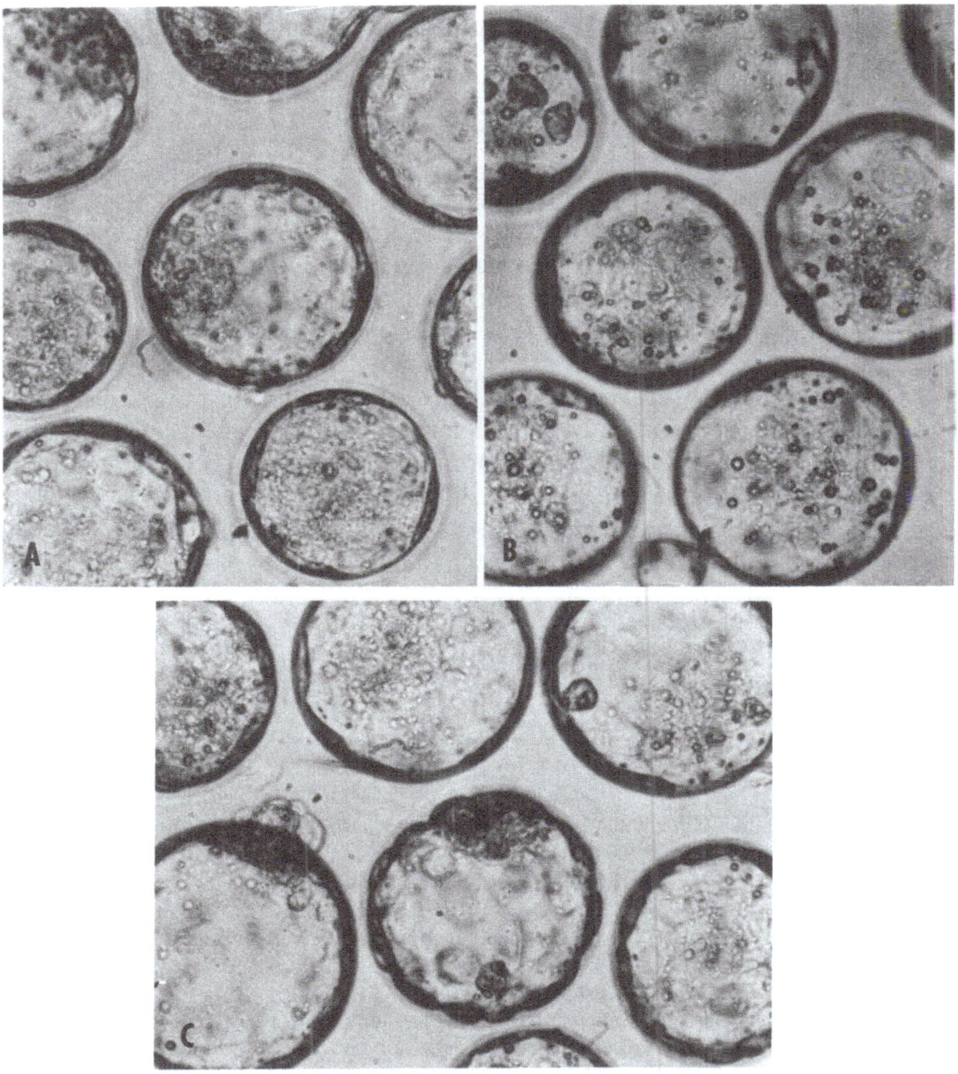

Figure 10. Blastocysts from (A) untreated control; (B) fused 2-cell stages, and (C) 2-cell embryos exposed to PEG but that did not fuse.

delay of cavitation of these embryos reflect a sensitivity to tetraploidy rather than an artifact of virus or PEG (Graham, 1971*b*; Eglitis, 1980).

Blastocysts that developed from fused blastomeres failed to develop into live young when transferred into appropriate pseudopregnant recipients (Graham, 1971*b*; B. J. Gulyas, 1984, unpublished; and B. J. Gulyas, M. Wood, and D. G. Whittingham, 1982, unpublished results). A portion gave decidual swelling, but no embryos could be found in them. The inability of tetraploid blastocytes to develop into live young is thought to be a consequence of tetraploidy itself (Graham, 1971*b*) and not of the fusion technique.

5. Concluding Remarks

This chapter reviews oocyte fusion as a means of manipulating early mammalian development. The intent was to better understand oocyte maturation, fertilization, and initiation of cleavage-stage development and to explore new avenues to transfer genetic information into oocytes. With further refinements and appropriate adaptions, cell fusion techniques should be instrumental in the future in such endeavors as: (1) studying nucleocytoplasmic regulatory mechanisms during cleavage stages; (2) further exploring the developmental significance of sperm membrane fusion with oolemma and the sperm fusion related activation of the oocytes; (3) transferring new genetic information into oocytes; and (4) producing offspring of exact genetic copies. Whereas we improved the blastocyst yield from OFP of two metaphase II oocytes, viable offspring from such embryo is still not in sight. The side effects of the various chemicals that the oocytes were exposed to in the present study can be circumvented in future studies through the application of more advanced electric field-induced cell fusion techniques.

References

Balakier, H., 1978, Induction of maturation in small oocytes from sexually immature mice by fusion with meiotic or mitotic cells, *Exp. Cell Res.* **112:**137–141.

Baranska, W., and Koprowski, H., 1970, Fusion of unfertilized mouse eggs with somatic cells, *J. Exp. Zool.* **174:**1–14.

Bedford, J. M., and Cooper, G. W., 1979, Membrane fusion events in the fertilization of vertebrate eggs, in: *Cell Surface Reviews*, Vol. 5: *Membrane Fusion* (G. Poste and G. L. Nicolson, eds.), pp. 65–125, Elsevier/North-Holland, Amsterdam.

Bromhall, J. D., 1975, Nuclear transplantation in the rabbit egg, *Nature (Lond.)* **258:**719–721.

Cuthbertson, K. S. R., 1983, Parthenogenetic activation of mouse oocytes in vitro with ethanol and benzyl alcohol, *J. Exp. Zool.* **226:**311–314.

Cuthbertson, K. S. R., Whittingham, D. G., and Cobbold, P. H., 1981, Free Ca^{2+} increases in exponential phases during mouse oocyte activation, *Nature (Lond.)* **294:**754–757.

Eglitis, M. A., 1980, Formation of tetraploid mouse blastocysts following blastomere fusion with polyethylene glycol, *J. Exp. Zool.* **213:**309–312.

Fulka, J., Jr., 1983, Nuclear maturation in pig and rabbit oocytes after interspecific fusion, *Exp. Cell. Res.* **146:**212–218.

Fulton, B., and Whittingham, D. G., 1981, Activation of mammalian oocytes by intracellular injection of calcium, *Nature (Lond.)* **273:**149–151.

Goddard, M. J., and Pratt, H. P. M., 1983, Control of events during early cleavage of the mouse embryo: An analysis of the "2-cell block," *J. Embryol. Exp. Morphol.* **73:**111–133.

Graham, C. F., 1969, The fusion of cells with one- and two-cell mouse embryos, in: *Wistar Institute Symposium Monographs*, Vol. 9: *Heterospecific Genome Interactions* (V. Defendi, ed.), pp. 19–33, The Wistar Institute Press, Philadelphia.

Graham, C. F., 1971*a*, Experimental early parthenogenesis in mammals, in: *Advances in the Biosciences*, Vol. 6: *Shering Symposium on Mammalian Development* (G. Raspe, ed.), pp. 87–97, Pergamon Press, Oxford.

Graham, C. F., 1971*b*, Virus assisted fusion of embryonic cells, in: *In Vitro Methods in Reproductive Cell Biology* (E. Diczfalusy, ed.), *Acta Endocrinol. (Suppl.)(Copenh.)* **153**:154–167.

Graham, C. F., 1974, The production of parthenogenetic mammalian embryos and their use in biological research, *Biol. Rev.* **49**:399–422.

Gulyas, B. J., and Yuan, L. C., 1985, Cortical reaction in parthenogenetically (Ethanol) activated mouse oocytes. *J. Exp. Zool.* **233**:269–276.

Gulyas, B. J., Wood, M., and Whittingham, D. G., 1984, Fusion of oocytes and development of oocyte fusion products in the mouse, *Dev. Biol.* **101**:246–250.

Gwatkin, R. B. L., 1976, Fertilization, in: *The Cell Surface in Animal Embryogenesis and Development* (G. Poste and G. L. Nicolson, eds.), Vol. 1, pp. 1–54, Elsevier/North-Holland, Amsterdam.

Harris, H., and Watkins, J. F., 1965, Hybrid cells derived from mouse and man: Artificial heterokaryons of mammalian cells from different species, *Nature (Lond.)* **205**:640–646.

Hoppe, P. C., and Illmensee, K., 1977, Microsurgically produced homozygous–diploid uniparental mice, *Proc. Natl. Acad. Sci. USA* **74**:5657–5661.

Kaufman, M. H., 1975, The experimental induction of parthenogenesis in the mouse, in: *The Early Development of Mammals* (M. Balls and A. E. Wild, ed.), pp. 25–44, Cambridge University Press, London.

Kaufman, M. H., 1982, The chromosome complement of single-pronuclear haploid mouse embryos following activation by ethanol treatment, *J. Embryol. Exp. Morphol.* **71**:139–154.

Kaufman, M. H., Barton, S. C., and Surani, M. A. H., 1977, Normal postimplantation development of mouse parthenogenetic embryos to the forelimb bud stage, *Nature (Lond.)* **265**:53–55.

Kaufman, M. H., Guc-Cubrilo, M., and Lyon, M. F., 1978, X chromosome inactivation in diploid parthenogenetic mouse embryos, *Nature (Lond.)* **271**:547–549.

Klebe, R. J., and Mancuso, M. G., 1981, Chemicals which promote cell hybridization, *Somat. Cell Genet.* **7**:473–488.

Knutton, S., and Pasternak, C. A., 1979, The mechanism of cell–cell fusion, *Trends Biochem. Sci.* **4**:220–223.

Lucy, J. A., 1975, Aspects of the fusion of cells *in vitro* without viruses, *J. Reprod. Fertil.* **44**:193–205.

Lucy, J. A., 1978, Mechanisms of chemically induced cell fusion, in: *Membrane Fusion* (G. Poste and G. L. Nicolson, eds.), Vol. 5, pp. 267–304, Elsevier/North-Holland, Amsterdam.

Mann, J. R., and Lovell-Badge, R. H., 1984, Inviability of parthenogenones is determined by pronuclei, not egg cytoplasm, *Nature (Lond.)* **310**:66–67.

Markert, C. L., and Seidel, G. E., Jr., Parthenogenesis, identical twins, and cloning in mammals, in: *New Technologies in Animal Breeding* (B. G. Brackett, G. E. Seidel, Jr., and S. M. Seidel, eds.), pp. 181–200, Academic Press, New York.

McGrath, J., and Solter, D., 1983, Nuclear transplantation in the mouse embryo by microsurgery and cell fusion, *Science* **220**:1300–1302.

McGrath, J., and Solter, D., 1984, Completion of mouse embryogenesis requires both the maternal and paternal genomes, *Cell* **37**:179–183.

Meldolesi, J., Borgese, N., De Camilli, P., and Ceccarelli, B., 1979, Cytoplasmic membranes and the secretory process, in: *Membrane Fusion* (G. Poste and G. L. Nicolson, eds.), Vol. 5, pp. 509–627, Elsevier/North-Holland, Amsterdam.

Mercer, W. E., and Baserga, R., 1982, Techniques for decreasing the toxicity of polyethylene glycol, in: *Techniques in Somatic Cell Genetics* (J. W. Shay, ed.), pp. 23–45, Plenum Press, New York.

Mintz, B., Gearhart, J. D., and Guymont, A. O., 1973, Phytohemagglutinin-mediated blastomere aggregation and development of allophenic mice, *Dev. Biol.* **31**:195–199.

Muggleton-Harris, A., Whittingham, D. G., and Wilson, L., 1982, Cytoplasmic control of preimplantation development *in vitro* in the mouse, *Nature (Lond.)* **299**:460–462.

Papahadjopoulos, D., Poste, G., and Vail, W. J., 1979, Studies on membrane fusion with natural and model membranes, in: *Methods in Membrane Biology*, (E. D. Korn, ed.), Vol. 10, pp. 1–121, Plenum Press, New York.

Poste, G., and Pasternak, C. A., 1979, Virus-induced cell fusion, in: *Membrane Fusion* (G. Poste and G. L. Nicolson, eds.), Vol. 5, pp. 305–367, Elsevier/North-Holland, Amsterdam.

Richter, H. -P., Scheurich, P., and Zimmermann, U., 1981, Electric field-induced fusion of sea urchin eggs, *Dev. Growth. Diff.* **23**:479–486.

Schneiderman, S., Farber, J. L., and Baserga, R., 1979, A simple method for decreasing the toxicity of polyethylene glycol in mammalian cell hybridization, *Somat. Cell Genet.* **5**:263–269.

Sheridan, J. D., 1976, Cell coupling and cell communication during embryogenesis, in: *The Cell Surface in Animal Embryogenesis and Development* (G. Poste and G. L. Nicolson, eds.), Vol. 1, pp. 409–447, Elsevier/North-Holland, Amsterdam.

Soupart, P., 1980, Initiation of mouse embryonic development by oocyte fusion, *Arch. And.* **5**:55–57.

Soupart, P., 1982, Initiation of mouse embryonic development by oocyte fusion, in: *In Vitro Fertilization and Embryo Transfer* (E. S. E. Hafez and K. Semm, eds.), pp. 51–63, MTP Press, Falcon House, Lancaster, England.

Spindle, A., 1981, Polyethylene glycol-induced fusion of two-cell mouse embryo blastomeres, *Exp. Cell Res.* **131**:465–470.

Szollosi, D., Balakier, H., Czolowska, R., and Tarkowski, A. K., 1980, Ultrastructure of cell hybrids between mouse oocytes and blastomeres, *J. Exp. Zool.* **213**:315–325.

Tarkowski, A. K., 1961, Mouse chimaeras developed from fused eggs, *Nature (Lond.)* **190**:857–860.

Tarkowski, A. K., and Balakier, H., 1980, Nucleo-cytoplasmic interactions in cell hybrids between mouse oocytes, blastomeres and somatic cells, *J. Embryol. Exp. Morphol.* **55**:319–330.

Whittingham, D. G., 1971, Culture of mouse ova, *J. Reprod. Fertil. (Suppl.)* **14**:7–21.

Whittingham, D. G., 1980, Parthenogenesis in mammals, in: *Oxford Reviews in Reproductive Biology* (C. H. Finn, ed.), Vol. 2, pp. 205–231, Oxford University Press, Oxford.

Zimmermann, U., Vienken, J., 1982, Electric field-induced cell-to-cell fusion, *J. Membr. Biol.* **67**:165–182.

Chapter 4

Animal *in Vitro* Fertilization and Embryo Development

BARRY D. BAVISTER

1. Introduction

There are considerable disparities in the quality and quantity of information available on fertilization and embryonic development in mammals compared with invertebrates (Bavister, 1980). In invertebrates, the availability of vast numbers of gametes and embryos for experimental studies and the natural occurrence of external fertilization greatly simplify duplication of normal events under laboratory conditions. By contrast, minute numbers of eggs and embryos are produced in each cycle by female mammals, and their development in the protected environment afforded by the female reproductive tract presents a formidable obstacle to research. It is remarkably difficult to elucidate the precise nature of the environment provided to the gametes and embryos by the female tract, and thus to duplicate these conditions in the laboratory. As a result, most attempts to examine mammalian gamete interactions and early embryo development under experimental conditions have utilized some form of culture solution that has been devised empirically or one that is appropriate for culture of somatic cells. Such conditions are most likely less than optimal for supporting normal fertilization and embryogenesis. This chapter examines some of the major problems inherent in experimental studies of mammalian early development and reviews some recent attempts to increase understanding of fertilization mechanisms using *in vitro* techniques. As much as possible, results of *in vitro* experiments are related to the natural situation in which fertilization and embryonic development take place.

It is generally believed that the use of *in vitro* techniques for the study of early mammalian development is easier, more informative, and more controllable than the alternative of attempting to study events occurring in the natural environment. While this is essentially true, there is much to be gained from the latter approach. The risk of introducing artifacts into an experiment should be

BARRY D. BAVISTER • Wisconsin Regional Primate Research Center, and Department of Veterinary Science, University of Wisconsin—Madison, Madison, Wisconsin 53715.

much less with the *in vivo* approach, and it avoids the whole problem of designing culture media suitable for gametes and embryos. The two most fundamental discoveries about mammalian fertilization that have been made to date, namely the need for spermatozoa to first undergo capacitation and then the acrosome reaction, were both accomplished by examination of eggs undergoing fertilization that were recovered from mated animals. By this means, Austin (1951) and Chang (1951) showed that a period of residence in the female reproductive tract was needed by spermatozoa before they could effect fertilization ("capacitation"); later, Austin and Bishop (1958) were able to examine the acrosomal status of rodent spermatozoa passing through the intact cumulus oophorus. The latter feat is hard to duplicate using the most highly developed *in vitro* fertilization techniques available. Observation of events occurring *in vivo* can provide invaluable confirmation of the physiological relevance of events taking place *in vitro*. For example, the phenomenon of hyperactivation (a change in sperm motility behavior associated with fertilizing ability) was first described in hamster spermatozoa incubated *in vitro* by Yanagimachi (1969*a*, 1970) and later in guinea pig spermatozoa (Yanagimachi, 1972). However, the normality of this motility behavior was established when it was seen in spermatozoa swimming within the oviductal contents by observation through the thin-walled oviductal ampulla of the hamster (Katz and Yanagimachi, 1980). Hyperactivated motility has now been reported in spermatozoa of numerous species, (recently) including nonhuman primates and cattle (Boatman and Bavister, 1984; Rose, 1984), and we may feel confident that it is a natural event that has some significance in fertilization (Cummins, 1982). Furthermore, much of our knowledge about mammalian fertilization and early development is derived from studies of material recovered from mated animals. Some of the most notable reports are the classic morphological observations made by Austin (1955) and by Hartman and colleagues at the Carnegie Foundation (e.g., Lewis and Hartman, 1933).

One of the most useful assets of the *in vivo* experimental approach is its ability to complement *in vitro* studies. As an example, the control of oocyte maturation is very poorly understood. Numerous attempts have been made to induce oocyte maturation *in vitro* in order to provide some insights, but the functional status of these oocytes needs to be assessed before any claims can be made about their normality. This could be achieved most easily using *in vitro* fertilization; for many species, however, appropriate techniques are unreliable or unavailable. Some investigators have circumvented this problem by transferring *in vitro* matured oocytes to oviducts of mated animals and later recovering embryos as evidence of normality of the oocytes. In some especially rigorous experiments, Staigmiller and Moor (1984) obtained lambs after transfer of *in vitro* matured oocytes to inseminated ewes, thereby providing incontrovertible evidence that normal maturation had taken place. Numerous other examples exist of *in vivo* studies on fertilization and associated events that have provided insights into mechanisms of early development. It may well be that the *in vivo* approach has been underutilized and may have been overshadowed by the popularity of *in vitro* fertilization procedures during the past 20 years or so.

Nevertheless, the *in vivo* approach has severe limitations. It is particularly difficult to obtain dynamic information on rapidly occurring events such as those associated with penetration of spermatozoa through the egg investments (e.g., onset of the acrosome reaction), although a number of attempts have been made (e.g., Cummins and Yanagimachi, 1982; Yanagimachi and Phillips, 1984). In addition, it is virtually impossible under natural conditions to analyze the manner in which the immediate environment (i.e., the oviductal secretions) regulates the functions of the gametes and early embryos. A fundamental drawback with the *in vivo* approach is that even if these limitations are accepted it can only provide information of a static type. Once removed from the natural milieu, the intimate relationship between the gametes or embryos and the oviductal secretions is lost. Further development along the pathway of fertilization and/or embryogenesis will not occur, unless the gametes or embryos are placed in a suitable culture environment or transferred to a recipient animal. In the latter case, they are obscured from further observation. It is these kinds of constraints which have stimulated development of *in vitro* techniques for the study of fertilization and embryogenesis in mammals. In spite of the technical difficulties and limitations involved with the *in vitro* approach, and the likelihood of introducing artifacts that complicate the interpretation of data, the advantages offered by examination of mammalian development *in vitro*, especially the prospects for observing the continuum of events as a dynamic process, underlie the ever-increasing interest in *in vitro* fertilization and *in vitro* embryogenesis, which is readily apparent from the vast literature published on these topics.

2. Techniques for *in Vitro* Fertilization

2.1. Methods

A variety of different methods have been described for accomplishing *in vitro* fertilization. Detailed discussion of these methods is beyond the scope of this review. Many of the procedures in common use as well as problem areas and results obtained by different laboratories are described in reviews by Gwatkin (1977) and Rogers (1978). The major areas of difference between these techniques lie in the physical conditions of the culture system (e.g., small drops of culture medium under an oil overlay vs. larger volumes in a humidified atmosphere without oil) and the type of culture medium used. Two major categories of culture medium are used for *in vitro* fertilization: (1) "simple" media, consisting of balanced salt solutions with various additives such as protein (serum albumin) and energy substrates (Bavister, 1981*a*); and (2) "complex" media, such as Ham's F10 or Medium 199, often with protein or blood serum supplements. Both types of culture medium have been shown to give satisfactory results. We prefer to use a simple medium formulation, TALP (Bavister and Yanagimachi, 1977), primarily because greater control can be exercised over its preparation (the medium is made entirely in the laboratory) and because modifications of the medium for different purposes can readily be made (e.g., see Bavister *et al.*, 1983*a*). Most of the simple culture media used for *in vitro* fertilization are based on either Tyrode's solution or Krebs-Ringer bicarbonate solution (Bavister, 1981*a*).

2.2. Standardization of Terms

Ambiguity exists in published work over the definition of a "defined" culture medium. Some investigators have used this term to denote a solution in which every constituent is chemically pure, while others have used it to signify only that no biological fluid component was used. Some standardization of terminology for gamete/embryo culture media is long overdue. The following classification is proposed:

1. *Undefined medium:* one containing biological fluid(s) such as blood serum as a component, regardless of the fact that the basic culture medium may be a salt solution of known composition
2. *Semi-defined medium:* one containing no biological fluids but with a protein component, which has only been partially purified, e.g., serum albumin (however, this represents a significant increment in standardization compared with the undefined medium)
3. *Chemically defined:* equivalent to "fully defined" or even "synthetic" medium, in which every ingredient is well characterized, if not of analytical grade

These definitions are preferable to ones such as "serum-free" medium, which could refer to either a semi-defined or a chemically defined solution. Description of a medium containing serum albumin as being "chemically defined" (e.g., Yanagimachi, 1982*a*; Iritani *et al.*, 1984) can be misleading.

2.3. Example of a Successful *in Vitro* Fertilization Procedure

One reason it is difficult to compare results of *in vitro* fertilization experiments between species and from one laboratory to another is that the culture conditions employed by investigators often vary quite widely. In my laboratory, we have attempted to minimize this variability by using essentially the same culture system for three species (hamster, rhesus monkey, and bovine), so that some consensus can be reached on the requirements for *in vitro* fertilization. In this way, we hope to demonstrate "real" differences and similarities between species (Table I). The method that we use for these species has been developed over the past 15 years or so using the golden hamster. Full technical details of our procedure for *in vitro* fertilization in this species are being prepared for publication. Using this procedure, 80–90% fertilization can be routinely achieved with hamster gametes in TALP medium, and about 70% of the inseminated eggs will cleave to the 2-cell stage (Juetten and Bavister, 1983*a*) (see Fig.

Table I. Comparison of Three Species for *in Vitro* Capacitation/AR Studies[a]

Species	Motility/viability factors needed	Stimulus for hyperactivation and fertilizing ability	Hyperactivation associated with fertilizing ability
Hamster[b]	Taurine/hypotaurine	Epinephrine	+
Rhesus monkey[c]	0	Caffeine/dbcAMP	+
Bovine[d]	0	0?	+?

[a]AR = acrosome reaction. Assessed by direct observation in hamster spermatozoa and by inference (fertilization) in rhesus and bovine.
[b]Meizel *et al.* (1980); Leibfried and Bavister (1982).
[c]Bavister *et al.* (1983*b*); Boatman and Bavister (1984).
[d]Hyperactivated motility observed but not yet correlated with penetrating ability (Rose, 1984).

1). Rhesus monkey gametes subjected to this procedure, with minor modifications, will yield about 80% fertilization, and about 70% of inseminated eggs have cleaved to the 8-cell stage or further *in vitro* (Bavister *et al.*, 1983*b*). Approximately 30% of bovine eggs can undergo sperm penetration and/or fertilization in TALP medium; in a pilot study we have obtained one or two cleavage divisions of *in vitro* fertilized bovine eggs (T. A. Rose and B. D. Bavister, unpublished observations).

An outline of our *in vitro* fertilization technique used for hamster gametes is given here as an example of a successful method. Hamster epididymal spermatozoa (washed or unwashed, depending on the experimental objectives) are incubated at 37°C in 100 μl drops of culture medium in plastic petri dishes under a layer of silicone or mineral oil. The culture medium (TALP) is a modified Tyrode's solution containing 10 mM sodium lactate, 0.1 mM sodium pyruvate, and 3 mg/ml bovine serum albumin (BSA) (Bavister *et al.*, 1983*a*). The sodium bicarbonate concentration is raised to 25 mM and the medium is equilibrated with 5% CO_2 in air. Sperm motility factors, consisting of 20 μM penicillamine, 100 μM hypotaurine, and 1 μM epinephrine (PHE) are added (see Leibfried and Bavister, 1982). Either cumulus-intact or cumulus-free hamster eggs obtained from superovulated animals are added at the beginning of incubation, or after approximately 4 hr when sperm capacitation has been completed. The final sperm concentration is 2–4 × 10^5/ml with cumulus-intact eggs or 2–4 × 10^4/ml with cumulus-free eggs (Bavister, 1982*a*). Incubation is carried out for 6 hr from the time of insemination of the culture drops (i.e., coincubation of spermatozoa with eggs is for either 6 or 2 hr). At the end of this time, each fertilized egg should display two well-formed pronuclei, together with a fertilizing sperm tail and two polar bodies, all of which can easily be visualized in hamster eggs under the phase-contrast microscope (Bavister, 1980). A variation on this technique that we use for more critical work is to preincubate hamster spermatozoa with motility factors at sperm concentrations of approximately 2 × 10^6/ml (see Section 3.2.2). This can conveniently be obtained by diluting 2 μl of epididymal contents in 2 ml of equilibrated culture

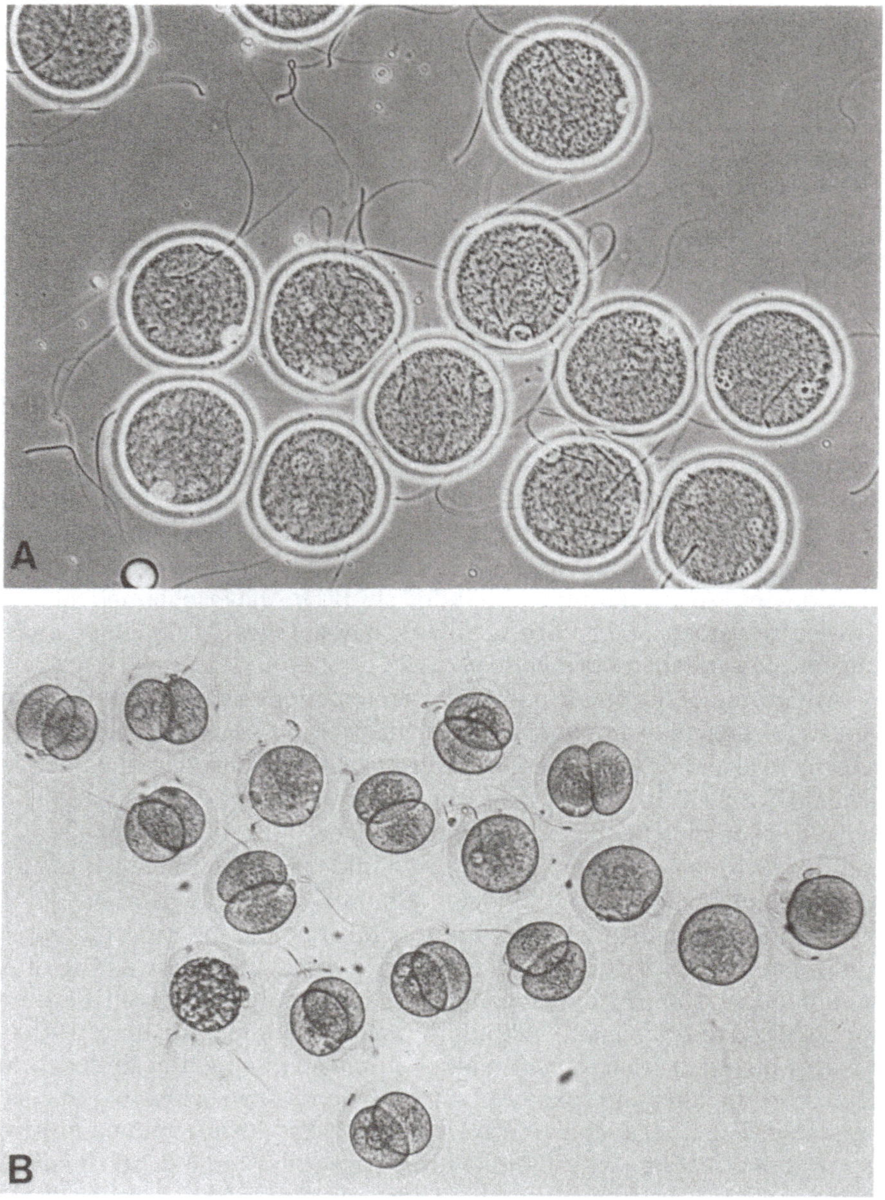

Figure 1. Typical results of hamster *in vitro* fertilization experiments. (A) Group of 12 hamster eggs incubated in same drop of culture medium with epididymal spermatozoa. Eleven are undergoing fertilization. In this focal plane, the fertilizing spermatozoon and/or one or more pronuclei can be seen in each egg. All 11 eggs had a sperm tail and two pronuclei in the egg cytoplasm and two polar bodies. Phase-contrast microscopy. Optical magnification : ×156. (B) Complete group of eggs from a single fertilization drop about 24 hr after insemination. Eleven of 17 eggs were fertilized and cleaved once in culture. Nuclei can be seen in some blastomeres. One egg fragmented and the remaining five failed to cleave. Interference-contrast microscopy. Optical magnification : ×66.

medium. After 4 hr, 1-μl or 10-μl aliquots of this primary sperm suspension are diluted into 100-μl or 90-μl drops of fresh equilibrated medium containing cumulus-free or cumulus-intact eggs, respectively; incubation is maintained for another 2 hr or longer.

Virtually the same procedures are used for bovine and for rhesus monkey gametes (Ball *et al.*, 1983; Bavister *et al.*, 1983*b*; Rose, 1984). Ejaculated spermatozoa are washed free from seminal plasma components (and semen extender ingredients in the case of the bovine) by centrifugation or by the swim-up technique (see Section 3.1). Bovine spermatozoa do not need any exogenous sperm motility factors or stimulants (Table I) and are inseminated directly into culture medium TALP containing eggs (Rose, 1984). Incubation of spermatozoa and eggs is maintained for 18–24 hr, after which eggs are fixed and examined for evidence of activation (presence of two polar bodies and/or pronuclei). Rhesus monkey spermatozoa require caffeine and dibutyryl cAMP (1 mM each) to stimulate the acquisition of fertilizing ability (Bavister *et al.*, 1983*b*; Boatman and Bavister, 1984) and are preincubated with these compounds before insemination of eggs. Rhesus eggs are either recovered from excised ovaries and matured *in vitro* or recovered by aspiration of follicles at laparoscopy in gonadotropin-stimulated animals (Bavister *et al.*, 1983*b*). Eggs from the latter source have a much higher capacity for embryonic development following *in vitro* fertilization (Boatman and Bavister, 1984). Evidence of fertilization is obtained either by examining fixed eggs or by assessing embryonic development during prolonged incubation. In species such as the bovine and rhesus monkey, in which it is difficult to visualize sperm tails in the egg cytoplasm, because of the presence of large amounts of lipid material and/or the rapid disintegration of sperm tail components (Bavister *et al.*, 1983*b*), activation of eggs is generally the criterion used to indicate that fertilization has occurred. In this situation, it is important to perform control incubations with killed spermatozoa to guard against the possibility of parthenogenetic activation of eggs (Bavister *et al.*, 1983*b*).

Since the *in vitro* fertilization procedures used in our laboratory are adequate for such different species as the golden hamster, bovine, and rhesus monkey, it seems likely that it will prove suitable for other mammals as well. The same techniques have in fact been used successfully with gametes of the mouse and rabbit (Bavister, 1972).

3. Techniques for *in Vitro* Fertilization: Advantages and Limitations

3.1. Recovery of Spermatozoa and Preparation for Fertilization

Most of the information on fertilization derived from *in vitro* studies has come from experiments using rodents and rabbits. Mouse, rat, and hamster spermatozoa are typically recovered by expressing the contents of the excised epididymides or vasa deferentia. In the golden hamster, this material consists of approximately equal volumes of epididymal plasma and spermatozoa (Bavister, 1974). Insemination of eggs *in vitro* with unwashed spermatozoa thus introduces a significant amount of epididymal plasma into the gamete culture environment. This contamination may be of little importance if the objective of an experiment is solely to examine details of fertilization (i.e., egg activation). For more critical experiments, however, epididymal plasma components may substantially affect the outcome. Both sperm-stimulating and sperm-inhibiting substances may be present. For example, epididymal plasma from golden hamsters contains components that have a pronounced sperm motility-stimulating effect that is apparent even at concentrations of epididymal contents as low as 0.1–0.4% (Bavister *et al.*, 1978). These components are probably taurine and related compounds (Meizel *et al.*, 1980). Their presence in the gamete culture environment may be beneficial, aiding the achievement of fertilization, but they can confound results of experiments such as those aimed at defining the role of egg–cumulus complex components in fertilization (Yanagimachi, 1969*a*; Bavister, 1982*a*). In addition, epididymal plasma from the hamster, and probably that of other species as well, contains inhibitors whose role may be to stabilize the sperm acrosome within the male reproductive tract. These components significantly inhibit the acrosome reaction *in vitro* even at epididymal plasma concentrations as low as 0.1% (Bavister *et al.*, 1978) and could thus reduce the incidence of fertilization as well as interfere with attempts to characterize the role of acrosome reaction stimulators.

The complications caused by epididymal plasma components can be avoided either by washing spermatozoa or by using very high dilutions of epididymal contents. For biochemical studies of the mechanism of the acrosome reaction, in which large numbers of spermatozoa are usually required at reasonably high concentrations, sperm washing seems mandatory. This procedure must be performed with considerable care, especially with delicate spermatozoa such as those of the golden hamster, which are easily damaged both by excessive centrifugation and by exposure to physiological salt solutions during washing. A solution that minimizes damage to hamster spermatozoa during washing can be prepared by mixing equal volumes of (isotonic) phosphate-buffered saline and sucrose (0.3 M), as described by Bavister and Yanagimachi (1977). The addition of 1 mg/ml polyvinylalcohol (PVA) to the washing solution provides further protection to the spermatozoa (B. D. Bavister, unpublished data). At least 80% of the spermatozoa remain motile after washing if this procedure is carried out carefully. Alternatively, hamster spermatozoa can be washed by passing them through a glass-bead column (Lui *et al.*, 1979). This method gives excellent results and has the added advantage that most of the immotile spermatozoa are removed, although the overall yield of spermatozoa is reduced compared with the centrifugation method. Another method that appears to be very promising for studies on sperm fertilizing ability consists of centrifuging spermatozoa on a continuous Percoll density gradient (Lessley and Garner, 1983). Not only does this technique achieve good separation of spermatozoa from seminal plasma, cytoplasmic droplets, and particulate matter, but it also yields a high recovery of sperm cells—about 80%. This separation is accomplished with relatively little dilution of the original sperm suspension and with virtually no mechanical or osmotic trauma to the spermatozoa.

Finally, mention should be made of the swim-up technique for preparing ejaculated spermatozoa for fertilization. In this technique, which was originally devised for human semen (Overstreet *et al.*, 1980), ejaculates are overlaid with a culture solution and incubated for an hour or more to permit vigorously motile spermatozoa to enter the upper layer, which is then recovered. This procedure simultaneously eliminates most of the seminal plasma and dead or poorly motile spermatozoa; at the same time it may initiate sperm capacitation. The technique has the additional advantages of being nontraumatic to the spermatozoa, and allowing time for other procedures (e.g., preparation of eggs for fertilization) during the swim-up period. We have used this technique with ejaculated spermatozoa of the rhesus monkey and bovine; it could probably be used advantageously with other species as well.

Epididymal spermatozoa, mostly collected from rodent species, have been used for convenience in many studies on capacitation and the acrosome reaction. It is tacitly inferred that such spermatozoa are physiologically comparable to those taking part in fertilization *in vivo*, yet this is not necessarily the case. Epididymal spermatozoa normally undergo changes resulting from contact with seminal plasma that modify their behavior and characteristics, which may well affect their eventual ability to fertilize eggs. For example, in the rat and pig, epididymal and ejaculated spermatozoa show distinct differences in their ability to fertilize eggs *in vitro* (Shalgi *et al.*, 1981; Nagai *et al.*, 1984). Proteins derived from seminal plasma become tightly bound to spermatozoa and may constitute as much as 35% of total sperm plasma membrane proteins (Russell *et al.*, 1984). The function of most of these seminal plasma-derived proteins is obscure. However, one such protein inhibits Ca^{2+} uptake by bovine spermatozoa (Babcock *et al.*, 1979), and removal of this protein during capacitation may be part of the mechanism leading to expression of the acrosome reaction (Rufo *et al.*, 1984). Calsemin, a protein found in ram seminal plasma, stimulates a Ca^{2+}-dependent Mg^{2+}-ATPase in the flagellar plasma membrane of epididymal spermatozoa and regulates Ca^{2+}-efflux (Bradley and Forrester, 1982). These and other studies clearly point to a role for seminal plasma proteins in regulating sperm activity and fertilizing ability *in vivo*. This regulatory function needs to be investigated by performing more comparative studies within species using both epididymal and ejaculated spermatozoa. Such studies can well be conducted in the rabbit and in cattle, with epididymal spermatozoa obtained from excised male reproductive tracts and with ejaculated sperm collected with an artificial vagina. Spermatozoa from these sources can be compared using *in vitro* fertilization procedures developed for these species (Brackett and Oliphant, 1975; Brackett *et al.*, 1982*a*,*b*; Ball *et al.*, 1983).

3.2. Capacitation and Acrosome Reactions of Spermatozoa *in Vitro*

Capacitation, originally defined as acquisition of the capacity to fertilize eggs (Austin, 1951), remains a poorly understood phenomenon. It appears to be a universal requirement for mammalian spermatozoa, but its functional significance is uncertain (Bedford, 1983). Although numerous changes in the properties of spermatozoa have been reported to be concomitant with capacitation (see reviews by Yanagimachi, 1981; O'Rand, 1982; Clegg, 1983), it is still not known which of these are directly associated with accomplishment of capacitation and which are simply expressions of increased activity of spermatozoa following their release from the male reproductive tract. A major reason for this uncertainty arises from the dichotomy between the requirements for biochemical studies of capacitation and those for physiological analyses of this process. The former usually necessitates examination of large numbers of spermatozoa, while the latter is optimally performed with as few spermatozoa as possible. The difference may reach several orders of magnitude in terms of sperm concentration. Thus, it may be difficult to relate biochemical and physiological correlates of capacitation within the same sperm suspension. This problem may be overcome in some respects by the introduction of monoclonal antibody techniques, e.g., to examine surface changes in spermatozoa during capacitation (Myles and Primakoff, 1983). Alternatively, the novel approach of examining sperm binding to lectin-coated beads seems very promising (Ahuja, 1984).

To illustrate this point, consider, for example, the functional status of a population of spermatozoa recovered several hours after mating from the uterus of a rabbit. Some of these spermatozoa are undoubtedly capacitated, as shown by their ability to fertilize eggs *in vitro* (or *in vivo*) after a short period of sperm : egg incubation (Brackett and Williams, 1965; Soupart, 1967). However, the proportion of spermatozoa that have undergone capacitation is hard to determine. Very few spermatozoa are required to fertilize a test sample of eggs, which probably number a few dozen at most. Yet the number of spermatozoa in the population being studied is probably many thousands or as many as a million. It cannot be assumed that the few spermatozoa shown by *in vitro* fertilization to be capacitated are representative of the sperm population as a whole. The vast majority of spermatozoa in a uterine sperm population would not normally participate in fertilization; many of them may not have undergone capacitation and may even be moribund. Thus, there is a possibility that characteristics evaluated in a large sperm population may be more representative of nonfertile than of capacitated spermatozoa. The same argument can be applied to large populations of spermatozoa incubated under *in vitro* conditions, leading to capacitation of a few cells as determined by the fertilization of a relatively very small number of eggs (see Section 3.3.1a). This argument does not imply that any changes observed in the majority of spermatozoa during *in vivo* or *in vitro* incubation are not relevant to the process of capacitation—only that it is difficult to unequivocally correlate biochemical and physiological events.

3.2.1. Events Associated with Capacitation

The problem described in Section 3.2 can be reduced, although not eliminated, in several ways that increase the precision of experimental studies on capacitation. The number of spermatozoa that can be analyzed in a population can be greatly increased by examination of two physiological correlates or sequelae of capacitation: (1) the development of a peculiar "whiplash" type of sperm motility, termed "hyperactivation," and (2) the acrosome reaction (Yanagimachi, 1970, 1981). Hyperactivated motility is known to occur in hamster and rabbit spermatozoa within the oviductal ampulla near to the time of fertilization (Cooper *et al.*, 1979; Katz and Yanagimachi, 1980). This phenomenon has been observed in numerous species of spermatozoa under *in vitro* conditions, including the mouse, bat, ram, dog, guinea pig, marmoset monkey, and bovine (Yanagimachi, 1972; Mahi and Yanagimachi 1976; Fraser, 1977; Lambert, 1981; Moore, 1981; Cummins, 1982; Rose, 1984). We recently showed a strong correlation between expression of hyperactivated motility in rhesus monkey spermatozoa and their ability to penetrate the zona pellucida of rhesus eggs *in vitro* (Boatman and Bavister, 1984). This study provided evidence of tight coupling between hyperactivation and capacitation.

Since there is good evidence that hyperactivation is normally a physiological expression of capacitation, it seems reasonable to use this motility behavior as a semiquantitative index of the capacitation status of sperm populations. Biochemical parameters of spermatozoa may then be related more precisely to the achievement of capacitation. However, by incubating golden hamster spermatozoa under some experimental conditions, it is possible to obtain hyperactivation without capacitation. These conditions include incubation of spermatozoa in a high potassium environment (Barros and Berrios, 1977) or in a special culture medium lacking serum albumin, a component required for capacitation of hamster spermatozoa (Bavister, 1981*b*,*c*). The latter observation has been used in our laboratory to devise a system for differentiating the changes occurring in hamster spermatozoa that are directly related to capacitation from those that are merely expressions of increased sperm activity (see Fig. 2). In this system, washed spermatozoa are incubated at 37°C in a chemically defined culture solution containing PVA instead of serum albumin (TLP–PVA) (Bavister, 1981*b*), together with the defined sperm motility-stimulating factors, hypotaurine and epinephrine (Leibfried and Bavister, 1982). Samples of the same washed sperm preparation are incubated in the same culture solution containing serum albumin. After about 5 hours, most of the spermatozoa in both suspensions are vigorously motile, and display hyperactivated motility. However, acrosome reactions, used as an indicator of capacitation (see below), are seen only in spermatozoa incubated with serum albumin. Between 60–90% of motile spermatozoa exhibit acrosome reactions by 5–6 hr. By contrast, spermatozoa incubated without albumin do not undergo acrosome reactions, nor do they appear to be capacitated, even after 8–10 hr (Bavister, 1981*b*). Thus, although PVA fails to support acrosome reactions, it is not inhibitory when spermatozoa are incubated in both PVA and serum albumin. By comparing aliquots of the two sperm suspensions at increasing time intervals (Fig. 2), differences in biochemical and cell membrane properties associated with capacitation can be examined.

A widely used method for assessing the capacitation status of sperm populations is to estimate the proportion of motile (live) cells showing the acrosome reaction. This assay method is predicated on the knowledge that fertilizing

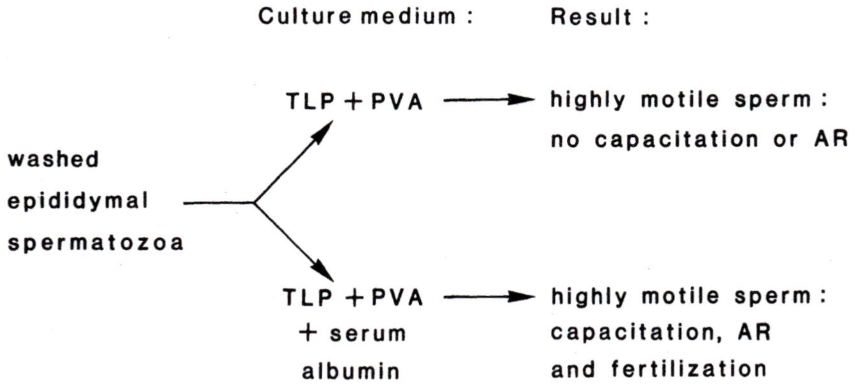

Figure 2. Scheme for analysis of changes associated with capacitation in hamster spermatozoa. Epididymal spermatozoa are washed in a chemically defined medium, then incubated in medium TLP, i.e., albumin-free TALP (see Bavister and Yanagimachi, 1977) with PVA (polyvinylalcohol), with or without bovine serum albumin (BSA) for up to 6 hr. Both treatments produce highly motile spermatozoa, but only BSA-containing medium (TALP) supports capacitation and acrosome reactions (AR). Biochemical correlates of capacitation and AR can be examined by comparative studies on spermatozoa incubated under the two culture conditions.

spermatozoa undergo acrosome reactions immediately before or during penetration of the egg envelopes (Austin and Bishop, 1958). Stated another way, the acrosome reaction is the physiological sequel to capacitation. The original definition of capacitation (Austin, 1951) did not make a distinction between these two events because the need for mammalian spermatozoa to undergo acrosome reactions was not recognized until several years later (Austin and Bishop, 1958). In the latter study, changes in the sperm acrosome leading to its detachment were considered part of capacitation. To avoid confusion, however, most investigators now consider capacitation an event, or a series of events, preceding the acrosome reaction. Throughout this chapter, capacitation is defined more rigorously as the process by which spermatozoa acquire the ability to undergo acrosome reactions in the presence of appropriate stimuli.

The acrosome reaction is a gross morphological change in which the acrosomal matrix, most of the outer acrosomal membrane, and the overlying sperm plasma membrane are lost due to fusion and vesiculation of these membranes (Franklin *et al.*, 1970; Yanagimachi, 1981). In those species having spermatozoa with large acrosomes that are visible in living cells under the light microscope (e.g., hamster and guinea pig), the percentage of motile spermatozoa that are acrosome reacted (Talbot and Franklin, 1976) can be used as a convenient, quantitative, and rapid indication of the proportion of spermatozoa in a sperm population that have completed capacitation. In most species, however, the acrosome is a relatively small structure. Various methods are used to assess acrosome reactions in these species, including the use of fluorescent probes such as chlortetracycline, e.g., with live mouse spermatozoa (Saling and Storey, 1979), and acrosomal staining methods that are applied after drying sperm suspensions on glass slides (Bryan and Akruk, 1977). The precision of the latter approach can be improved by combining it with a vital stain (trypan blue) before air-drying and differential staining (Talbot and Chacon, 1981). This technique reduces complications in the assay by eliminating dead sperm, which frequently undergo acrosomal loss, from the acrosome reaction analysis.

A more refined approach has been described in which changes in capacitated spermatozoa are evaluated by chlortetracycline fluorescence before the onset of acrosome reactions. Mouse spermatozoa showing a banded fluorescent pattern were capable of undergoing acrosome reactions induced by solubilized zonae pellucidae (see Section 4.7.3), whereas sperm with fluorescence patterns equivalent to fresh epididymal sperm were not able to do so. Therefore, it was concluded that "an operational criterion for capacitation is taken to be the zonae-induced acrosome reaction. The fluorescent probe method appears to provide a direct means for monitoring the capacitation of mouse sperm *in vitro*" (Ward and Storey, 1983). It will be most interesting to discover whether this technique can be applied to other species.

A new multiple sperm penetration assay for measuring capacitation is described in Section 3.3.1a. This assay effectively examines the ability of capacitated sperm to undergo physiological acrosome reactions.

3.2.2. Analysis of Responses to Experimental Treatments

Regardless of how acrosome reactions are detected, there is the question of how to relate the data to the efficiency of different experimental treatments in supporting capacitation. Simply assessing the highest percentage of acrosome reactions produced can be misleading, as represented diagrammatically in Figure 3. The results of a hypothetical experiment are shown using two different conditions for capacitation. Both treatments eventually yield the same percentage of acrosome-reacted spermatozoa (percentage of the motile sperm population). However, the maximum response is obtained in about 6.5 hr with treatment A but not until more than 8 hr have elapsed with treatment B. Such treatment differences can be physiologically very meaningful, since most eggs may be penetrated within a short time following insemination by a population of capacitated sperm showing only 1–10% acrosome reactions (Bavister, 1973; Barros *et al.*, 1984).

To make acrosome reaction assays more meaningful and precise, the time required to obtain acrosome reactions in 50% of the sperm (AR_{50}) can be used (point b in Fig. 3). Treatment A has an AR_{50} of 4.75 hours whereas B has an AR_{50} of 7.4 hr. Additional information can be derived by measuring the slope of the rising phase of the acrosome reaction curve. In the example shown in Figure 3, the slope of A is greater than the slope of B, indicating that treatment A is more efficient in supporting capacitation (and acrosome reactions) than B. Finally, the time elapsed between the intercepts of the AR slope and the AR_{50} with the time axis (a to b in Fig. 3) provides another indication of the efficiency of the two treatments: in treatment A this time is only about 0.7 hr, whereas in B it is 1.5 hr. Use of one or more of these acrosome reaction measurements

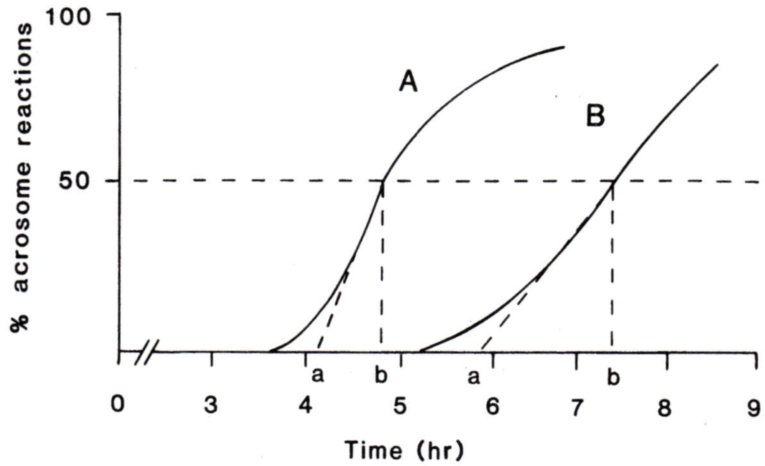

Figure 3. Results of a hypothetical experimental study on sperm acrosome reactions. Data curves typical for hamster spermatozoa incubated in good (A) or poor (B) conditions for stimulating capacitation and/or acrosome reactions. Aliquots of sperm suspensions are examined at regular intervals, and the percentages of motile spermatozoa showing acrosome reactions are determined, usually by examining 100 motile spermatozoa. Similar curves might be obtained using staining methods for acrosome reactions. The times required to achieve 50% acrosome reactions (AR_{50}) are shown by points b; times for onset of AR (points a) are taken as the intercepts of the initial slope of the AR curve with the X axis. The time interval between a and b may also give useful information about the ability of different treatments to stimulate the AR.

could substantially improve the quality of information obtained from *in vitro* experiments.

Apart from determinations of hyperactivated motility and of acrosome reactions in the sperm population, another practical approach that can help relate biochemical and physiological events has been used in our laboratory. This procedure consists of incubating hamster spermatozoa in primary suspensions (about 10 ml) at a high concentration, e.g., 2×10^6/ml (see Section 2.3). Biochemical analyses can be performed on aliquots removed at intervals during the incubation period or on the entire suspensions at an appropriate time point. The proportion of motile spermatozoa should not drop below 80% for approximately 6 hr, and hyperactivated motility should be apparent after incubation for about 4 hr. Small aliquots (e.g., 10 μl) of each suspension are used (1) to determine the percentage of acrosome-reacted spermatozoa, and (2) to inseminate eggs in drops of culture medium to evaluate sperm fertilizing ability.

This approach has several advantages. First, enough spermatozoa are incubated under each treatment condition in the primary suspensions to permit biochemical analyses to be performed. Second, biochemical events can be correlated with several physiological changes (motility, acrosome reactions, fertilizing ability) using the same sperm suspension. Third, the final assessment of fertilizing ability is performed using a very small and representative (randomly selected) sample of the primary sperm suspension. This procedure goes a long way toward addressing the sperm sampling difficulties described in Section

3.2. The quality of data that can be obtained using this experimental approach can be increased substantially if comparative studies are performed with noncapacitating conditions as controls, such as in the system described in Figure 2. Additional ways of increasing the sensitivity of the *in vitro* fertilization component of these assays are described in the following section.

3.3. *In Vitro* Fertilization

The technique of *in vitro* fertilization has been applied to a large number of species during the past 25 years with varying degrees of success (Gwatkin, 1977; Rogers, 1978). In general, the technology for obtaining sperm penetration of eggs *in vitro* is well established, although a bewildering variety of procedures are currently in use. Less well established, sometimes by far, is the ability to obtain development of *in vitro* fertilized eggs, without which it is difficult to ascertain the "normality" of fertilization obtained in this way (see Section 5). Moreover, the mechanisms involved in fertilization are still rather obscure in spite of the enormous literature available on *in vitro* fertilization. In part, this discrepancy results from the technical problems involved in devising methods for consistently achieving *in vitro* fertilization in some species. In addition, the variety of procedures used with different species makes the achievement of a consensus of information very difficult. Undoubtedly there are also some profound species differences in the characteristics of spermatozoa and of gamete interactions *in vitro* that further confuse the overall picture. This section describes some of the ways in which more meaningful information might be derived from *in vitro* fertilization experiments and highlights some of the advances that have been made in recent years.

In vitro fertilization of animal eggs is performed for three principal reasons: to gain information about sperm fertilizing ability and its regulation, to obtain eggs in very early stages of activation so that the fertilization process itself can be examined, and to provide a convenient supply of early cleavage stage embryos of certain species so that events in embryonic development can be studied. The technique has proved invaluable in all these respects. Section 3.3.1 deals with the use of *in vitro* fertilization as a means to assess the fertilizing capacity of spermatozoa.

3.3.1. Assessment of Sperm Fertilizing Ability

3.3.1a. Homologous (Homospecific) Fertilization. The ability of a spermatozoon to penetrate the zona pellucida of an egg and to fuse with the egg plasma membrane requires completion of capacitation and the acrosome reaction (Austin and Bishop, 1958). From a conceptual viewpoint, accomplishment of *in vitro* fertilization provides unequivocal evidence that these events have occurred. Moreover, this technique can provide a more sensitive indication of the fertilizing ability of a sperm population than can the simpler procedure of assessing acrosome reactions (see Section 3.2.1). The time course for penetration of intact eggs by hamster spermatozoa may precede the time course for expression of acrosome reactions (in the bulk sperm suspension) by 2 hr or more, and a high percentage of acrosome reactions is associated with loss of fertilizing ability (Barros *et al.*, 1973, 1984; Gwatkin *et al.*, 1976).

Paradoxically, from a technical viewpoint, *in vitro* fertilization as it is usually performed may be less than satisfactory as an indicator of sperm fertilizing ability. One reason for this relates to the variability and technical problems inherent in most *in vitro* fertilization procedures, while another stems from the sperm population sampling problems already referred to in the context of analyzing hyperactivation and acrosome reactions (Section 3.2). Even after technical difficulties with *in vitro* fertilization procedures have been worked out with a particular species, there remain the biological variables inherent in both spermatozoa and eggs. If only the spermatozoa are evaluated, e.g., for occurrence of hyperactivation or of acrosome reactions, there is only one set of variables to contend with. By carefully selecting sperm donors, variability between replicate experiments can be substantially reduced. However, when egg differences are also introduced, the results of *in vitro* fertilization experiments can vary remarkably. Variability can be contained within acceptable limits when using laboratory rodents for *in vitro* fertilization experiments. Animals stimulated with exogenous gonadotropins will produce large numbers (e.g., 40–50 each) of ovulated eggs that usually yield consistently uniform results, i.e., sperm penetration, fertilization, and cleavage, *in vitro* and *in vivo* (e.g., Toyoda and Chang, 1974; Liebfried and Bavister, 1982; Fraser, 1983*a*; Juetten and Bavister, 1983*a*,*b*; Yanagimachi, 1984). However, *in vitro* fertilized eggs may not be capable of normal development (see Section 5.1). A major advantage of using rodents is that the eggs are translucent, permitting identification of the fertilizing sperm tail, which often persists beyond the first cleavage division (Toyoda and Chang, 1974; Bavister, 1980). This feature provides evidence of fertilization. This is not the case with gametes of domesticated animals and primates (see Section 2.3). An additional problem with eggs of these species is that they are usually matured (or complete maturation) *in vitro*, since they are most conveniently collected from excised ovaries or recovered from preovulatory follicles. *In vitro* matured eggs frequently exhibit anomalies, and their ability to take part in fertilization is very often impaired (see Section 5.1). There are some notable exceptions, e.g. the studies by Staigmiller and Moor (1984) and by Schroeder and Eppig (1984).

A second problem relating to the use of *in vitro* fertilization concerns sampling of sperm populations (see Section 3.2). It may seem perfectly straightforward to incubate spermatozoa *in vitro* under different (capacitating) conditions and then to assess the efficiency of the treatments by inseminating eggs and evaluating the incidence of penetration and/or fertilization. However, unless experiments are very carefully designed, there are some inherent flaws in this approach that derive from the abnormally high sperm : egg ratios conventionally used to accomplish fertilization *in vitro*. Table II shows data obtained from a hypothetical *in vitro* fertilization experiment, using two treatments to support sperm fertilizing ability. Treatment A has 10× the efficiency of treat-

Table II. Numerical Limitations with *in Vitro* Assays for Sperm-Fertilizing Ability[a,b]

Sperm concentration	No. of sperm per 100 μl[c]	Total sperm : egg ratio[d]	Sperm treatment[e]	Capacitated sperm : egg ratio	IVF result[f] (%)
2 × 10⁵/ml	2 × 10⁴	1000 : 1	A	500 : 1	95
			B	50 : 1	90
2 × 10⁴/ml	2 × 10³	100 : 1	A	50 : 1	90
			B	5 : 1	25
2 × 10³/ml	2 × 10²	10 : 1	A	5 : 1	25
			B	0.5 : 1	1

[a]See also Figure 4.
[b]Hypothetical data for a typical IVF experiment.
[c]Typical 100-μl drop of culture medium (fertilization drop).
[d]Assumes 20 eggs per fertilization drop.
[e]Treatment A supports capacitation in 50% of sperm population, treatment B in 5% of sperm population.
[f]IVF = *in vitro* fertilization. For simplicity, all penetrated eggs are assumed to be fertilized.

ment B. In the first example, a sperm concentration of 2×10^5/ml in a volume of 100 μl is assumed, which is typical of many experimental studies. At this high sperm : egg ratio (1000 : 1), there is virtually no difference in the fertilization response. Such an experimental result would lead to erroneous conclusions about the sperm treatments. This type of experimental design flaw has probably undermined many *in vitro* fertilization studies. It should be emphasized that this problem can arise with both excessive sperm concentrations and sperm numbers, i.e., the final sperm : egg ratio is a critical factor.

By reducing the total sperm : egg ratio, the sensitivity of the *in vitro* fertilization test system is increased, revealing the difference in efficiency of the two sperm treatments (Table II). In this hypothetical case, either 100 : 1 or 10 : 1 total sperm : egg ratios should be used, although the former would be easier to use from a practical point of view. The data shown in Table II are presented diagrammatically in Figure 4, which graphically illustrates the difference between

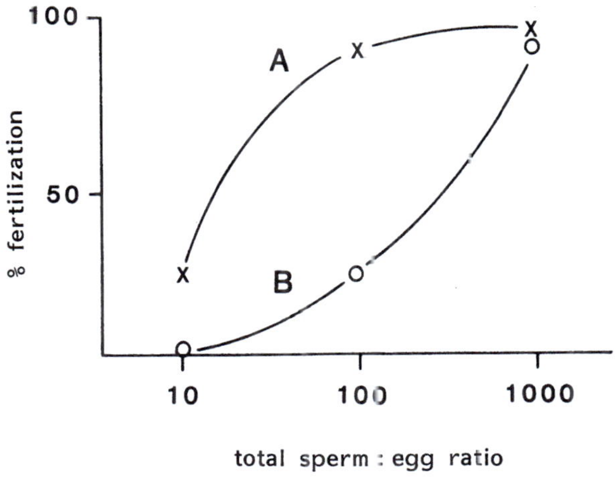

Figure 4. Results of a hypothetical study on sperm-fertilizing ability. Because of the usual discrepancy between numbers of spermatozoa and eggs in a fertilization test system, treatments with differing abilities to support sperm-fertilizing ability (A,B) may give similar results at high sperm : egg ratios. By reducing these ratios, differences between the treatments become more apparent.

the two sperm treatments. The same experimental approach could be used to demonstrate differences in sperm fertilizing ability between males of the same or different strains.

In an actual study using hamster gametes, 50-μl drops of culture medium were used containing $1–2 \times 10^4$ spermatozoa, with a final sperm : egg ratio of approximately 100 : 1 (Bavister, 1982*a*). Under these conditions, the optimal treatment supported about 80% penetration and 20% fertilization of eggs. This system served as a basis for examining the ability of egg–cumulus complex components to support sperm capacitation and fertilizing ability.

A second way to increase the sensitivity of *in vitro* fertilization assays is to minimize the period of sperm : egg co-incubation (e.g., Fraser, 1983*a*). A treatment that produces a high percentage of spermatozoa capable of penetrating eggs only after prolonged incubation should clearly give a different response from one that accomplishes this in a shorter time. However, in experimental studies using *in vitro* fertilization as a measure of the ability of different treatments to effect sperm capacitation and acrosome reactions, it is common practice to compare the percentages of eggs fertilized after a prolonged period of sperm : egg co-incubation—usually several hours. The rationale for this procedure seems to be to obtain the maximum possible percentages of fertilized eggs. This type of study is the simplest way to compare different sperm treatments, but it is not necessarily the most informative method. Examination of the time course of *in vitro* fertilization will usually reveal a sigmoid curve, i.e., the slope of the relationship between the percentage of fertilized eggs and the duration of sperm : egg co-incubation decreases after a period of time (e.g., Aonuma *et al.*, 1982). This may result in part from variability in egg quality, as well as heterogeneity of fertilizing ability in the sperm population (see Section 3.2). A more stringent test of sperm capacitation is to use the time for 50% fertilization (actually, penetration*) to be accomplished (P_{50}) as the relative index of sperm-fertilizing ability, analogous to the AR_{50} proposed for acrosome reaction studies (Section 3.2.2). This analysis may provide more meaningful data than the absolute percentage of eggs fertilized. For example, in one study (Aonuma *et al.*, 1982), mouse spermatozoa preincubated with BSA reached the P_{50} level about 20 min after insemination of ova, whereas fresh epididymal spermatozoa or spermatozoa preincubated without albumin required 55–60 min to reach P_{50} after insemination of ova. By contrast, examination of ova 1.25 hr after insemination would have revealed essentially no difference between any of the sperm treatments—about 60% of eggs penetrated in all cases.

A third approach is to increase the sensitivity of the eggs as indicators of sperm penetration. This approach is equivalent to increasing the number of

*The distinction here between penetration and fertilization is often ignored. Strictly, an egg is not fertilized until syngamy has occurred. However, the terms "fertilized" or "undergoing fertilization" are frequently used to indicate the presence of a sperm nucleus in the egg cytoplasm. In the present context, penetration of eggs is a more accurate term, although it may be used to denote presence of spermatozoa in the perivitelline space or in the egg cytoplasm; in either case, sperm must have undergone capacitation and the acrosome reaction in order to penetrate the intact zona pellucida.

eggs used in the test system, which is usually not feasible in species such as the rhesus monkey, in which relatively few eggs can be obtained. Normally, the block to polyspermy is initiated when a single spermatozoon fuses with the vitelline membrane (Wolf, 1981), so that each egg can serve as an indicator of the penetrating ability of only one spermatozoon. However, if the block to polyspermy is defective or absent, spermatozoa can continue to penetrate the zona pellucida and enter the perivitelline space. In this way, dozens of spermatozoa may penetrate each zona, thereby increasing the resolving power of the test system by the same numerical factor. It should be mentioned that this is the normal condition with rabbit eggs, in which the block to polyspermy is mostly at the level of the egg plasma membrane, unlike most other species. This property of rabbit eggs was recently used to advantage in an ingenious study by Kuzan *et al.* (1984). Numerous perivitelline spermatozoa were recovered by splitting the zonae pellucidae of *in vivo* fertilized rabbit eggs; the released motile sperm were able to fertilize fresh zona-intact eggs *in vitro*. This experiment shows that eggs could be used to selectively collect many spermatozoa capable of fertilizing from a heterogeneous population. The numbers of spermatozoa penetrating each egg would then give a quantitative index of the efficiency of different treatments in supporting sperm-fertilizing ability.

Eggs of other species can be treated to render the block to multiple zona penetration inoperative. One approach is to inhibit the block to polyspermy by chemical methods, e.g., by using local anesthetics (Ahuja, 1982). Yanagimachi *et al.* (1979) used hypertonic salt treatment to preserve zona-intact hamster eggs. This procedure caused shrinkage of the vitellus and an increase in the perivitelline space. Numerous spermatozoa were able to penetrate the zonae pellucidae of these eggs, but the chemical and physical characteristics of the zonae were unchanged, as was their biological specificity (i.e., ability to exclude uncapacitated spermatozoa).

We have used Yanagimachi's salt-storage procedure to devise a quantitative sperm penetration assay, in which treated hamster eggs are exposed to spermatozoa that have been incubated under different conditions. Because the block to polyspermy is eliminated, many sperm can penetrate the zona pellucida provided that they have undergone capacitation and are capable of undergoing acrosome reactions. Between 100 and 200 sperm have been observed in the perivitelline space of some eggs but for convenience of counting sperm, the sperm concentration and length of sperm : egg coincubation are adjusted to give a maximum of about 50 sperm per egg. A group of only 10 eggs can thus confirm the capacitated condition of up to 500 sperm, an increase of 50-fold over the information gained with conventional monospermic egg penetration as the measure of capacitation. This multiple sperm penetration assay, which essentially combines features of acrosome reaction and egg penetration assays, should be helpful in elucidating conditions favoring sperm capacitation. Details of the procedure are being prepared for publication.

We have achieved a similar effect with rhesus monkey eggs that failed to mature *in vitro* and that lacked the block to polyspermy. In some of these eggs, the cytoplasm became shrunken, thereby greatly enlarging the perivitelline space and facilitating observation and counting of the penetrating spermatozoa (D. E. Boatman and B. D. Bavister, unpublished data). Also in this species, the block to polyspermy appears to operate both at the level of the egg plasma membrane (unlike the golden hamster) and at the innermost region of the zona pellucida. Some eggs can be observed following *in vitro* insemination with 20–50 spermatozoa deeply embedded in the zona pellucida, with their heads about two-thirds of the way through. It may be assumed that these spermatozoa had all undergone capacitation and acrosome reactions, although this could not be assessed directly. Finally, rhesus spermatozoa incubated under culture conditions that did not stimulate hyperactivated motility, and thus by inference were not capacitated, failed even to attach to the zonae pellucidae of eggs, although frequent collisions of spermatozoa with eggs were observed (Boatman and Bavister, 1984). These examples illustrate that useful data concerning the functional status of spermatozoa can be obtained by careful observation in species in which large numbers of eggs cannot be obtained.

3.3.1b. Heterospecific Fertilization. A different approach to the evaluation of sperm fertilizing ability is to employ the technique of heterospecific *in vitro* fertilization (Yanagimachi *et al.*, 1976; Rogers, 1978). Most of the comments made in Section 3.3.1a apply equally here. Heterospecific fertilization nearly always involves the use of zona pellucida-free hamster eggs, which have the remarkable (and perhaps unique) property of being able to undergo fusion with spermatozoa of a large number of other species (Yanagimachi, 1981). Among the common laboratory animals, the hamster is unusual in having the block to polyspermy operate almost entirely at the level of the zona pellucida. Perhaps these two properties are functionally related. In order to fuse with the plasma membranes of naked (zona-free) hamster eggs, spermatozoa must undergo acrosome reactions, so the heterospecific fertilization assay is widely assumed to be a satisfactory test of sperm fertilizing ability. This is true only to a limited extent. The absence of the egg envelopes (cumulus cells and zona pellucida), which represent selective barriers to sperm penetration in the normal situation, eliminates several of the normal constraints to accomplishment of fertilization. In particular, the role of sperm acrosomal enzymes and of vigorous sperm motility (hyperactivation) in sperm penetration may be reduced or even eliminated. At least in some species, the mechanisms of fertilization may be profoundly different in the zona-free test compared with the situation with intact homologous eggs. For example, rhesus monkey spermatozoa incubated under conditions that support fertilization of intact eggs *in vitro* (e.g., stimulation with cyclic nucleotide mediators) rarely penetrate zona-free hamster eggs (Boatman and Bavister, 1984). This discrepancy may imply a functional role for the cumulus oophorus and/or the zona pellucida in supporting sperm-fertilizing ability. Such a role of these components is indeed indicated by experiments with mouse and hamster eggs (Saling and Storey, 1979; Bavister, 1982*a*; Bleil and Wassarman, 1983; Jacobs *et al.*, 1984) (see also Section 4.7.3).

Conversely, guinea pig and hamster spermatozoa remain able to fertilize zona-free hamster eggs for several hours after losing their ability to fertilize intact homologous eggs (Fleming and Yanagimachi, 1982; Barros *et al.*, 1984). This discrepancy shows that different aspects of sperm fertilizing ability are discriminated by the zona-intact and zona-free egg assays.

3.4. Summary of Approaches for Improving *in Vitro* Assays of Sperm Fertilizing Ability

From the foregoing discussion, it is clear that there are a variety of ways in which the quality and/or quantity of useful data obtained from *in vitro* fertilization experiments can be increased. These approaches, which are not all applicable to every species or to every *in vitro* technique, are summarized below.

1. Use of epididymal spermatozoa is not strictly physiological. Attempts can be made to compare characteristics of epididymal and ejaculated spermatozoa during *in vitro* incubation in order to gain a deeper understanding of mechanisms of sperm capacitation and of the acrosome reaction. Spermatozoa recovered from the uteri of mated animals should also be studied to provide comparative data (e.g., Shalgi *et al.*, 1981; Fraser, 1983*b*).
2. Whenever possible, male reproductive tract secretions should be removed from spermatozoa. This may be done using centrifugation/washing, glass bead columns, or the swim-up technique.
3. An appreciation of the problems inherent in sampling a sperm population is needed. Efforts should be made to ensure as far as possible that spermatozoa used for fertilization and other assays are representative of the population as a whole.
4. Information on the acrosome-reaction status of a sperm population can be enhanced by using more rigorous determinations than simple percentages, e.g., use of the AR_{50}.
5. The time allowed for development of capacitation/acrosome reactions and/or for estimation of sperm-fertilizing ability (sperm : egg co-incubation time) should be kept to a minimum to enhance differences in responses between treatments.
6. Fertilization assays should be performed using minimal sperm : egg ratios. The sensitivity of assays may be increased by use of the P_{50} instead of waiting for maximum fertilization percentages to be achieved.
7. Attempts should be made to relate biochemical and physiological parameters of a sperm population by performing multiple assays on the same population.
8. The resolving power of *in vitro* fertilization assays for testing sperm capacitation can be increased by using eggs lacking an effective block to polyspermy. This is especially useful in species in which few eggs can be obtained.
9. The zona-free (heterologous) *in vitro* fertilization technique can provide useful information on sperm capacitation and acrosome reactions, es-

pecially with species in which homologous eggs are difficult to obtain. However, there are some inherent limitations with this technique, and the results of heterologous fertilization assays may not be comparable to the homologous situation in terms of sperm fertilizing ability.

4. Regulation of Sperm-Fertilizing Ability

Over the past several years, there has been an enormous increase in the amount of published information on fertilization and associated events in animals. In general, this has led to greater awareness of the complexity of fertilization processes, but in few instances has a clear picture emerged of the underlying mechanisms. For reasons outlined in Section 1, it is not surprising that major advances have been achieved more readily by investigators using invertebrate material. Nevertheless, insights into mammalian fertilization are forthcoming with increasing frequency due to the resurgence of interest in this topic and to the current tendency towards relating observations made at the whole-animal, cellular, membrane, and molecular levels.

It would be redundant to attempt a survey of recent developments in fertilization research using *in vitro* (and *in vivo*) techniques. Several thorough, if not exhaustive, major reviews have recently been published on this topic, including a monumental one by Yanagimachi (1981) on fertilization mechanisms and an entire book on the same topic edited by Hartmann (1983). Both sources provide a wealth of stimulating information and ideas.

This section focuses on the regulation of sperm-fertilizing ability by the immediate environment of the gametes. It is still far from clear how interactions of spermatozoa with their environmental constituents influence acquisition of fertilizing ability, either *in vitro* or *in vivo*. Nevertheless, understanding is increasing as more techniques are brought into play and as different analytical approaches are integrated. The emphasis in the following discussion is not another review of the mechanisms of capacitation and the acrosome reaction *per se*, but rather a brief survey of the factors and conditions that appear to favor acquisition of sperm-fertilizing ability. It is hoped that emphasising research on the regulation of sperm fertilizing ability by extrinsic factors will provide greater insights into the underlying mechanisms of fertilization.

4.1. Distinction between Capacitation and the Acrosome Reaction

There are two principal reasons for confusion about the regulation of sperm fertilizing ability. First, it may be very difficult experimentally to separate the occurrence of acrosome reactions from the completion of capacitation; this is evident in many *in vitro* studies. In the guinea pig, separation of these events can be accomplished by withholding exogenous calcium from the spermatozoa. Capacitation can occur under these conditions, but not the acrosome reaction. The latter event occurs almost synchronously in the sperm population upon the addition of calcium (Yanagimachi and Usui, 1974). This useful technique is not feasible with spermatozoa of some other species,such as the golden hamster, which require exogenous calcium for viability (Yanagimachi, 1982*b*). In the majority of *in vitro* studies, capacitation and the acrosome reaction have occurred as a continuum, at least in the bulk sperm suspension. This is no doubt because all the factors needed for acquisition of fertilizing ability by spermatozoa have usually been present throughout the incubation period. This continuity of events makes it difficult to analyze the mechanisms and regulation of capacitation separately from the acrosome reaction. The situation *in vivo* appears to be different in that acquisition of fertilizing ability by spermatozoa seems to be a discontinuous process (see Section 4.7.4).

Second, caution needs to be exercised in interpretation of data concerning alterations in sperm properties (e.g., membrane characteristics) during *in vitro* incubation. Again, the question arises whether such changes constitute part of the process of capacitation (Section 3.2), or whether they should be considered early events in the acrosome reaction (Section 3.2.1). This distinction is important because the mechanisms of these two events are probably quite distinct (Yanagimachi, 1981; Clegg, 1983).

The fertilizing capacity of spermatozoa can be influenced *in vitro* by a wide variety of constituents of their environment. Some of these factors may also play a role *in vivo*. Because of the difficulty at present of clearly differentiating between events involved in capacitation and those concerned with the acrosome reaction, the influence of these putative regulatory factors will be considered from the broad viewpoint of their actions on sperm motility and/or fertilizing ability. Fertilizing ability is judged by percentages either of spermatozoa showing acrosome reactions or of penetrated eggs (see Sections 3.2.1 and 3.3.1). However, in a very few cases, regulatory factors have been shown to exert effects either on early events (i.e., capacitation) or on terminal events (the acrosome reaction).

In order to provide a conceptual framework for discussion of factors regulating sperm fertilizing ability, the changes undergone by spermatozoa during and after preparation for fertilization are diagrammed schematically in Figure 5. This outline is divided into two major phases. Phase A is the transition of the epididymal or ejaculated spermatozoon to a state in which it is potentially able to undergo acrosome reactions; this is the process designated as capacitation. In phase B, the capacitated spermatozoon, presumably influenced by some extracellular factor or conditions, undergoes the acrosome reaction and may also penetrate the zona pellucida of an egg. Between these two phases, under natural conditions, capacitated spermatozoa may be held for a time in a relatively quiescent state by environmental conditions existing within the female reproductive tract, perhaps including those discussed in Section 4.3. After completion of phase B, spermatozoa enter the final phase of fertilization, the fusion of sperm and egg plasma membranes, sperm nuclear decondensation, and formation of pronuclei (phase C). The events of this final phase are not discussed here, as they have been thoroughly reviewed by Moore and Bedford (1983) and by Wolgemuth (1983). In addition, recent data obtained by Perreault and col-

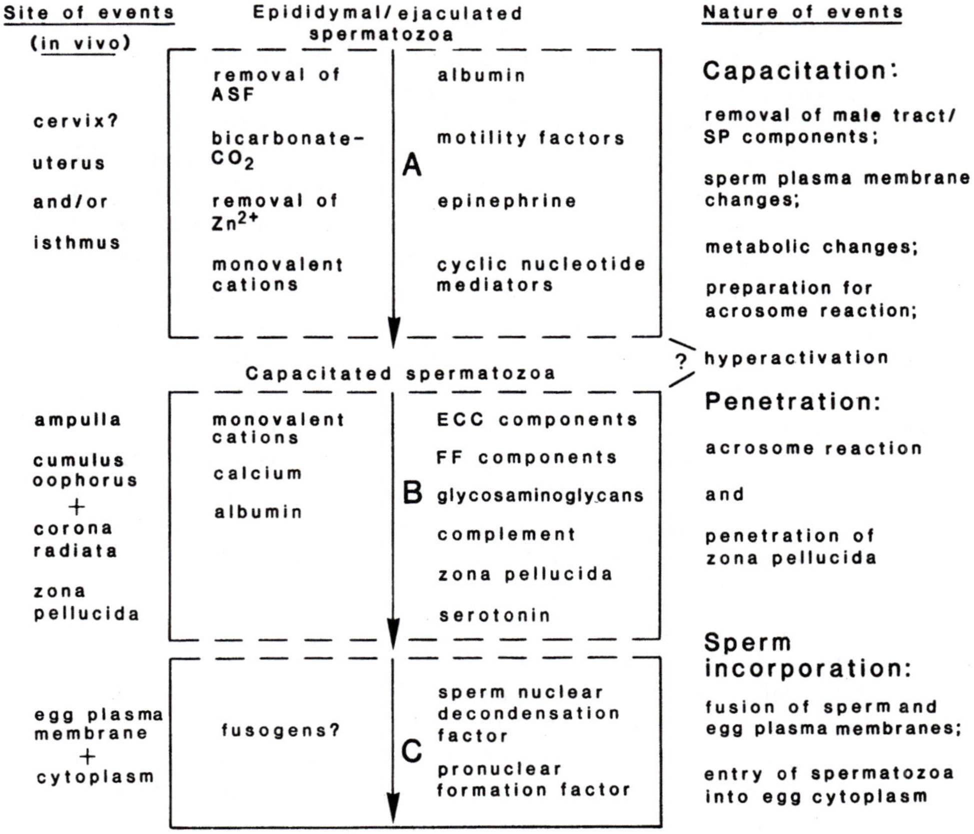

Figure 5. Factors that may be involved in the regulation of sperm-fertilizing ability. Diagram showing the three principal phases of sperm fertilizing ability (A, B, C), listing some of the constituents of the sperm environment that have been implicated as regulators of the acquisition of fertilizing ability.

leagues on the mechanism of sperm nuclear decondensation should receive attention (Perreault and Zirkin, 1982; Perreault *et al.*, 1984).

4.2. Removal of Sperm Membrane Components

An early event in capacitation of rabbit spermatozoa appears to be the removal of membrane-bound surface components, such as the acrosome stabilizing factor (ASF) studied by Oliphant and colleagues (Eng and Oliphant, 1978). Removal of this component, a large (360,000-M_r) glycoprotein, is needed to permit spermatozoa ultimately to undergo acrosome reactions. Since capacitation undoubtedly involves more than the removal of this inhibitor, it seems best to avoid use of the term "decapacitation factor" to designate this or similar substances; ASF is preferable as a functionally descriptive term. Presumably,

ASF and functionally similar components are applied to spermatozoa during epididymal maturation and/or ejaculation to prevent acrosomal changes from occurring prematurely (see Section 3.1). If so, removal of such components from spermatozoa either in the female reproductive tract or *in vitro* would appear to be one of the few events that can presently be claimed as an integral part of the process of capacitation (Oliphant and Singhas, 1979). The difficulty of correlating the biochemical and physiological (functional) events that take place during capacitation (see Section 3.2) is illustrated by experiments in which seminal plasma antigens were shown to be removed from the surface of rabbit spermatozoa (or modified) by treatment with hypertonic salt solution, but not by treatment with an isotonic solution (Brackett and Oliphant, 1975). However, spermatozoa from individual bucks were equally capable of fertilizing eggs *in vitro* regardless of which salt treatment was used to wash them. This observation has been confirmed by Hosoi *et al.* (1981); a similar result was obtained in cattle (Bondioli and Wright, 1983).

While the relevance of these particular antigens to capacitation is therefore questionable, there is a substantial amount of evidence that removal of sperm surface components occurs during capacitation (Oliphant and Singhas, 1979; Yanagimachi, 1981; O'Rand, 1982); the problem is to determine the functional significance of these changes. It appears to be important to remove from the vicinity of sperm some components eluted during brief incubation in isotonic salt solution; this was accomplished by washing. Failure to do so resulted in inability of spermatozoa to fertilize eggs (Brackett and Oliphant, 1975; Hosoi *et al.*, 1981; Bondioli and Wright, 1983). Again, this observation indicates that some sperm-bound inhibitors or stabilizers exert a regulatory action on capacitation. Experiments with mouse and guinea pig spermatozoa support this suggestion (Oliphant and Brackett, 1973; Aonuma *et al.*, 1973, 1982). In the hamster, factors present in epididymal plasma prevent spermatozoa from undergoing acrosome reactions but do not interfere with capacitation (Bavister, 1973; Bavister *et al.*, 1978), indicating that the regulatory role of these factors is exerted between these two events. A primary reason for continuing these studies on ASF and similar compounds is that they may provide important clues to the mechanism of the acrosome reaction by interference with the activity of factors that can stimulate this event.

4.3. Regulation by Ionic Components of the Environment

Compared with the amount of work done with recognized stimulators of metabolic pathways, such as catecholamines and cyclic nucleotide mediators (Sections 4.5.1 and 4.6), the possible roles of simple ionic components of the environment (except for H^+ and Ca^{2+}) in regulating capacitation has been virtually ignored until very recently. Now it has been shown that other ionic constituents can play a major regulatory role *in vitro;* in some species, they may do so *in vivo* as well.

4.3.1. Bicarbonate, Carbon Dioxide, and pH

Even the principal pH-buffering system (bicarbonate–carbon dioxide) exerts a surprisingly powerful effect on sperm capacitation, as shown with hamster spermatozoa. By varying bicarbonate and carbon dioxide proportionally, we were able to maintain the pH of sperm suspensions constant at 7.4. Under these conditions, the incubation time needed to onset of hyperactivated motility, the time to the beginning of acrosome reactions, and the final percentage of hyperactivated spermatozoa (all used as indices of capacitation) varied inversely with the bicarbonate–carbon dioxide concentration. Hyperactivation and acrosome reactions showed a dose–response curve with increasing bicarbonate concentration (Boatman and Bavister, 1983). The effects of varying bicarbonate–carbon dioxide were primarily on the regulation of capacitation; effects on acrosome reactions appeared to occur secondarily. In these experiments, responses were near-maximal with 6 mM bicarbonate (and 1.25% carbon dioxide). Bicarbonate also stimulates calcium-dependent increases in cAMP in guinea pig sperm; this effect was maximal at about 25 mM bicarbonate (Garbers *et al.*, 1982). It is improbable that bicarbonate or carbon dioxide directly regulate sperm-fertilizing ability *in vivo*, in view of the very high concentrations of bicarbonate, i.e., 30 mM or more (Hamner and Williams, 1965; Maas *et al.*, 1977) in oviductal secretions. Nevertheless, these experimental data implicate some fundamental bicarbonate- and/or carbon dioxide-dependent events in sperm function.

The pH of the culture medium is an important factor regulating occurrence of the acrosome reaction *in vitro*. Guinea pig spermatozoa readily undergo acrosome reactions *in vitro* at pH 7.8 or greater, but at pH 7.4 biological stimuli (serum factors) are required (Hyne and Garbers, 1980). This is similar to the situation in the sea urchin (*Arbacia*), in which egg jelly is needed to stimulate acrosome reactions at physiological pH (8.6), but is unnecessary at a pH greater than 9.6 (Decker *et al.*, 1976). Increased extracellular pH may aid in the elevated intra-acrosomal pH associated with the acrosome reaction of hamster spermatozoa (Working and Meizel, 1983). In the hamster and the dog, an increase in the proportion of spermatozoa undergoing acrosome reactions was observed when the pH of the culture medium was raised above 7.4 (Mahi and Yanagimachi, 1973, 1978). Acrosome reactions of guinea pig spermatozoa were also reversibly inhibited by low pH (pH 6.1) and were maximal at pH 7.5 or above (Murphy and Yanagimachi, 1984).

An increased pH of the ampullary oviduct secretions around the time of ovulation could well facilitate occurrence of the acrosome reaction *in vivo*. However, in spite of a number of attempts to measure the ampullary fluid pH either directly or indirectly by examining its bicarbonate concentration, there is still no consensus on this important parameter of the natural milieu for fertilization. It would be of interest to know whether the cumulus oophorus provides a high pH for the acrosome reaction to occur; this, together with, for example, zona pellucida factors (see Section 4.7.3), could provide a physiological "trigger" for the acrosome reaction *in vivo*.

The pH of the extracellular environment can also influence events in fertilization other than the acrosome reaction. Acrosome–reacted hamster spermatozoa were unable to fuse with zona-free eggs at low pH (pH 6.0), although sperm binding to egg plasma membranes was unaffected. Raising the pH above 7.1 permitted sperm : egg membrane fusion to occur (Yanagimachi *et al.*, 1980).

4.3.2. Monovalent Cations

The monovalent cations (sodium and potassium ions) should be considered as possible regulators of sperm capacitation (and of the acrosome reaction). In this context, they have received scant attention, because the great majority of experimental studies on sperm capacitation and fertilizing ability have been carried out *in vitro*, usually by diluting spermatozoa in a culture solution with an ionic composition resembling that of blood serum (Bavister, 1981*a*). Only a few studies have been performed to examine effects of altering the monovalent ion composition of culture solutions on sperm capacitation and/or fertilizing ability. However, it is becoming increasingly clear that, at least in some species, the concentrations and/or ratios of monovalent cations found *in vivo* may be important regulators of sperm capacitation, acrosome reactions, and fertilizing ability (Rogers *et al.*, 1981; Fraser, 1983*b*).

The Na^+/K^+ ion ratios measured in male and female reproductive tract fluids increase steadily from a very low level in epididymal plasma through the female tract secretions to the ampulla (see Fig. 6). The very unusual ionic composition of the male reproductive tract fluid (cauda epididymal plasma) (Jones, 1978) may be functionally associated with the quiescence of spermatozoa in this location. Dilution of spermatozoa into culture media or into female tract fluids having a more "normal" ionic composition stimulates sperm motility ("activation") and permits capacitation to take place. Thus, the Na^+/K^+ ratio in the environment of spermatozoa may be one regulator of the transition from the uncapacitated to the capacitated state (phase A in Fig. 5). When rat spermatozoa were flushed from the cauda epididymidis with a solution containing 4.7 mM K^+ but no Na^+, their initial motility was suppressed within 30 min; Na^+ was required to reactivate them (Wong and Lee, 1983). Reactivation was largely inhibited in 20 mM KCl, which is interesting in view of the high K^+ concentration of uterine fluids reported for some species, such as, pig (16 mEq/liter) and mouse (32 mEq/liter) (Iritani *et al.*, 1974; Borland *et al.*, 1977). The activation or inactivation of rat sperm by monovalent cations may be paralleled by the initiation or inhibition of capacitation. This possibility needs to be verified experimentally.

Several groups of investigators have examined the effects of altering Na^+/K^+ ratios and concentrations on capacitation and/or sperm-fertilizing ability. A dramatic increase in this ratio occurs in mouse bursal fluid (Fig. 6), which may approximate the environment in which acrosome reactions and sperm–egg interactions occur *in vivo* in this species. It is most interesting that a culture solution with a Na^+/K^+ ratio of 4.5, close to that occurring in the

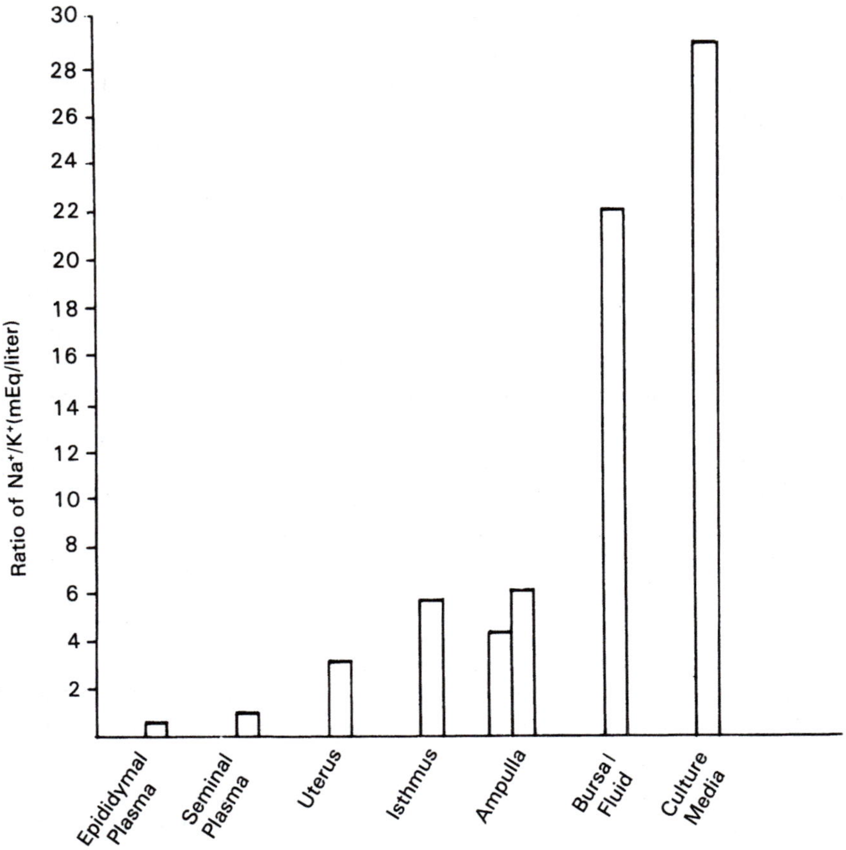

Figure 6. Comparison of Na^+/K^+ ratios in male and female reproductive tract fluids and in culture media for *in vitro* fertilization. Values for epididymal plasma are mean values of data from hamster and rat (Jones, 1978). Values for seminal plasma and secretions of uterus, isthmus, ampulla, and bursal fluid are mean values for mouse (Borland *et al.*, 1977). Lower value for ampulla is for fluid collected 0–3 hr p.c., higher value is for 11–13 hr p.c. Value for culture media is mean of ratios in 14 media used for *in vitro* studies on sperm-fertilizing ability (Bavister, 1981*a*). (p.c., post coitus).

uterus and isthmus, did not adequately support acquisition of fertilizing ability in mouse spermatozoa. This was not due to a failure of sperm capacitation, but rather to suppression of acrosome reactions in this medium (Fraser, 1983*b*). Expression of hyperactivated sperm motility was also greatly reduced in the high K^+ medium. It was proposed that similar changes in Na^+/K^+ ratios *in vivo* could serve to modulate sperm-fertilizing ability, i.e., acrosome reactions and hyperactivation (Fraser, 1983*b*). The increased Na^+/K^+ ratio found in the uterus and oviductal isthmus may permit capacitation to occur yet block development of acrosome reactions even in the presence of appropriate stimuli of the latter event (see Section 4.7). Thus, under natural conditions, capacitated spermatozoa may be held, perhaps for several or many hours, within reservoirs such as the uterus and/or isthmus, in a relatively quiescent state by inhibitory ratios of Na^+/K^+ (transition phase between events in phase A and phase B, Fig 5). This may serve as a mechanism both to prolong the fertile life of spermatozoa and also to afford better synchronization of acrosome reactions and hyperactivation with appearance of eggs in the ampulla. Then, when capacitated spermatozoa are released into the ampulla at ovulation (Cummins and Yanagimachi, 1982), they may encounter an environment with a higher Na^+/K^+ ratio, such as the bursal fluid (similar to follicular fluid?), the ampullary fluid (possibly modified by follicular fluid*), or even the microenvironment within the egg–cumulus complex. Acrosome reactions may then occur, enabling egg penetration to be accomplished. Some support for the scheme just outlined can be found with other species. Johnson and Overstreet (1982) found rabbit oviduct fluid to contain 30 mEq/ℓ K^+. Rabbit spermatozoa at the time of recovery from the isthmus have depressed motility (Johnson *et al.*, 1981; Cooper *et al.*, 1979; Johnson and Overstreet, 1982). It is tempting to speculate that monovalent ion ratios are involved in holding the isthmic spermatozoa in this suppressed condition.

Using high K^+ (27 mEq/ℓ) culture medium, we have found that acrosome reactions in hamster spermatozoa were reduced to about one-sixth the level obtained in normal K^+ medium (3 mEq/ℓ); motility was also depressed (D. E. Boatman and B. D. Bavister, unpublished observations). These results were obtained using a low concentration (0.1 mg/ml) of hamster serum albumin to capacitate spermatozoa. However, when we used 3 mg/ml albumin, no differences were found in the percentages of acrosome-reacted spermatozoa obtained with 3 or 27 mEq/ℓ K^+. The latter result is consistent with the data of Mrsny and Meizel (1981). We have also examined the regulation of sperm-fertilizing ability by K^+ ions using an *in vitro* egg penetration assay (see Section 2.3). Using an egg penetration index, as follows:

$$\text{Penetration index} = (\text{total no. of penetrating sperm})/(\text{total no. eggs inseminated}) \times 100$$

the index obtained with culture medium containing 3mM KCl was reduced by 64% in 27mM KCl (D. E. Boatman, J. C. Westrick, and B. D. Bavister, unpublished observations). These results of *in vitro* experiments may indicate a regulatory role for K^+ ions on fertilization *in vivo*.

Informative experiments have also been performed with guinea pig spermatozoa. In this species, influences of different factors on capacitation can be analyzed separately from those on the acrosome reaction by withholding Ca^{2+} (Yanagimachi and Usui, 1974). Using this technique, Hyne *et al.* (1984) showed that Na^+ concentrations greater than 45 mM are required for capacitation. Since the cauda epididymal plasma in the guinea pig contains only 17 mEq/ℓ (17 mM) Na^+ (Jones, 1978), whereas female reproductive tract fluids generally

*Some modification of this concept is necessary in view of the important data of Moore and Bedford (1978), who found that fertilization of cumulus-free hamster eggs could be accomplished within the ampulla in the absence of products released by the ovary. It may be that the Na^+/K^+ ratio of ampullary fluid changes dramatically, albeit transiently, at ovulation, permitting spermatozoa to undergo acrosome reactions in response to appropriate stimulatory molecules.

contain in excess of 100 mEq/ℓ, the Na^+ concentration could well be a major factor regulating capacitation. While K^+ ions were found to be unnecessary for guinea pig sperm capacitation *in vitro* (Hyne *et al.*, 1984), it is uncertain whether a high K^+ concentration such as that found in the epididymal plasma can suppress capacitation, although, curiously, K^+ concentrations similar to or less than those found in oviduct fluid of several species (5 mM or more) inhibit acrosome reactions in guinea pig spermatozoa *in vitro* (Rogers *et al.*, 1981).

4.3.3. Divalent Cations

4.3.3a. Calcium and Magnesium. Divalent cations, especially calcium, play a critical role in the acrosome reaction; a voluminous number of data have been gathered on this topic (Rogers and Yanagimachi, 1976; Yanagimachi, 1982*b*). The classic study by Yanagimachi and Usui (1974) demonstrated that extracellular calcium is unnecessary for capacitation of guinea pig spermatozoa *in vitro*, but its functional importance in other species is more difficult to evaluate (Yanagimachi, 1982*b*). Spermatozoa are undoubtedly exposed to calcium ions during capacitation *in vivo*, so regulation of capacitation *in vivo* is probably not exercised through calcium ions whether calcium is necessary for this event or not.

An influx of Ca^{2+} ions is a key event in initiation of the acrosome reaction. Numerous Ca^{2+}-dependent processes have been implicated in the mechanism of the acrosome reaction (Yanagimachi, 1981), including elevation of sperm cAMP (Hyne and Garbers, 1979). We would expect the Ca^{2+} permeability of spermatozoa to be highly regulated. One possible mechanism, the presence of proteins inhibiting Ca^{2+} uptake by sperm, has already been mentioned (see Section 3.1). The available amounts of divalent cations may play a role as well. In epididymal plasma of the guinea pig, rabbit, rat, and hamster, the Ca^{2+}/Ma^{2+} ratio is 0.05–0.08 (Jones, 1978) which is much lower than the ratio found in blood serum and in standard culture media, i.e., 1.4–5.6 (Bavister, 1981*a*). A Ca^{2+}/Mg^{2+} ratio of 0.17 is sufficient to completely inhibit acrosome reactions in guinea pig spermatozoa *in vitro*, while a ratio of 0.3 produces almost a fourfold reduction in acrosome reactions compared with the control Ca^{2+}/Mg^{2+} ratio of 1.7 (Rogers and Yanagimachi, 1976). Data from the mouse indicate that not until spermatozoa reach the oviductal isthmus would they encounter Ca^{2+}/Mg^{2+} ratios comparable to those present in blood serum (Borland *et al.*, 1977). It is therefore likely that the ratios of these divalent ions help regulate the occurrence of acrosome reactions *in vivo*.

4.3.3b. Zinc. An interesting candidate for a physiological regulator of sperm capacitation is zinc, which is found in high concentrations in male reproductive tract fluids, e.g., up to 2.0 mM in seminal plasma (Mann and Lutwak-Mann, 1981). Lower concentrations of zinc than this are found to inhibit fertilization of mouse eggs *in vitro* (Aonuma *et al.*, 1978, 1981, 1982). Zinc ions appear to inhibit sperm capacitation. Most interestingly, the complete inhibition of sperm-fertilizing ability by zinc ions could be partially or completely reversed with increasing concentrations of serum albumin (Aonuma *et al.*, 1982). Serum albumin is an effective chelator of zinc (Charlwood, 1979; Aonuma *et al.*, 1982). Spermatozoa of the starfish (*Asterias*) and other invertebrates release considerable amounts of zinc when first activated by dilution in seawater (Fujii *et al.*, 1955). Mammalian spermatozoa contain substantial amounts of zinc (Lindholmer and Eliasson, 1972; Arver and Eliasson, 1980; Srivastava *et al.*, 1983). Chelation of zinc by serum albumin could thus be a key event regulating capacitation (Johnsen and Eliasson, 1978*a*).

Zinc ions could exert a powerful regulatory influence on spermatozoa by acting at several different levels. Not only does zinc stabilize biological membranes, possibly by interacting with SH groups, but this element also inhibits lipid peroxidation, mitochondrial electron transport, sperm respiration, calmodulin and Na^+/K^+-ATPase (Chvapil, 1973; Johnsen and Eliasson, 1978*a*,*b*; Brewer *et al.*, 1979). Effects of zinc such as these and their functional significance in regulating acquisition of sperm-fertilizing ability could very readily be tested with existing *in vitro* techniques. It would be interesting to discover whether the effectiveness of albumin treated in various ways, e.g., to remove fatty acids (Lui and Meizel, 1977), and of other proteins (Bavister and Morton, 1974) in stimulating sperm-fertilizing ability can be correlated with their zinc-chelating efficiency.

Data from our laboratory are consistent with a role of zinc in regulating sperm-fertilizing ability. Hamster spermatozoa (8×10^5/ml) were incubated with different concentrations of zinc chloride. Acrosome reactions were examined in highly motile spermatozoa after 5 hr. The percentage of sperm showing acrosome reactions with 75 μM zinc chloride was only 25% that of controls; there was no significant depression of sperm motility (J. C. Westrick, D. E. Boatman, and B. D. Bavister, unpublished observations). However, we have not yet shown whether the inhibitory effect of zinc is exerted on capacitation, on the acrosome reaction, or on both events. Similar results have been obtained by Meizel and Lui (1976).

4.4. Sperm-Motility Factors

The golden hamster is the only species in which a role for "sperm motility factors" (hypotaurine or taurine) in supporting motility and capacitation *in vitro* has been unequivocally shown (Meizel *et al.*, 1980) (Table I). This does not preclude the possibility that these β-amino acids have some functional role in sperm capacitation in other species; it may be that only hamster spermatozoa require exogenous motility factors. This possibility warrants further study. It has been suggested that hamster spermatozoa are unusually "leaky" (Bavister and Yanagimachi, 1977); they may readily lose endogenous β-amino acids during sperm washing and dilution. The observation that guinea pig and human spermatozoa contain sperm motility factors (Meizel *et al.*, 1980), possibly β-amino acids in view of the effects of sperm extracts on hamster spermatozoa (Bavister and Yanagimachi, 1977), lends support to this idea. Taurine protects chimpanzee spermatozoa against damage *in vitro* (Ozasa and Gould, 1982), and appears to enhance the fertilizing ability of bovine spermatozoa *in vitro* (Ball *et al.*, 1983), so there may be a more subtle role of β-amino acids in species other than the hamster.

Neither the mechanism of action nor the functional role of these substances in supporting sperm-fertilizing ability is known. However, taurine and hypotaurine protect rabbit spermatozoa against loss of motility *in vitro* by inhibiting lipid peroxidation (Alvarez and Storey, 1983). Analogues of hypotaurine, including taurine, are less effective in stimulating hamster sperm motility; there is some evidence that uptake of these compounds is necessary for their action (Gwatkin, 1983). It is especially interesting from a functional point of view that the cumulus oophorus contains taurine and that a taurine uptake inhibitor blocks the ability of the cumulus to stimulate hamster sperm motility (Gwatkin, 1983).

4.5. Biogenic Amines

4.5.1. Catecholamines

The hamster is also the only species in which it has been demonstrated that spermatozoa are dependent on exogenous epinephrine (or other catecholamines) for achievement of fertilizing ability *in vitro*. The action of catecholamines may be exerted during the early part of capacitation (Meizel *et al.*, 1984), although other evidence suggests an effect of epinephrine during later stages of this event (Leibfried and Bavister, 1982). Epinephrine is certainly needed at some point in the capacitation phase; otherwise hyperactivated sperm motility and subsequent acrosome reactions are poorly developed (D. E. Boatman and B. D. Bavister, unpublished observations). All these experiments were done using BSA in the culture medium. We have recently found, using 3 mg/ml of *hamster* serum albumin (HSA), that no exogenous (added) epinephrine is needed; after 5 hr of incubation, more than 60% of hamster spermatozoa had undergone acrosome reactions (D. E. Boatman and B. D. Bavister, unpublished observations). Quite possibly, HSA contains epinephrine or some other biogenic amine (Meizel *et al.*, 1984) perhaps in larger amounts than BSA, in view of its greater efficiency in capacitating spermatozoa. Epinephrine has recently been shown to significantly enhance the fertilizing ability of mouse spermatozoa *in vitro* when they were incubated at suboptimal concentrations (Stanger, 1983). The mechanism of action of epinephrine (and of other biogenic amines) in stimulating sperm-fertilizing ability is unknown. It is possible that epinephrine stimulates adenylate cyclase activity in hamster spermatozoa; alternatively (or additionally), these biogenic amines may be acting by altering cellular calcium fluxes (Rasmussen and Goodman, 1977).

4.5.2. Serotonin

Meizel and Turner (1983) showed that, unlike catecholamines, serotonin or its agonist 5-methoxytryptamine can stimulate acrosome reactions within 15 min after addition to a suspension of capacitated hamster spermatozoa. In view of this rapid effect by serotonin and its action at the end of the capacitation process, this biogenic amine must be considered a possible effector of acrosome reactions *in vivo*. Appreciable amounts of serotonin are bound to albumin (Meizel *et al.*, 1984) and thus are likely to be present at the site of fertilization. The binding of serotonin to albumin could also explain, at least in part, the stimulatory effects of this protein on acrosome reactions *in vitro*. The presence of serotonin in follicular fluid (Meizel *et al.*, 1984) is also consistent with the notion that this biogenic amine is the naturally occurring ARIF in the egg–cumulus complex (see Section 4.7.4).

4.6. Cyclic Nucleotide Mediators

Rhesus monkey spermatozoa require stimulation by cyclic nucleotide mediators (caffeine and dibutyryl cAMP) for expression of hyperactivated motility *in vitro* (Bavister *et al.*, 1983b; Boatman and Bavister, 1984). Spermatozoa treated in this way can readily penetrate the zonae of cumulus-free homologous eggs but will very rarely fertilize hamster zona-free eggs (Boatman and Bavister, 1984; and unpublished observations). These observations may indicate that these cyclic nucleotide mediators are assisting capacitation of spermatozoa in this species, while acrosome reactions appear to require the presence of the zona pellucida. Enhancement of *in vitro* fertilization in the squirrel monkey by treating spermatozoa with these mediators was reported by Chan *et al.* (1982), but it is not known whether this effect was restricted to capacitation of spermatozoa, and there was substantial fertilizing ability with untreated spermatozoa, which was not the case with rhesus monkey spermatozoa (Boatman and Bavister, 1984). Capacitation of mouse spermatozoa is accelerated by caffeine and by dbcAMP (Fraser, 1979, 1981). However, it is not known whether these substances are acting on spermatozoa by elevating intracellular cAMP or by some other mechanism such as increasing calcium permeability (Kopf *et al.*, 1981). This needs to be studied, since some fundamental regulatory pathway in sperm capacitation may be involved. It is difficult nonetheless to relate these observations to sperm capacitation *in vivo*. Perhaps stimulatory substances with similar actions on spermatozoa occur naturally in the female reproductive tract secretions. Alternatively, the need to stimulate spermatozoa *in vitro* with these drugs may simply reflect a requirement to compensate for some abnormality of the culture environment provided to the spermatozoa. It is interesting that the rhesus monkey egg–cumulus complex does not seem able to provide stimulators of sperm capacitation with comparable actions to caffeine and dbcAMP. Rhesus spermatozoa that were incubated in culture medium without these substances were unable to effect fertilization of cumulus-intact eggs *in vitro* (Boatman and Bavister, 1984).

4.7. Macromolecules

4.7.1. Serum Albumin

The role of serum albumin in supporting sperm-fertilizing ability is not at all clear. Since albumin is the major protein constituent of female reproductive tract secretions (Shapiro *et al.*, 1971; Shalgi *et al.*, 1977), it could play an important role *in vivo*. In virtually all *in vitro* studies using animal spermatozoa, serum albumin (usually BSA) has been routinely included in the culture solution used for acrosome reaction or fertilization experiments. Uncertainty about the role of serum albumin *in vitro* can arise for three principal reasons.

First, it may be technically feasible to demonstrate a role for albumin in supporting fertilization *in vitro*, e.g., in the mouse and rat (Miyamoto and Chang, 1973; Davis, 1976), although possible contributions from the cumulus oophorus and oviduct fluid should be eliminated (Hoppe and Whitten, 1974). However, it is much more difficult to show whether albumin is required only for sperm capacitation, or for acrosome reactions, or for both events. In the great majority of *in vitro* experiments, capacitation and the acrosome reaction have occurred as a continuous process. It is likely that very different events take place during capacitation (phase A in Fig. 5) and the acrosome reaction (phase B). Unless a clear distinction can be made, both conceptually and experimentally, between these two phases of sperm fertilizing ability (see Section 4.1), the actions of albumin on spermatozoa will be difficult to assess. A requirement for serum albumin in capacitation of golden hamster and mouse spermatozoa has been unequivocally demonstrated (Bavister, 1981*b,c;* Aonuma *et al.*, 1982).

Second, albumin may not be required in some species. Guinea pig spermatozoa do not require albumin in the culture medium for capacitation or acrosome reactions, although under some conditions a dramatic increase in the proportion of acrosome-reacted spermatozoa can occur in the presence of serum albumin (Yanagimachi, 1972). Critical data are lacking for other species.

The third, and perhaps most vexing, problem associated with the use of serum albumin for *in vitro* experiments is to actually elucidate its mechanism of action on spermatozoa. There are a variety of quite different ways in which albumin could interact with spermatozoa, due to its diverse properties (Peters, 1975; Meizel, 1978), but in spite of valiant attempts to analyze this interaction by several investigators, notably Davis and Meizel, its role in the achievement of sperm fertilizing ability is still uncertain (see Meizel, 1978).

Numerous mechanisms have been proposed to explain the effects of serum albumin in stimulating sperm-fertilizing ability. The difficulties referred to in the preceding paragraphs should be kept in mind when evaluating these proposed mechanisms.

A strong possibility is that albumin capacitates spermatozoa by chelating zinc ions, thus destabilizing sperm membranes and stimulating sperm activity

(Section 4.3.3b). It has also been suggested that catecholamines bound to albumin facilitate capacitation (Meizel *et al.*, 1984), although epinephrine in the absence of albumin does not effectively capacitate hamster spermatozoa (Bavister, 1981*b*). It has been proposed that albumin removes cholesterol from sperm membranes, reducing their cholesterol : phospholipid ratio, which correlates with the occurrence of capacitation (Davis, 1976, 1980, 1981; Davis *et al.*, 1979). This change could initiate sperm membrane fusion leading to the acrosome reaction (Davis, 1981). Alternatively, a phospholipase contaminant of BSA has been suggested as the active component involved in stimulating acrosome reactions (Singleton and Killian, 1983), although other evidence indicates the participation of sperm endogenous phospholipase but not exogenous phospholipase in stimulation of the acrosome reaction (Llanos *et al.*, 1982). Serotonin bound to albumin could be the active acrosome reaction-stimulating substance, both *in vitro* and *in vivo* (see Section 4.5.2).

Chemical treatment of BSA, e.g., to remove associated fatty acids, greatly increases its ability to stimulate acrosome reactions in hamster spermatozoa (Lui and Meizel, 1977; Lui *et al.*, 1977). Another interesting fact that has emerged is that serum albumin is almost unique in its ability to stimulate sperm-fertilizing ability; a variety of other proteins, including globulins and chicken egg albumin, have little or no effect (Bavister and Morton, 1974; Davis, 1976; Aonuma *et al.*, 1982). There are also differences between animal sources of serum albumin in its ability to stimulate acrosome reactions and sperm-fertilizing ability. We have found that maximal acrosome reactions (about 85%) are achieved in about 5 hr when hamster spermatozoa are incubated with 3 mg/ml BSA, but this response can be obtained in 4 hr using the same concentration of HSA. Moreover, 20–30% of *in vitro* fertilized eggs are polyspermic when the culture medium (TALP) contains HSA, compared with 0–5% with BSA, all other conditions being equal (J. C. Westrick, D. E. Boatman, and B. D. Bavister, unpublished observations). These results indicate that HSA is considerably more efficient in stimulating capacitation and/or acrosome reactions than an equivalent amount of BSA.

Continued research is clearly needed to elucidate the nature of the interactions between serum albumin and spermatozoa. Since albumin is implicated in the stimulation of sperm-fertilizing ability *in vitro* and *in vivo* (see Section 4.7.4), much useful information stands to be gained about the mechanisms of capacitation and of the acrosome reaction.

4.7.2. Complement and Glycosaminoglycans

Apart from serum albumin, two other macromolecular components of follicular fluid (and the egg–cumulus complex) have been proposed as being involved in induction of the acrosome reaction: complement and glycosaminoglycans (GAGs). Complement contained in follicular fluid has been implicated in stimulation of rabbit sperm acrosome reactions (Cabot and Oliphant, 1978; Clegg, 1983). Acrosome reactions in bovine and rabbit spermatozoa are stimulated by GAGs (Handrow *et al.*, 1982; Lenz *et al.*, 1983).

We have tested these possibilities *in vitro* using hamster spermatozoa. The ability of hamster follicular fluid to support both capacitation and acrosome reactions in a defined medium was not significantly affected by heating at 56°C for 30 min (D. E. Boatman, J. C. Westrick, and B. D. Bavister, unpublished observations), which contraindicates a role for complement in induction of acrosome reactions in the hamster. We have also tested five GAGs (heparin, chondroitin 4-sulfate, chondroitin 6-sulfate, dermatan sulfate, and hyaluronic acid) at concentrations of 1, 10, and 100 μM. Hamster spermatozoa were incubated with a low concentration of hamster serum albumin (0.1 mg/ml), which capacitates spermatozoa but has low acrosome reaction-inducing ability. No significant increase over controls in the proportion of acrosome reacted spermatozoa was found with any of these GAGs at any of the concentrations tested (D. E. Boatman and B. D. Bavister, unpublished observations). However, another study did show a stimulatory effect of GAGs on acrosome reactions in hamster sperm (Meizel and Turner, 1984). Sperm were incubated for 4.5 hr with BSA to capacitate them, then either hyaluronic acid or heparin was added. Within 15 min there was a substantial increase in the percentage of acrosome-reacted sperm compared with controls. The discrepancy between the results of these studies shows a need for further investigation of the role of GAGs in capacitation and/or the acrosome reaction.

4.7.3. The Zona Pellucida

Evidence is accumulating from *in vitro* experiments that the zona pellucida is the site of the acrosome reaction, or at least the final stages of this event, in fertilizing spermatozoa, and that the zona actually stimulates the acrosome reaction (Saling and Storey, 1979). This property of the zona pellucida was attributed to a glycoprotein component of this envelope, designated ZP3 (Wassarman and Bleil, 1982; Bleil and Wassarman, 1983). This protein apparently serves as both the receptor for (mouse) spermatozoa and the inducer of the acrosome reaction; following fertilization, it becomes modified so that both of these activities are lost (Bleil and Wassarman, 1983). Solubilized zonae pellucidae of the hamster (but not mouse) are potent stimulators of acrosome reactions in hamster spermatozoa (Cherr *et al.*, 1984*a*). We have found that the acrosome reaction-stimulating component of the hamster zona pellucida has an apparent molecular weight of >30,000 (D. E. Boatman, J. C. Westrick, and B. D. Bavister, unpublished observations). Interpretation of all these experiments on the role of the zona pellucida in stimulating acrosome reactions is complicated by the fact that they were performed in culture media containing serum albumin. This protein is also an effective inducer of acrosome reactions (see Section 4.7.1). Accomplishment of *in vitro* fertilization in the absence of albumin, other proteins, or any biological fluid may help resolve this difficulty.

We have recently achieved this result in the hamster. Washed epididymal spermatozoa and cumulus-free eggs were incubated in a chemically defined culture medium containing PVA in place of serum albumin. Hypotaurine was used to support sperm motility and epinephrine to stimulate fertilizing ability (see Sections 4.4 and 4.5.1). Penicillamine was also present, together with a small amount (10 μM) of EDTA. The proportion of fertilized eggs increased with increasing concentrations of penicillamine. A higher amount of EDTA did not mimic this effect (Jacobs *et al.*, 1984). In these experiments, the zona pellucida may be implicated in the stimulation of acrosome reactions in fertilizing spermatozoa; no acrosome reactions were detected in spermatozoa in the bulk sperm suspension, i.e., sperm unattached to eggs. In addition, these data are consistent with the idea that capacitation involves removal of zinc from spermatozoa (see Section 4.3.3b), since penicillamine is a chelator of heavy metals (Klaassen, 1980). This chemically defined system promises to be of value in elucidating the regulation of sperm-fertilizing ability by environmental constituents.

4.7.4. The ARIF Concept

Once spermatozoa have completed capacitation, they are ready to undergo the acrosome reaction, an event that occurs *in vivo* during or immediately before penetration of the egg envelopes (see Section 4.8). However, there may be a substantial delay before the acrosome reaction occurs (Bryan, 1974; Suarez *et al.*, 1983). It is not known whether the acrosome reaction occurs *in vivo* as the result of removal of some inhibitory influence (see Section 4.3), and/or from the action of stimulatory molecules such as the putative acrosome reaction-inducing factor (ARIF) associated with the ovulated egg–cumulus complex (Bavister, 1982*a*). This factor may "trigger" the acrosome reaction, in a comparable manner to the mechanism in some invertebrates (*Strongylocentrotus* and *Asterias*). In these animals, the egg jelly coats contain macromolecules (characterized as fucosylated glycoproteins) that trigger the acrosome reaction in spermatozoa upon contact (Uno and Hoshi, 1978; SeGall and Lennarz, 1979, 1981) and also stimulate production of cAMP in spermatozoa (Garbers *et al.*, 1983). Eggs of the teleost *Acipenser* may have similar properties (Cherr and Clark, 1983).

Some evidence consistent with the existence of a similar mechanism in mammals has been derived from *in vitro* experiments. The egg–cumulus complex may be functionally similar in this respect to the jelly coat of invertebrate eggs. Hamster follicular fluid and the fluid component of the egg–cumulus complex can stimulate acrosome reactions in hamster spermatozoa *in vitro* (Barros and Austin, 1967; Yanagimachi, 1969*a*; Bavister, 1982*a*). We have shown that the active substance in hamster follicular fluid has a molecular weight of more than 30,000, and this activity is abolished by heating at 100°C (Westrick *et al.*, 1985). Bovine follicular fluid effectively stimulates hamster sperm acrosome reactions (Yanagimachi, 1969*b*), and the active substance appears to be serum albumin (Lui *et al.*, 1977). Rabbit spermatozoa capacitated *in utero* need to interact with follicular fluid *in vitro* in order to undergo acrosome reactions (Bedford, 1969).

Taken together, these and similar observations all indicate that serum albumin as a component of follicular fluid, presumably present as well in the ovulated egg–cumulus complex, is a naturally occurring ARIF that "triggers" the occurrence of acrosome reactions in capacitated spermatozoa during interaction with ovulated egg–cumulus complexes. The ARIF would thus synchronize the occurrence of acrosome reactions in fertilizing spermatozoa with the presence of the eggs in the ampulla. Whether the albumin molecule itself or some associated ligand is responsible is unknown (see Sections 4.5 and 4.7.1). However, there are some problems with this interpretation of events.

First, albumin is present in female reproductive tract secretions (Shapiro *et al.*, 1971; Shalgi *et al.*, 1977), so that spermatozoa would have been exposed to its actions long before encountering the supposed "trigger" or ARIF within the egg–cumulus complex. A counterargument can be put forward. The albumin in the egg–cumulus complex may be "different" from that found in the female reproductive tract secretions in some way that makes it much more potent in stimulating acrosome reactions. However, the most potent form of albumin *in vitro* is that which has been rendered fatty acid free (Lui and Meizel, 1977; Lui *et al.*, 1977); this form of albumin is hardly likely to be found *in vivo*.

A second problem with the ARIF concept as stated above is that the experiments of Moore and Bedford (1978) show conclusively that neither follicular fluid nor any of the components of the cumulus oophorus is essential for supporting sperm-fertilizing ability *in vivo*. While this does not preclude a role

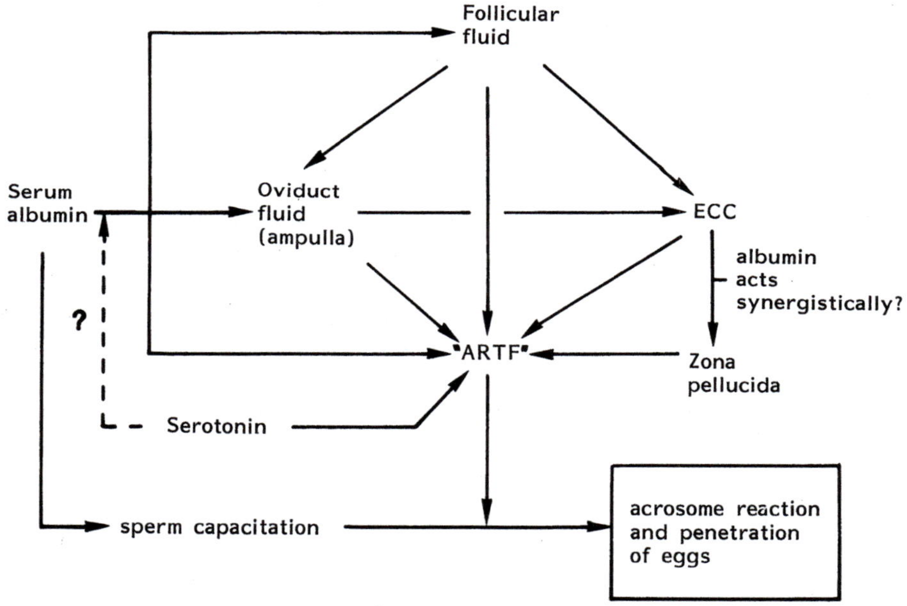

Figure 7. Possible relationships between different sources of ARIF (acrosome reaction-inducing factor). Serum albumin is represented as being a component of follicular fluid, of oviduct fluid, and of the egg–cumulus complex (ECC), as well as having direct ARIF activity and taking part in sperm capacitation. Follicular fluid is a component of the ECC and (transiently at least) of postovulatory ampullary fluid. The amount of follicular fluid at the site of fertilization may be very small or negligible in nonrodent species i.e., those lacking a periovarian bursa.

for these materials under normal conditions, it certainly means that other mechanisms are capable of operating. Under the experimental conditions used by Moore and Bedford (1978), a crucial role of ampullary oviduct fluid and/or of the zona pellucida is indicated (see Section 4.7.3). One way to reconcile these apparently conflicting data from *in vitro* and *in vivo* experiments is to postulate that the physiological ARIF is present in all these locations, i.e., in oviduct secretions, in the fluid component of the egg–cumulus complex, and in the zona pellucida. This situation may serve to provide backup mechanisms for stimulating acrosome reactions in fertilizing spermatozoa, which could be important if the cumulus oophorus has dispersed by the time of arrival of spermatozoa in the ampulla. This may be the normal situation in some animals (e.g., the cow). An important possibility is that different species rely more on one or other of these sources of ARIF to trigger acrosome reactions in fertilizing spermatozoa; this could help explain some of the discrepancies observed concerning the site of functional acrosome reactions in different species (see Section 4.8).

The various sources of ARIF which have been reported are shown diagrammatically in Figure 7. This diagram attempts to show possible relationships (i.e., homology) between these different ARIF sources in order to provide a more coherent picture of events. Sources of ARIF to which spermatozoa are not normally exposed (e.g., blood serum) are not shown. More work is urgently needed to elucidate the nature of ARIFs from different sources and on their mechanism of action.

4.8. Site of the Acrosome Reaction in Fertilizing Spermatozoa and the Role of the Cumulus Oophorus

Few capacitated spermatozoa appear in the oviductal ampulla before the arrival of eggs in this location (Bryan, 1974; Cummins and Yanagimachi, 1982; Suarez *et al.*, 1983). Acrosome-reacted spermatozoa are found mostly associated with the egg–cumulus complex in the mouse and rabbit (Bedford, 1969; Zamboni, 1974; Suarez *et al.*, 1983) although not always in the hamster and ram (Cummins, 1982; Cummins and Yanagimachi, 1982). It is still controversial precisely where "functional" acrosome reactions, i.e., those of the fertilizing spermatozoa, are initiated and where they are completed. Possibly, functional acrosome reactions are initiated at the periphery or just within the cumulus oophorus and are completed by the time the fertilizing spermatozoon reaches the zona pellucida (Bedford, 1968; Cummins and Yanagimachi, 1982; Yanagimachi and Phillips, 1984). Alternatively, the fertilizing spermatozoon may penetrate the cumulus/corona cell layers with its acrosome essentially intact and is then induced to undergo the acrosome reaction by the zona pellucida. However, the progression of unequivocally fertilizing spermatozoa through the intact egg investments has rarely been studied as a dynamic process, and even then (Blandau and Odor, 1952; Yang *et al.*, 1972), the acrosomal condition of this class of spermatozoa was not recorded. Resolution of this controversy has important implications for understanding the functional role of the acrosome reaction, the role of acrosomal enzymes, and the location and chemical nature of the physiological ARIF (see Section 4.7.4).

In the great majority of *in vitro* fertilization experiments, cumulus-free eggs have been used, so that the progression of fertilizing spermatozoa through the primary barrier to penetration (the cumulus–corona cell layer) cannot be observed.† *In vitro* experiments using cumulus-free eggs can give helpful, but not absolutely conclusive, information on the normal site of the acrosome reaction in fertilizing spermatozoa. On the basis of data from such experiments, it seems that there may be two variations on the strategy for accomplishing sperm penetration of the zona pellucida. In one variation, spermatozoa may undergo acrosome reactions before reaching the zona surface, since already reacted spermatozoa can penetrate the zona, i.e., in guinea pig and rabbit (Huang *et al.*, 1981; Kuzan *et al.*, 1984). In another variation, intact spermatozoa may bind to the zona pellucida and are then induced to react, i.e., in mouse and golden hamster (Gwatkin *et al.*, 1976; Saling and Storey, 1979; Bleil and Wassarman, 1983). Recent evidence indicates that mouse sperm do indeed penetrate the cumulus oophorus with acrosomes intact, leaving little room for doubt that in this species functional acrosome reactions occur on the zona pellucida (Storey *et al.*, 1984).

One experimental approach to provide critical information on this question is to study penetration of the intact egg–cumulus complex and fertilization *in vitro* by individual spermatozoa while continuously monitoring their acrosomal condition. This is remarkably difficult to accomplish. I have performed some preliminary experiments using this technique.

After inseminating freshly ovulated hamster eggs with a (heterogeneous) population of capacitated spermatozoa, using a microchamber *in vitro* fertilization technique, the ability of a few spermatozoa to penetrate the intact cumulus oophorus was examined by continuous observation with Nomarski optics (B. D. Bavister, unpublished data). Between 10 to 20 spermatozoa were used to inseminate each egg–cumulus complex. Three categories of spermatozoa were recognized: (1) those with intact acrosomes (presumably not fully capacitated), which did not succeed in penetrating the cumulus; (2) those with reacted acrosomes, which could sometimes penetrate a little way into the cumulus but invariably became permanently stuck to cumulus cells, in spite of vigorous flagellar activity; and (3) those (presumably capacitated) that showed intact acrosomes and were able to penetrate the cumulus oophorus with ease and reach the zona pellucida. More observations are needed before conclusive statements can be

†It should be realized that in many *in vitro* fertilization studies, cumulus-intact eggs were added to sperm suspensions (or *vice-versa*). However, the very high concentrations of spermatozoa that have nearly always been used rapidly disperse the cumulus and corona cells *in vitro* (e.g., Yanagimachi, 1969*a*; Gwatkin *et al.*, 1972). Under these conditions, it is true that components of the cumulus oophorus/corona radiata are present in the fertilization environment, but they are not a barrier limiting sperm access to the zona pellucida.

made, but it appears from these results that, in the hamster, functional acrosome reactions *in vitro* occur on the surface of the zona pellucida. The same conclusion was reached in two other studies in which video-micrography was used to examine progression of hamster spermatozoa through the cumulus oophorus (Cherr *et al.*, 1984*b*; Corselli and Talbot, 1986). However, Cummins and Yanagimachi (1982) cited their unpublished data, which are quite the opposite of these, namely that hamster spermatozoa with "modified" acrosomes could enter the cumulus oophorus *in vitro* and rapidly reach the zona pellucida. It should be mentioned that penetration of eggs, proving that the sperm being monitored are indeed fertilizing sperm, is difficult to achieve consistently under conditions of continuous observation with intact cumulus oophorus. Other investigators studying cumulus-intact hamster eggs have noted the difficulty of obtaining sperm penetration through the zona pellucida *in vitro* (Cummins and Yanagimachi, 1972; Yang *et al.*, 1982). Our laboratory is working on technical refinements to improve this situation.

Considering the data from all the experiments cited above, it is apparent that the functional role of the cumulus oophorus/corona cell layer in fertilization is not clear. The physiological ARIF may be located in this layer. However, in view of observations that the cumulus oophorus and corona radiata are not essential for accomplishment of fertilization *in vitro* or *in vivo* (Yanagimachi and Noda, 1970; Fraser *et al.*, 1971; Bedford, 1974; Moore and Bedford, 1978), an interesting alternative role of the cumulus oophorus in fertilization can be suggested. It may be that the function of the cumulus oophorus is not simply to aid the development of acrosome reactions in fertilizing spermatozoa (as in the case of the jelly coat of invertebrates), but rather to block access to the zona pellucida of spermatozoa that are not appropriately conditioned to effect fertilization. This interpretation is consistent with the unpublished observations from my laboratory that were described above. Another function of the cumulus–corona cell layer may be to reduce the total numbers of spermatozoa reaching the zona pellucida. The experiments conducted by Moore and Bedford (1978) demonstrated a striking increase in the incidence of polyspermy when the cellular investments of the eggs were removed. Yet another possible role of the cumulus oophorus is indicated by observations that egg–cumulus complexes contain substances capable of stimulating sperm motility (Yanagimachi, 1969*a*; Bradley and Garbers, 1983; Gwatkin, 1983). This property of the egg–cumulus complex could have a functional role in the accomplishment of sperm penetration of the egg investments.

4.9. Mechanism of the Acrosome Reaction —Electrochemical Initiation?

Finally, some reference should be made to the mechanism of the acrosome reaction itself. This is another complex and confusing topic that has been addressed in numerous reviews (e.g., Meizel, 1978; Yanagimachi, 1981). The regulation of the acrosome reaction, briefly outlined in the foregoing discussion, is described in much greater detail in these and other reviews. However, one intriguing possibility that appears to have received little or no attention is that a late event in the mammalian sperm acrosome reaction could be an action potential occurring in the sperm plasma membrane, or perhaps a substantial depolarization. This idea seems consistent with much of the evidence gathered to date on the influence of extracellular monovalent cations on sperm-fertilizing ability and/or acrosome reactions, and with postulated ion fluxes and changes in ion conductance in the sperm plasma membrane close to the time of the acrosome reaction. At least one report has demonstrated that depolarization of the membrane potential (in sea urchin sperm) is associated with induction of the acrosome reaction by egg jelly (Schackmann *et al.*, 1981). This may be analogous to membrane depolarization triggering exocytosis (e.g., at nerve terminals), a process with which the acrosome reaction is frequently compared.

More data on membrane potential changes are urgently needed, especially for mammalian spermatozoa. It is tempting to speculate that an action potential occurring across the sperm plasma membrane serves to "detonate" the acrosome reaction, perhaps by radically altering the membrane permeability to ions such as calcium. This hypothesis should be amenable to further experimental verification using *in vitro* techniques such as ion probes and ion conductance blockers (see Peterson *et al.*, 1978; Schackmann *et al.*, 1978, 1981; Mrsny and Meizel, 1981; Hyne *et al.*, 1984). In this context, the data reported by Hyne *et al.* (1984) are intriguing. These workers found that capacitated guinea pig spermatozoa could not undergo acrosome reactions (membrane fusion was inhibited) in the absence of extracellular Na^+; however, in the presence of a monovalent cation ionophore (monensin), addition of Na^+ and Ca^{2+} produced acrosome reactions very rapidly. The use of amiloride, a Na^+ channel blocker, retarded acrosome reactions. Hyne *et al.*, (1984) stated, "Based on these data, it appears that an increase in intracellular Na^+ has a functional role in acquisition of guinea pig sperm fertilizing ability." The similarities between these results and those of classic physiological experiments on action potentials in excitable cells, such as nerve and smooth muscle cells, are striking. Perhaps further studies on the ionic mechanisms involved in the acrosome reaction could profitably be conducted with these classic experiments in mind.

4.10. Summary

The regulation of sperm-fertilizing ability has been discussed in Section 4. Emphasis has been given to discussion of differences between the *in vitro* and *in vivo* situations. Some of the concepts and experimental data are brought together in Figure 8, which attempts to provide a coherent scheme, no doubt greatly oversimplified, for the regulation of capacitation and acrosome reactions *in vivo* and *in vitro*. These ideas are tentative and will need modification as more critical information is obtained; however, they may help stimulate some productive research in this highly complex area.

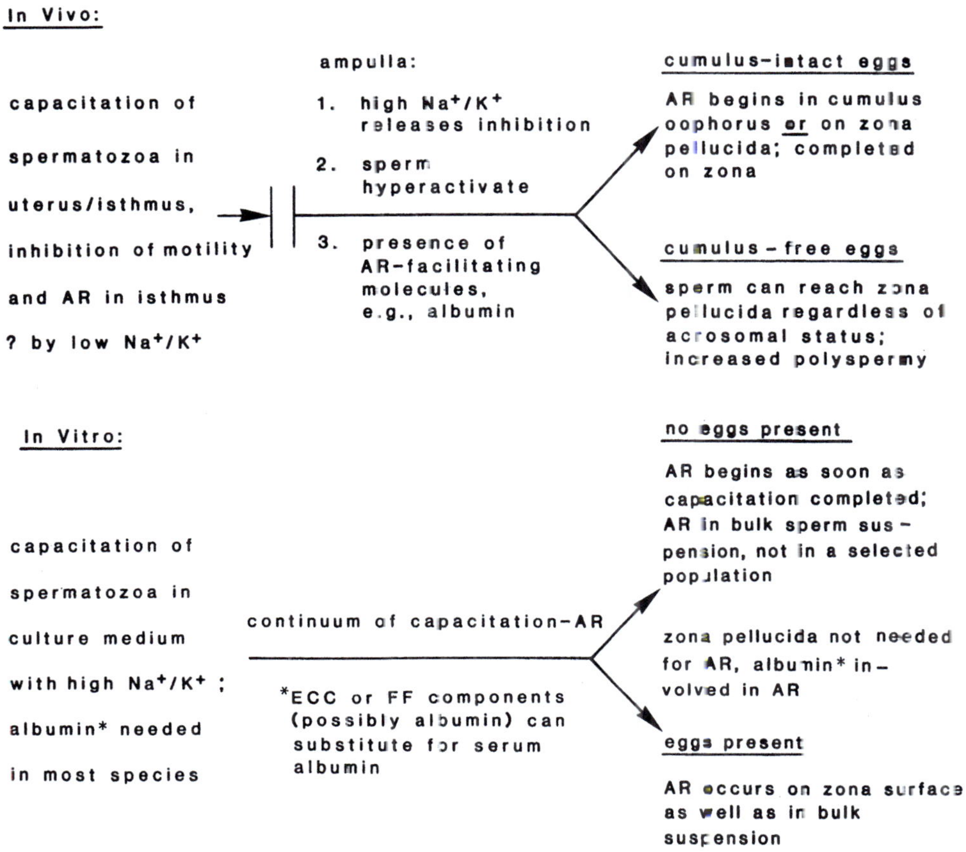

Figure 8. Scheme for the regulation of sperm-fertilizing ability both *in vivo* and *in vitro*. This scheme outlines only some of the potential regulatory factors described in Section 4. AR, acrosome reaction; ECC, egg–cumulus complex; FF, follicular fluid.

5. Embryonic Development following *in Vitro* Fertilization

As mentioned in Section 3.3, the three principal reasons for conducting experiments on fertilization *in vitro* are to evaluate sperm-fertilizing ability, to examine details of the fertilization process itself (sperm penetration and egg activation), and to provide a source of early embryos. For the latter purpose, it is important to know whether eggs fertilized *in vitro* are comparable to those fertilized *in vivo*. While a variety of criteria can be applied for this purpose (Bavister, 1981*a*), the ultimate proof of similarity between fertilization under *in vitro* and *in vivo* conditions rests on the ability to demonstrate continued normal development of eggs that are activated following sperm penetration *in vitro*. This proof is discussed in more detail in this section.

5.1. Rationale for Culturing *in Vitro* Fertilized Eggs

Continued embryonic development of *in vitro* fertilized eggs has been accomplished in several species, although with widely varying degrees of success (Table III). The simplest way to evaluate the normality of *in vitro* fertilized eggs is to perform embryo transfer (oviductal) with zygotes or with very early cleavage stages. This has been done most often in the mouse, rat, and rabbit. However, there are at least two major incentives for attempting to sustain *in vitro* fertilized eggs in culture until the 8-cell stage or later. First, much more information can be derived about the growth requirements of early embryos and about the etiology of abnormal embryogenesis. This rationale depends very much on the species. In laboratory animals such as the mouse and rabbit, studies on growth requirements are most conveniently performed with 1- or 2-cell embryos recovered from mated animals, avoiding the problems associated with *in vitro* fertilization. Large numbers of embryos can be obtained in this way following superovulation with gonadotropins. However, very early embryos are more difficult to obtain by this means in reasonable numbers from animals such as cattle and nonhuman primates. This involves flushing the oviducts either *in situ* or after excising reproductive tracts. Moreover, problems arise with obtaining embryos showing synchronous development, especially following gonadotropin stimulation of donors. In nonhuman primates, occurrence of normal fertilization and embryonic development *in vivo* following superovulation has not been reported.* In these species, *in vitro* fertilization offers an attractive alternative for producing early embryos for experimental studies.

A second motivation for sustaining *in vitro* fertilized eggs in culture concerns the development of animal models for obtaining information relevant to human clinical applications of this technique. In the clinical situation, *in vitro* fertilized human embryos are invariably transferred back to the egg donors by nonsurgical means (cervical transfers). The development of suitable animal models, particularly with nonhuman primates, for this purpose implies an ability to sustain *in vitro* fertilized eggs in culture at least to stages of embryonic growth that are appropriate for uterine transfers, i.e., about the 8-cell stage. It is remarkable that human embryos at the 4-cell stage or even earlier can survive in the uterine environment (Trounson, 1983); most animal embryos seem to be more sensitive, and it is unlikely that such early-stage embryos will tolerate such "ectopic" transfers (Adams, 1982).

It is unfortunate that, in general, very poor success has been obtained in supporting cleavage of *in vitro* (or *in vivo*) fertilized animal eggs in culture, with the exception of the laboratory animal species, i.e., mouse, rabbit, and rat (Table III). The most success by far has been obtained with the mouse, but even

*One exception to this statement is the recent demonstration by Kenigsberg *et al.* (1984) that several cynomolgus monkeys (*Macaca fascicularis*) mated following gonadotropin stimulation produced multiple early cleavage stage embryos. No data were provided on the viability of these embryos.

Table III. Status of *in Vitro* Fertilization and Subsequent Embryonic Development in Animal Species (Excluding Nonhuman Primates)[a,b]

Species	Fertilization *in vitro*	Cleavage of IVF embryos *in vitro*[c]	Differentiation of IVF embryos *in vitro*[d]	Birth of IVF young after embryo transfer[e]
Laboratory animals				
Mouse (1)	+	+	+	+
Rabbit (2)	+	+	+	+
Rat (3)	+	+[f]	−	+
Hamster (4)	+	+[f]	−	−
Guinea pig (5)	+	+	−	−
Farm animals				
Cow (6)	+	+	−	+/−
Pig (7)	+/−	+/−	−	+/−
Sheep (8)	+/−	+/−	−	−
Horse, goat	−	−	−	−
Carnivores				
Cat (9)	+/−	+/−	+/−	−
Ferret (10)	+/−	+/−	+/−	−
Dog (11)	+/−	−	−	−

[a]+, numerous published reports; +/−, few published reports; −, no published reports.
[b]References: (1) Whittingham (1968), Mukherjee and Cohen (1970); (2) Chang (1959), Brackett *et al.* (1982*b*); (3) Toyoda and Chang (1974); (4) Yanagimachi and Chang (1964), Juetten and Bavister (1983*b*); (5) Yanagimachi (1972); (6) Brackett *et al.* (1982*a*), Brackett *et al.* (1984), Ball *et al.* (1983), Sirard *et al.* (1984); (7) Cheng and Polge (1983), Nagai *et al.* (1984); (8) Bondioli and Wright (1982); (9) Hamner *et al.* (1970), Bowen (1977); (10) Whittingham (1975); (11) Mahi and Yanagimachi (1976, 1978).
[c]Cleavage of *in vitro* fertilized eggs to between 2- and 8-cell stage.
[d]Development of *in vitro* fertilized eggs to morula or blastocyst stage.
[e]In nearly all cases, cleavage stage embryos were transferred.
[f]Zygotes will only cleave to the 2-cell stage.

in this species, there are dramatic differences between inbred and random bred strains (Table IV). It is clear that satisfactory culture media for sustaining embryonic development in the great majority of species have not yet been devised. This topic is discussed in the following sections.

Quite apart from this difficulty, there is the question of how to obtain an adequate supply of mature eggs for *in vitro* fertilization. This presents no difficulty in laboratory animal species (see Section 3.3.1a) but can cause a severe problem in domesticated animals and in nonhuman primates. Viable ovulated eggs are difficult to collect in these species. Large numbers of immature oocytes can be obtained from excised ovaries, but at present, our ability to support normal maturation *in vitro* is quite inadequate (Tsafriri *et al.*, 1983). *In vitro* matured oocytes may appear superficially normal but will rarely undergo normal fertilization (but see Staigmiller and Moor, 1984). Anomalies include failure to effect decondensation of the penetrating sperm nucleus and reduced postfertilization development (Thompson and Zamboni, 1975; Thibault, 1977; Evans and Armstrong, 1983; Leibfried and Bavister, 1983). Eggs recovered by aspiration of immediately preovulatory follicles in gonadotropin-stimulated

Table IV. *In Vitro* Development of NMRI and C57Bl Mouse Preimplantation Embryos[a,b]

Stage of embryos in culture[c]		Time in culture	Percentage of embryos developing	
Begin	End	(hr)	NMRI	C57BL
1-cell	2-cell	24	72	96
2-cell	Blastocyst	72	85	96
1-cell	Morula	72	0	76
1-cell	Blastocyst	96	0	24

[a]Data from Spielmann *et al.* (1980). (Reproduced by permission of the authors and the editors of *Acta Morphol. Acad. Sci. Hung.*)
[b]NMRI is a random-bred strain; C57BL is an inbred strain.
[c]Embryos cultured in WMOC medium; at least 75 embryos in each group.

rhesus monkeys vary widely both within and between animals in terms of fertilizability and subsequent developmental capability (Bavister *et al.*, 1983*b*; Boatman *et al.*, 1986*a*).

Even with ovulated eggs, trauma can readily occur during recovery, e.g., temperature shock or delay in removal from excised oviducts, which is detrimental to their fertilizability (Keefer *et al.*, 1983). Particularly in rodents, the postovulatory viability of eggs is short, and aging for only a few hours in the oviduct can lead to anomalies of fertilization and cleavage (Juetten and Bavister, 1983*a*).

There are three species or groups in which techniques for sustaining embryonic growth *in vitro* are most urgently needed: (1) nonhuman primates, which may provide information that is most relevant to humans; (2) domesticated animals, some of which are commercially valuable; and (3) the golden hamster.

Rapid progress has taken place within the past 2 years in the ability to obtain development in culture of *in vitro* fertilized nonhuman primate eggs. These advances are discussed in more detail in Section 5.4.

5.2. Domesticated Animals

Among the domesticated animals, most interest has focused on cattle. The ability to fertilize eggs and to sustain early embryo development *in vitro* is of importance in these commercially valuable animals for several reasons. First, it would provide the most convenient means to study very early embryogenesis and thereby to provide insights into the reasons for the high-pregnancy failure rate found in cattle, especially in dairy cows (Ayalon, 1978). Second, preliminary evaluation of semen quality (sperm fertilizing ability) could be carried out in the laboratory instead of by field trials. Third, early cleavage stage embryos could be obtained in quantities for development and application of experimental procedures such as embryo sexing and splitting, which are potentially useful for increasing the quality and/or quantity of offspring obtained from valuable donors (Willadsen, 1979; Seidel, 1982; Lambeth *et al.*, 1983). Early cleavage stage embryos may represent the most convenient, and perhaps the most appropriate, material for the application of these techniques. *In vitro* fertilization of bovine eggs has become a reliable procedure and normal offspring have been produced after transfer of *in vitro* fertilized embryos (Brackett *et al.*, 1982*a*, 1984). However, the great majority of efforts to sustain *in vitro* growth of cleavage stage embryos of cattle (and the other domesticated species) have met with little or no success (Wright and Bondioli, 1981). One of several exceptions to this statement is the demonstration that sheep embryos at the 1–8-cell stages will develop successfully to the morula or blastocyst stage in a simple culture medium based on the ionic composition of oviduct fluid. A few cattle ova placed in this medium developed as far as the morula stage, but were mostly retarded (Tervit *et al.*, 1972). More emphasis needs to be placed on this approach, and perhaps less on the use of commercial, complex media and those based on the composition of blood serum (Bavister, 1981*a*). Recently, improved success has been reported in obtaining cleavage of *in vitro* fertilized bovine oocytes to 4- and 8-cell stages (Brackett *et al.*, 1982*a;* Sirard *et al.*, 1984). In both studies, the semen donor appeared to have been a major variable affecting the outcome of *in vitro* fertilization and/or cleavage of embryos. An exciting new area of research is opened up by discoveries of embryotrophic factors. Trophoblast from day 13 bovine embryos secretes substances that stimulate growth of early cleavage stage embryos *in vitro* (Camous *et al.*, 1984). A low molecular weight substance associated with BSA stimulates growth of rabbit embryos *in vitro* (Kane, 1985).

5.3. Laboratory Animals

In contrast to the situation with domesticated animals, early embryos of the mouse and rabbit can easily be grown, at least from the 2-cell stage, to the hatched blastocyst stage in culture (Brinster, 1963; Kane and Headon, 1980). Much has been learned about the growth requirements of these embryos, especially with the mouse. This progress is made possible by the ready availability of large numbers of embryos obtained by flushing oviducts of superovulated mated donors. It should therefore be very easy to obtain comparable embryonic development in culture with eggs fertilized *in vitro*. However, this is not necessarily true. Two distinct experimental approaches have evolved with these (and other) species during the past 20 years or so. For obvious reasons, one research direction has focused on the exploitation of *in vitro* fertilization, with relatively little emphasis on extended development of the fertilized eggs *in vitro* (e.g., Gwatkin, 1977; Rogers, 1978; Hartmann, 1983). For example, we have recently used our hamster *in vitro* fertilization procedure (see Section 2.3) to examine the effects of postovulatory egg aging on fertilization and the first cleavage division (Juetten and Bavister, 1983*a*) and to study the competence of *in vitro* matured oocytes to take part in normal fertilization (Leibfried and Bavister, 1983). A quite different thrust (almost entirely by a different group of investigators) has been to solve the problems of supporting embryonic growth *in vitro*, in order to study events in normal embryogenesis. Such studies have nearly always utilized *in vivo* fertilized embryos, usually at the 2-cell stage (e.g., Biggers *et al.*, 1971).

These two approaches are entirely valid and logical. However, a basic discrepancy emerges between these areas when attempts are made to encourage *in vitro* fertilized eggs to develop into blastocysts in culture. The complete spectrum of preimplantation development, from fertilization of the egg to the blastocyst stage, has not been obtained *in vitro* with the great majority of animal species studied to date (Table III). In part, this may result from abnormalities of the fertilization process under *in vitro* conditions (Shalgi, 1984; Vanderhyden *et al.*, 1984), but experiments with *in vivo* fertilized 1-cell embryos show that inadequate culture conditions must make a major contribution.

As an example of the problems with culture media, Table IV summarizes data from experiments with mouse embryos (*in vivo* fertilized). One-cell embryos readily cleaved to the 2-cell stage *in vitro*, while most embryos recovered at the 2-cell stage formed blastocysts after culture for 72 hr, regardless of whether inbred or random-bred mouse strains were used. However, most 1-cell embryos from the inbred strain failed to develop into blastocysts, and none of the outbred strain embryos reached even the morula stage. The conclusion to be drawn from these and similar experiments is that, even in the mouse, the first cleavage division is a very sensitive stage of development. It is therefore hardly surprising that attempts to sustain growth of *in vitro* fertilized eggs in culture have often met with limited success. One approach to resolving this problem is to direct more attention toward devising culture conditions that more closely resemble the natural environment in which fertilization and early cleavage (2–4-cell stages) take place, i.e., the oviductal fluid (see Section 5.2). This will be tedious but in the long run may greatly enhance our ability to fertilize eggs *in vitro* and to support normal embryonic development, at least through the critical early cleavage stages.

5.3.1. Development of Golden Hamster Embryos *in Vitro*

As mentioned in Section 3.3.1a, golden hamster gametes provide excellent material for *in vitro* fertilization experiments. This species has been used extensively in our laboratory. However, studies of early embryonic development using this species are severely limited because of a failure of 2-cell embryos to continue to divide in culture. This is true whether the embryos are obtained by *in vitro* or *in vivo* fertilization (Whittingham and Bavister, 1974). We regard this "2-cell block" to development as an opportunity to examine the regulation of embryo cleavage, rather than as a reason to reject the hamster as a model for embryogenesis studies (Farrell and Bavister, 1984). It is interesting that *in vivo* fertilized 8-cell embryos can continue to develop in culture (Whittingham, 1975; Bavister *et al.*, 1983*a*), supporting the argument presented in Section 5.3

that oviductal stage embryos (1–4 cells) are especially sensitive to culture conditions. We hope that elucidation of the 2-cell block in the hamster will provide information that can be applied to embryos of other species and thus improve the capability of *in vitro* fertilized eggs to develop in culture through all the preimplantation stages. Embryos of the rat, another species commonly used for *in vitro* fertilization studies, also fail to develop much beyond the 2-cell stage in culture.

I have recently identified a morphological concomitant of loss of viability in hamster 2-cell embryos. Embryos obtained from mated animals, within an hour or so of being placed into TALP culture medium, undergo swelling, resulting in a large area of contact between the two blastomeres. In extreme cases, the perivitelline space may be virtually eliminated. Two-cell embryos resulting from cleavage of *in vitro* fertilized eggs, which have no capacity for further development *in vitro* or *in vivo*, invariably show considerable swelling compared with freshly collected *in vivo* fertilized embryos (B. D. Bavister, unpublished observations); this effect is apparent in Figure 1. This blastomere swelling effect does not appear to be the result of a simple osmotic imbalance in the culture medium. We have also noted blastomere swelling in some rhesus monkey and bovine embryos derived from *in vitro* fertilized eggs, although we have not been able to correlate this effect with embryo viability in these species. We are presently trying to identify the cause of *in vitro* blastomere swelling in the hamster in order to gain information on the 2-cell block.

We have shown that hamster 2-cell embryos are damaged by very brief exposure to simple culture media (Farrell and Bavister, 1984). This discovery may serve to alter the focus of studies aimed at improving embryo culture conditions. In this study, hamster embryos were held in TALP-Hepes (Bavister *et al.*, 1983*a*) or PBl media (Whittingham, 1971) for up to 20 min before transferring them to oviducts of pseudopregnant recipients. Most of these embryos failed to reach the 8-cell stage *in vivo*, whereas a majority of 2-cell embryos exposed to these media for 5 min or less before transfer retained their developmental capacity. Further experiments along these lines should lead to identification of detrimental components of the culture environment that contribute to the 2-cell block. By eliminating this block, we expect to be able to support development of *in vitro* fertilized hamster eggs in culture at least up to the 8-cell stage, at which point uterine embryo transfers can be performed (Bavister *et al.*, 1983*a*). This would open the way to exploitation of hamster *in vitro* fertilization as a model for studies on (e.g.) the consequences for embryonic development of delayed fertilization (see Juetten and Bavister, 1983*a*), which is believed to be a major reason for the occurrence of defective embryogenesis and birth defects.

5.4. *In Vitro* Fertilization and Development in Nonhuman Primates

The history of initial attempts to achieve *in vitro* fertilization in nonhuman primates has been described by Brackett (1978) and by Gould (1979), together with accounts of the major difficulties inherent in working with gametes of

Table V. *In Vitro* Fertilization and Development in Nonhuman Primates[a,b]

Species	Fertilization *in vitro*	Cleavage of IVF embryos *in vitro*[c]	Differentiation of IVF embryos *in vitro*[d]	Birth of IVF young after embryo transfer[e]
Rhesus monkey (*Macaca mulatta*) (1)	+/−	+/−	+/−[f]	+/−[g]
Baboon (*Papio cynocephalus*) (2)	+/−	+/−	−	+/−[h]
Cynomolgus monkey (*Macaca fascicularis*) (3)	+/−	+/−	+/−	+/−[h]
Squirrel monkey (*Saimiri sciureus*) (4)	+	+/−	−	−
Chimpanzee (*Pan troglodytes*) (5)	+/−	+/−	−	−

[a]+, +/−, −: see legend to Table III.
[b]*References:* (1) Bavister *et al.* (1983*b*, 1984), Morgan *et al.* (1984); (2) Clayton and Kuehl *et al.* (1984); (3) Kreitmann *et al.* (1982), Balmaceda *et al.* (1984); (4) Cline *et al.* (1972), Gould *et al.* (1973), Kuehl and Dukelow (1979); (5) Gould (1983).
[c–e]See legend to Table III.
[f]Includes hatched blastocysts (see Table VI and Fig. 9).
[g]Two offspring born.
[h]One offspring born.

these animals. Several groups of investigators have used the squirrel monkey (*Saimiri sciureus*) for *in vitro* fertilization experiments. Sperm penetration of nonhuman primate eggs *in vitro* was first reported using this species (Cline *et al.*, 1972; Gould *et al.*, 1973). *In vitro* fertilization in the human was first reported several years earlier (Edwards *et al.*, 1969). It is ironic that progress with human *in vitro* fertilization has preceded developments with nonhuman primate species. In a large series of experiments with *S. sciureus* gametes, *in vitro* fertilization was consistently achieved and some of the fertilized eggs cleaved as far as the 8-cell stage, but no pregnancies were established after transfer of embryos (zygotes) to recipients (Kuehl and Dukelow, 1979). Other investigators have attempted *in vitro* fertilization using Old World monkeys exhibiting menstrual cycles, in order to provide information on early development that may be more applicable to humans. Apart from a very few sporadic reports of *in vitro* fertilization using rhesus monkey (*Macaca mulatta*) gametes, the first convincing report of success with a higher nonhuman primate species appears to be that of Kreitmann *et al.* (1982), who obtained fertilization *in vitro* of 22 cynomolgus monkey (*M. fascicularis*) eggs. Some of these eggs cleaved as far as the early morula stage. The present status of *in vitro* fertilization and development with nonhuman primates is shown in Table V.

5.4.1. Development of *in Vitro* Fertilized Rhesus Monkey Eggs in Culture

In our laboratory, we have used techniques derived from *in vitro* fertilization experiments with golden hamsters to accomplish fertilization of rhesus monkey eggs and to sustain embryonic development *in vitro* (Bavister *et al.* 1983*b*). Mature eggs were obtained by aspiration of follicles at laparoscopy in PMSG-stimulated monkeys. A critically important factor in the success of these experiments was the stimulation of sperm fertilizing ability with caffeine and dbcAMP: without this stimulation, neither hyperactivated sperm motility, attachment to the zona pellucida, nor penetration of the zona was observed (Boatman and Bavister, 1984). The "hamster" fertilization medium (TALP) was used (Bavister and Yanagimachi, 1977), supplemented with a very small amount (2%) of heated rhesus blood serum. Although *in vitro* fertilization of rhesus monkey eggs could be obtained in this solution, only sporadic success was achieved with cleavage of these eggs until we incorporated four amino acids (glutamine, isoleucine, methionine, and phenylalanine) in the TALP medium used for embryo culture (Bavister *et al.*, 1983*b*). The rationale for this supplementation was that these amino acids are required for cleavage of hamster embryos *in vitro* (Bavister *et al.*, 1983*a*; Juetten and Bavister, 1983*b*). This is but one example of several benefits derived from ongoing studies of fertilization and embryonic development in several species—hamster, rhesus, and bovine—in our laboratory.

Under the conditions described, 79% (23 of 29) of rhesus eggs inseminated *in vitro* were fertilized and cleaved in culture; 11 reached the 8-cell or early morula stages, and their cleavage timing was comparable to that reported for primate development *in vivo* (Bavister *et al.*, 1983*b*). At this stage, a block to development seemed to have occurred, and we suggested that a change in requirements of the embryos might occur at the 8-cell stage, restricting growth in such a limited culture environment. This explanation was most probably correct, since we have recently cultured *in vitro* fertilized rhesus eggs to the hatched blastocyst stage (Morgan *et al.*, 1984). The results of these experiments are summarized in Table VI. In these experiments, medium TALP with amino acids was again used for *in vitro* fertilization; eggs were removed 24 hr after insemination, by which time some had already undergone the first cleavage division, and transferred to medium CMRL-1066 containing 20% human fetal

Table VI. Development of *in Vitro* Fertilized Rhesus Monkey Eggs in Culture

			Blastocysts		
			Zonal		
No. of eggs inseminated[a]	Eggs fertilized (%)[b]	Morulae (%)[c,d]	Early (%)[c,d]	Expanded (%)[c,d]	Hatched (%)[c,e]
112	64 (57)	15 (23)	12 (19)	5 (8)	11 (17)

[a]Data from nine experiments (Boatman *et al.*, 1986*b*). Only type I eggs (Bavister *et al.*, 1983*b*) included.
[b]Occurrence of fertilization judged by cleavage (Bavister *et al.*, 1983*b*).
[c]Percentage of fertilized eggs: Categories are independent.
[d]Number of embryos progressing no further than stage shown.
[e]All hatched blastocysts continued development for several days after hatching.

cord serum. This culture method was devised by Hsu (1979) to support prolonged growth and differentiation of mouse blastocysts [see also Chen and Hsu, (1982)]. Recently, Pope *et al.*, (1982) used the same method to obtain growth of baboon embryos *in vitro.* We have found it to be suitable for sustaining development of *in vitro* fertilized rhesus monkey embryos. Very recently, we have been able to replace human cord serum with selected animal blood sera from a commercial supplier. In our experiments, 11 out of a total of 64 *in vitro* fertilized eggs developed into hatched blastocysts after prolonged culture (Table VI). The complete spectrum of preimplantation embryo development in the rhesus monkey, from fertilization of the egg to hatched blastocyst, can thus be studied *in vitro.* However, deterioration of the inner cell mass occurred in these embryos after hatching, although trophectoderm development was progressing well (Enders *et al.*, 1986).

Some of the blastocysts obtained from *in vitro* fertilized eggs hatched within 9 days after insemination (Table VI), which is within the normal range for the rhesus monkey. Information on the morphology and timing of embryonic development was obtained by making photographic records of the progress of each embryo at frequent intervals, using Nomarski differential-interference contrast microscopy (examples in Fig. 9). Measurements of embryo diameters made from these records were used to construct growth curves, which provide comparative data that are useful for predicting the normality of cleavage rates of individual embryos. Further studies using this *in vitro* embryogenesis system should provide valuable data on both normal and abnormal development in primates.

5.4.2. Birth of Offspring after *in Vitro* Fertilization

The ultimate proof of the "normality" of fertilization achieved under *in vitro* conditions lies in the ability to obtain healthy offspring after transfer of embryos to recipients. As mentioned in Section 5.1, this has been accomplished in only a few animal species (Table III). The first nonhuman primate offspring resulting from *in vitro* fertilization was born in July 1983 at the Southwest Foundation for Research and Education, San Antonio, Texas (Clayton and Kuehl, 1984). Four *in vitro* fertilized baboon embryos (three pronucleate stage and one 2-cell) were transferred surgically to the oviduct of a surrogate recipient. One healthy female infant was produced.

The first rhesus monkey derived from an *in vitro* fertilized egg was born at the Wisconsin Regional Primate Research Center 3 weeks after the baboon infant (Bavister *et al.*, 1984). This egg was one of a number recovered by follicular aspiration from a PMSG-stimulated donor. We transferred one 4-cell and one 6-cell embryo nonsurgically into the uterus of a surrogate rhesus monkey. Establishment of pregnancy was confirmed by uterine enlargement and by detection of monkey chorionic gonadotropin in maternal blood 16–20 days after embryo transfer. The presence of a normal fetus was shown by ultrasound scans. A male offspring, "Petri," was delivered by cesarean section 176 days from the date of fertilization *in vitro.* The parentage of this animal was

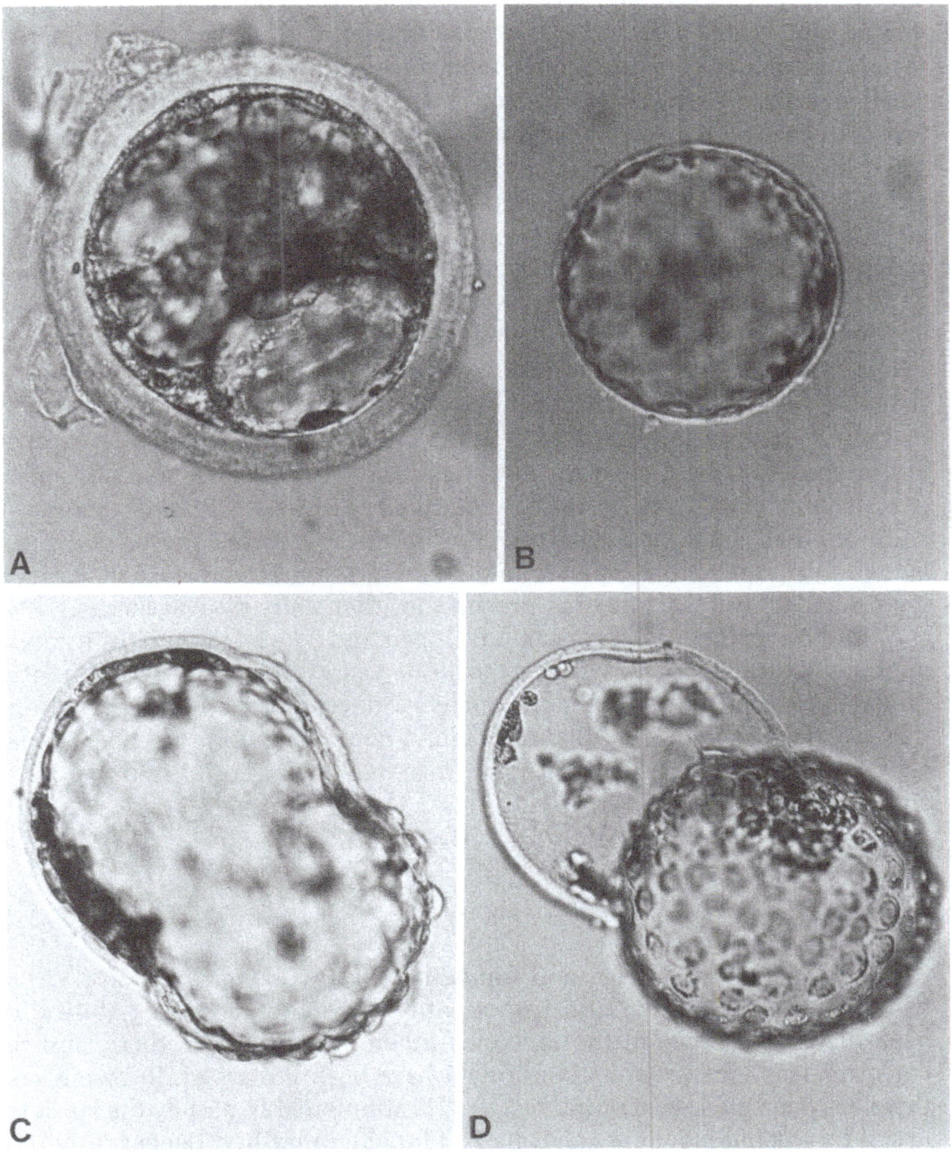

Figure 9. Stages of embryonic development *in vitro* of a rhesus monkey egg fertilized *in vitro*. Successive stages in development of the same embryo grown *in vitro*. This embryo grew more slowly than normal. Nomarski optics. (A) Mid-blastocyst stage, 221 hr after insemination. Two large blastocoel cavities can be seen that are not yet confluent. Blastocyst expansion has not begun, as shown by normal thickness (15–19 μm) of the zona pellucida. Optical magnification : ×342. (B) Expanded blastocyst showing blastocoel cavity and very thin (stretched, perhaps partially digested) zona pellucida. Shown 257 hr after insemination. Optical magnification : ×166. (C) Hatching blastocyst. The embryo is now about halfway out of the ruptured zona pellucida. Shown 265 hours after insemination. Optical magnification : ×184. (D) Fully hatched blastocyst, still attached to zona pellucida containing debris from the embryo. Shown 268 hr after insemination. Optical magnification : ×166.

conclusively demonstrated by genetic typing of blood cells taken from the infant, from the sperm and egg donors and from the surrogate recipient. A second pregnancy was established after transferring a single *in vitro* fertilized 8-cell embryo nonsurgically to a surrogate recipient. A male infant, "Orwell," was delivered naturally in April 1984. These successful pregnancies have validated the *in vitro* fertilization procedures used in our laboratory.

A male cynomolgus (*M. fascicularis*) infant was born after transfer of an *in vitro* fertilized embryo to a rhesus (*M. mulatta*) recipient (Balmaceda *et al.*, 1984). This result illustrates the practicality of combining *in vitro* fertilization with cross-species transfer, which could eventually prove immensely valuable to attempts at preserving endangered species.

In our laboratory, current experiments are aimed at producing more consistent development of *in vitro* fertilized rhesus embryos into viable blastocysts in culture and sustaining growth through early "postimplantation" stages. These studies should provide new data on normal and abnormal embryogenesis in primates, and on culture conditions needed for primate embryos. In addition, early blastocysts derived from *in vitro* fertilized rhesus oocytes are being transferred into recipients to determine if these later-stage embryos are more successful in establishing pregnancies than early cleavage stages.

In vitro fertilization has been developed and proved successful in several nonhuman primate species during the past 3 years. Much useful basic information stands to be gained by more intensive development and exploitation of these models. In addition, this information could be helpful in increasing the success of *in vitro* fertilization for the treatment of infertility in humans.

6. Summary

6.1. Overview

Some of the advantages and limitations of *in vitro* fertilization (IVF) as a technique for obtaining critical information on sperm-fertilizing ability have been discussed. Some of the problems inherent in IVF procedures and their consequences for interpretation of data have been described. In recent years, there has been a renewal of interest in IVF, stimulated in part by the successful use of this technique for the treatment of human infertility. Deeper insights are being gained into the way that spermatozoa acquire fertilizing ability (capacitation and the acrosome reaction), and into the molecular mechanisms of sperm : egg interaction. Details of these events are, however, still lacking in most areas.

Section 4 surveys the topic of regulation of sperm-fertilizing ability by constituents of the environment. This overview has of necessity been superficial because of the great volume of published information on this topic and because in many respects it is not certain how data obtained from experiments performed *in vitro* relate to the situation *in vivo*. Nevertheless, I have attempted to compare data obtained under *in vitro* and *in vivo* conditions in order to emphasize some potentially productive lines of investigation. The analysis of the ways in which capacitation and the acrosome reaction are regulated by the environment of spermatozoa *in vitro* and *in vivo* is a productive approach for addressing some of the outstanding questions about mammalian fertilization. Furthermore, if we knew how these events are controlled by extrinsic factors, we could design culture conditions for IVF that more closely mimic the natural environment of the gametes. We could then have greater confidence that events studied *in vitro* (e.g., changes in the properties of spermatozoa during capacitation) are physiological. The case of the "sperm motility factors" is one example of this experimental approach. The discovery that the hamster egg–cumulus complex contains sperm motility-stimulating substances (Yanagimachi, 1969*a*), which are also found in follicular fluid and blood serum (Yanagimachi, 1969*b*; Bavister, 1975) led to the characterization of the defined sperm motility factors, taurine (or hypotaurine) and catecholamines (see Meizel *et al.*, 1980). The use of these factors in a semi-defined culture medium supports *in vitro* fertilization of hamster eggs at very low sperm : egg ratios, approaching those found at the natural site of fertilization (J. C. Westrick and B. D. Bavister, unpublished observations). Preliminary data on this subject were reported by Bavister (1979). This capability not only supplies basic information on the regulation of sperm fertilizing ability, but also allows us to examine the behavior of individual fertilizing spermatozoa and their interactions with the egg investments. More intensive research into the relationship between spermatozoa and constituents of their environment is certainly warranted, and should lead to significant advances in knowledge about mammalian fertilization.

The potential of IVF techniques for gaining information about fertilization cannot be fully exploited if normal embryogenesis cannot be sustained in culture, as in the hamster and cattle (Bavister, 1982*b*). Present knowledge of culture requirements for supporting early embryonic development is inadequate for most mammalian species, the rabbit being an outstanding exception. Most of our information on preimplantation embryo growth requirements comes from studies with inbred mice, which are not likely to be typical of other mammals in this respect. Available evidence shows that cleavage stage embryos of other species are much more exacting in their needs for development in culture than are mouse embryos (Kane and Foote, 1970; Whittingham, 1975; Wright and Bondioli, 1981; Bavister *et al.*, 1983*a*,*b*). More intensive efforts are sorely needed to devise adequate culture environments for the embryos of species that are useful in basic research and/or are commercially valuable. It is self-evident that we must learn how to sustain these embryos *in vitro* before we can fully use them to examine mechanisms of normal and abnormal embryogenesis or to exploit their potential for improving productivity.

6.2. Outstanding Questions

Following are some of the major questions about mammalian fertilization and embryonic development that are particularly in need of more critical investigation. This list is biased by my own research interests. Some of these questions can be addressed using IVF or embryo culture techniques, others may require *in vivo* experimental designs, and some may be best answered using a combination of these approaches.

6.2.1. Capacitation

What is the functional significance of capacitation?
What are the functional changes undergone by spermatozoa during capacitation?
How is the process of capacitation regulated *in vitro* and *in vivo?*
What is the functional significance of changes in sperm motility (hyperactivation) associated with capacitation (or the acrosome reaction)?
How is the expression of hyperactivation regulated *in vivo?*

6.2.2. The Acrosome Reaction

What is the functional role of the acrosome reaction in fertilization?
Where does the acrosome reaction occur in fertilizing spermatozoa in relationship to penetration of the egg investments?
Where is the natural ARIF located (in the cumulus oophorus, zona pellucida, and/or oviductal secretions)?
What is the molecular identity of the natural ARIF?
What is the mechanism of the acrosome reaction and how is it induced by ARIF?
How is induction of acrosome reactions in fertilizing spermatozoa synchronized temporally and spatially with proximity of the products of ovulation?

6.2.3. Penetration of Egg Investments and Early Events in Fertilization

How does the fertilizing spermatozoon accomplish penetration of the egg investments? To what extent does this involve sperm enzymes; sperm motility; a combination of these? What is the nature of the so-called zona lysin?
What molecular components are involved in sperm : egg recognition and in sperm binding to the zona pellucida?
What is the mechanism of sperm : egg membrane fusion?
What is the molecular explanation for the "block to polyspermy" (in zona pellucida and/or egg plasma membrane)?
What are the mechanisms involved in sperm nuclear decondensation and male and female pronuclear formation?

6.2.4. Continued Development of *in Vitro* or *in Vivo* Fertilized Eggs in Culture

What are the reasons for the "blocks to development" during early stages of embryogenesis (e.g., hamster, bovine)?

Why is embryonic growth in vitro generally slower than *in vivo*?

Why are 1-cell mouse embryos generally more difficult to sustain in culture than 2-cell embryos?

What is the cause of the "blastomere swelling" phenomenon observed in cultured embryos of the hamster and other species?

What are the factors present in blood sera that stimulate growth of primate embryos *in vitro*?

Can a "universal" culture system be devised that will support preimplantation embryonic development, especially in the cleavage stages, in a wide variety of mammalian species? Alternatively, will it be necessary to devise different culture conditions for each species?

What are the critically important characteristics of such culture systems?

6.3. Conclusions

The performance of experiments with the objectives of directly answering these and other fundamental questions is expected to provide significant advances in our knowledge of the mechanisms and regulation of mammalian fertilization and embryonic development within the next several years. Conducting studies at both the cellular and molecular levels, and combining *in vitro* and *in vivo* experimental designs, will most rapidly lead to answers to these questions.

ACKNOWLEDGMENTS. I am grateful to the Department of Veterinary Science, College of Agricultural and Life Sciences, and to the Wisconsin Regional Primate Research Center, University of Wisconsin—Madison for providing support during the writing of this chapter. I am indebted to Dorothy Boatman for her assistance in researching information and in editing the text. I am grateful to Dorothy Boatman, Patricia Morgan, and Jane Westrick for providing unpublished data cited in this work. I thank Cec Warfield and Linda Endlich for their skillful assistance in preparing the manuscript and figures.

References

Adams, C. E. 1982, Egg transfer in the rabbit, in: *Mammalian Egg Transfer* (C. E. Adams, ed.), pp. 29–48, CRC Press, Boca Raton.

Ahuja, K. K., 1982, In-vitro inhibition of the block to polyspermy of hamster eggs by tertiary amine local anaesthetics, *J. Reprod. Fertil.* **65**:15–22.

Ahuja, K. K., 1984, Lectin-coated agarose beads in the investigation of sperm capacitation in the hamster, *Dev. Biol.* **104**:131–142.

Alvarez, J. G., and Storey, B. T., 1983, Taurine, hypotaurine, epinephrine and albumin inhibit lipid peroxidation in rabbit spermatozoa and protect against loss of motility, *Biol. Reprod.* **29**:548–555.

Aonuma, S., Mayumi, T., Suzuki, K., Noguchi, T., Iwai, M., and Okabe, M., 1973, Studies on sperm capacitation. I. The relationship between a guinea pig sperm-coating antigen and a sperm capacitation phenomenon, *J. Reprod. Fertil.* **35**:425–432.

Aonuma, S., Okabe, M., and Kawaguchi, M., 1978, The effect of zinc ions on fertilization of mouse ova in vitro, *J. Reprod. Fertil.* **53**:179–183.

Aonuma, S., Okabe, M., Kawaguchi, M., and Kishi, Y., 1981, Zinc effects on mouse spermatozoa and in-vitro fertilization, *J. Reprod. Fertil.* **63**:463–466.

Aonuma, S., Okabe, M., Kishi, Y., Kawaguchi, M., and Yamada, H., 1982, Capacitation inducing activity of serum albumin in fertilization of mouse ova in vitro, *J. Pharm. Dyn.* **5**:980–987.

Arver, S., and Eliasson, R., 1980, Zinc and magnesium in bull and boar spermatozoa, *J. Reprod. Fertil.* **60**:481–484.

Austin, C. R., 1951, Observations on the penetration of the sperm into the mammalian egg, *Aust. J. Sci. Res.* **4**:581–596.

Austin, C. R., 1955, Ovulation, fertilization and early cleavage in the hamster (*Mesocricetus auratus*), *J. R. Microsc. Soc.* **75**:141–157.

Austin, C. R., and Bishop, M. W. H., 1958, Role of the rodent acrosome and perforatorium in fertilization, *Proc. R. Soc. Ser. B* **149**:241–248.

Ayalon, N., 1978, A review of embryonic mortality in cattle, *J. Reprod. Fertil.* **54**:483–493.

Babcock, D. F., Singh, J. P., and Lardy, H. A., 1979, Alteration of membrane permeability to calcium ions during maturation of bovine spermatozoa, *Dev. Biol.* **69**:85–93.

Ball, G. D., Leibfried, M. L., Lenz, R. W., Ax, R. L., Bavister, B. D., and First, N. L., 1983, Factors affecting successful in vitro fertilization of bovine follicular oocytes, *Biol. Reprod.* **28**:717–725.

Balmaceda, J. P., Pool, T. B., Arana, J. B., Heitman, T. S., and Asch, R. H., 1984, Successful in vitro fertilization and embryo transfer in cynomolgus monkeys, *Fertil. Steril.* **42**:791–795.

Barros, C., and Austin, C. R., 1967, In vitro fertilization and the sperm acrosome reaction in the hamster, *J. Exp. Zool.* **166**:317–323.

Barros, C., and Berrios, M., 1977, Is the activated spermatozoon really capacitated?, *J. Exp. Zool.* **201**:65–72.

Barros, C., Fujimoto, M., and Yanagimachi, R., 1973, Failure of zona penetration by hamster spermatozoa after prolonged preincubation in a blood serum fraction, *J. Reprod. Fertil.* **35**:89–95.

Barros, C., Jedlicki, A., Bize, I., and Aguirre, E., 1984, Relationship between the length of sperm preincubation and zona penetration in the golden hamster: A scanning electron microscopy study, *Gamete Res.* **9**:31–43.

Bavister, B. D., 1972, A study of in vitro fertilization and capacitation in the hamster, Ph. D. dissertation, Cambridge University, England.

Bavister, B. D., 1973, Capacitation of golden hamster spermatozoa during incubation in culture medium, *J. Reprod. Fertil.* **35**:161–163.

Bavister, B. D., 1974, The effect of variations in culture conditions on the motility of hamster spermatozoa, *J. Reprod. Fertil.* **38**:431–440.

Bavister, B. D., 1975, Properties of the sperm motility-stimulating component derived from human serum, *J. Reprod. Fertil.* **43**:363–366.

Bavister, B. D., 1979, Fertilization of hamster eggs in vitro at sperm : egg ratios close to unity, *J. Exp. Zool.* **210**:259–264.

Bavister, B. D., 1980, Recent progress in the study of early events in mammalian fertilization, *Dev. Growth Diff.* **22**:385–402.

Bavister, B. D., 1981*a*, Analysis of culture media for in vitro fertilization and criteria for success, in: *Fertilization and Embryonic Development in Vitro* (L. Mastroianni and J. D. Biggers, eds.), pp. 41–60, Plenum Press, New York.

Bavister, B. D., 1981*b*, Substitution of a synthetic polymer for protein in a mammalian gamete culture system, *J. Exp. Zool.* **217**:45–51.

Bavister, B. D., 1981c, Bovine serum albumin is required for efficient capacitation of hamster spermatozoa in vitro, *Biol. Reprod.* **24**(Suppl. 1):37A.

Bavister, B. D., 1982a, Evidence for a role of post-ovulatory cumulus components in supporting fertilizing ability of hamster spermatozoa, *J. Androl.* **3**:365–372.

Bavister, B. D., 1982b, In vitro fertilization: Principles, practice and potential, in: *In Vitro Fertilization and Embryo Transfer* (E. S. E. Hafez and K. Semm, eds.), pp. 13–29, MTP Press, Lancaster, England.

Bavister, B. D., and Morton, D. B., 1974, Separation of human serum components capable of inducing the acrosome reaction in hamster spermatozoa, *J. Reprod. Fertil.* **40**:495–498.

Bavister, B. D., and Yanagimachi, R., 1977, The effects of sperm extracts and energy sources on the motility and acrosome reaction of hamster spermatozoa in vitro, *Biol. Reprod.* **16**:228–237.

Bavister, B. D., Rogers, B. J., and Yanagimachi, R., 1978, The effects of cauda epididymal plasma on the motility and acrosome reaction of hamster and guinea pig spermatozoa in vitro, *Biol. Reprod.* **19**:358–363.

Bavister, B. D., Leibfried, M. L., and Lieberman, G., 1983a, Development of preimplantation embryos of the golden hamster in a defined culture medium, *Biol. Reprod.* **28**:235–247.

Bavister, B. D., Boatman, D. E., Leibfried, L., Loose, M., and Vernon, M. W., 1983b, Fertilization and cleavage of rhesus monkey oocytes in vitro, *Biol. Reprod.* **28**:983–999.

Bavister, B. D., Boatman, D. E., Collins, K., Dierschke, D. J., and Eisele, S. G., 1984, Birth of rhesus monkey infant after in vitro fertilization and nonsurgical embryo transfer, *Proc. Natl. Acad. Sci. USA* **81**:2218–2222.

Bedford, J. M., 1968, Ultrastructural changes in the sperm head during fertilization in the rabbit, *Am. J. Anat.* **123**:329–358.

Bedford, J. M., 1969, Morphological aspects of capacitation, in: *Advances in Biosciences (Schering Symposium on Mechanisms Involved in Conception*, G. Raspe, ed., Vol. 4) pp. 35–50, Pergamon Press, Oxford.

Bedford, J. M., 1974, Mechanisms involved in penetration of spermatozoa through the vestments of the mammalian egg, in: *Physiology and Genetics of Reproduction* (E. M. Coutinho and F. Fuchs, eds.), Part B, pp. 55–68, Plenum Press, New York.

Bedford, J. M., 1983, Significance of the need for sperm capacitation before fertilization in eutherian mammals, *Biol. Reprod.* **28**:108–120.

Biggers, J. D., Whitten, W. K., and Whittingham, D. G., 1971, The culture of mouse embryos in vitro, in: *Methods in Mammalian Embryology* (J. C. Daniel, ed.), pp. 86–116, W. H. Freeman, San Francisco.

Blandau, R. J., and Odor, D. L., 1952, Observations on sperm penetration into the ooplasm and changes in the cytoplasmic components of the fertilizing spermatozoon in rat ova, *Fertil. Steril.* **3**:13–26.

Bleil, J. D., and Wassarman, P. M., 1983, Sperm–egg interactions in the mouse: Sequence of events and induction of the acrosome reaction by a zona pellucida glycoprotein, *Dev. Biol.* **95**:317–324.

Boatman, D. E., and Bavister, B. D., 1983, Regulation of hamster sperm capacitation by bicarbonate ion–carbon dioxide, *J. Cell Biol.* **97**:12a.

Boatman, D. E., and Bavister, B. D., 1984, Stimulation of rhesus monkey sperm capacitation by cyclic nucleotide mediators, *J. Reprod. Fertil.* **71**:357–366.

Boatman, D. E., Morgan, P. M., and Bavister, B. D., 1986a, Variables affecting the yield and developmental potential of embryos following superstimulation and *in vitro* fertilization in rhesus monkeys, *Gamete Res.* (in press).

Boatman, D. E., Morgan, P. M., Collins, K., and Bavister, B. D., 1986b, Development of in vitro fertilized rhesus monkey oocytes to the hatched blastocyst stage in culture (in preparation).

Bondioli, K. R., and Wright, R. W., Jr., 1982, In vitro fertilization of ovine ovulated and ovarian oocytes, *Theriogenology* **17**:79.

Bondioli, K. R., and Wright, R. W., Jr., 1983, In vitro fertilization of bovine oocytes by spermatozoa capacitated in vitro, *J. Anim. Sci.* **57**:1001–1005.

Borland, R. M., Hazra, S., Biggers, J. D., and Lechene, C. P., 1977, The elemental composition of the environments of the gametes and preimplantation embryo during initiation of pregnancy, *Biol. Reprod.* **16**:147–157.

Bowen, R. W., 1977, Fertilization in vitro of feline ova by spermatozoa from the ductus deferens, *Biol. Reprod.* **17**:144–147.

Brackett, B. G., 1978, Experimentation involving primate embryos, in: *Methods in Mammalian Reproduction* (J. C. Daniel, ed.), pp. 333–357, Academic Press, New York.

Brackett, B. G., and Oliphant, G., 1975, Capacitation of rabbit spermatozoa in vitro, *Biol. Reprod.* **12**:260–274.

Brackett, B. G., and Williams, W. L., 1965, In vitro fertilization of rabbit ova, *J. Exp. Zool.* **160**:271–282.

Brackett, B. G., Bousquet, D., Boice, M. L., Donawick, W. J., Evans, J. F., and Dressel, M. A., 1982*a*, Normal development following in vitro fertilization in the cow, *Biol. Reprod.* **27**:147–158.

Brackett, B. G., Bousquet, D., and Dressel, M. A., 1982*b*, In vitro sperm capacitation and in vitro fertilization with normal development in the rabbit, *J. Androl.* **3**:402–411.

Brackett, B. G., Keefer, C. L., Troop, C. G., Donawick, W. J., and Bennett, K. A., 1984, Bovine twins resulting from in vitro fertilization, *Theriogenology* **21**:224.

Bradley, M. P., and Forrester, I. T., 1982, Calsemin: A novel Ca^{2+}-dependent regulator of mammalian sperm functions, *Arch. Androl.* **9**:78 (abst.).

Bradley, M. P., and Garbers, D. L., 1983, The stimulation of bovine caudal epididymal sperm forward motility by bovine cumulus–egg complexes in vitro, *Biochem. Biophys. Res. Commun.* **115**:777–787.

Brewer, G. J., Aster, J. C., Knutsen, C. A., and Kruckeberg, W. C., 1979, Zinc inhibition of calmodulin: A proposed molecular mechanism of zinc action on cellular functions, *Am. J. Hematol.* **7**:53–60.

Brinster, R. L., 1963, A method for in vitro cultivation of mouse ova from two-cell to blastocyst, *Exp. Cell Res.* **32**:205–208.

Bryan, J. H. D., 1974, Capacitation in the mouse: The response of murine acrosomes to the environment of the female reproductive tract, *Biol. Reprod.* **10**:414–421.

Bryan, J. H., and Akruk, S. R., 1977, A naphthol yellow S and erythrosin B staining procedure for use in studies of the acrosome reaction of rabbit spermatozoa, *Stain Technol.* **52**:47–51.

Cabot, C. L., and Oliphant, G., 1978, The possible role of immunological complement in the induction of rabbit sperm acrosome reaction, *Biol. Reprod.* **19**:666–672.

Camous, S. , Heyman, Y., Meziou, W., and Menézo, Y., 1984, Cleavage beyond the block stage and survival after transfer of early bovine embryos cultured with trophoblastic vesicles, *J. Reprod. Fertil.* **72**:479–485.

Chan, P. J., Hutz, R. J., and Dukelow, R. W., 1982, Nonhuman primate in vitro fertilization: Seasonality, cumulus cells, cyclic nucleotides, ribonucleic acid, and viability assays, *Fertil. Steril.* **38**:609–615.

Chang, M. C., 1951, Fertilizing capacity of spermatozoa deposited into the fallopian tubes, *Nature (Lond.)* **168**:697–698.

Chang, M. C., 1959, Fertilization of rabbit ova in vitro, *Nature (Lond.)* **184**:466–467.

Charlwood, P. A., 1979, The relative affinity of transferrin and albumin for zinc, *Biochim. Biophys. Acta* **581**:260–265.

Chen, L. T., and Hsu, Y. C., 1982, Development of mouse embryos in vitro: Preimplantation to the limb bud stage, *Science* **218**:66–68.

Cheng, W., and Polge, C., 1983, In vitro fertilization of pig oocytes, in: *Winter Meeting of the Society for the Study of Fertility, London, England,* p. 20, abst. 30.

Cherr, G. N., and Clark, W. H., 1983, Characterization of the natural inducers of the acrosome reaction in sturgeon sperm, *J. Cell Biol.* **5**:180a.

Cherr, G. N., Meizel, S., and Katz, D. F., 1984*a*, Homologous soluble zonae pellucidae elicit acrosome reactions in hamster sperm, in: *International Cell Biology* (S. Seno and Y. Okada, eds.), *Proceedings of the Third International Congress on Cell Biology,* p. 401(abst.)

Cherr, G. N., Lambert, H., and Katz, D. F., 1984*b*, Completion of the hamster sperm acrosome reaction on the zona pellucida in vitro, *J. Cell Biol.* **99**:261(#960).

Chvapil. M., 1973, New aspects in the biological role of zinc: A stabilizer of macromolecules and biological membranes, *Life Sci.* **13**:1041–1049.

Clayton, O., and Kuehl, T. J., 1984, The first successful in vitro fertilization and embryo transfer in a non-human primate, *Theriogenology* **21**:228(abst.).

Clegg, E. D., 1983, Mechanisms of mammalian sperm capacitation, in: *Mechanism and Control of Animal Fertilization* (J. F. Hartmann, ed.), pp. 177–206, Academic Press, New York.

Cline, E. M., Gould, K. G., and Foley, C. W., 1972, Regulation of ovulation, recovery of mature ova and fertilization in vitro of mature ova of the squirrel monkey (*Saimiri sciureus*), *Fed. Am. Soc. Exp. Biol.* **31**:360 (abst.).

Cooper, G. W., Overstreet, J. W., and Katz, D. F., 1979, The motility of rabbit spermatozoa recovered from the female reproductive tract, *Gamete Res.* **2**:35–42.

Corselli, J., and Talbot, P., 1986, An in vitro technique to study penetration of hamster oocyte-cumulus complexes using physiological numbers of sperm, *Gamete Res.* (in press).

Cummins, J. M., 1982, Hyperactivated motility patterns of ram spermatozoa recovered from the oviducts of mated ewes, *Gamete Res.* **6**:53–63.

Cummins, J. M., and Yanagimachi, R., 1982, Sperm–egg ratios and the site of the acrosome reaction during in vivo fertilization in the hamster, *Gamete Res.* **5**:239–256.

Davis, B. K., 1976, Influence of serum albumin on the fertilizing ability in vitro of rat spermatozoa *Proc. Soc. Exp. Biol. Med.* **151**:240–243.

Davis, B. K., 1980, Interaction of lipids with the plasma membrane of sperm cells. I. The antifertilization action of cholesterol, *Arch. Androl.* **5**:249–254.

Davis, B. K., 1981, Timing of fertilization in mammals: Sperm cholesterol/phospholipid ratio as a determinant of the capacitation interval, *Proc. Natl. Acad. Sci. USA* **78**:7560–7564.

Davis, B. K., Byrne, R., and Hungund, B., 1979, Studies on the mechanism of capacitation. II. Evidence for lipid transfer between plasma membrane of rat sperm and serum albumin during capacitation in vitro, *Biochim. Biophys. Acta* **558**:257–266.

Decker, G. L., Joseph, D. B., and Lennarz, W. J., 1976, A study of factors involved in induction of the acrosomal reaction in sperm of the sea urchin, *Arbacia punctulata*, *Dev. Biol.* **53**:115–125.

Edwards, R. G., Bavister, B. D., and Steptoe, P. C., 1969, Early stages of fertilization in vitro of human oocytes matured in vitro, *Nature (Lond.)* **221**:632–635.

Enders, A. C., Schlafke, S., Boatman, D. E., Morgan, P. M., and Bavister, B. D., 1986, Differentiation of blastocysts derived from in vitro fertilized rhesus monkey oocytes (in preparation).

Eng, L. A., and Oliphant, G., 1978, Rabbit sperm reversible decapacitation by membrane stabilization with a highly purified glycoprotein from seminal plasma, *Biol. Reprod.* **19**:1083–1094.

Evans, G., and Armstrong, D. T., 1983, In vitro fertilization and embryo transfer following in vitro vs. in vivo maturation of rat oocytes, *Biol. Reprod.* **28**(Suppl. 1):73.

Farrell, P. S., and Bavister, B. D., 1984, Short-term exposure of 2-cell hamster embryos to collection media is detrimental to viability, *Biol. Reprod.* **31**:109–114.

Fleming, A. D., and Yanagimachi, R., 1982, Fertile life of acrosome-reacted guinea pig spermatozoa, *J. Exp. Zool.* **220**:109–115.

Franklin, L. E., Barros, C., and Fussell, E. N., 1970, The acrosomal region and the acrosome reaction in sperm of the golden hamster, *Biol. Reprod.* **3**:180–200.

Fraser, L. R., 1977, Motility patterns in mouse spermatozoa before and after capacitation, *J. Exp. Zool.* **202**:439–444.

Fraser, L. R., 1979, Accelerated mouse sperm penetration in vitro in the presence of caffeine, *J. Reprod. Fertil.* **57**:377–384.

Fraser, L. R., 1981, Dibutyryl cyclic AMP decreases capacitation time in vitro in mouse spermatozoa, *J. Reprod. Fertil.* **62**:63–72.

Fraser, L. R., 1983*a*, Mouse sperm capacitation assessed by kinetics and morphology of fertilization in vitro, *J. Reprod. Fertil.* **69**:419–428.

Fraser, L. R., 1983*b*, Potassium ions modulate expression of mouse sperm fertilizing ability, acrosome reaction and hyperactivated motility in vitro, *J. Reprod. Fertil.* **69**:539–553.

Fraser, L. R., Dandekar, P. V., and Vaidya, R. A., 1971, In vitro fertilization of tubal rabbit ova partially or totally denuded of follicular cells, *Biol. Reprod.* **4**:229–233.

Fujii, L., Utida, S., and Mizuno, L., 1955, Reaction of starfish spermatozoa to histidine and certain other substances considered in relation to zinc, *Nature (Lond.)* **176**:1068–1069.

Garbers, D. L., Tubb, D. J., and Hyne, R. V., 1982, A requirement of bicarbonate for Ca^{2+}-induced elevations of cyclic AMP in guinea pig spermatozoa, *J. Biol. Chem.* **257**:8980–8984.

Garbers, D. L., Kopf, G. S., Tubb, D. J., and Olson, G., 1983, Elevation of sperm adenosine 3′:5′-monophosphate concentrations by a fucose-sulfate-rich complex associated with eggs. I. Structural characterization, *Biol. Reprod.* **29**:1211–1220.

Gould, K. G., 1979, Fertilization in vitro of nonhuman primate ova: Present status and rationale for further development of the technique, in: *Report to the Ethics Advisory Board, Appendix: HEW Support of Research Involving Human In Vitro Fertilization and Embryo Transfer, No. 14*, pp. 1–25, U.S. Government Printing Office, Washington, D.C.

Gould, K. G., 1983, Ovum recovery and in vitro fertilization in the chimpanzee, *Fertil. Steril.* **40**:378–383.

Gould, K. G., Cline, E. M., and Williams, W. L., 1973, Observations on the induction of ovulation and fertilization in vitro in the squirrel monkey *(Saimiri sciureus)*, *Fertil. Steril.* **24**:260–268.

Gwatkin, R. B. L., 1977, *Fertilization Mechanisms in Man and Mammals*, Plenum Press, New York.

Gwatkin, R. B. L., 1983, Effect of compounds structurally related to taurine and of taurine uptake inhibitors on the motility of hamster sperm in vitro, *Gamete Res.* **4**:347–350.

Gwatkin, R. B. L., Andersen, O. F., and Hutchinson, C. F., 1972, Capacitation of hamster spermatozoa in vitro: The role of cumulus components, *J. Reprod. Fertil.* **30**:389–394.

Gwatkin, R. B. L., Carter, H. W., and Patterson, H., 1976, Association of mammalian sperm with the cumulus cells and the zona pellucida studied by scanning electron microscopy, in: *Scanning Electron Microscopy* (H. Johari and R. P. Becker, eds.), Vol., 2, Part 2, pp. 379–384, IIT Research Institute, Chicago, Illinois.

Hamner, C. E., and Williams, W. L., 1965, Composition of rabbit oviduct secretions, *Fertil. Steril.* **16**:170–176.

Hamner, C. E., Jennings, L. L., and Sojka, N. J., 1970, Cat *(felis catus L.)* spermatozoa require capacitation, *J. Reprod. Fertil.* **73**:477–480.

Handrow, R. R., Lenz, R. W., and Ax, R. L., 1982, Structural comparisons among glycosaminoglycans to promote an acrosome reaction in bovine spermatozoa, *Biochem. Biophys. Res. Commun.* **107**:1326–1332.

Hartmann, J. K., 1983, Mammalian fertilization: Gamete surface interactions in vitro, in: *Mechanism and Control of Animal Fertilization* (J. K. Hartmann, ed.), pp. 325–364, Academic Press, New York.

Hoppe, P. C., and Whitten, W. K., 1974, An albumin requirement for fertilization of mouse eggs in vitro, *J. Reprod. Fertil.* **39**:433–436.

Hosoi, Y., Niwa, K., Hatanaka, S., and Iritani, A., 1981, Fertilization in vitro of rabbit eggs by epididymal spermatozoa capacitated in a chemically-defined medium, *Biol. Reprod.* **24**:637–642.

Hsu, Y., 1979, In vitro development of individually cultured whole mouse embryos from blastocyst to early somite stage, *Dev. Biol.* **68**:453–461.

Huang, T. T. F., Fleming, A. D., and Yanagimachi, R., 1981, Only acrosome-reacted spermatozoa can bind to and penetrate zona pellucida: A study using the guinea pig, *J. Exp. Zool.* **217**:287–290.

Hyne, R. V., and Garbers, D. L., 1979, Calcium-dependent increase in adenosine 3′, 5′-monophosphate and induction of the acrosome reaction in guinea pig spermatozoa, *Proc. Natl. Acad. Sci. USA* **76**:5699–5703.

Hyne, R. V., and Garbers, D. L., 1980, Induction of an in vitro acrosome reaction of guinea pig spermatozoa by biological factors at pH 7.4, *Biol. Reprod.* **22**(Suppl. 1):9a.

Hyne, R. V., Higginson, R. E., Kohlman, D., and Lopata, A., 1984, Sodium requirement for capacitation and membrane fusion during the guinea pig sperm acrosome reaction, *J. Reprod. Fertil.* **70**:83–94.

Iritani, A., Sato, E., and Nishikawa, Y., 1974, Secretion rates and chemical composition of oviduct and uterine fluids in sows, *J. Anim. Sci.* **39**:582–588.

Iritani, A., Kasai, M., Niwa, K., and Song, H. B., 1984, Fertilization in vitro of cattle follicular oocytes with ejaculated spermatozoa capacitated in a chemically defined medium, *J. Reprod. Fertil.* **70**:487–492.

Jacobs, J., Boatman, D., and Bavister, B. D., 1984, Fertilization of hamster ova in vitro using a completely chemically-defined culture medium, *Biol. Reprod.* **30**(Suppl. 1):43a.

Johnsen, O., and Eliasson, R., 1978*a*, Destabilization of human sperm membranes by albumin, EDTA and histidine, *Int. J. Androl* **1**:485–488.

Johnsen, O., and Eliasson, R., 1978*b*, Studies on the succinate induced respiration of human spermatozoa in relation to zinc ions, *Int. J. Androl. (Suppl.)* **1**:175.

Johnson, L. L., and Overstreet, J. W., 1982, Potassium and pyruvate may be regulators of rabbit oviductal sperm motility, *Biol. Reprod.* **26**(Suppl. 1):145A.

Johnson, L. L., Katz, D. F., and Overstreet, J. W., 1981, The movement characteristics of rabbit spermatozoa before and after activation, *Gamete Res.* **4**:275–282.

Jones, R., 1978, Comparative biochemistry of mammalian epididymal plasma, *Comp. Biochem Physiol.* **61B**:365–370.

Juetten, J., and Bavister, B. D., 1983*a*, Effects of egg aging on in vitro fertilization and first cleavage division in the hamster, *Gamete Res.* **8**:219–230.

Juetten, J., and Bavister, B. D., 1983*b*, The effects of amino acids, cumulus cells, and bovine serum albumin on in vitro fertilization and first cleavage of hamster eggs, *J. Exp. Zool.* **227**:487–490.

Kane, M. T., 1985, A low molecular weight extract of bovine serum albumin stimulates rabbit blastocyst cell division and expansion in vitro, *J. Reprod. Fertil.* **73** 147–150.

Kane, M. T., and Foote, R. H., 1970, Culture of two- and four-cell rabbit embryos to the expanding blastocyst stage in synthetic media, *Proc. Soc. Exp. Biol. Med.* **133**:921–925.

Kane, M. T., and Headon, D. R., 1980, The role of commercial bovine serum albumin preparations in culture of one-cell rabbit embryos to blastocysts, *J. Reprod. Fertil.* **60**:469–475.

Katz, D. F., and Yanagimachi, R., 1980, Movement characteristics of hamster spermatozoa within the oviduct, *Biol. Reprod.* **22**:759–764.

Keefer, C. L., Bennett, K. A., and Brackett, B. G., 1983, Rabbit fertilization in vitro after delayed ovum recovery, *Biol. Reprod.* **28**(Suppl. 1):37.

Kenigsberg, D., Littman, B. A., Williams, R. F., and Hodgen, G. D., 1984, Medical hypophysectomy. II. Variability of ovarian response to gonadotropin therapy, *Fertil. Steril.* **42**:116–126.

Klaassen, C. D., 1980, Heavy metals and heavy-metal antagonists, in: *The Pharmacological Basis of Therapeutics* (A. Gilman, L. S. Goodman, and A. Gilman, eds.), pp. 1615–1637, Macmillan, New York.

Kopf, G. S., Lewis, C. A., and Vacquier, V. D., 1981, Regulation of abalone sperm cyclic nucleotide (CN) metabolism and the acrosome reaction (AR) by extracellular calcium and methylxanthines (MX), *J. Cell Biol.* **91**:178a.

Kreitmann, O., Lynch, A., Nixon, W. E., and Hodgen, G. D., 1982, Ovum collection, induced luteal dysfunction, in vitro fertilization, embryo development and low tubal ovum transfer in primates, in: *In Vitro Fertilization and Embryo Transfer* (E. S. E. Hafez and K. Semm, eds.), pp. 303–324, MTP Press, Lancaster, England.

Kuehl, T. J., and Dukelow, W. R., 1979, Maturation and in vitro fertilization of follicular oocytes of the squirrel monkey *(Saimiri sciureus)*. *Biol. Reprod.* **21**:545–556.

Kuzan, F. B., Fleming, A. D., and Seidel, G. E., Jr., 1984, Successful fertilization in vitro of fresh intact oocytes by perivitelline (acrosome reacted) spermatozoa of the rabbit *Fertil. Steril.* **41**:766–770.

Lambert, H., 1981, Temperature dependence of capacitation in bat sperm monitored by zona-free hamster ova, *Gamete Res.* **4**:525–533.

Lambeth, V. A., Looney, C. R., Voelkel, S. A., Jackson, D. A., Hill, K. G., and Godke, R. A., 1983, Microsurgery on bovine embryos at the morula stage to produce monozygotic twin calves, *Theriogenology* **20**:85–95.

Leibfried, M. L., and Bavister, B. D., 1982, Effects of epinephrine and hypotaurine on in vitro fertilization in the golden hamster, *J. Reprod. Fertil.* **66**:87–93.

Leibfried, M. L., and Bavister, B. D., 1983, Fertilizability of in vitro matured oocytes from golden hamsters, *J. Exp. Zool.* **226**:481–485.

Lenz, R. W., Bellin, M. E., and Ax, R. L., 1983, Rabbit spermatozoa undergo an acrosome reaction in the presence of glycosaminoglycans, *Gamete Res.* **8**:11–19.

Lessley, B. A., and Garner, D. L., 1983, Isolation of motile spermatozoa by density gradient centrifugation in Percoll, *Gamete Res.* **7**:49–61.

Lewis, W. H., and Hartmann, C. G., 1933, Early cleavage stages of the egg of the monkey *(Macacus rhesus)*, *Contrib. Embryol.* **24**:189–201.

Lindholmer, C., and Eliasson, R., 1972, Zinc and magnesium in human spermatozoa, *Int. J. Fertil.* **17**:153–160.

Llanos, M. N., Lui, C. W., and Meizel, S.,1982, Studies of phospholipase A_2 related to the hamster sperm acrosome reaction, *J. Exp. Zool.* **221**:107–117.

Lui, C. W., and Meizel, S., 1977, Biochemical studies of the in vitro acrosome reaction inducing activity of bovine serum albumin, *Differentiation* **9**:59–66.

Lui, C. W., Cornett, L. E., and Meizel, S., 1977, Identification of the bovine follicular fluid protein involved in the in vitro induction of the hamster sperm acrosome reaction, *Biol. Reprod.* **17**:34–41.

Lui, C. W., Mrsny, R. J., and Meizel, S., 1979, Procedures for obtaining high percentages of viable in vitro capacitated hamster sperm, *Gamete Res.* **2**:207–211.

Maas, D. H. A., Storey, B. T., and Mastroianni, L., 1977, Hydrogen ion and carbon dioxide of the oviductal fluid of the rhesus monkey *(Macaca mulatta)*, *Fertil. Steril.* **28**:981–984.

Mahi, C. A., and Yanagimachi, R., 1973, The effects of temperature, osmolality and hydrogen ion concentration on the activation and acrosome reaction of golden hamster spermatozoa, *J. Reprod. Fertil.* **35**:55–66.

Mahi, C. A., and Yanagimachi, R., 1976, Maturation and sperm penetration of canine ovarian oocytes in vitro, *J. Exp. Zool.* **196**:189–194.

Mahi, C. A., and Yanagimachi, R., 1978, Capacitation, acrosome reaction, and egg penetration by canine spermatozoa in a simple defined medium, *Gamete Res.* **1**:101–109.

Mann, T., and Lutwak-Mann, C., 1981, *Male Reproductive Function and Semen*, pp. 272–280, Springer-Verlag, Berlin.

Meizel, S., 1978, The mammalian sperm acrosome reaction, a biochemical approach, in: *Development in Mammals* (M. H. Johnson, ed.), Vol. 3, pp. 1–64, North-Holland Publishing Co., Amsterdam.

Meizel, S., and Lui, C. W., 1976, Evidence for the role of a trypsin-like enzyme in the hamster sperm acrosome reaction, *J. Exp. Zool.* **195**:137–144.

Meizel, S., and Turner, K. O., 1983, Serotonin or its agonist 5-methoxytryptamine can stimulate hamster sperm acrosome reactions in a more direct manner than catecholamines, *J. Exp. Zool.* **226**:171–174.

Meizel, S., and Turner, K. O., 1984, Glycosaminoglycans stimulate the acrosome reaction of previously capacitated hamster sperm, *J. Cell Biol.* **99**:261 (Abstract #959).

Meizel, S., Lui, C. W., Working, P. K., and Mrsny, R. J., 1980, Taurine and hypotaurine: Their effects on motility, capacitation and the acrosome reaction of hamster sperm in vitro and their presence in sperm and reproductive tract fluids of several mammals, *Dev. Growth Diff.* **22**:483–494.

Meizel, S., Turner, K. O., and Thomas, P., 1984, The stimulation of hamster sperm capacitation and acrosome reactions by biogenic amines, in: *Catecholamines as Hormone Regulators* (N. Ben-Jonathon, J. M. Bahr, and R. I. Weiner, eds.), pp. 329–343, Raven Press, New York.

Miyamoto, H., and Chang, M. C., 1973, The importance of serum albumin and metabolic intermediates for capacitation of spermatozoa and fertilization of mouse eggs in vitro, *J. Reprod. Fertil.* **32**:193–205.

Moore, H. D. M., 1981, An assessment of the fertilizing ability of spermatozoa in the epididymis of the marmoset monkey *(Callithrix jacchus)*, *J. Androl.* **4**:321–330.

Moore, H. D. M., and Bedford, J. M., 1978, An in vivo analysis of factors influencing the fertilization of hamster eggs, *Biol. Reprod.* **19**:879–885.

Moore, H. D. M., and Bedford, J. M., 1983, The interaction of mammalian gametes in the female, in: *Mechanism and Control of Animal Fertilization* (J. K. Hartmann, ed.), pp. 453–497, Academic Press, New York.

Morgan, P. M., Boatman, D. E., Collins, K., and Bavister, B. D., 1984, Complete preimplantation development in culture of in vitro fertilized rhesus monkey oocytes, *Biol. Reprod.* **30**(Suppl. 1):96a.

Mrsny, R. J., and Meizel, S., 1981, Potassium ion influx and Na^+,K^+-ATPase activity are required for the hamster sperm acrosome reaction, *J. Cell Biol.* **91**:77–82.

Mukherjee, A. B., and Cohen, M. M., 1970, Development of normal mice by in vitro fertilization *Nature (Lond.)* **228**:472–473.

Murphy, S. J., and Yanagimachi, R., 1984, The pH dependence of motility and the acrosome reaction of guinea pig spermatozoa, *Gamete Res.* **10**:1–8.

Myles, D. G., and Primakoff, P., 1983, A guinea pig sperm surface antigen that migrates during capacitation, *Biol. Reprod.* **28**(Suppl. 1):70.

Nagai, T., Niwa, K., and Iritani, A., 1984, Effect of sperm concentration during preincubation in a defined medium on fertilization in vitro of pig follicular oocytes, *J. Reprod. Fertil.* **70**:271–275.

Oliphant, G., and Brackett, B. G., 1973, Immunological assessment of surface changes of rabbit sperm undergoing capacitation, *Biol. Reprod.* **9**:404–414.

Oliphant, G., and Singhas, C. A., 1979, Iodination of rabbit sperm plasma membrane: Relationship of specific surface proteins to epididymal function and sperm capacitation, *Biol. Reprod.* **21**:937–944.

O'Rand, M. G., 1982, Modification of the sperm membrane during capacitation, *Ann. N.Y. Acad. Sci.* **383**:392–404.

Overstreet, J. W., Yanagimachi, R., Katz D. F., Hayashi, K., and Hanson, F. W., 1980, Penetration of human spermatozoa into the human zona pellucida and the zona-free hamster egg: A study of fertile donors and infertile patients, *Fertil. Steril.* **33**:534–542.

Ozasa, H., and Gould, K. G., 1982, Protective effect of taurine from osmotic stress on chimpanzee spermatozoa, *Arch. Androl.* **9**:121–126.

Perreault, S. D., and Zirkin, B. R., 1982, Sperm nuclear decondensation in mammals: Role of sperm-associated proteinase in vivo, *J. Exp. Zool.* **224**:253–257.

Perreault, S. D., Wolff, R. A., and Zirkin, B. R., 1984, The role of disulfide bond reduction during mammalian sperm nuclear decondensation in vivo, *Dev. Biol.* **101**:160–167.

Peters, T., Jr., 1975, Serum albumin, in: *The Plasma Proteins: Structure, Function, and Genetic Control* (F. W. Putnam, ed.), Vol. 1, pp. 133–181, Academic Press, New York.

Peterson, R. N., Bundman, D., and Freund, M., 1978, Use of a fluorescent dye to measure drug-induced changes in the membrane potential of boar spermatozoa, *Life Sci.* **22**:659–666.

Pope, C. E., Pope, V. Z., and Beck, L. R., 1982, Development of baboon preimplantation embryos to post-implantation stages in vitro, *Biol. Reprod.* **27**:915–923.

Rasmussen, H., and Goodman, D. B. P., 1977, Relationships between calcium and cyclic nucleotides in cell activation, *Physiol. Rev.* **57**:421–508.

Rogers, B. J., 1978, Mammalian sperm capacitation and fertilization in vitro: A critique of methodology, *Gamete Res.* **1**:165–223.

Rogers, B. J., and Yanagimachi, R., 1976, Competitive effect of magnesium on the calcium-dependent acrosome reaction in guinea pig spermatozoa, *Biol. Reprod.* **15**:614–619.

Rogers, B. J., Ueno, M., and Yanagimachi R., 1981, Fertilization by guinea pig spermatozoa requires potassium ions, *Biol. Reprod.* **25**:639–648.

Rose, T. A., 1984, Assessment of bovine oocyte maturation and its control with a simple in vitro fertilization procedure, *Biol. Reprod.* **30**(Suppl. 1):95a.

Rufo, G. A., Jr., Schoff, P. K., and Lardy, H. A., 1984, Regulation of calcium content in bovine spermatozoa, *J. Biol. Chem.* **259**:2547–2552.

Russell, L. D., Peterson, R. N., Hunt, W., and Strack, L. E., 1984, Posttesticular surface modifications and contributions of reproductive tract fluids to the surface polypeptide composition of boar spermatozoa, *Biol. Reprod.* **30**:959–978.

Saling, P. M., and Storey, B. T., 1979, Mouse gamete interactions during fertilization in vitro: Chlortetracycline as a fluorescent probe for the mouse sperm acrosome reaction, *J. Cell. Biol.* **83**:544–555.

Schackmann, R. W., Eddy, E. M., and Shapiro, B. M., 1978, The acrosome reaction of *Strongylocentrotus purpuratus* sperm, *Dev. Biol.* **65**:483–495.

Schackmann, R. W., Christen, R., and Shapiro, B. M., 1981, Membrane potential depolarization and increased intracellular pH accompany the acrosome reaction of sea urchin sperm, *Proc. Natl. Acad. Sci. USA* **78:**6066–6070.

Schroeder, A. C., and Eppig, J. J., 1984, The developmental capacity of mouse oocytes that matured spontaneously in vitro is normal. *Dev. Biol.* **102:**493–497.

SeGall, G. K., and Lennarz, W. J., 1979, Chemical characterization of the component of the jelly coat from sea urchin eggs responsible for induction of the acrosome reaction, *Dev. Biol.* **71:**33–48.

SeGall, G. K., and Lennarz, W. J., 1981, Jelly coat and induction of the acrosome reaction in echinoid sperm, *Dev. Biol.* **86:**87–93.

Seidel, G. E., Jr., 1982, Applications of microsurgery to mammalian embryos, *Theriogenology* **17:**23–34.

Shalgi, R., 1984, Developmental capacity of rat embryos produced by in vivo or in vitro fertilization, *Gamete Res.* **10:**77–82.

Shalgi, R., Kaplan, R., and Kraicer, P. F., 1977, Proteins of follicular, bursal and ampullar fluids of rats, *Biol. Reprod.* **17:**333–338.

Shalgi, R., Kaplan, R., Nebel, L., and Kraicer, P. F., 1981, The male factor in fertilization of rat eggs in vitro, *J. Exp. Zool.* **217:**399–402.

Shapiro, S. S., Jentsch, I. R., and Yard, A. S., 1971, Protein composition of rabbit oviductal fluid, *J. Reprod. Fertil.* **24:**403–408.

Singleton, C. L., and Killian, G. J., 1983, A study of phospholipase in albumin and its role in inducing the acrosome reaction of guinea pig spermatozoa *in vitro*, *J. Androl.* **4:**150–156.

Sirard, M. A., Lambert, R., and Beland, R., 1984, Evaluation of bovine capacitation by in vitro fertilization, *Theriogenology* **21:**261.

Soupart, P., 1967, Studies on the hormonal control of rabbit sperm capacitation, *J. Reprod. Fertil.* **Suppl. 2:**49–63.

Spielmann, H., Eibs, H. -G., and Jacob-Muller, U., 1980, In vitro methods for the study of the effects of teratogens on preimplantation embryos, *Acta Morphol. Acad. Sci. Hung.* **28:**105–115.

Srivastava, A., Chowdhury, A. R., and Setty, B. S., 1983, Zinc content of maturing spermatozoa in oestrogen treated rats, *Int. J. Androl.* **6:**103–108.

Staigmiller, R. B., and Moor, R. M., 1984, Effect of follicle cells on the maturation and developmental competence of ovine oocytes matured outside the follicle, *Gamete Res.* **9:**221–229.

Stanger, J. D., (1983) The effect of catecholamines and their antagonists on the fertilization of cumulus-free mouse ova in vitro at a suboptimal spermatozoal density, *Gamete Res.* **7:**111–122.

Storey, B. T., Lee, M. A., Muller, C., Ward, C. R., and Wirtschafter, D. G., 1984, Binding of mouse spermatozoa to the zonae pellucidae of mouse eggs in cumulus: evidence that the acrosomes remain substantially intact. *Biol. Reprod.* **31:**1119–1128.

Suarez, S. S., Katz, D. F., and Overstreet, J. W., 1983, Movement characteristics and acrosomal status of rabbit spermatozoa recovered at the site and time of fertilization, *Biol. Reprod.* **29:**1277–1287.

Talbot, P., and Chacon, R. S., 1981, A triple-stain technique for evaluating normal acrosome reactions of human sperm, *J. Exp. Zool.* **215:**201–208.

Talbot, P., and Franklin, L. E., 1976, Morphology and kinetics of the hamster sperm acrosome reaction, *J. Exp. Zool.* **198:**163–176.

Tervit, H. R., Whittingham, D. G., and Rowson, L. E. A., 1972, Successful culture in vitro of sheep and cattle ova, *J. Reprod. Fertil.* **30:**493–497.

Thibault, C., 1977, Are follicular maturation and oocyte maturation independent processes?, *J. Reprod. Fertil.* **51:**1–15.

Thompson, R. S., and Zamboni, L., 1975, Anomalous patterns of mammalian oocyte maturation and fertilization, *Am. J. Anat.* **142:**233–264.

Toyoda, Y., and Chang, M. C., 1974, Fertilization of rat eggs in vitro by epididymal spermatozoa and the development of eggs following transfer, *J. Reprod. Fertil.* **36:**9–22.

Trounson, A., 1983, Factors controlling normal embryo development and implantation of human oocytes fertilized in vitro, in: *Fertilization of the Human Egg in Vitro* (H. M. Beier and H. R. Lindner, eds.), pp. 235–250, Springer-Verlag, Berlin.

Tsafriri, A., Bar-Ami, S., and Lindner, H. R., 1983, Control of the development of meiotic competence of oocyte maturation in mammals, in: *Fertilization of the Human Egg in Vitro* (H. M Beier and H. R. Lindner, eds.), pp. 3–17, Springer-Verlag, Berlin.

Uno, Y., and Hoshi, M., 1978, Separation of the sperm agglutinin and the acrosome reaction-inducing substance in egg jelly of starfish, Science **200**:58–59.

Vanderhyden, B., Rouleau-Coulter, A. Walton, E. A., and Armstrong, D. T., 1984, Increased embryonic mortality following in vitro fertilization of rat oocytes *Biol. Reprod.* **30**(Suppl. 1):45a.

Ward, C. R., and Storey, B. T., 1983, Capacitation of mouse spermatozoa in vitro: Chlortetracycline as fluorescent probe for its time course, *J. Cell Biol.* **5**:473a.

Wassarman, P. M., and Bleil, J. D., 1982, The role of zona pellucida glycoproteins as regulators of sperm-egg interactions in the mouse, in: *Cellular Recognition* (W. Fraser, L. Glaser, and D. Gottlieb, eds.), pp. 845–863, Alan R. Liss, New York.

Westrick, J. C., Boatman, D. E., and Bavister, B. D., 1985, Characteristics of acrosome reaction-inducing factor from hamster cumulus cophorus and follicular fluid, *Biol. Reprod.* **32**(Suppl. 1):213a.

Whittingham, D. G., 1968, Fertilization of mouse eggs in vitro, *Nature (Lond.)* **220**:592–593.

Whittingham, D. G., 1971, Survival of mouse embryos after freezing and thawing, *Nature (Lond.)* **233**:125.

Whittingham, D. G., 1975, Fertilization, early development and storage of mammalian ova in vitro, in: *The Early Development of Mammals* (M. Balls and A. Wild, eds.), pp. 1–24, Cambridge University Press, Cambridge.

Whittingham, D. G., and Bavister, B. D., 1974, Development of hamster eggs fertilized in vitro or in vivo, *J. Reprod. Fertil.* **38**:489–492.

Willadsen, S. M., 1979, A method for culture of micromanipulated sheep embryos and its use to produce monozygotic twins, *Nature (Lond.)* **277**:298–300.

Wolf, D. P., 1981, The mammalian egg's block to polyspermy, in: *Fertilization and Embryonic Development in Vitro* (L. Mastroianni and J. D. Biggers, eds.), pp. 185–197, Plenum Press, New York.

Wolgemuth, D. J., 1983, Synthetic activities of the mammalian early embryo: Molecular and genetic alterations following fertilization, in: *Mechanism and Control of Animal Fertilization* (J. K. Hartmann, ed.), pp. 415–452, Academic Press, New York.

Wong, P. Y. D., and Lee, W. M., 1983, Potassium movement during sodium-induced motility initiation in the rat caudal epididymal spermatozoa, *Biol. Reprod.* **28**:206–212.

Working, P. K., and Meizel, S., 1983, Correlation of increased intraacrosomal pH with the hamster sperm acrosome reaction, *J. Exp. Zool.* **227**:97–107.

Wright, R. W., Jr., and Bondioli, K. R., 1981, Aspects of in vitro fertilization and embryo culture in domestic animals, *J. Anim. Sci.* **53**:702–729.

Yanagimachi, R., 1969*a*, In vitro capacitation of hamster spermatozoa by follicular fluid, *J. Reprod. Fertil.* **18**:275–286.

Yanagimachi, R., 1969*b*, In vitro acrosome reaction and capacitation of golden hamster spermatozoa by bovine follicular fluid and its fractions, *J. Exp. Zool.* **170**:269–280.

Yanagimachi, R., 1970, The movement of golden hamster spermatozoa before and after capacitation, *J. Reprod. Fertil.* **23**:193–196.

Yanagimachi, R., 1972, Fertilization of guinea pig eggs in vitro, *Anat. Rec.* **174**:9–20.

Yanagimachi, R., 1981, Mechanisms of fertilization in mammals, in: *Fertilization and Embryonic Development in Vitro* (L. Mastroianni and J. D. Biggers, eds.), pp. 81–182, Plenum Press, New York.

Yanagimachi, R., 1982*a*, In vitro sperm capacitation and fertilization of golden-hamster eggs in a chemically defined medium, in: *In Vitro Fertilization and Embryo Transfer* (E. S. E. Hafez and K. Semm, eds.), pp. 65–76, MTP Press Limited, Lancaster, England.

Yanagimachi, R., 1982*b*, Requirement of extracellular calcium ions for various stages of fertilization and fertilization-related phenomena in the hamster, *Gamete Res.* **5**:323–344.

Yanagimachi, R., 1984, Developmental ability of precociously superovulated golden hamster eggs, *Gamete Res.* **9**:231–237.

Yanagimachi, R., and Chang, M. C., 1964, In vitro fertilization of golden hamster ova, *J. Exp. Zool.* **156**:361–376.

Yanagimachi, R., and Noda, Y. D., 1970, Ultrastructural changes in the hamster sperm head during fertilization, *J. Ultrastruct. Res.* **31**:465–485.

Yanagimachi, R., and Phillips, D. M., 1984, The status of acrosomal caps of hamster spermatozoa immediately before fertilization in vivo, *Gamete Res.* **9**:1–19.

Yanagimachi, R., and Usui, N., 1974, Calcium dependence of the acrosome reaction and activation of guinea pig spermatozoa, *Exp. Cell Res.* **89**:161–174.

Yanagimachi, R., Yanagimachi, H., and Rogers, B. J., 1976, The use of zona-free animal ova as a test-system for the assessment of the fertilizing capacity of human spermatozoa, *Biol. Reprod.* **15**:471–476.

Yanagimachi, R., Lopata, A., Odom, C. B., Bronson, R. A., Mahi, C. A., and Nicolson, G. L., 1979, Retention of biologic characteristics of zona pellucida in highly concentrated salt solution: The use of salt-stored eggs for assessing the fertilizing capacity of spermatozoa, *Fertil. Steril.* **31**:562–574.

Yanagimachi, R., Miyashiro, L. H., and Yanagimachi, H., 1980, Reversible inhibition of sperm-egg fusion in the hamster by low pH, *Dev. Growth Diff.* **22**:281–288.

Yang, W. H., Lin, L. L., Wang, J. R., and Chang, M. C., 1972, Sperm penetration through zona pellucida and perivitelline space in the hamster, *J. Exp. Zool.* **179**:191–206.

Zamboni, L., 1974, Morphological aspects of mammalian fertilization, in: *Modern Perinatal Medicine* (L. Gluck, ed.), pp. 11–36, Year Book Medical Publishers, Chicago.

Chapter 5

Recent Progress in Human *in Vitro* Fertilization and Embryo Transfer

ALAN TROUNSON

1. Introduction

The present technique for human *in vitro* fertilization (IVF) has been derived from clinical and basic science components which, when welded together, form a complicated but successful procedure for the treatment of a number of aspects of human infertility. An interesting historical synthesis of these components was made by Biggers (1984). The initial report by Steptoe and Edwards (1978) of a completed pregnancy obtained by IVF procedures will remain one of the milestones of medicine, even though most of the significant scientific observations were made many years earlier by the same workers (Edwards *et al.*, 1970; Steptoe *et al.*, 1971). As required in any new scientific or medical advance, the confirmation of their success followed rapidly (Lopata *et al.*, 1980*a*); IVF began spectacular evolvement into an accepted clinical procedure and advanced into new areas of application as more groups entered the field. The escalation of research and clinical use of IVF has brought public attention and ethical concerns to focus on the procedure and associated technologies. The community's sensitivity to the intrusion of science and technology into conception has been provoked by the notion that artificial breeding and manipulation may progress in an uninhibited cascade. The response has been varied and often unpredictable with government moritoria in the United States and Australia, numerous government-appointed committee reviews, and the initiation of specific legislation to control IVF procedures and experimentation. This is a unique situation for science and medicine, and there is widespread concern and confusion as to whether any further developments will be allowed in the application or improvement of human IVF.

In this condensed review of human IVF, I have attempted to place in perspective the momentum of cell biology and the current state of IVF technology. Whether any further significant developments of human IVF are per-

ALAN TROUNSON • Centre for Early Human Development, Monash University and Queen Victoria Medical Centre, Melbourne, Australia 3000.

mitted will largely depend on community attitude and legal restrictions. I have the feeling that human medicine may become an observer of developments in science for some time to come, as it was before Edwards, Steptoe, Wood, and the others began applying the principles of cell biology to the problem of human infertility.

2. Collection of Mature Oocytes

The recovery of mature oocytes has always been considered a central requirement for successful human IVF (Trounson, 1983*a*). Indeed, it was thought that oocytes should be aspirated from the follicle as close to ovulation as possible. However, at least one IVF group (Veeck *et al.*, 1983) has reported that apparently immature oocytes recovered after human menopausal gonadotropin (hMG) stimulation are capable of maturation *in vitro* and may subsequently develop to normal fetuses when replaced *in utero*. Additional support for this claim has been obtained by A. Templeton and colleagues in Edinburgh, who found that oocytes obtained from women undergoing elective sterilization, before the LH surge, or injection of human chorionic gonadotropin (hCG) spontaneously matured during 42 hr *in vitro* in the complete absence of gonadotropins. These oocytes could be fertilized and the embryos developed *in vitro* to at least the blastocyst stage. These observations suggest that the oocyte may reach the stage of competance to mature at some point during follicle growth rather than at the end of follicular growth. If human oocytes can be spontaneously matured *in vitro*, there is little need for precision in timing oocyte recovery close to ovulation, except that it may be easier to aspirate fully mature oocytes loosely attached by the cumulus oophorus to the follicle wall.

The control of inhibition of oocyte maturation and process that governs the initiation and completion of oocyte maturation is complex, and the interrelationships between the sequence of events are yet not fully elucidated (Crosby and Moor, 1984; Eppig and Downs, 1984; Eppig *et al.*, 1984). It was considered essential that the cumulus cells remain attached to the oocyte to permit completion of cytoplasmic maturation, the normal sequence of protein synthetic changes, and the establishment of full fertilizing capacity and embryonic viability. However, in the mouse there is now one report (Schroeder and Eppig, 1984) that oocytes denuded of cumulus that spontaneously mature *in vitro* fertilize and develop to live young at rates comparable to those matured *in vivo* or to those surrounded by cumulus cells *in vitro*. In the sheep the presence of cumulus cells appears to be essential for normal maturation *in vitro* (Crosby *et al.*, 1981; Crosby and Moor, 1984). The requirements in the human have not yet been established.

The search for follicular fluid components that can identify the mature and viable oocyte has been disappointing. Numerous reports associating steroids, protein hormones and inhibitors, and other proteins (Trounson and Lutjen, 1984) of follicular fluid with the success of IVF have been either not specific, not repeatable, or too variable to be of prospective value in IVF. If the human oocyte readily undergoes spontaneous maturation when cultured *in vitro*, it may not be surprising that strict association with follicular fluid components is difficult to demonstrate, particulary when 4–12 hr preincubation of oocytes *in vitro* before insemination is almost a universal procedure in clinical IVF (Trounson *et al.*, 1982*a*).

A recent report by Trapp *et al.* (1984) on the presence of biozides (chlorinated hydrocarbons and pesticides) in human follicular fluid at concentrations that raise concerns as to their cytotoxic and teratogenic effects requires further investigation. The exposure of primordial oocytes to such chemicals over very long periods of time and exposure of the maturing oocyte undergoing resumption of meiosis, including DNA replication, has important implications for embryology and perhaps the community in general. It is important that any associations with embryo viability or genetic abnormality and biozide concentrations be identified by careful research; human IVF provides the opportunity to do this.

There is considerable variance in methods used for ovarian stimulation and the criteria used to administer hCG to time the final sequence of changes of oocyte maturation. Some IVF programs operate very effectively by inducing follicles to grow with low doses of hMG and administer hCG when the follicles are still in the rapid growth phase (size of the largest follicle by ultrasound about 1.5 cm diameter). Other IVF programs administer hCG close to the time of the spontaneous luteinizing hormone (LH) surge (Trounson and Calabrese, 1984). Probably more oocytes are recovered with "immature" or unactivated cumulus if hCG is given earlier in the follicular phase; it is possible, however, that these oocytes have already developed maturational competence, and they may be matured *in vitro* (Veeck *et al.*, 1983), whereas very few oocytes with "immature" or unactivated cumulus are recovered if hCG is given later in the follicular phase. Prolonged incubation of these oocytes already surrounded by activated cumulus may produce quite a different result (Trounson and Webb, 1984) to that for oocytes surrounded by inactivated cumulus. It is not known why some follicles fail to respond to hCG, neither producing any changes in the cumulus nor inducing the initiation of oocyte maturation, when given earlier in the follicular phase. Doses of hCG as high as 10,000 IU (H. W. Jones *et al.*, 1983*a*) are associated with the recovery of those immature oocytes. It would appear that either LH receptors are not present on the cells of these follicles or that they are inadequate in number. Alternatively, follicular inhibition, which is presiding over cumulus activation and oocyte maturation, predominates.

The requirement for a period of preincubation of human oocytes before insemination to enable completion of oocyte maturation (Sathananthan and Trounson, 1982*a*; Trounson *et al.*, 1982*a*) has been challenged by the data reported by Edwards *et al.* (1984). Other reports confirm a significant advantage in the preincubation oocytes before insemination (Johnston *et al.*, 1984), although proper controlled trials to retest this strategy have not been reported. There appears at least to be no disadvantage of preincubation in fertilization or pregnancy rates. The requirement for preincubation, like maturation *in vitro*, may depend on the superovulation procedures used and the method for timing oocyte recovery. The risk taken with prolonged preincubation before insemination resides in the possibility of oocyte aging, which may lead to an increase in defects of fertilization and reduced embryo development to blastocysts (Gianfortoni and Gulyas, 1985).

2.1. Follicular Maturation in the Spontaneous Ovulatory Cycle

The initial success in human IVF was obtained by collecting the single oocyte from women just prior to spontaneous ovulation (Edwards *et al.*, 1980; Lopata *et al.*, 1980*a*). This led Edwards and others to conclude that the natural ovulatory cycle was mandatory for successful IVF. In retrospect it seems more likely that the timing of the collection of oocytes by accurate prospective LH assays, which may predict very precisely the time of ovulation, resulted in more completely mature oocytes than had previous procedures.

Allowing the spontaneous growth and development of a single follicle and the maturation of a single oocyte under endogenous hormonal control *in vivo* did not require manipulation of any endocrine or physiological events—rather just the accurate prediction of the time of ovulation. Some interesting observations were made concerning the events surrounding ovulation:

1. The time of the onset of the LH surge appears to have a diurnal distribution, more frequently occurring early in the morning (5:00 to 10:00 hr) than at other times of day (Edwards, 1981; Testart *et al.*, 1982; Edwards *et al.*, 1982).
2. Plasma progesterone concentrations begin to elevate 3–12 hr before the onset of the LH surge (Trounson, 1983*b*; Hoff *et al.*, 1983), and the slope of the progesterone rise is parallel to the ascending arm of the LH surge.
3. Ultrasonography studies (O'Herlihy *et al.*, 1980; Kerin *et al.*, 1981; Bryce *et al.*, 1982) have shown that in the spontaneous ovulatory cycle there is almost a linear day-to-day increase in follicle size from 12 to 23 mm over the 5 days preceeding ovulation. A similar linear relationship between follicle diameter and plasma estradiol levels was found by Bryce *et al.* (1982), although the variation within and between the two parameters was too large to predict ovulation to within less than 1.4 ± 1.2 days.

The hormonal events underlying and regulating the recruitment, selection, and growth of the ovulating follicle have been reviewed extensively by Hodgen (1982) and Healy and Hodgen (1984). The elegant hypothesis derived by Hodgen and colleagues for the selection and dominance of the single follicle destined for ovulation centers around a delicate balance between the availability of pituitary follicle-stimulating hormone (FSH) to the cells of the rapidly growing follicles and the inhibition of FSH secretion exerted by increasing levels of a follicular factor, which is probably inhibin.

Additional support for this hypothesis has been observed in our own studies on women with primary ovarian failure receiving steroid replacement therapy (Lutjen *et al.*, 1984). In these women, normal ovulatory changes in plasma levels for estrogen, progesterone, and LH are observed, but follicular phase levels of FSH remain elevated, presumably due to the absence of any production of follicular inhibin. Any manipulation of this delicate balance between the circulating levels of FSH and inhibin will result in multiple follicular development due to increased availability of FSH or to follicular atresia caused by decreased availability of FSH. The follicular control of LH secretion is less clear. Rising levels of follicular estrogen secretion are associated with a decrease in peripheral plasma LH levels, which can be demonstrated by estrogen replacement therapy in women with primary ovarian failure. The LH surge is accompanied by increased pulsatility of gonadotropin-releasing hormone (GnRH) secretion from the hypothalamus (Clarke and Cummins, 1982). Progesterone secretion of follicular origin also begins to elevate before the onset of the LH surge (Hoff *et al.*, 1983) and, although the exact sequence of events controlling the surge of LH secretion has not been established, high levels of estrogen, increasing progesterone, increasing pulse rate of GnRH secretion, and the onset of the LH surge are closely linked. GnRH antagonists, potent analogues, and constant infusion of GnRH will reduce or abolish LH and FSH secretion; these approaches have been successfully used in primates to manipulate the ovulatory cycle (Sandow, 1983; Ory, 1983). The use of GnRH and analogues of GnRH for suppression of endogenous gonadotropins and antral follicle growth is attractive for human IVF because there is a significant proportion of women who undergo IVF treatment but who are insensitive to clomiphene citrate, hMG, and purified FSH. These women have characteristically elevated gonadotropins, and often only one antral follicle develops, even with large doses of clomiphene or hMG. Suppression of endogenous gonadotropin for 10–20 days may enable a cohort of follicles to reach the gonadotropin-sensitive stage rather than the single follicle, which responds immediately to the elevated FSH present under normal circumstances in these patients.

The administration of pulsatile high-dose GnRH results in multiple follicular development (Liu *et al.*, 1983); studies using this method of superovulation for IVF are being examined in our own laboratory. Whether this method provides any advantage over the established methods of superovulation for human IVF will be determined in time.

2.2. Manipulation of Follicular Growth Using Clomiphene and Human Menopausal Gonadotropin

The initial success of ovarian stimulation in IVF, using a very simple regimen of clomiphene citrate (Trounson *et al.*, 1981*a*), has been followed by many varied protocols based on combinations of clomiphene and hMG and hMG alone. The IVF group at Norfolk have progressively adjusted their hMG-alone protocols (Garcia *et al.*, 1984) and now include purified FSH in addition to hMG to induce multiple follicular growth (Jones *et al.*, 1984; Porter *et al.*, 1984). Considering the many different stimulation protocols used successfully by different IVF groups (Kerin and Seamark, 1984), it is surprising that the final success rates are so comparable. It is possible, however, that components of the IVF process may in fact differ quite markedly so that, as suggested by Rogers *et al.* (1986), increased embryo viability from one ovarian stimulation protocol may be accompanied by reduced uterine receptivity for implantation, when compared with another stimulation protocol.

Clomiphene citrate is a chlorotrianisene analogue that successfully competes for 17β-estradiol receptor sites and is thought to interfere in the negative feedback of estradiol on the hypothalamic–pituitary axis. This results in increased gonadotropin secretion, resulting in a mild dose-dependent increase in multiple follicular development. Clomiphene has a low clearance rate from the body, with only 50% of an oral dose eliminated in 5 days. Radiolabeled clomiphene is still present in excretions (feces) 6 weeks after administration (Schreiber *et al.*, 1966). This has led to a concern about the physiological effects of clomiphene in other tissues that are estrogen dependent for normal function. In particular, antiestrogenic effects on the uterine endometrium may adversely affect uterine receptivity for implantation.

Clomiphene does enable the development of a relatively synchronous cohort of follicles as compared with hMG alone. For this reason, the initial administration of clomiphene followed by hMG has been developed (McBain and Trounson, 1984) and used effectively (Trounson *et al.*, 1983*a*; McBain *et al.*, 1984) for ovarian stimulation in IVF. The hMG component enables an increased number of follicles to continue growth and maturation. Recent studies in our own laboratory involving the enzymatic removal of cumulus from the oocytes at laparoscopy show that the vast majority of oocytes obtained on this superovulation schedule have undergone nuclear maturation and expulsion of the first polar body. This is apparently not the case when hMG alone is used for ovarian stimulation (Veeck *et al.*, 1983; P. Rogers, unpublished data), and oocytes are frequently recovered at the germinal vesicle stage. It is much more difficult to superovulate older women (>40 years of age) with both clomiphene and hMG; reduced numbers of follicles, oocytes, and pregnancies occur in this group (see Table I).

Monitoring and development in the stimulated ovulatory cycle are usually done by tracking daily plasma or urine estrogen levels (Kerin and Seamark,

Table I. Effect of Maternal Age on *in Vitro* Fertilization Success Rates

Maternal age (years)	Mean number of follicles aspirated per patient	Mean number of mature oocytes recovered per patient	Fertilization rate (%)	Birth rate per treatment cycle (%)
>40	2.21	1.49	86	2.8
<40	3.03	2.60	79	10.3
Significance	$p < 0.005$	$p < 0.001$	N.S.	$p < 0.05$

1984) and ultrasound (Tarlatzis *et al.*, 1984; Richie, 1985). Problems in the use of ultrasound tracking of stimulated ovaries (Johnston *et al.*, 1981; Buttery *et al.*, 1983; Prins and Vogelzang, 1984) raise difficulties in the use of the parameter as the only monitor. However, Smith *et al.* (1984) reported that when changes in follicle diameter are combined with ultrasonically observed changes in endometrial thickening, ultrasonic parameters alone may be used successfully to monitor follicular growth and to determine the appropriate time to administer hCG. The use of an endometrial parameter is attractive because uterine factors appear to have a significant influence on pregnancy rate (Rogers *et al.*, 1986).

2.3. Manipulation of Ovulation with Human Chorionic Gonadotropin

The use of hCG to control the time of oocyte recovery (Trounson *et al.*, 1981*a*) is almost universal. The criteria for timing administration of hCG varies from simply ultrasonic follicle size (the leading follicle >1.5 cm diameter), or the sixth day of sustained plasma estrogen rise, to more complicated criteria involving a statistical determination of the expected day of the endogenous LH surge, cervical mucus changes, or indexes computed from vaginal cell and mucus changes, the plateau of plasma estrogen concentrations, and the observation of increasing plasma progesterone levels. In view of the lack of universal agreement on the optimum criteria for timing hCG injection, perhaps the best advice for IVF groups is to match the criteria used to the stimulation protocol and monitoring methods. Some of the stimulation protocols, monitoring methods, and criteria for hCG currently being used successfully were recently reviewed by H. W. Jones *et al.* (1983*a*), Seegar Jones (1983), Trounson *et al.* (1983*a*), Kemeter *et al.* (1984), and Kerin and Seamark (1984).

2.4. The Spontaneous LH Surge

Few IVF groups now rely on the detection of the spontaneous LH surge (Edwards, 1981; Testart *et al.*, 1981; Trounson, 1983*b*; Trounson *et al.*, 1983*a*) for the timing of oocyte recovery in stimulated ovulatory cycles. The onset of the spontaneous LH surge has a diurnal variation (Edwards, 1981, 1984; Testart *et al.*, 1982), occurring more frequently early in the morning than at other times during the day. Plasma progesterone levels also appear to have a diurnal nadir early in the morning (Trounson and Calabrese, 1984). Changes in urinary and plasma LH levels can be used to time ovulation accurately (Edwards, 1984), although sampling intervals (3 hr) for urinary LH need to be shorter than for plasma LH (Testart *et al.*, 1981; Trounson and Calabrese, 1984). In our own IVF program, patients preferred three times daily blood sampling to restricted fluid intake and 3-hourly urine collection.

Plasma progesterone levels rapidly increase at the same time that the LH surge begins, and changes in progesterone may be used successfully to determine the time of laparoscopy in the absence of a clearly defined LH surge (Trounson and Calabrese, 1984). In some patients, plasma progesterone levels may show a minor elevation during the 24 hr before the onset of the LH rise, which may be taken as a likely indicator of an approaching LH surge.

The IVF group at Norfolk, Virginia, consider that exogenous hMG suppresses the endogenous LH surge (Seegar Jones, 1984); experiments in monkeys given hMG early in the menstrual cycle (diZerega and Hodgen, 1980) support this view. Our own studies with clomiphene and hMG show no indication of suppression of the endogenous LH surge, even when hMG is given to within 24 hr of the endogenous LH surge. In fact, there is a close correlation between the expected day of the LH surge (McIntosh *et al.*, 1980) and the spontaneous LH surge observed for the same patients treated with clomiphene and hMG or hMG alone (Trounson *et al.*, 1985*a* (see Fig. 1). High plasma estradiol levels certainly occur for many days before the LH surge (Trounson, 1984*a*), indicating that high estrogen levels alone do not induce the LH surge. The Norfolk group usually administer hCG during the endogenous estrogen rise, so that patients do not have the opportunity to release their own LH. It is not infrequent that premature cessation of hMG administration may result in a rapid decline of estrogen levels (Seegar Jones, 1984; Trounson, 1984*a*) and a poor prognosis for continuing IVF in that cycle. The reason may be a decline in endogenous FSH, although this did not appear to be the case in diZerega and Hodgen's (1980)

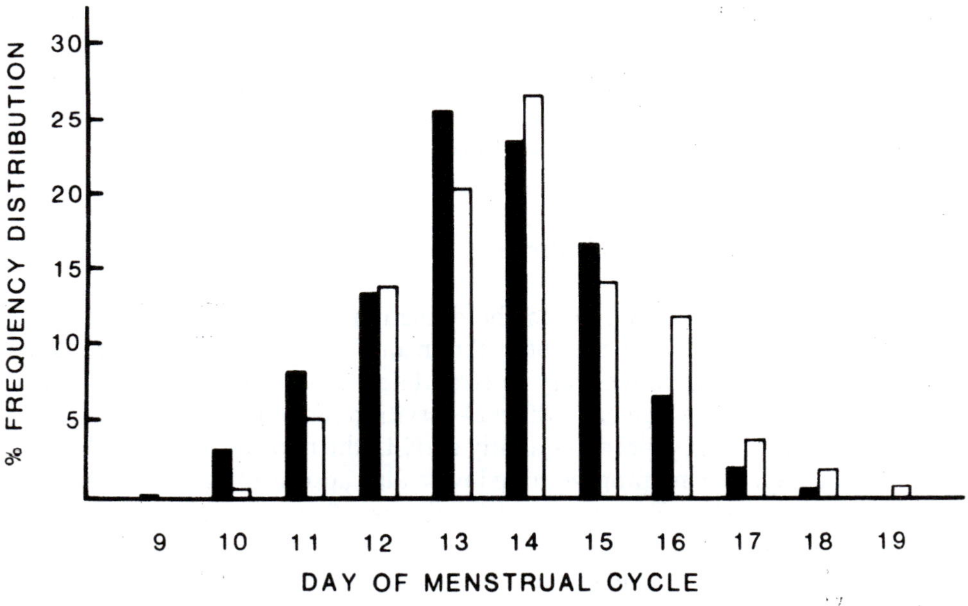

Figure 1. Relationship between the expected mean day (□) of the LH surge, calculated by the method of McIntosh *et al.* (1980), from the natural menstrual cycle length data and the observed day (■) of the LH surge after superovulatory treatment of the same women with clomiphene and hMG for IVF. (From Trounson *et al.*, 1985.)

experiments. One way to overcome a premature decline in estrogen levels is to fit the superovulation schedule to the patients natural ovulatory pattern (Trounson, 1984*a*) or to administer an ovulating dose of hCG earlier in the cycle when estrogen levels are still rising.

2.5. Follicular Aspiration

Laparoscopy is a safe and effective method for oocyte recovery (Wood *et al.*, 1981). The laparoscopy procedure was extensively reviewed recently by Kovacs *et al.* (1984) and Downing (1984). Ultrasonically guided oocyte recovery, percutaneously through the fluid-filled bladder or transvaginally, is rapidly gaining acceptance as a preferred alternative to laparoscopy. Recent experience with this technique has been reviewed by Wikland and Hamberger (1984), Lenz (1984), and Feichtinger and Kemeter (1984). The necessity of gaining experience with this technique in an effort to obtain oocyte recovery rates approaching that of laparoscopy is frequently recommended by those experienced in the procedure. Variations exist in the fluid used to fill the bladder, including urine, normal saline, culture media, and Hartman's solution. If the aspiration needle is cleared by aspiration of fluid in the bladder, physiological solutions should be used. The technology for ultrasonic oocyte recovery will continue to improve, as will the recovery rates and pregnancy rates associated with this technique.

A number of advantages are evident for ultrasonic oocyte recovery, including (1) the use of local anesthesia, requiring an outpatient environment instead of an operating room; (2) the fact that gas is not required for pneumoperitoneum, reducing the risk of exposing the oocyte to acid pH, which may occur with carbon dioxide; and (3) reduced cost of the IVF procedure and less stress and anxiety for the patient.

3. Fertilization *in Vitro*

The techniques of fertilization *in vitro* were developed primarily in mice, hamsters, and other animal species (Brackett and Bousquet, 1984) to study the process of fertilization and early embryonic development. Understanding of sperm capacitation requirements and the detailed interactions of the sperm and oocyte and its investments were established well before successful human IVF was obtained. It has indeed been fortunate that the degree of difficulty in the preparation of human sperm for fertilization *in vitro* is less than any other species studied; the relatively nonspecific requirements for embryo cleavage *in vitro* enables human embryos to develop under a wide range of culture media and conditions (Trounson, 1984*a*).

Human sperm spontaneously capacitate following removal of the seminal plasma components, require no preincubation prior to insemination (Trounson *et al.*, 1982*a*), and can be detected in the ooplasm within 3 hr of insemination

(Lopata *et al.*, 1978; McMaster *et al.*, 1978), do not require protein in the insemination medium for fertilization (C. Caro and A. Trounson, unpublished data), and do not require the presence of the cumulus on the oocyte for fertilization (Mahadevan and Trounson, 1985). In fact, the semen preparation and insemination techniques required for semen with quality parameters in the normal range is exceedingly simple.

3.1. Sperm Preparation

The routine methods used for sperm preparation (Purdy, 1982), frequently described in IVF publications, have been reviewed by Mahadevan and Baker (1984). Seminal plasma is usually separated from sperm and other cellular material by gentle centrifugation and a concentrated solution of motile sperm obtained by an overlay technique. A variation of these procedures utilizes a discontinuous or continuous Percol gradient to separate abnormal and normal motile sperm by gentle centrifugation (Braude and Bolton, 1984; Berger *et al.*, 1985; Dravland and Mortimer, 1985). This procedure may have an application in the preparation of sperm from men with abnormal semen profiles.

There is considerable variation in the number of motile sperm added to the oocyte at the time of insemination. As few as 10,000 sperm/ml culture medium has been used successfully (Craft *et al.*, 1981), and the number normally used varies from 20,000 to 100,000 sperm/ml. Increasing sperm concentrations may increase the incidence of polyspermy, and reducing the concentration to 10,000 sperm/ml may reduce the rate of normal fertilization *in vitro* (Wolf *et al.*, 1984).

It is essential that sperm be prepared in culture media containing protein (serum or serum albumin) to prevent agglutination during centrifugation. However, sperm added to protein-free culture media containing washed oocytes fertilize at the same rate as do those added to serum containing media (C. Caro and A. Trounson, unpublished data).

3.2. Factors Affecting Fertilization Rate

A number of factors are known to affect fertilization rate *in vitro* including the following.

3.2.1. Oocyte Maturation

The immature human oocyte at the germinal vesicle stage is rarely penetrated by spermatozoa (Edwards *et al.*, 1969; Overstreet and Hembree, 1976). Incomplete cytoplasmic maturation may also lead to polyspermy (Sathananthan and Trounson, 1984*a*,*b*) due to failure of the cortical granule reaction. It has also been shown by Ahuja (1985) that exposure of capacitating hamster sperm to inhibitors of glycoprotein synthesis prevents sperm binding to the zona. It is not known whether the human zona undergoes alteration in glycoprotein composition during maturation, but it is readily observable (Trounson *et al.*, 1980) that the zona of some human oocytes do not bind sperm, or bind few sperm, when placed in the same insemination droplet or tube as other oocytes which are covered with firmly bound sperm. It is possible that defects in zona glycoprotein maturation or incomplete maturation may also reduce fertilization rate *in vitro*.

The requirements for, and importance of, complete nuclear, cytoplasmic, and membrane maturation for normal fertilization and embryonic viability were reviewed recently by Crosby and Moor (1984) and Eppig and Downs (1984). Despite apparent differences between species in the regulation of oocyte maturation, there is agreement that defects arising through incomplete maturation or the exposure of oocytes to an abnormal environment that gives rise to defects in such crucial events as the biosynthesis and phosphorylation of new proteins (Warnes *et al.*, 1977; Crosby and Moor, 1984) will almost certainly result in abnormal fertilization and radically reduced embryo viability (Moor and Warnes, 1978; Trounson *et al.*, 1982*b*).

3.2.2. Semen Quality

In our own IVF program, we have been able to demonstrate that the quality of semen used for insemination can have a major effect on fertilization rate (Mahadevan and Trounson, 1984*a*). Partial correlation analyses showed that the percentage of sperm with forward progressive motility in the semen, or after preparation for insemination, as well as the proportion of morphologically abnormal sperm, were correlated to fertilization rate, whereas sperm concentration alone was not. Accordingly, reduced sperm motility and increased abnormal forms in semen used for insemination were found to reduce the fertilization rate. Cohen *et al.* (1984) claimed that abnormalities of semen quality do not have a major effect on the success rate of IVF, although a detailed analysis of data to support this conclusion was not presented. Reduced fertilization rate of oocytes due to oligozoospermia ($<15 \times 10^6$ sperm/ml) asthenozoospermia ($<50\%$ motility) and teratozoospermia ($>50\%$ abnormal sperm) has been reported by Testart *et al.* (1983), and a reduced fertilization rate was observed by Wolf *et al.* (1984) when sperm motility used for insemination dropped below 40%.

There is some suggestion that fertilization may be reduced with high concentrations of sperm used for insemination (Mahadevan and Trounson, 1984*a*; Wolf *et al.*, 1984) and an accompanying increase in the incidence of polyspermic fertilization with increasing concentration of sperm (Wolf *et al.*, 1984). We have not observed such an association in our own studies. It is possible that both defects of oocyte maturation (Sathananthan and Trounson, 1982*a*,*b*) and sperm concentration influence the rate of polyspermic fertilization. It is of interest to note that oocytes with multiple pronuclei frequently undergo normal embryonic cleavage (Van Blerkom *et al.*, 1984). Maternal age does not affect fertilization rate (Table I).

3.2.3. Category of Infertility

A detailed analysis of fertilization rates obtained in various classifications of infertility was reported by Mahadevan *et al.* (1983*a*). Fertilization rate was highest in patients with tubal infertility (80%) and endometriosis (88%). In patients with infertility of unknown cause (idiopathic infertility) with husbands who have apparently normal semen and those treated unsuccessfully by artificial insemination with donor semen (AID) fertilization rates were decreased (66%). This report showed there was a substantial difference in fertilization rate for men with persistently low-quality semen (sperm concentration $<20 \times 10^6$/ml, sperm motility <50%, or >40% abnormal sperm forms), depending on whether their wives had normal (38%) or blocked tubes (67%). Presumably in the latter case the husbands were in fact fertile despite having lower-quality semen than this arbitrarily chosen range.

These observations indicate that fertilization is not a major component of idiopathic infertility, and some other causes need to be investigated. In our own laboratory, studies on biochemical infertility due to defects in the sialic acid components of glycoprotein secretions (Lutjen *et al.*, 1985*a*) have shown that relatively minor biochemical alterations can have profound effects on the reproductive process. It is also apparent that some aspects of male infertility may be treated by IVF, but the pregnancy rates obtained in some conditions may be very low because of the reduction in expected fertilization rate (de Kretser *et al.*, 1985).

3.3. Sperm Penetration of the Zona Pellucida

Sathananthan *et al.* (1982*a*) presented a detailed account of the ultrastructure of sperm binding to, and penetration of, the human zona, and Sundstrom (1982) showed some visually esthetic and impressive views of sperm–zona interaction. The sperm most frequently found that are firmly bound to the zona have intact acrosomes. The sperm head lies parallel to the zona surface and the plasma membrane in contact with the zona swells and disrupts on this side (Sathananthan, 1984). It is of interest to note that the acrosome reaction takes place after sperm binding to the zona but prior to passage through the zona. Most sperm seen in transit through the zona have partially or fully reacted acrosomes. The sperm head enters the zona tangentially to the surface (Sundstrom, 1982) because the initial binding is along the flat side of the sperm head. During the initial phase of sperm penetration, the side of the sperm head initially in contact with the zona vesiculates and, often on the opposite side, the plasma membrane may be observed to be intact. Sperm penetrating the inner half of the zona are fully reacted.

The zona has an important role in sperm–oocyte recognition, sperm binding, and the prevention of polyspermy. Very little information is available on the specific nature of glycoproteins involved in human sperm–oocyte interaction. Species specificity of this interaction would suggest that particular oligosaccharides, proteins, or combinations of the two are involved. In some polyspermic oocytes, acrosome-reacted sperm enter the zona without any indication of prior binding (Sathananthan *et al.*, 1982*a*), which may indicate that the normal sequence of events can be altered by modification of the specific zona glycoproteins. The elucidation of these biochemical interactions in the human is important for our evaluation of oocyte maturity, sperm-fertilizing capacity, and biochemical infertility.

Motility and morphological abnormality of sperm are both related to fertilization rate (Mahadevan and Trounson, 1984*a*). Immotile sperm or those with no forward progression are unable to pass through the zona. Sperm with tail abnormalities usually have abnormal motility, and those with head abnormalities are infrequently found attached or penetrating the zona (Sathananthan *et al.*, 1982*a*; Sundstrom, 1982). It may be necessary to consider microsurgical methods for placing sperm within the oocyte (Uhera and Yanigimachi, 1976), providing preliminary screening of the initial embryos and pregnancies obtained with this technique shows that there is no dramatic increase in chromosomal abnormalities.

3.4. Sperm Fusion with the Oocyte and the Cortical Granule Reaction

The fusion of human sperm with the oocyte plasma membrane during polyspermic fertilization of zona-free oocytes was described by Soupart and Strong (1975). It has also recently been observed in monospermic fertilization of zona intact oocytes (A. H. Sathananthan, unpublished data) and involves the initial fusion of the intact plasma membrane of the postacrosomal segment with the oocyte vitelline membrane; the entire sperm is then gradually incorporated into the ooplasm. The fusion of gametes is generally species specific, except in the case of the hamster oocyte. Membrane fusion requires drainage of aqueous solution from between membrane bilayers to permit intimate contact (Fisher and Parker, 1984). This may be achieved by the presence of an osmotic gradient established at the time of the initial binding of the vitellus to the postacrosomal remnant of the sperm plasma membrane. It is established that drastically increased levels of intracellular calcium ions occur at fertilization (Jaffe, 1983) due to release of intracellular calcium ion stores and increased calcium ion permeability of the vitelline membrane. The establishment of these ionic gradients may be sufficient to produce osmotic flow through membrane bilayers, reducing the film of aqueous solution between the bilayers to some critical thickness, at which time the thin film collapses, permitting contact and fusion (Fisher and Parker, 1984). This may be the mechanism for gamete fusion; of equal importance, it may also be the mechanism that initiates cortical granule exocytosis.

For the cortical granule reaction, it is necessary that they be in close proximity to the vitelline membrane. In mature oocytes they are very close to the vitellus (Sathananthan and Trounson, 1982*a*). The cortical granules are surrounded by a membrane that fuses with vitelline membrane at the time of dehiscence. Cortical granules may fuse together forming caverns that enable deeper-seated granules to fuse with the vitellus. These observations have been documented in the human by Sathananthan and Trounson (1982*b*). The cortical granule contents diffuse into the zona, altering the sperm penetrability of the inner zona (zona reaction). The cortical granules are believed to contain a trypsinlike protease (Gwatkin, 1977), but the actual chemical nature of these granules in the human has not been established.

3.5. Formation of Pronuclei

The formation of pronuclei in human oocytes has been described by Sathananthan and Trounson (1985). The fine structure of the completion of meiosis (late telophase II) and extrusion of the polar body were reported by Lopata *et al.* (1980*b*); the complete process was reviewed by Sathananthan (1984). Pronuclei may be identified within 3 hr of insemination (McMaster *et al.*, 1978) and are clearly visible by dissecting microscopy, 12–16 hr after insemination (Trounson *et al.*, 1982*a*). At this time it is possible to identify polyspermic fertilization by the presence of three or more pronuclei (Wolf *et al.*, 1984). At 19–24 hr after insemination, the pronuclear envelopes are dismantled and the pronuclei disappear from view as the chromosomes condense at prophase of syngamy. At 22–28 hr, the zygote undergoes the first cleavage division, and the embryonic phase of development begins.

3.6. Errors of Fertilization

Abnormalities that may arise through chromosomal errors in gamete maturation, fertilization, and culture *in vitro* have been reviewed by Pike (1984). A study of the chromosomal constitution of human IVF embryos reported by Angell *et al.* (1983) identified monoploidy and trisomy in almost one-third of embryos examined. Further studies by R. Angell and colleagues (unpublished data) have confirmed that chromosomal errors, particularly monosmy and trisomy of the D group chromosomes, occur in about 30% of embryos karyotyped. When this information is considered together with the incidence of irregular numbers of pronuclei, 2–7% (Pike, 1984), chromosomal aberrations constitute a significant component of embryo abnormality and wastage. Monoploidy has not been identified in surveys of spontaneously aborted material and must arise through parthenogenetic activation, failure of male or female pronucleus formation (Pike, 1984), or failure of both pronuclei to associate at syngamy. Oocytes with multiple pronuclei will undergo apparently normal cleavage and development to at least the morula stage (Van Blerkom *et al.*, 1984). Karyotyping of some cleaving embryos in which three pronuclei were clearly visible in our own studies has revealed diploidy, which probably means that one pronucleus was excluded from participation in syngamy. Other cleaving embryos with three pronuclei contained a triploid quantity of DNA (R. Angell and A. Trounson, unpublished data).

Chromosomal deletion and addition errors arising from chromosome nondysjunction during gametogenesis have been identified in the female pronucleus of oocytes with multiple pronuclei at fertilization (Rudak *et al.*, 1984). Chromosomal errors were also identified in male pronuclei (Rudak *et al.* 1984), demonstrating the contribution made by both gametes to defects in the embryonic genome. Abnormalities of oocyte and sperm nuclear maturation would appear to be a significant problem in human IVF. Since these defects are found in both superovulated oocytes (Rudak *et al.*, 1984) and oocytes obtained from nonstimulated women (Angell *et al.*, 1983), it is likely that this may also be a general problem in natural human reproduction.

Fragmentation of fertilized oocytes into anucleate cells of variable size is also frequently observed in human IVF (Mohr, 1984). Fragmentation may be initiated by fertilization but is likely to be an abnormality of cytoplasmic or membrane maturation of the oocyte. Very little is known about the underlying events or the cause of this abnormality in embryogenesis.

3.7. Frozen Semen in IVF

Human semen may be frozen and thawed for use in human IVF (Mahadevan *et al.*, 1983*b*). However, semen quality should be in the normal quality range and the freezing, thawing, and insemination methods closely controlled. Semen with low initial motility cannot be easily cryopreserved to retain motility, and semen with other defects, including teratozoospermia and oligozoospermia, needs to be checked carefully after thawing to determine whether a sufficient number of normal motile sperm can be harvested. Methods for semen freezing have been described in detail by Mahadevan and Trounson (1983; 1984*b*).

4. Culture of the Embryo *in Vitro*

Human embryos develop from the 1-cell stage to the hatched blastocyst stage (Fig. 2) in a wide variety of culture media and conditions (Trounson, 1984*a*). The fine structure of apparently normal and abnormal human embryo development has been described from the 2-cell to the hatched blastocyst stage (Sundstrom *et al.*, 1981; Dvorak *et al.*, 1982; Lopata *et al.*, 1982; Mohr and Trounson, 1982; Sathananthan *et al.*, 1982*b*; Trounson and Sathananthan, 1984). Fine structural details and functional implications have been reviewed by Sathananthan (1984), Mohr (1984), and Sathananthan *et al.* (1985). There are

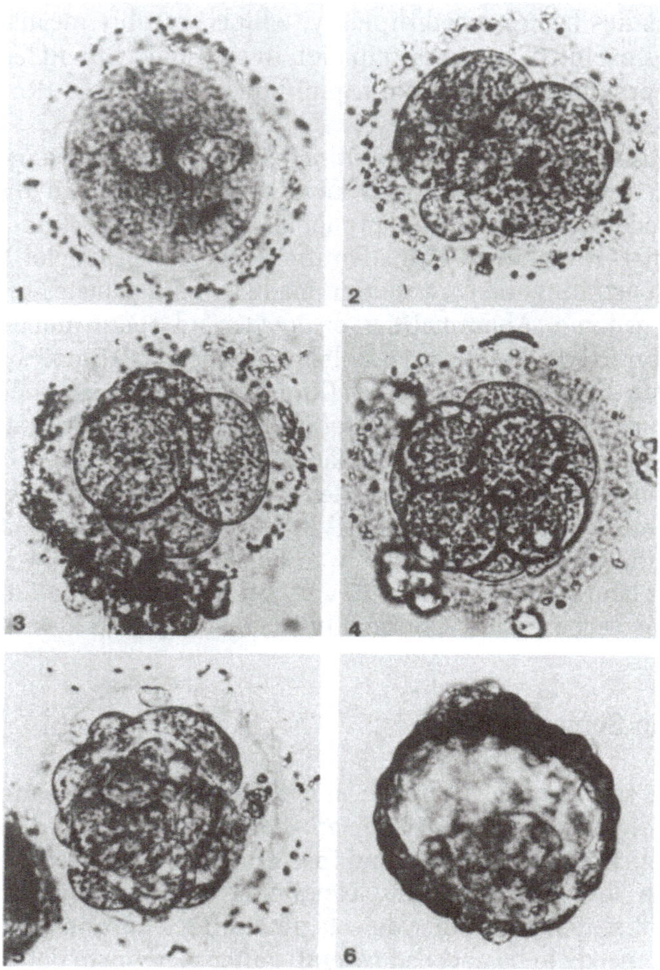

Figure 2. Stages of development of the human preimplantation embryo. (1) Pronuclear 1-cell, 2. 2-cell, 3. 4-cell, 4. 8-cell, 5. Morula, 6. Hatched blastocyst. (From Trounson *et al.*, 1982*a*.)

many similarities in the ultrastructure of human embryos to those of other mammalian species, but some characteristics are specifically primate or human (Sathananthan *et al.*, 1985). Some of the comparative observations have enabled more informed decisions to be made in the development of new techniques. For example, the early cleavage stage human embryo has cytoplasmic yolk characteristics similar to those of the mouse. This observation led us to freeze early cleavage stage human embryos rather than those at the late morula or early blastocyst stage. Mouse embryos can be frozen at the early cleavage stages, whereas late morulae and early blastocysts are preferred for freezing in most other species. The nature of the cytoplasm at different stages of development has a major effect on embryo survival rates after cooling and freezing (Mohr and Trounson, 1981).

4.1. Embryo Cleavage

Regular cleavage is a normal feature of preimplantation embryo development, and a correlation exists between the rate of cleavage and pregnancy after embryo transfer. More rapidly dividing embryos are more likely to result in pregnancy than are those with retarded cleavage (Trounson *et al.*, 1982*a*; Mohr, 1984). However, the range in postinsemination age of the various cleavage stages that result in pregnancy is substantial (Mohr, 1984).

There is considerable variation in the appearance of early cleavage stage embryos (Mohr, 1984; Mohr and Trounson, 1984); they may be graded according to criteria relating to shape and size of blastomeres, number and location of nuclei within the blastomeres, number of anucleate fragments, and appearance of the cytoplasm. However, the grade of embryos does not absolutely determine their developmental potential, although poor-quality embryos are less frequently associated with pregnancy.

Cleavage rate and gross morphological appearance may indicate whether the *in vitro* procedures and maturity of oocytes collected are in the range expected for successful IVF. Failure of embryos to continue from one cleavage stage to another at high rates, i.e., >90% (Trounson, 1983*c*), at least for the first three cleavage divisions, may indicate suboptimal culture conditions or some abnormality of oocyte maturation relating to the procedures used.

It may be argued that the capacity of embryos to develop to morphologically normal blastocysts is closely related to their potential to develop to term (Moor and Trounson, 1977). Human embryos may develop to blastocysts with clearly differentiated trophectoderm, inner cell mass cells and primitive endoderm (Mohr and Trounson, 1982) and may begin to produce hCG *in vitro* (Fishel *et al.*, 1983). While these observations indicate that human embryos are capable of complete preimplantation development *in vitro*, the proportion that develop to apparently normal blastocysts may give a better indication of embryo viability than presumed from the transfer of early cleavage stages. Many abnormalities may be observed in embryos grown to the blastocyst stage (Lopata *et al.*, 1983), but it is not known whether these are due to prolonged culture *in vitro* or to spontaneous developmental abnormalities. No adequately detailed data have been published on the proportion of morphologically normal blastocysts that develop from fertilized oocytes *in vitro*, nor is there any indication that pregnancy rate increases with postinsemination age of the embryo (Edwards *et al.*, 1984).

4.2. Factors Affecting Embryo Development *in Vitro*

Culture media and conditions *in vitro* have been considered crucial for normal human embryo cleavage and development (Quinn *et al.*, 1984). However, many conditions purported to be important have apparently little or no effect on cleavage or pregnancy rate including the following:

1. A concentration of 5% oxygen in the culture gas phase appears to be no different from 20% oxygen in our own laboratory.
2. A controlled trial of four basically different culture media showed no difference in fertilization, cleavage, or pregnancy rates (Trounson, 1984*a*), although more pregnancies were obtained in the simpler media developed for mouse embryo culture (Whittingham's T6 and modified Whitten's medium). The widely used B2 medium was not available for testing in this trial.
3. Mouse embryos have been shown to develop normally in the absence of protein, amino acid, or large-molecular-weight supplementation of culture medium (Caro and Trounson, 1984). We have also found in protein-free medium (T6) that human fertilization, embryo cleavage, and pregnancy rate are the same as in medium containing 10% patient's serum (C. Caro and A. Trounson, unpublished data) (see Table II). In fact, it is possible that some protein supplements may be detrimental for embryo cleavage *in vitro*, particulary at high levels (Caro and Trounson, 1984). Detrimental effects of supplementing Hams F10 culture medium with maternal or fetal cord serum and plasma on mouse embryo development have also been reported by Shirely *et al.* (1985).
4. The washing of fertilized oocytes and transfer from the insemination medium containing sperm to a fresh droplet or tube of culture medium have no significant effect on cleavage or pregnancy rate as compared with oocytes left in the insemination medium until the time of transfer (Table III). This simplifies the laboratory procedures for IVF but does not permit detection of multiple pronuclei (potentially polyploid embryos).

There are no convincing reports on factors that positively improve the rate of human embryo development *in vitro*. This implies that either embryo development *in vitro* is determined by gamete quality or by conditions that have not

Table II. Development and Viability of Human Embryos Grown in Protein-Free Culture Medium

	Insemination and culture medium[a]	
	T6 + 10% maternal serum	Protein-free T6[b]
Number of patients treated	197	178
Number of patients pregnant (%)	36 (18.2%)	25 (14.0%)
Number of patients aborting	11 (30.5%)	6 (24.0%)
Number of births	25 (12.7%)	19 (10.6%)

[a]Oocytes were fertilized and cultured in the medium described.
[b]Whittingham T6 medium containing no protein or amino acids.

Table III. Effect on Human Embryo Viability of Prolonged Incubation in Insemination Medium

Treatment	Number of patients treated	Number of patients pregnant (%)
Oocytes changed to fresh medium[a]	196	28 (14.3%)
Oocytes left in insemination medium[b]	195	36 (18.5%)

[a]12–18 hr after insemination
[b]30–50 hr after insemination

been examined. It would probably be erroneous to assume that fertilization and culture conditions are optimized at present because the calculated viability of human embryos (Rogers *et al.*, 1986) is about one-half that calculated by Short (1979) for embryos fertilized *in vivo*.

4.3. Embryo Viability

The techniques used to assess human embryo viability have been reviewed by Mohr (1984). These techniques fall within two broad categories, invasive and noninvasive tests. Invasive tests result in the destruction of all or part of the embryo, but these tests are a more informative than noninvasive tests. There are ethical objections to the destruction of embryos or to tests that may themselves reduce embryo viability. This raises major difficulties in determining the underlying mechanisms that reduce embryo viability and prevent the development of more acceptable noninvasive tests of viability. There are many morphological and ultrastructural abnormalities in embryos classified by routine IVF assessment as normal and abnormal (Trounson and Sathananthan, 1984). Chromosomal defects may also occur in otherwise apparently normal 8-cell embryos (Angell *et al.*, 1983), and there may be defects in gene expression, as assessed by protein synthetic capacity, of otherwise normally cleaving embryos. An example of such defects in otherwise apparently normal cattle embryos is shown in Figure 3 (G. M. Warnes and A. Trounson, unpublished data). The development of noninvasive tests that identify or correlate with these defects should be a priority for IVF research. If we are unable to address these matters, we shall have to accept the relatively low human embryonic viability apparent in human IVF (Wood *et al.*, 1985; Rogers *et al.*, 1986).

The noninvasive fluorescein diacetate test (Mohr and Trounson, 1980) does have some relevance for the assessment of human embryo viability (Mohr, 1984) but relates only to intracellular esterase enzyme activity. Other noninvasive tests, which are primarily based on light microscopic assessment, have only a minor association with embryo viability (Mohr, 1984).

Microsurgical biopsy of early cleavage embryos, which will permit retention of embryonic viability and provide cells for detailed evaluation of function and normality, may be the ideal compromise for assessing embryonic viability.

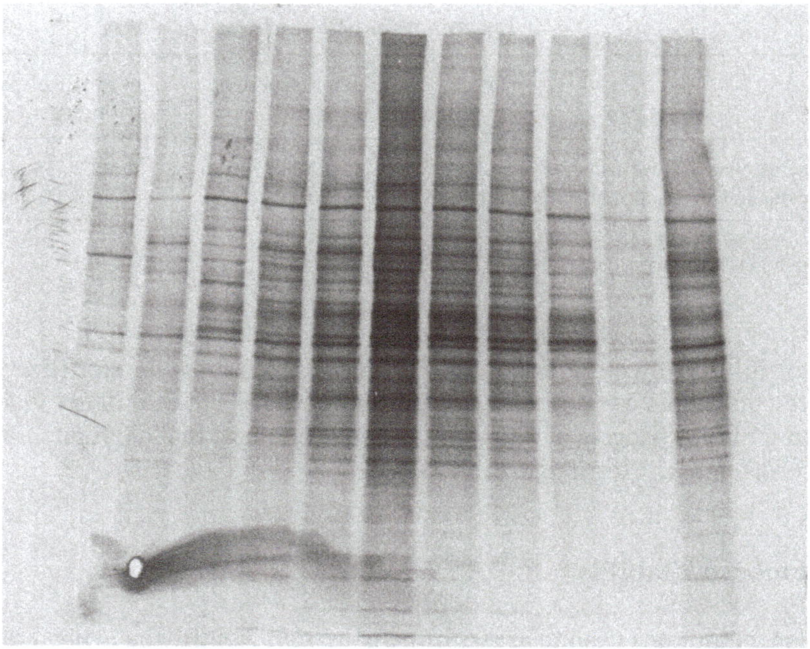

Figure 3. Autoradiograph of an SDS gel of [^{35}S]labeled proteins from 11 8-cell cow embryos flushed from a single donor animal. There are marked differences in the protein profiles and degree of [^{35}S]label incorporation between individual embryos that all appeared morphologically normal. (From G. M. Warnes and A. O. Trounson, unpublished data.)

Invasive techniques could be used on the biopsied cells and the developmental capacity of the biopsied embryo determined by replacement *in utero*.

5. Embryo Replacement *in Utero*

The replacement of embryos *in utero* is probably the least defined and least understood process in human IVF. Most patients' transferred embryos (70–80%) do not become pregnant; the reasons for this have not been elucidated. Low embryonic viability (20–30%) and varying states of uterine receptivity (38–60%) are thought to interact to produce the overall success rate of establishing pregnancy (Rogers *et al.*, 1986). The true values for both factors would be altered if miscarriage were considered (could be either embryonic or uterine) and the efficiency of the embryo replacement procedure were known. Some embryos may not be placed properly *in utero* or may be expelled from the uterus because of the transfer procedures or the state of the uterus.

5.1. Stage of Embryo Development

There appears to be very little effect of the stage of development (2–8-cell) on the success rate of embryo transfer (Trounson, 1983*a*; Edwards *et al.*, 1984). A reduced pregnancy rate appears to occur if pronuclear embryos (Trounson, 1983*a*; Edwards *et al.*, 1984) and oocytes (Craft *et al.*, 1982) are transferred to the uterus. Information on the success rate of transfer of later-stage embryos (16 cells to blastocysts) is sparse since the demonstration of successful replacement of 8-cell embryos (Edwards *et al.*, 1980) and 2–4-cell embryos (Trounson *et al.*, 1981*a*,*b*).

It should be possible to be more selective about the morphological quality of embryos as cleavage progresses. Embryos that are retarded in cleavage or those arrested in cleavage can be discarded because they are not viable or are of low viability. The assessment of morulae may be very difficult because the human embryo compacts at the late 8-cell or 16-cell stage. However, at least up until compaction, pregnancy rate should rise with the replacement of successive cleavage stages. This does not appear to be the case when confounding factors such as the number of embryos transferred are taken into account (Edwards *et al.*, 1984). This observation is in agreement with our own data. The failure to observe increasing pregnancy rate with advancing cleavage stage, or time after fertilization, may be due to decreased viability of embryos with prolonged culture *in vitro*, as observed in some other species (Trounson, 1984*a*). These two factors may act to balance each other. Embryos are generally replaced *in utero* at the 2–8-cell stage.

5.2. Number of Embryos Replaced

The number of embryos replaced is undoubtedly one of the most influential factors in determining pregnancy rate in IVF (see Table IV). Almost every study reported in IVF concludes that pregnancy rate increases with the number

Table IV. Pregnancy Rate from Multiple Embryo Replacement

Number of embryos replaced	Pregnancy rate	
	Helsinki[a]	Cargese[b]
1	9.5	8.2
2	14.6	13.9
3	19.3	16.1
4	24.1	19.4
5	—	21.5

[a]Unpublished data, International IVF Congress, Helsinki (1984).
[b]Unpublished data, INSERM IVF Congress, Cargese (1984).

of embryos replaced. Wood *et al.* (1985) have proposed that this factor, more than any other factor analyzed, controls the final pregnancy rate. The increase in pregnancy rate with increasing numbers of embryos transferred varies from a minor increase of 2–4% with each additional embryo transferred (Garcia *et al.*, 1984) to about 10% for each additional embryo (Edwards *et al.*, 1984; Wood *et al.*, 1985).

Some IVF groups, particularly those in the United States, replace all the fertilized embryos on the same occasion. This can result in up to 10 embryos being replaced on a single occasion. It is perhaps extremely fortunate that quintuplet and higher-order multiple pregnancies have not occurred or at least have not been reported as yet. Two sets of quadruplets have occurred from the transfer of four and six embryos. In our own IVF program (Wood *et al.*, 1984), four sets of triplets have been delivered from the transfer of three (three cases) and four embryos (one case). We believe that the documented risks to the mother and the infants, as well as the social difficulties encountered with triplet and larger multiple pregnancies, should restrict the numbers of embryos replaced at one time to no more than three embryos. The preservation of embryos by freezing, enabling replacement on another occasion (Trounson and Mohr, 1983; Mohr *et al.*, 1985), has been found to be an attractive alternative to replacement of large numbers of embryos (Wood *et al.*, 1984). This procedure effectively increases pregnancy rate per laparoscopy by about another 10% (Trounson *et al.*, 1985*a*) (see Figure 4).

5.3. Method of Embryo Replacement

A variety of procedures and instrumentation have been used for embryo transfer, most of which have been reviewed by Johnston (1982); Leeton and

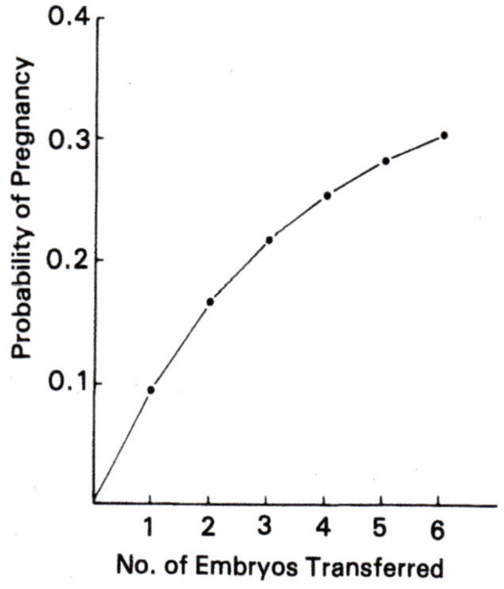

Figure 4. The probability of pregnancy according to the number of embryos replaced *in utero*, calculated by Rogers *et al.* (1986). The single spot above this curve is the calculated pregnancy rate when five embryos are obtained but three embryos are replaced in the cycle of treatment and the remaining two frozen-thawed and replaced in a subsequent natural ovulatory circle. (From Trounson *et al.*, 1985*a*.)

Kerin (1984), Leeton (1984), and Edwards (1984). This is no universal agreement on the optimum procedure or instrumentation for embryo replacement. The variety of instruments and assertions as to their effacacy, are largely based on personal experience. With a few exceptions (Leeton *et al.*, 1982), controlled trials are almost nonexistent in this area. Until such time as properly controlled trials comparing instruments and procedures are reported, firm conclusions cannot be made. Some of the factors that vary considerably include patient position at replacement, relaxants and drugs used at replacement, type and material composition of catheters, method of passing through the cervical os, volume of fluid and air used to deliver the embryos, site of expulsion of embryos, and postembryo replacement treatment of the patient. Embryo replacement under ultrasonic guidance is being evaluated in our own laboratory in an attempt to standardize the site of deposition of embryos in the uterus.

There was some speculation that the transfer of embryos in high serum (75 or 98%) concentrations (Leung *et al.*, 1984) may be of benefit to pregnancy rate. Edwards *et al.* (1984) found no significant difference in pregnancy rate when embryos were transferred in 15 or 75% serum. Similar results were found in our own laboratory.

6. Early Pregnancy

When compared with other techniques for treatment of infertility, such as induction of ovulation, artificial insemination with donor semen, and tubal microsurgery, IVF is relatively inefficient. The actual birth rate per treatment cycle is on the order of 9–15%. When it is remembered that two or more embryos are normally replaced at the one time, rather than the single oocyte ovulated under normal circumstances, there is a considerable potential to increase pregnancy rates in IVF. Many factors have been considered in reviews on the progress of IVF, but there are very few hard data to indicate where the research effort should be concentrated.

6.1. Uterine Factors in the Establishment of Pregnancy

Adverse conditions in the uterus may prevent normal embryo development or implantation. Calculations by Rogers *et al.* (1986) certainly suggest that uterine receptivity for continued embryo development may be a major factor preventing the attainment of higher pregnancy rates in IVF. Maternal age, for example, has a major effect on pregnancy rate of embryos transferred (Table V), probably due to a small increase in embryonic genetic defects and a substantial reduction in uterine receptivity. The physiological basis for decreased uterine receptivity has not been elucidated. In some patients there appears to be a considerable variation in endometrial histology inconsistent with the stage of the ovulatory cycle (Jones, 1983; Seegar-Jones, 1984). Relating changes in uterine morphology to actual pregnancy in IVF is a major difficulty because endo-

Table V. Maternal Age and Pregnancy Rate per Embryo Replacement[a]

Maternal age (years)	Pregnancy rate per embryo transfer (%)
<30	14.3
30–34	12.8
35–39	12.2
>40	4.3

[a]Unpublished data, INSERM IVF Congress, Cargese (1984).

metrial biopsy and embryo replacement in the same ovulatory cycle are considered undesirable for the establishment of pregnancy. In the only reported study to date of simultaneous endometrial biopsy and embryo replacement (Abate *et al.*, 1984), pregnancies occurred only when the histological findings were consistent with the expected stage of the ovulatory cycle. More detail of the structure and function of the uterine endometrium is needed under various conditions in IVF to determine the receptive and nonreceptive states of the uterus for embryo growth, development and implantation. Findlay (1984) recently reviewed our present understanding of human uterine receptivity and implantation.

The effect of the antiestrogen, clomiphene citrate, on uterine function is of some concern in IVF stimulation protocols that employ this drug. There is no obvious effect on the surface morphology of the endometrium (Sundstom *et al.*, 1983), but the reduced estrogen and progestin cytosol receptor concentrations reported in menopausal women receiving estrogen replacement therapy (Kokko *et al.*, 1981) raise some questions as to the possible interference of clomiphene in implantation. The apparent difference in calculated uterine receptivity between IVF programs using hMG alone and in combination with clomiphene has been suggested to be due to the antiestrogenic qualities of clomiphene (Rogers *et al.*, 1986).

The levels of peripheral plasma estrogen in both the follicular phase and luteal phase and progesterone in the luteal phase are markedly elevated in stimulated IVF patients (Trounson, 1984*a*) because of multiple follicular growth and multiple corpora lutea. The effect of such elevated steroid levels on uterine function has not been addressed. Elevation of progesterone levels in the 36 hr before the onset of the LH surge does not appear to affect the establishment of pregnancy (Trounson and Calabrese, 1984) but may hasten endometrial changes, producing advanced morphological changes in this tissue (Jones, 1983). An advanced state of the uterine endometrium has been argued to be beneficial for IVF success rate (Seegar-Jones, 1984), although little evidence is available to support this view.

6.2. Exogenous Pregesterone Support of the Luteal Phase

The use of progesterone supplementation of the luteal phase in IVF therapy is based on reported corpus luteum dysfunction in nonhuman primates (Kreitmann *et al.*, 1981) following aspiration of the single dominant follicle. However, reduced progesterone or estrogen secretions in the luteal phase could not be demonstrated by Kerin *et al.* (1984) in IVF patients receiving clomiphene or clomiphene and hMG. These researchers were unable to detect any effect of ovarian stimulation on cycle or luteal phase length (Kerin and Seamark, 1984). However, reports that IVF cycles induced with hMG alone may be shorter (Olson *et al.*, 1983) and of some degree of luteal insufficiency seem to be based on very flimsy evidence (Dlugi *et al.*, 1984). Garcia *et al.* (1984) were unable to detect any insufficiency in the luteal phase of their patients, except in a few individual cases. This may indicate that progesterone supplementation could be advantageous in ovarian cycles of the occasional patient stimulated with hMG, but not when clomiphene is included in the superovulation schedule. A randomized controlled trial of daily progesterone supplementation during the luteal phase of patients stimulated with clomiphene and hMG for IVF showed no significant improvement in pregnancy rate (Leeton *et al.*, 1985). In this trial, progesterone supplementation was from the seventh to sixteenth day after laparoscopy, as endogenous progesterone levels for the first 7 days after laparoscopy are very high (Trounson, 1983*a*). The intention of this study was to ensure adequate progesterone coverage during the implantation and postimplantation period (days 8–14), enabling the initiation of embryonic hCG secretions. In patients with previously documented luteal phase defects, supplementation with progesterone is probably beneficial (Leeton *et al.*, 1985).

6.3. Detection of Early Pregnancy

One of the most disappointing areas of human IVF is the high rate of the failure to detect any indication of pregnancy in patients' transferred embryos (Table VI). The initial detection of embryonic hCG is usually 10–16 days after embryo replacement *in utero*. It was suggested that the enigmatic early pregnancy factor (EPF) could be detected in the culture solutions containing the dividing human embryos (Smart *et al.*, 1981) and that EPF may then be detected in the patient's blood after embryo transfer. This suggestion has not received any further interest in the literature and can only be classed as speculative.

Another early pregnancy test has been reported by O'Neil *et al.* (1985), which is based on a reduction in platelet counts in the blood of patients' transferred embryos. This interesting development needs confirmation in properly controlled studies. Should this approach be confirmed by other workers, we may be able to determine whether embryos continue to develop *in utero* but fail to implant, because of embryonic abnormality or reduced uterine recep-

Table VI. Pregnancy Statistics in the Monas[redacted]—IVF Program, 1983–1984[a]

	1983	1984
Number of laparoscopies	518	607
Number of embryo replacements	439 (85%)	529 (87%)
Number of pregnancies		
Biochemical	3	9
Abortions	13	26
Ectopic	2	5
Deliveries	56 (13%)	63 (12%)
Total	74 (17%)	103 (19%)

[a]1979–1982 data have been reported by Trounson and Wood (1984).

tivity. We have thus far been unable to ascertain whether embryo wastage is due to failure to replace embryos properly *in utero*, failure of continued embryo development, or failure of implantation.

6.4. Biochemical Pregnancy

The close monitoring of endocrine changes following embryo replacement in IVF provide the opportunity to identify pregnancies that do not proceed to ultrasonically identifiable clinical pregnancies. These biochemical pregnancies are variably included or excluded from reported IVF success rates. This has led H. W. Jones *et al.* (1983*b*) to question the validity of some reporting on IVF pregnancies and success rates. The general trend is now to report biochemical pregnancies separately to miscarriage and ongoing pregnancies or births, or they are excluded from pregnancy and success-rate data. The former is probably a more informative way to present IVF data because the proportion of biochemical pregnancies, miscarriages, and births permits an assessment of the state and progress of an IVF program. Frequently IVF programs begin with a high proportion of biochemical pregnancies, which then decreases as the technical procedures are optimized. The occurrence of an unusually large number of biochemical pregnancies may indicate an unforeseen alteration in either procedures or the quality of materials used in IVF.

Biochemical pregnancy is characterized by a prolonged luteal phase and the appearance of hCG in blood or urine. The concentration of hCG may initially follow the increase seen in normal pregnancy but normally plateaus well below maximum levels observed in normal pregnancy and then rapidly declines. This is followed by a rapid decline in progesterone levels and the initiation of menstruation. It is not known whether these pregnancies represent lethal genetic abnormalities or embryonic developmental abnormalities, such as the classic trophoblastic vesicle or blighted ovum, in which the embryonic component fails to develop but the trophoblast component does.

The sensitivity of the hCG assay used and the monitoring schedule employed will affect the proportion of biochemical pregnancies detected. This leads to a discrepancy in the proportion of these pregnancies observed between IVF programs. All the IVF programs operating in Australia provide the National Perinatal Statistics Unit with data on all aspects of IVF pregnancies, including biochemical pregnancies. The incidence of biochemical pregnancies as a percentage of total IVF pregnancies from this source (1980–1983) was 18% (41 of 233). In our own IVF program, the percentage has varied year to year from 4% to 9% (Trounson and Wood, 1984) (see Table VI).

6.5. Miscarriage

Spontaneous abortion occurs most frequently at 7–10 weeks gestation. The Australian Perinatal Statistics Unit report (1980–1983) that this occurs in 20% of all IVF pregnancies, or 24% if biochemical pregnancies are excluded from the calculations. There is very little information on the chromosomal and structural abnormalities of the abortuses. In our own program almost one-half the successfully karyotyped abortuses were chromosomally abnormal, with trisomy and monosomy of the D group chromosomes being the most frequent abnormality detected.

The spontaneous abortion rate may be slightly higher than that expected in natural conception, possibly related to the fact that these are the initial data obtained in the early development phase for most IVF clinics or to other factors such as infertility or age.

6.6. Ectopic Pregnancy

Ectopic pregnancy is a concerning side effect of IVF that occurs at an alarming incidence. In the Australian National Perinatal Statistics Report (1980–1983), the incidence was 5.1% or 6.3% when biochemical pregnancies were excluded. In some reports ectopic pregnancy has been as high as 11% (Lopata, 1983). In our own program, 4% of pregnancies have been ectopic (Table VI). The reasons for ectopic pregnancy occurring in IVF are unclear. Either embryos must be expelled into the tube or they migrate from the uterus into the tube. A detailed analysis of 10 ectopic pregnancies in our own IVF program has revealed no endocrine factor, infertility category, or procedure consistently related to the occurrence of this condition. However, in 9 of the 10 cases, the patients had previously diagnosed tubal occlusion or had pathological evidence of tubal infection at the time of embryo replacement. In one patient, no tubal pathology was found.

It is necessary to define the predisposition of IVF patients to ectopic pregnancy and to design treatments to minimize the occurrence. Normally functioning tubes would deliver any embryos back into the uterus if they were accidently placed in the tubes. Multiple ovulation and multiple corpora lutea may affect tubal function through increased relaxin secretion. In addition, tubal infection prevents normal tubal function. Prevention of embryo migration into the tubes by the correct placement of the transfer catheter and the judicious use of appropriate drugs to prevent uterine contraction may need to be considered. The removal of tubes has been recommended in the past as a solution to this problem but has not been adopted clinically in IVF.

6.7. Birth

According to the Australian National Perinatal Statistics Unit Report (1980–1983), equal numbers of male and female infants were born as a result of IVF. One-third of the babies were delivered by cesarean section. The percentages of single, twin, and triplet births were 78, 19, and 4%, respectively. A higher incidence of preterm births was reported (22%), and even in single pregnancies preterm delivery (19%) was about three to four times higher than in non-IVF pregnancies (5–8%). Of 175 live and stillborn babies, 57 (32.6%) were low birth weight (less than 2500 g)—a much higher incidence than is normally recorded in non-IVF pregnancies (6%). In single pregnancies 19% were of low birth weight, in twin pregnancy 42% were of low birth weight, and all triplets were in this category. Of the 180 births, four were stillborn (2.2%), which is about three times higher than expected. Two congenital abnormalities were noted (transposition of the great vessels and dislocation of the hip), which is no higher than other pregnancies.

As a total of all IVF pregnancies, 56% resulted in live births, or 68% excluding biochemically induced pregnancies. The increased incidence of premature delivery and of low birth weight deserve closer scrutiny. Factors that need to be considered include higher maternal age, cause of infertility, maternal endocrine and social factors, obstetrical complication, induction of labor, maternal stress, and multiple pregnancy. The concerns of increased congenital malformations raised as criticism of the development of IVF do not appear to be occurring. However, the occurrence of ectopic pregnancy, premature delivery, and low birth weight, and the possible increase in stillborn infants, requires the attention of IVF and prenatal researchers.

7. Egg Donation

The development of procedures to enable egg or embryo donation has been an important new application of IVF in the treatment of human infertility. Many women are unable to provide their own oocytes because of primary ovarian failure or inaccessible ovaries or do not wish to pass on inheritable genetic disorders to their offspring. These women have no other infertility treatment available to them and would normally choose not to have a family or adopt children if this were possible. IVF with donated oocytes enables the husband to participate genetically and enables the wife to carry a normal pregnancy. This degree of participation in family formation is often very attractive for such infertile couples. The donation of oocytes and embryos in IVF was recently reviewed (Trounson, 1984b; Trounson *et al.*, 1985*a*,*b*).

7.1. Conditions of Infertility Relevant to Egg Donation

These may be broadly classified into two main treatment types: (1) women with functional ovaries who ovulate and menstruate without exogenous hormone therapy (see Fig. 5), and (2) women with primary ovarian failure who are unable to ovulate or who rarely ovulate and who menstruate only with exogenous hormone therapy (see Fig. 6).

The first category includes a number of women with a familial history of inheritable genetic disease, particularly those disorders that are difficult or that cannot be diagnosed by prenatal techniques (see Fig. 5) Conditions include Huntington's chorea, Tay-Sachs disease, and fibrocystic disease of the pancreas. Many patients have chosen elective sterilization to prevent pregnancy involving their own oocytes. Other indications in this category include inaccessible ovaries, where pelvic disease or prior surgery has resulted in adhesions and scar tissue or close proximity of the ovaries to the bowel or other vital organs and major blood vessels. In these patients, the risk to the patient of attempted follicular needle aspiration is too high. The introduction of ultrasonically guided oocyte recovery has reduced the number of patients in this category. There may also be patients who have developmental abnormalities of

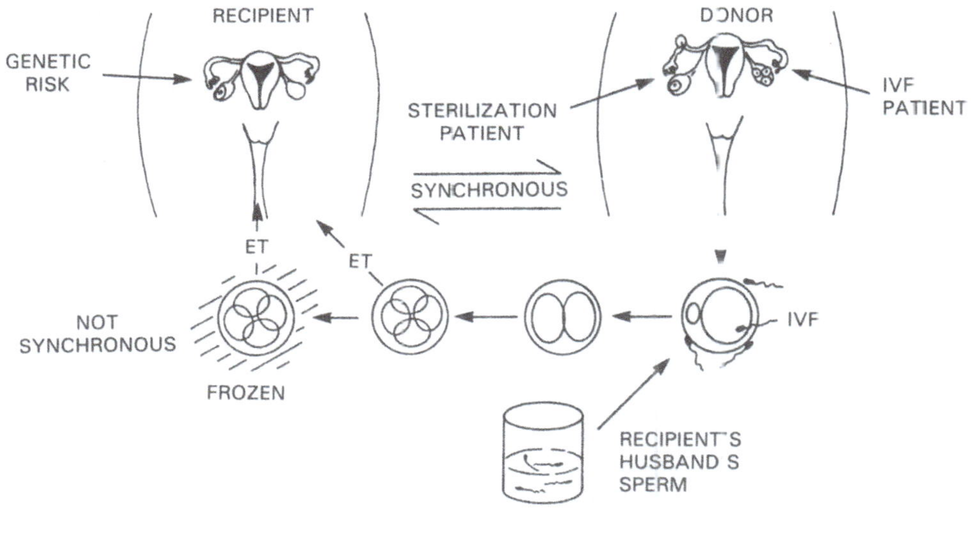

Figure 5. Egg donation to women with functional ovaries. If the donor and recipient are ovulating within 24–36 hr of each other, embryos are transferred (ET) directly. Embryo freezing may be used to enable embryo transfer at the correct time if ovulation in the donor and recipient is not synchronous.

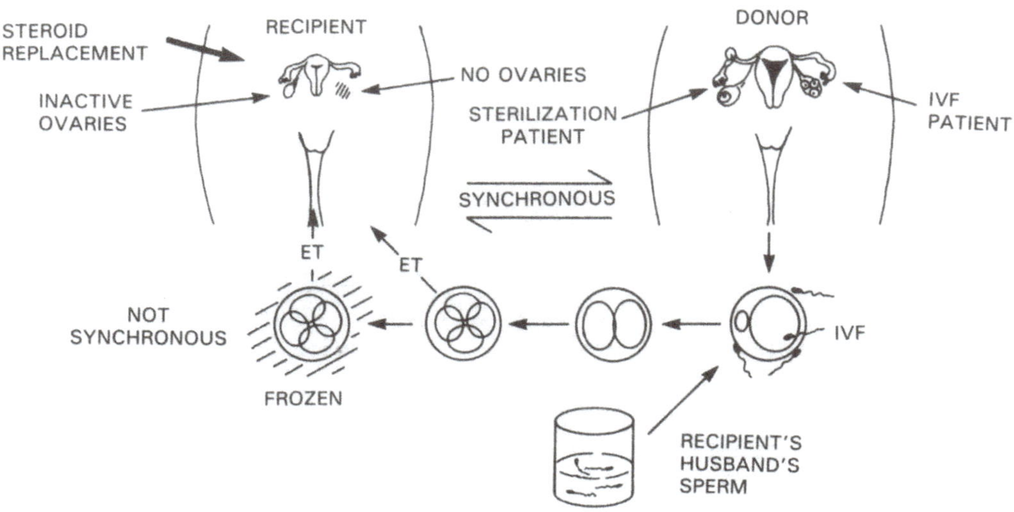

Figure 6. Egg donation to women who have primary ovarian failure. Steroid replacement therapy is required to imitate a uterus receptive to the transformed embryos (ET). Embryo freezing may also be used to enable ET at the correct stage of the artificial menstrual cycle, although steroid replacement can more easily be manipulated to achieve synchrony with the donor.

their oocytes. For example, we observed a patient who had a congenital absence of cortical granules in her oocytes, which resulted in polyspermic fertilization *in vitro*, and presumably *in vivo*. Another patient has oocytes with an abnormal zona pellucida, which dissolves when exposed to sperm *in vitro*, and presumably *in vivo*. Other developmental abnormalities in embryo development that prevent the establishment of pregnancy may exist as well, including genetic and chromosomal abnormalities derived from the germ cells. However, these abnormalities are more difficult to determine with the present embryological techniques and the growing government attitudes toward the prevention of research on human embryos.

In the second category of patients with primary ovarian failure (Fig. 6), indications would include premature menopause, congenital absence of ovaries, surgical ovariectomy, loss of ovarian function following radiotherapy or chemotherapy, severe endometriosis involving the ovary, and patients with Turner's syndrome and streak ovaries arising from other causes. A potentially normal uterus is necessary, which can respond to steroid replacement therapy.

7.2. Synchronization of Ovulation for Embryo Transfer

Procedures used for synchronization of a donor and recipient necessary to establish pregnancy have been described by Trounson *et al.* (1983*b*). Approximate synchronization of ovulation may be achieved by the extension of the recipient's or donor's luteal phase with progesterone in the ovulatory cycle before the cycle of transfer (Templeton *et al.*, 1984). The time of expected ovulation may be calculated for both the donor and recipient (McIntosh *et al.*, 1980), and the time of the onset of menstruation for both may be determined to achieve ovulation on about the same day. The actual time of ovulation can then be controlled by administration of hCG.

Alternatively, recipients who are tracked for the day of their spontaneous LH surge (Trounson *et al.*, 1983*b*) may coincide by chance with the time of recovery of an anonymously donated oocyte. A frozen sample (Mahadevan *et al.*, 1983*b*) of the recipient's husband's semen is normally available to inseminate these donated oocytes.

Other methods of synchronization of ovulation could involve the use of GnRH analogues (Kemeter *et al.*, 1984; Porter *et al.*, 1984), exogenous progesterone, or complete steroid replacement (Lutjen *et al.*, 1984). Progress in our research on these procedures, and on egg donation in general, has been hampered by a government moratorium (1983), the almost complete change of attitude toward donating oocytes by normal IVF patients following the development of successful embryo freezing techniques, and the requirement to keep on record in the Victorian Health Commission the donor's name and other particulars. In the State of Victoria the latter has been made mandatory by law.

Alternative sources of oocytes for donation are being explored, including family or friends of the recipient couple who volunteer to donate oocytes, women undergoing elective sterilization who consent to donate oocytes, and women who volunteer to donate oocytes anonymously because of the plight of these patients. Specific counseling and guidance are needed for donors and recipients under these circumstances. We believe that counseling should also extend to the children and families resulting from these associations.

7.3. Steroid Hormone Replacement

Two completed pregnancies have now been described by Lutjen *et al.* (1984; 1985*b*) in which total ovarian agenesis required oocyte donation and embryo transfer as well steroid replacement therapy to establish and maintain pregnancy. A third patient has also completed pregnancy following the same treatment in our IVF clinic. The success rate of this procedure (23% pregnancy and birthrate) is at least as good as, or possibly better than, that achieved in the routine IVF program. Single embryos are transferred to these patients.

This remarkable program was based on research carried out into the establishment and maintenance of pregnancy in the sheep (Trounson and Moore, 1974; Moore *et al.*, 1983). The steroid replacement schedule that was chosen closely mimics normal ovulatory cycle changes in endogenous 17β-estradiol, progesterone, and LH (P. Lutjen *et al.*, 1986). By contrast, FSH remained above the normal range even after repeated cycles of steroid therapy. This observation supports the notion that ovarian control of FSH secretion arises from the follicular components of the ovary. This is probably the peptide inhibin, which is found in follicular fluid (Tsonis *et al.*, 1983). Other interesting observations have also been made on these patients, including evidence to support a role for progesterone in facilitating the midcycle FSH surge. It has also been noted in pregnant patients that there is a complete absence of changes that normally occur at the time of expected labor (Lutjen *et al.*, 1985*b*). No cervical softening or elongation of the inner uterine segment occurred, no contractions were reported, and all the patients had undeveloped pelvices. We are examining human relaxin and other endocrine levels in these patients, particulary around the time of expected parturition. Since the possibility of vaginal delivery for these patients is remote, caesarean section will probably always be necessary.

We have come to several conclusions on the basis of these studies:

1. Of the ovarian hormones, only estradiol and progesterone are required for the establishment and maintenance of pregnancy.
2. Ovarian estrogen and progesterone are only required for a maximum of the first 12–19 weeks of pregnancy, respectively. Thereafter, placental estrogen and progesterone are produced in sufficient quantities to maintain pregnancy.
3. Any other ovarian derived factors, e.g., early pregnancy factor (Cavanagh *et al.*, 1982), are not required for embryo growth and development, implantation, or pregnancy. However, ovarian-derived factors may be necessary to achieve normal labor and parturition.

This research program has much to offer in terms of clinical therapy for primary ovarian failure and in terms of valuable insight into the control of the normal reproductive process.

8. Embryo Cryopreservation

The principles of human embryo freezing were recently reviewed by Trounson (1984*a*) and Mohr and Trounson (1985). The organization required to integrate embryo freezing into a routine IVF program was described by Trounson and Freemann (1986). In our own IVF clinic, embryo freezing and thawing is now a routine clinical component (Mohr *et al.*, 1985; Trounson and Freemann, 1986), although we are continuing our research program to find more efficient and technically simpler procedures.

Having been instrumental in the introduction of superovulation techniques in human IVF (Trounson *et al.*, 1981*a*), we presented the argument to our hospital ethics committee that the development of a cryopreservation technique for patients is an ethical obligation (Trounson *et al.*, 1982*a*). Freezing would enable the patients to preserve embryos in excess of their immediate requirements for their own use at a later time. This removes the necessity to (1) dispose of embryos; (2) replace large numbers of embryos on the one occasion, which risks the chance of quadruplet or larger numbers of fetuses; (3) restrict the number of eggs collected or inseminated; and (4) restrict the availability of embryos for research purposes. Our ethics committee approved the project, and more recently the Victorian government approved the technique and set forth the requirements for operating a freezing program. They also detailed the alternatives for disposition of frozen embryos for patients who decline to have them returned or who divorce or die while embryos are still in cryostorage.

8.1. Techniques for Freezing and Thawing Human Embryos

The initial research program investigated two cryoprotectants (glycerol and dimethyl sulfoxide), two freezing regimens (rapid and slow), and three groups of cleavage stages (2-cell, 4-cell, 8-cell, and late morula to early blastocyst). Preliminary results of these studies have been reported by Trounson (1984*a*) and the morphological details of some of these thawed embryos described by Mohr *et al.* (1983). Embryo survival was obtained in all treatment groups, with high survival rates observed for morulae and early blastocysts. Nevertheless, the good results for 4- and 8-cell embryos deserved further investigation because it was believed that prolonged culture *in vitro* may reduce embryo viability, as reported in animal investigations (Trounson, 1984*a*).

The next series of experiments were reported by Trounson and Mohr (1983). These experiments investigated 21 4-cell and 8-cell embryos, again using glycerol and dimethyl sulfoxide (DMSO) and two freezing rates (slow and rapid). The best results were obtained with embryos frozen both slowly and rapidly in DMSO. Two pregnancies were reported in this paper following replacement of these embryos *in utero*. One pregnancy was described in detail and the other as a note added in proof. Both embryos were frozen at the 8-cell stage in DMSO cryoprotectant using the slow freezing and thawing method (Trounson and Mohr, 1983).

The results of these freezing experiments were expanded and updated by Mohr and Trounson (1985) and by Mohr *et al.* (1985). These studies showed that the slow freezing and thawing method with DMSO as the cryoprotectant gave the highest embryo survival rate after thawing. Survival rates of 8-cell embryos were higher than 4-cell embryos, but irregularities of blastomere shape and size had no effect on embryo survival. Pregnancies were obtained with 4-cell and 8-cell freeze-thawed embryos. A biochemical pregnancy occurred in one 3-cell embryo that had only two blastomeres intact after thawing.

Pregnancies were also established when 8-cell embryos had lost three to five of their eight original blastomeres after thawing. Despite apparently good survival of 4-cell embryos frozen in glycerol using the rapid freezing and thawing technique, no pregnancies have been initiated with this procedure. Zeilmaker *et al.* (1984) reported embryo survival rates of 53% using DMSO and a method very close to the rapid freeze-thawing technique described by Trounson and Mohr (1983). Our own results (Mohr *et al.*, 1985) showed a significantly higher embryo survival rate when 4-cell and 8-cell embryos are frozen in DMSO by the slow freezing method. However, the subtle differences in technique, cooling rates, and thawing rates used by Zeilmaker *et al.* (1984) and ourselves may have a significant effect on the outcome of freezing and thawing and invalidate any comparison.

A new generation of very simple freezing procedures are now evolving in animal studies; these are being examined in our own laboratory for the freezing of unfertilized oocytes and early cleavage stage human embryos. It is likely that these methods will replace current freezing procedures (Trounson and Mohr, 1983). Freezing the human oocyte prior to fertilization is attractive because of the ethical attitude that prefers cryostorage of eggs rather than of embryos. This would also be of more practical benefit in IVF because of patients' own preferences (donors and recipients) for donated eggs rather than embryos. This would apply in cases in which IVF patients are unable, or do not wish, to use all of the stored embryos themselves.

8.2. Success Rates of Embryo Cryopreservation

Survival of human embryos after freezing and thawing has been arbitrarily chosen as retention of 50% or more of the original number of blastomeres. Embryo survival rate after freezing and thawing under this definition is approximately 50% (Trounson and Mohr, 1983; Zeilmaker *et al.*, 1984; Mohr *et al.*, 1985). Pregnancy rate of transferred embryos was 20% (9 pregnancies from 45 transfers (Mohr *et al.*, 1985) to 25% (2 pregnancies from 8 transfers); (Zeilmaker *et al.*, 1984). The embryo viability computed from the number of embryos originally frozen was 7.3% and 13.3%, respectively, in the two studies. The addition of this figure to the probability of pregnancy from an IVF treatment cycle is substantial. This improvement in the success rate of IVF would be further increased if multiple embryos were frozen for the patient. The data presented by Trounson and Freemann (1986) show that pregnancy rate was 10%, 6%, 25%, and 17% for 1, 2, 3, and 4+ embryos frozen, respectively (21, 17, 4, and 6 patients in each respective group). As the number of patients increased in each group we would expect a progressive increase in pregnancy rate with each additional embryo transferred.

Few other developments, with the exception of superovulation, have had such an effect on the increase in success rate of IVF. Seven healthy babies have been born in our own program to date, and one infant was delivered stillborn after premature rupture of membranes and the development of intrauterine infection (Trounson and Mohr, 1983).

9. Male Factor Infertility and IVF

IVF has been considered a technique that could be therapeutic in the treatment of male infertility, particularly in the conditions of oligozoospermia, asthenozoospermia, and teratozoospermia, which are the least responsive male factor conditions to other forms of therapy. Pregnancies have certainly been obtained through IVF in such cases of male infertility (Mahadevan *et al.*, 1983*a*;

1985; Testart *et al.*, 1983; Cohen *et al.*, 1984; Yovich and Stanger, 1984; de Kretser *et al.*, 1985). However, there are major discrepancies in the conclusions reached as to fertilization rates and the success of IVF for treatment of male factor infertility. For example, Cohen *et al.* (1984) reported pregnancy initiation with semen from men with very low quality semen but provided very little information on their partners' infertility status, which has a major effect on fertilization rate (Mahadevan *et al.*, 1983*a*), or information on the total male factor population treated. The implication of this report is that male factor infertility is very successfully treated by IVF. Testart *et al.* (1983) reported reduced IVF success rates in oligozoospermia, asthenozoospermia, and teratozoospermia, while Mahadevan and Trounson (1984*a*) reported that when other factors were controlled, the percentage sperm motility and sperm abnormality affected fertilization rate but sperm concentration had no effect. Yovich and Stanger (1984) found that severe oligozoospermia ($<5 \times 10^6$ motile sperm/ml semen) reduced fertilization rates but moderate oligozoospermia ($6-2 \times 10^6$ motile sperm/ml semen) and teratozoospermia had no effect.

These divergent conclusions have produced considerable confusion as to the expectations of IVF in male factor infertility. The variance in opinion may be due to differences in the technique for preparing sperm from poor quality semen but is probably due mainly to the lack of consistent definition in the establishment of the various categories of male factor infertility. This has also been a problem in reporting other therapies employed in the treatment of male infertility and is the main reason that the World Health Organization drew up very specific guidelines (WHO, 1980) to permit consistent reporting between centers, particularly for multicenter trials. de Krestser *et al.* (1985) used these criteria and definitions to establish categories of male infertility prior to IVF treatment, from repeated single and multiple defects of semen quality. Their preliminary findings, updated by C. Yates (unpublished data) show that semen with combined defects of sperm motility, morphology, and concentration have the lowest fertilization rate (39–42%). Single defects of reduced motility (0–40% and 40–70%) had reduced a fertilization rate (64%) as compared with the fertilization rate in cases of normal males with wives who had blocked tubes (80%). A single defect of reduced sperm concentration ($<20 \times 10^6$ sperm/ml) also had a reduced fertilization rate (60%). These preliminary results show that within any category there is marked individual patient variation and that pre-assessment of male factor infertility may lead to different conclusions from those computed from classifications established from semen quality at the time of insemination (Mahadevan and Trounson, 1984*a*). However, these studies in the long term will provide a clearer assessment for the clinical indications and expectations for the treatment of different types of male factor infertility.

Yovich and Stanger (1984) reported that teratozoospermia is associated with retardation of embryo cleavage. This observation requires confirmation because of the possible association of sperm defects with embryo development, viability, and genetic normality.

Stimulation of fertilization rate of low-quality sperm with motility and capacitation stimulants has yet to be investigated in human IVF. A preliminary trial with hypotaurine and epinephrine (Mahadevan and Trounson, 1985) had no effect on fertilization rate when added to the insemination medium of either normal-quality or low-quality sperm.

9.1. Donor Sperm and Reinsemination in IVF

Donor sperm is frequently used in cases of azoospermia in which the wife has tubal infertility or has failed to conceive by artificial insemination with donor semen. The use of donor sperm for a diagnostic fertilization test, proposed by Trounson *et al.* (1981*b*), was studied by Mahadevan *et al.* (1983*a*) for idiopathic and male factor infertility as an alternative to the hamster-egg penetration test in an effort to establish the fertilizing capacity of sperm samples. In this test, oocytes obtained from the patient are divided into two groups and are inseminated with either husband's or donor sperm. In the case of couples with idiopathic infertility, both donor and husband's sperm succeeded in fertilizing oocytes on 67% of occasions. Only donor sperm fertilized oocytes 17% of occasions, and only husband's sperm on 3% of occasions. We concluded that this test should not be used in cases of idiopathic infertility because of the high success rate with husband's sperm. In the case of the male infertility cases chosen, fertilization of oocytes by both husband's and donor sperm occurred on only 9% of occasions, whereas on 82% of occasions only donor sperm fertilized the inseminated oocytes. These were severe cases of male factor infertility, and the test can be considered definitive for these patients.

An alternative to the diagnostic fertilization test, reinsemination of oocytes with either husband's or donor sperm, 12–14 hr after the original insemination with husband's sperm, was examined by Trounson and Webb (1984). Reinsemination with donor sperm gave a significantly increased fertilization than that with husband's sperm. However, in a control group of oocytes that were not reinseminated, spontaneously delayed fertilization occurred in 21% of all oocytes. The increase in fertilization rate with reinsemination of donor sperm (35%) was not significantly higher than spontaneously delayed fertilization. Moreover, only one pregnancy resulted from the transfer of 34 embryos following reinsemination, and this pregnancy aborted at 12 weeks gestation. No pregnancies occurred in oocytes with spontaneously delayed fertilization (18 patients). We concluded that this test was not as definitive as the diagnostic fertilization test and that reinsemination of oocytes is unlikely to increase IVF pregnancy rates significantly.

9.2. Epididymal Sperm and IVF

There are infertile men who have normal testicular function but who have obstructive azoospermia. Vasoepididymostomy has been disappointing as a procedure to overcome this type of infertility; we have investigated the use of

IVF for the fertilization of oocytes with epididymal sperm obtained from men with obstructive azoospermia.

Increased fertilizing capacity, as measured by penetration of zona-free hamster eggs, appears to occur as the source of epididymal sperm increases in distance from the testis (Moore *et al.*, 1983). In the first case attempted in our laboratory, sperm obtained from the body of the epididymis fertilized three of the six oocytes inseminated. However, fertilization was delayed, cleavage was retarded in one embryo, and the other two oocytes were arrested at the 1-cell pronuclear stage of development (Mahadevan and Trounson, 1985). In another two cases, fertilization did not occur. We recently described a successful pregnancy in a patient in whom one of five oocytes fertilized and a normal 2-cell embryo was returned to the wife's uterus (Temple-Smith *et al.*, 1985). A normal baby boy was delivered in this case.

There is a very reduced fertilization rate in this group of patients and in those who produce sperm after vasoepididymostomy (de Kretser *et al.*, 1985). We are investigating ways to increase this fertility rate through maturation of epididymal sperm and microsurgical insemination.

10. The Scope of Human IVF

The technique of IVF has already had profound effects on the treatment of human infertility and in the understanding of many aspects of the human reproductive process. The technique will continue to widen in its clinical application and to elucidate the endocrine, biochemical, genetic, and physiological processes governing ovarian function, conception, and pregnancy. It is difficult to predict the progress that is rapidly occurring, but it would seem that a number of advances will certainly occur in time. Some of these advances are presented in Figure 7 and may include the following:

1. IVF success rates have shown improvement. The isolated example of a 45–50% pregnancy rate reported in 25 patients by Fishel *et al.* (1984), which many IVF programs occasionally observe, may become the norm.
2. Genetic defects in human embryos may be routinely diagnosed at the preimplantation stage. Couples at risk of producing familial genetic defects may choose IVF in preference to prenatal diagnosis and therapeutic abortion.
3. The many and varied aspects of male infertility that do not respond to other therapies may be treated by IVF as chemical stimulants are located and as insemination and microsurgical procedures improve. The need for artificial insemination with donor semen will decline in parallel with these developments.
4. Techniques will be developed to cryopreserve the unfertilized human oocyte, and frozen egg banks will develop in parallel with sperm banks.
5. Superovulation techniques will simplify and improve as our under-

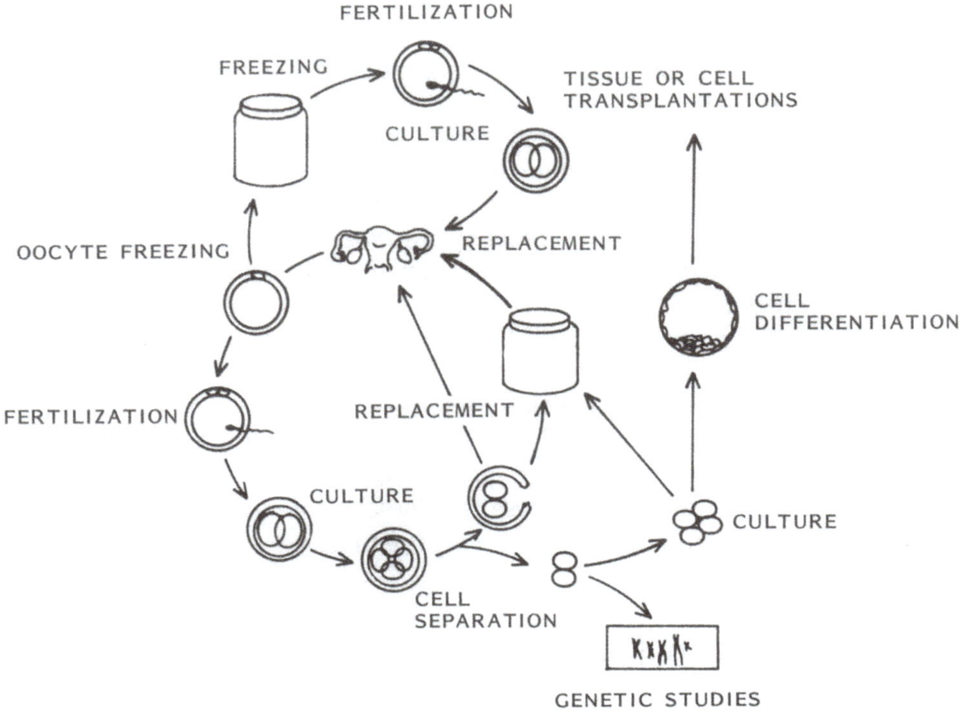

Figure 7. Future possible advances in IVF technology may include embryo biopsy for genetic analyses, egg preservation, and cell or tissue transplantation.

standing of follicle stimulation and growth improves, and as new drugs are developed and tested.

6. The basis for human fetal developmental defects will be elucidated, and it is possible that abnormalities such as neural tube defects may be prevented. At least some of this information is likely to derive from IVF and related technologies.
7. Edwards (1982, 1984) has argued that the primitive stem cells arising during the first few weeks of embryo development could be used for transplantation therapy. This proposal would seem unlikely to occur in the near future because this would need to be established in animal models first.

References

Abate, V., Call, A., and Stanchi, A., 1984, Endometrial biopsy at the time of embryo transfer: Correlation of histologic diagnosis with therapy and pregnancy rates, *Fertil. Steril.* **41:**43S (abst).

Ahuja, K. K., 1984, Inhibitors of glycoprotein biosynthesis block fertilization in the hamster, *Gamete Res.* **11:**179–189.

Angell, R. R., Aitken, R. J., van Look, P. F. A., Lumsden, M. A., and Templeton, A. A., 1983, Chromosome abnormalities in human embryos after in vitro fertilization *Nature (Lond.)* **303**:336–338.

Berger, T., Marrs, R. P., and Moyer, D. L., 1985, Comparison of techniques for selection of motile spermatozoa, *Fertil. Steril.* **43**:268–273.

Biggers, J. D., 1984, In vitro fertilization and embryo transfer in historical perspective, in: *In Vitro Fertilization and Embryo Transfer* (A. Trounson and C. Wood, eds.), pp. 3–15, Churchill Livingstone, Edinburgh.

Brackett, B. G., and Bousquet, D., 1984, Comparative aspects of fertilization in vitro, in: *In Vitro Fertilization and Embryo Transfer* (A. Trounson and C. Wood, eds.), pp. 32–43, Churchill Livingstone, Edinburgh.

Braude, P. R., and Bolton, V. N., 1984, The preparation of spermatozoa for in vitro fertilization by buoyant centrifugation, in: *Recent Progress in Human in Vitro Fertilization* (W. Feichtinger and P. Kemeter, eds.), pp. 125–134, Cofese, Palermo.

Bryce, R. L., Shutter, B., Sinosich, M. J., Stiel, J. M., Picker, R. H., and Saunders, D. M., 1982, The value of ultrasound, gonadotrophin, and estradiol measurements for precise ovulation prediction, *Fertil. Steril.* **37**:42–45.

Buttery, B., Trounson, A., McMaster, R., and Wood, C., 1983, Evaluation of diagnostic ultrasound as a parameter of follicular development in an in vitro fertilization program, *Fertil. Steril.* **39**:458–463.

Caro, C. M., and Trounson, A., 1984, The effect of protein on preimplantation mouse embryo development in vitro, *J. Vitro Fertil. Embryo Transfer* **1**:183–187.

Cavanagh, A., Morton, H., Rolfe, B., and Gidley-Baird, A. A., 1982, Ovum factor; a first signal of pregnancy? *Am. J. Reprod. Immunol.* **2**:97–101.

Clarke, I. J., and Cummins, J. T., 1982, The temporal relationship between gonadotropin releasing hormone (GnRH) and Luteinizing Hormone (LH) secretion in ovariectomized ewes, *Endocrinology* **111**:1737–1739.

Cohen, J., Fehilly, C. B., Fishel, S. B., Edwards, R. G., Hewitt, J., Rowland, G. F., Steptoe, P. C., and Webster, J., 1984, Male infertility successfully treated by in vitro fertilization, *Lancet* **1**:1239–1240.

Craft, I., McLeod, F., Bernard, A., Green, S., and Twigg, H., 1981, Letter. Sperm numbers and in-vitro fertilization, *Lancet* **2**:1165–1166.

Craft, I. L., Djahanbakheh, O., McLeod, F., Bernard, A., Green, S., Twigg, H., Smith, W., Edmonds, D. K., and Lindsay, K. S., 1982, Human pregnancies following intrauterine transfer of pre-ovulatory oocytes and sperm, *Lancet* **2**:1031–1033.

Crosby, I. M., and Moor, R. M., 1984, Oocyte maturation, in: *In Vitro Fertilization and Embryo Transfer* (A. Trounson and C. Wood, eds.), pp. 19–31, Churchill Livingstone, Edinburgh.

Crosby, I. M., Obsborn, J. C., and Moor, R. M., 1981, Follicle cell regulation of protein synthesis and developmental competance in sheep oocytes, *J. Reprod. Fertil.* **62**:575–582.

de Kretser, D. M., Yates, C. A., Leeton, J. F., Southwick, G. J., Temple-Smith, P. D., Trounson, A. O., and Wood, E. C., 1985, The use of in vitro fertilization in the management of male infertility, in: *Gamete Quality and Fertility Regulation* (R. Rolland, M. J. Heinemann, S. G. Hillier, and H. Vemer, eds.), pp. 213–223, Excerpta Medica, Amsterdam.

diZerega, G. S., and Hodgen, G. D., 1980, The primate ovarian cycle: Suppression of human menopausal gonadotropin-induced follicular growth in the presence of the dominant follicle, *J. Clin. Endoncrinol. Metab.* **50**:819–825.

Dlugi, A. A., Laufer, N., DeCherney, A. H., MacLusky, N. J., Haseltine, F. P., Polan, M. L., Mezer, H. C., Tarlatzis, B., and Naftolin, F., 1984, The periovulatory and luteal phase of conception cycles following in vitro fertilization and embryo transfer, *Fertil. Steril.* **41**:530–537.

Downing, B., 1984, Oocyte.pick-up, in: *Clinical in Vitro Fertilization* (C. Wood and A. Trounson, eds.), pp. 67–82, Springer-Verlag, Berlin.

Dravland, J. E., and Mortimer, D., 1985, A simple discontinuous Percoll gradient procedure for washing human spermatozoa, *IRCS Med. Sci.* **13**:16–17.

Dvorak, M., Tesarik, J., Pilka, L., and Travinik, P., 1982, Fine structure of human 2-cell ova fertilized and cleaved in-vitro, *Fertil. Steril.* **37**:661–667.

Edwards, R. G., 1981, Test-tube babies, *Nature (Lond.)* **293**:253–256.

Edwards, R. G., 1982, The case for studying human embryos and their constituent tissues in vitro, in: *Human Conception in Vitro* (R. G. Edwards and J. M. Purdy, eds.), pp. 371–388, Academic Press, London.

Edwards, R. G., 1984, Human conception in vitro: New opportunities in medicine and research, in : *In Vitro Fertilization and Embryo Transfer* (A. Trounson and C. Wood, eds.), pp. 217–250, Churchill Livingstone, Edinburgh.

Edwards, R. G., Bavister, B. D., and Steptoe, P. C., 1969, Early stages of fertilization in vitro of human oocytes matured in vitro, *Nature (Lond.)* **221**:632–635.

Edwards, R. G., Steptoe, P. C., and Purdy, J. M., 1970, Fertilization and cleavage in vitro of preovulatory human oocytes, *Nature (Lond.)* **227**:1307–1309.

Edwards, R. G., Steptoe, P. C., and Purdy, J. M., 1980, Establishing full-term human pregnancies using cleaving embryos grown in vitro, *Br. J. Obstet. Gynaecol.* **87**:737–756.

Edwards, R. G., Anderson, G., Pickering, J., and Purdy, J. M., 1982, A rapid assay for urinary LH in women using a simplified method of HI-Gonavis, in: *Human Conception in Vitro* (R. G. Edwards and J. M. Purdy, eds.), pp. 19–34, Academic Press, London.

Edwards, R. G., Fishel, S. B., Cohen, J., Fehilly, C. B., Purdy, J. M., Slater, J., Steptoe, P. C., and Webster, J. M., 1984, Factors influencing the success of in-vitro fertilization and embryo transfer, *J. In Vitro Fertil. Embryo Transfer* **1**:3–23.

Eppig, J. J., and Downs, S. M., 1984, Chemical signals that regulate mammalian oocyte maturation, *Biol. Reprod.* **30**:1–11.

Eppig, J. J., Freter, R. R., Ward-Bailey, P. F., and Schultz, R. M., 1984, Inhibition of oocyte maturation in the mouse: Participation of cAMP, steroid hormones, and a putative maturation-inhibiting factor, *Dev. Biol.* **100**:158–165.

Feichtinger, W., and Kemeter, P., 1984, Laparoscopic or ultrasonically guided follicle aspiration for in vitro fertilization?, *J. Vitro Fertil. Embryo Transfer* **1**:244–249.

Findlay, J. K., 1984, Implantation and early pregnancy, in: *In Vitro Fertilization and Embryo Transfer* (A. Trounson and C. Wood, eds.), pp. 57–72, Churchill Livingstone, Edinburgh.

Fishel, S. B., Edwards, R. G., and Evans, C. S., 1983, Human chorionic gonadotrophin secreted by preimplantation embryos cultured in vitro, *Science* **223**:816–818.

Fishel, S. B., Edwards, R. G., and Purdy, J. M., 1984, Analysis of 25 infertile patients treated consecutively by in vitro fertilization at Bourn Hall, *Fertil. Steril.* **42**:191–197.

Fisher, L. R., and Parker, N. S., 1984, Osmotic control of bilayer fusion, *Biophys. J.* **46**:253–258.

Garcia, J. E., Jones, G. S., Acosta, A. A., and Wright, G. L., 1981, Corpus luteum function after follicle aspiration for oocyte retrieval, *Fertil. Steril.* **36**:565.

Garcia, J., Acosta, A., Andrews, M. C., Jones, G. S., Jones, H. W., Jr., Mantzavinos, T., Mayer, J., McDowell, J., Sandow, B., Veeck, L., Whibley, T., Wilkes, C., and Wright, G., Jr., 1984, In vitro fertilization in Norfolk, Virginia, 1980–1983, *J. In Vitro Fertil. Embryo Transfer* **1**:24–28.

Gianfortoni, J. G., and Gulyas, B. J., 1985, The effects of short-term incubation (aging) of mouse oocytes on in vitro fertilization, zona solubility, and embryonic development, *Gamete Res.* **11**:59–68.

Gwatkin, R. B. L., 1977, *Fertilization Mechanisms in Man and Mammals.* Plenum Press, New York.

Healy, D. L., and Hodgen, G. D., 1984, Endocrinology of follicle growth, in: *In Vitro Fertilization and Embryo Transfer* (A. Trounson and C. Wood, eds.), pp. 75–93, Churchill Livingstone, Edinburgh.

Hodgen, G. D., 1982, The dominant ovarian follicle, *Fertil. Steril.* **38**:281–300.

Hoff, J. D., Quigley, M. E., and Yen, S. S. C., 1983, Hormonal dynamics at midcycle: A re-evaluation, *J. Clin. Endocrinol. Metab.* **57**:792–796.

Jaffe, L. F., 1983, Sources of calcium in egg activation: A review and hypothesis, *Dev. Biol.* **57**:364–374.

Johnson, I. W. H., 1982, The luteal uterus: Cellular interactions with blastocysts, and the damaging effects of catheters, in: *Human Conception in Vitro* (R. G. Edwards and J. M. Purdy, eds.), pp. 311–314, Academic Press, London.

Johnston, I., Lopata, A., Speirs, A., Hoult, T., Kellow, G., and Duplessis, Y., 1981, In Vitro fertilization—The challenge of the eighties, *Fertil. Steril.* **36**:699–706.

Johnston, I., Lopata, A., Nayudu, P., Speirs, A., McBain, J., Leung, P., Martin, M., and Gronow, M., 1984, Assessment of oocyte maturation and its application for in vitro fertilization, in: *Recent Progress in Human in Vitro Fertilization* (W. Feichtinger and P. Kemeter, eds.), pp. 207–230 Cofese, Palermo.

Jones, G. S., Garcia, J. E., and Rosenwaks, Z., 1984, The role of pituitary gonadotropins in follicular stimulation and oocyte maturation in the human, *J. Clin. Endocrinol. Metab.* **59**:178–180.

Jones, H. W., 1983, Factors influencing implantation and maintenance of pregnancy following embryo transfer, in: *Fertilization of the Human Egg in Vitro* (H. M. Beier and H. R. Lindner, eds.), pp. 293–306, Springer-Verlag, Berlin.

Jones, H. W., Acosta, A. A., Andrews, M. C., Garcia, J. E., Jones, G. S., Mantzavinos, T., McDowell, J., Sandow, B., Veeck, L., Whibley, T., Wilkes, C., and Wright, G., 1983*a*, The importance of the follicular phase to success and failure in in vitro fertilization, *Fertil. Steril.* **40**:317.

Jones, H. W., Jr., Acosta, A. A., Andrews, M. C., Garcia, J. E., Jones, G. S., Mantzavinos, T., McDowell, J., Sandow, B. A., Veeck, L., Whibley, T. W., Wilkes, C. A., and Wright, G. L., Jr., 1983*b*, What is pregnancy? A question for programs of in vitro fertilization, *Fertil. Steril.* **40**:728–733.

Kemeter, P., Bohnet, H. G., and Feichtinger, W., 1984, The control of feedback mechanisms during ovarian stimulation therapy, in: *Recent Progress in Human in Vitro Fertilization* (W. Feichtinger and P. Kemeter, eds.), pp. 33–44, Cofese, Palermo.

Kerin, J., and Seamark, R., 1984, Stimulation of ovulation for in vitro fertilization, in: *In Vitro Fertilization and Embryo Transfer* (A. Trounson and C. Wood, eds.), pp. 94–110, Churchill Livingstone, Edinburgh.

Kerin, J. F., Edmonds, D. K., Warnes, G. M., Cox, L. W., Seamark, R. F., Mathews, C. D., Yound, G. B., and Baird, D. T., 1981, Morphological and functional relations of Graafian follicle growth to ovulation in women using ultrasonic, laparoscopic and biochemical measurements, *Br. J. Obstet. Gynaecol.* **88**:81–90.

Kerin, J. F. P., Warnes, G. M., Quinn, P., Jeffrey, R., Godfrey, B., Hourigan, L., Broom, T. J., McEvoy, M., Johnson, M., and Cox, L. W., 1984, The effect of clomid induced superovulation on human follicular and luteal function for extra corporeal fertilization and embryo transfer, *J. Clin. Reprod. Fertil.* **2**:129–141.

Kokko, E., Janne, O., Kauppila, A., and Vihko, R., 1981, Cyclic clomiphene citrate treatment lowers cytozol estrogen and progestin receptor concentrations in the endometrium of post men-opausal women on estrogen replacement therapy, *J. Clin. Endocrinol. Metab.* **52**:345–349

Kovacs, G. T., Craft, I., and McLeod, F., 1984, Oocyte recovery, in: *In Vitro Fertilization and Embryo Transfer* (A. Trounson and C. Wood, eds.), pp. 189–196, Churchill Livingstone Edinburgh.

Kreitmann, O., Nixon, W. E., and Hodgen, S. D., 1981, Induced corpus luteum dysfunction after aspiration of the preovulatory follicle in monkeys, *Fertil. Steril.* **35**:671–675.

Leeton, J. F., 1984, Embryo transfer, in: *in Vitro Fertilization and Embryo Transfer* (A. Trounson and C. Wood, eds.), pp. 197–204, Churchill Livingstone, Edinburgh.

Leeton, J., and Kerin, J., 1984, Embryo transer, in: *Clinical In Vitro Fertilization* (C. Wood and A. Trounson, eds.), pp. 117–136, Springer-Verlag, Berlin.

Leeton, J., Trounson, A., Jessup, D., and Wood, C., 1982, The technique for human embryo transfer, *Fertil. Steril.* **38**:156–161.

Leeton, J., Trounson, A., and Jessup, D., 1985, Support of the luteal phase in in vitro fertilization programs: Results of a controlled trial with intramuscular Proluton, *J. Vitro Fertil. Embryo Transfer* **2**:166–169.

Lenz, S., 1984, Ultrasonic-guided follicle puncture under local anesthesia, *J. Vitro Fertil. Embryo Transfer* **1**:239–253.

Leung, P. C. S., Gronow, M. J., Kellow, G. N., Lopata, A., Speirs, A. L., McBain, J. C., du Plessis, Y. P., and Johnston, I., 1984, Sera supplement in human in vitro fertilization and embryo development, *Fertil. Steril.* **41**:36–39.

Liu, J. H., Durfee, R., Muse, K., and Yen, S. S. C., 1983, Induction of multiple ovulation by pulsatile administration of gonadotropin-releasing hormone, *Fertil. Steril.* **40**:18–22.

Lopata, A., 1983, Concepts in human in vitro fertilization and embryo transfer, *Fertil. Steril.* **40:**289–301.

Lopata, A., McMaster, R., McBain, J. C., and Johnston, W. I. H., 1978, In vitro fertilization of preovulatory human eggs, *J. Reprod. Fertil.* **52:**339–342.

Lopata, A., Johnston, W. I. H., Hoult, I. J., and Speirs, A. L., 1980*a*, Pregnancy following intrauterine implantation of an embryo obtained by in vitro fertilization of a preovulatory egg, *Fertil. Steril.* **33:**117–120.

Lopata, A., Sathananthan, A. H., McBain, J. C., Johnston, W. I. H., and Speirs, A. L., 1980*b*, Ultrastructure of the preovulatory human egg fertilized in vitro, *Fertil. Steril.* **33:**12–20.

Lopata, A., Kohlman, D. L., and Kellow, G., 1982, The fine structure of human blastocysts developed in culture, in: *Embryonic Development.* Part B: *Cellular Aspects* (M. M. Burger and R. Weber, eds.), pp. 69–86, Alan R. Liss, New York.

Lopata, A., Kohlman, D., and Johnston, I., 1983, The fine structure of normal and abnormal Human embryos developed in culture, in: *Fertilization of the Human Egg in Vitro* (H. M. Beir and H. R. Lindner, eds.), pp. 189–210, Springer-Verlag, Berlin.

Lutjen, P., Trounson, A., Leeton, J., Findlay, J., Wood, C., and Renou, P., 1984, The establishment and maintenance of pregnancy using in vitro fertilization and embryo donation in a patient with primary ovarian failure, *Nature (Lond.)* **307:**174–175.

Lutjen, P. J., Handley, C. J., de Witt, M. T., Trounson, A. O., and McBain, J. C., 1985*a*, Biochemical changes in cervical mucus factor infertility, *Gamete Res.* **12:**265–274.

Lutjen, P., Leeton, J., Trounson, A., Renou, P., Wood, C., and Findlay, J., 1985*b*, Pregnancy without ovarian function, *J. Vitro Fertil. Embryo Transfer* **2:**107–108.

Lutjen, P. J., Findlay, J. K., Trounson, A. O., Leeton, J. F., and Chan, L. K., 1986, Effect on plasma gonadotropins of cyclic steroid replacement prior to embryo transfer in women with premature ovarian failure, *J. Clin. Endocrinol. Metab.* (in press).

Mahadevan, M. M., and Baker, G., 1984, Assessment and preparation of semen for in vitro fertilization, in: *Clinical in Vitro Fertilization* (C. Wood and A. Trounson, eds.), pp. 83–97, Springer-Verlag, Berlin.

Mahadevan, M. M., and Trounson, A. O., 1983, Effect of cryoprotective media and dilution methods on the preservation of human spermatozoa, *Andrologia* **15:**355–366.

Mahadevan, M. M., and Trounson, A. O., 1984*a*, The influence of seminal characteristics on the success rate of human in vitro fertilization, *Fertil. Steril.* **42:**400–405.

Mahadevan, M. M., and Trounson, A. O., 1984*b*, Effect of cooling, freezing and thawing rates and storage conditions on preservation of human spermatozoa, *Andrologia* **16:**52–60.

Mahadevan, M. M., and Trounson, A. O., 1985, Removal of the cumulus oophorus from the human oocyte for in vitro fertilization, *Fertil. Steril.* **43:**263–267.

Mahadevan, M. M., Trounson, A. O., and Leeton, J. F., 1983*a*, Relationship of tubal blockage, infertility of unknown cause, suspected infertility and endometriosis to success of in vitro fertilization and embryo transfer, *Fertil. Steril.* **40:**755–762.

Mahadevan, M. M., Trounson, A. O., and Leeton, J. F., 1983*b*, Successful use of human semen banking for in vitro fertilization, *Fertil. Steril.* **40:**340–343.

Mahadevan, M. M., Trounson, A. O., Leeton, J. F., and Wood, C., 1985, Use of in vitro fertilization for patients with persisting low quality semen, *Ann. N.Y. Acad. Sci.* **442:**293–300.

McBain, J. C., and Trounson, A., 1984, Patient management-treatment cycle, in: *Clinical in Vitro Fertilization* (C. Wood and A. Trounson, eds.), pp. 49–66, Springer-Verlag, Berlin.

McBain, J. C., Gronow, M. J., Bayly, C. M., and Martin, M. J., 1984, Controlled ovarian stimulation with clomiphene and hMG for in vitro fertilization, in: *Recent Progress in Human in Vitro Fertilization* (W. Feichtinger and P. Kemeter, eds.), pp. 25–32, Cofese, Palermo.

McIntosh, J. E. A., Mathews, C. D., Crocker, J. M., Broom, T. J., and Cox, L. W., 1980, Predicting the luteinizing hormone surge: Relationship between the duration of the follicular and luteal phases and the length of the human menstrual cycle, *Fertil. Steril.* **34:**125–130.

McMaster, R., Yanagimachi, R., and Lopata, A., 1978, Penetration of human eggs by human spermatozoa in vitro, *Biol. Reprod.* **19:**212–216.

Mohr, L. R., 1984, Assessment of human embryos, in: *In Vitro Fertilization and Embryo Transfer* (A. Trounson and C. Wood, eds.), pp. 159–171, Churchill Livingstone, Edinburgh.

Mohr, L. R., and Trounson, A. O., 1980, The use of fluorescein diacetate to assess embryo viability in the mouse, *J. Reprod. Fertil.* **58**:189–196.

Mohr, L. R., and Trounson, A. O., 1981, Structural changes associated with freezing of bovine embryos, *Biol. Reprod.* **25**:1009–1025.

Mohr, L. R., and Trounson, A. O., 1982, Comparative ultrastructure of the hatched human, mouse and bovine blastocysts, *J. Reprod. Fertil.* **66**:499–504.

Mohr, L. R., and Trounson, A., 1984, In vitro fertilization and embryo growth, in: *Clinical in Vitro Fertilization* (C. Wood and A. Trounson, eds.), pp. 99–115, Springer-Verlag, Berlin.

Mohr, L. R., and Trounson, A. O., 1985, Cryopreservation of human embryos, *Ann. N.Y. Acad. Sci.* **442**:536–543.

Mohr, L. R., Trounson, A. O., Leeton, J. F., and Wood, C., 1983, Evaluation of normal and abnormal human embryo development during procedures in vitro, in: *Fertilization of the Human Egg in Vitro* (H. M. Beier and H. R. Lindner, eds.), pp. 211–222, Springer-Verlag, Berlin.

Mohr, L. R., Trounson, A., and Freemann, L., 1985, Deep-freezing and transfer of human embryos, *J. Vitro Fertil. Embryo Transfer* **2**:1–10.

Moor, R. M., and Trounson, A. O., 1977, Hormonal and follicular factors affecting maturation of sheep oocytes in vitro and their subsequent development capacity, *J. Reprod. Fertil.* **51**:321–327.

Moor, R. M., and Warnes, G. M., 1978, Regulation of oocyte maturation in mammals, in: *Control of Ovulation* (D. B. Crighton, G. R. Foxcroft, N. B. Haynes, and G. E. Lamming, eds.), pp. 159–176, Butterworths, London.

Moore, H. D. M., Hartman, T. D., and Pryor, J. P., 1983, Development of the oocyte-penetrating capacity of spermatozoa in the human epididymis, *Int. J. Androl.* **6**:310–318.

Moore, N. W., Miller, B. G., and Trappl, M. N., 1983, Transport and development of embryos transferred to the oviducts and uteri of entire and ovariectomized ewes, *J. Reprod. Fertil.* **68**:129–135.

National Perinatal Statistics Unit Report, 1984, In vitro fertilization pregnancies, Australia 1980–1983, Sydney University Press, Sydney.

O'Herlihy, C., de Crespigny, L. Ch., Lopata, A., Johnston, I., Hoult, I., and Robinson, H., 1980, Preovulatory follicle size: A comparison of ultrasound and laparoscopic measurements, *Fertil. Steril.* **34**:24–26.

Olson, J. L., Rebar, R. W., Schreiber, J. R., and Vaitukaitis, J. L., 1983, Shortened luteal phase after ovulation induction with human menopausal gonadotropin and human chorionic gonadotropin, *Fertil. Steril.* **39**:284–291.

O'Neil, C., Gidley-Baird, A. A., Pike, I. L., Porter, R. N., Sinosich, M. J., and Saunders, D. M., 1985, Maternal blood platelet physiology and luteal phase endocrinology as a means of monitoring pre and post implantation embryo viability following in vitro fertilization, *J. Vitro Fertil. Embryo Transfer* **2**:87–93.

Ory, S. J., 1983, Clinical uses of luteinizing hormone-releasing hormone, *Fertil. Steril.* **39**:577–591.

Overstreet, J. W., and Hembree, W. C., 1976, Penetration of the zona pellucida of nonliving human oocytes by human spermatozoa in vitro, *Fertil. Steril.* **27**:815–831.

Pike, I. L., 1984, Biological risks of in vitro fertilization and embryo transfer, in: *Clinical in Vitro Fertilization* (C. Wood and A. Trounson, eds.), pp. 137–146, Springer-Verlag, Berlin.

Porter, R. N., Smith, W., Craft, I. L., Abdulwahid, N. A., and Jacobs, H. S., 1984, Induction of ovulation for IVF using Buserelin and gonadotrophins, *Lancet* **1**:1284–1285.

Prins, G. S., and Vogelzang, R. L., 1984, Inherant sources of ultrasound variability in relation to follicular measurements, *J. Vitro Fertil. Embryo Transfer* **1**:221–225.

Purdy, J. M., 1982, Methods for fertilization and embryo culture in vitro, in: *Human Conception in Vitro* (R. G. Edwards and J. M. Purdy, eds.), pp. 135–148. Academic Press, London.

Quinn, P., Warnes, G. M., Kerin, J. F., and Kirby, C., 1984, Culture factors in relation to the success of human in vitro fertilization and embryo transfer, *Fertil. Steril.* **41**:202–209.

Richie, W. G. M., 1985, Ultrasound in the evaluation of normaland induced ovulation, *Fertil. Steril.* **43**:167–181.

Rogers, P. A. W., Milne, B. J., and Trounson, A. O., 1986, A model to show human uterine and embryo viability following ovarian stimulation for in vitro fertilization, *J. Vitro Fertil. Embryo Transfer* (in press).

Rudak, E., Dor, J., Mashiach, S., Nebel, L., and Goldman, B., 1984, Chromosome analysis of multipronuclear human oocytes fertilized in vitro, *Fertil. Steril.* **41**:538–545.

Sandow, J., 1983, Clinical applications of LHRH and its analogues, *Clin. Endocrinol.* **18**:571–592.

Sathananthan, A. H., 1984, Ultrastructural morphology of fertilization and early cleavage in the human, in: *In Vitro Fertilization and Embryo Transfer* (A. Trounson and C. Wood, eds.), pp. 131–158, Churchill Livingstone, Edinburgh.

Sathananthan, A. H., and Trounson, A. O., 1982*a*, Ultrastructural observations on cortical granules in human follicular oocytes cultured in vitro, *Gamete Res.* **5**:191–198.

Sathananthan, A. H., and Trounson, A. O., 1982*b*, Ultrastructure of cortical granule release and zona interaction in monospermic and polyspermic human ova fertilized in vitro, *Gamete Res.* **6**:225–234.

Sathananthan, A. H., and Trounson, A. O., 1985, The human pronuclear ovum: Fine structure of monospermic and polyspermic fertilization in vitro, *Gamete Res.* **12**:385–398.

Sathananthan, A. H., Trounson, A. O., Wood, C., and Leeton, J. F., 1982*a*, Ultrastructural observations on the penetration of human sperm into the zona pellucida of the human egg in vitro, *J. Androl.* **3**:356–364.

Sathananthan, A. H., Wood, C., and Leeton, J., 1982*b*, Ultrastructural evaluation of 8–16-cell human embryos developed in vitro, *Micron* **13**:193–203.

Sathananthan, A. H., Trounson, A. O., and Wood, C., 1985, *Atlas of Fine Structure of Human Sperm, Eggs and Embryos Cultured in Vitro*, Praeger Scientific, Philadelphia.

Schreiber, E., Johnson, J. E., Plotz, E. J., and Weiner, M., 1966, Studies with ^{14}C-labelled clomiphene citrate (cc), *Clin. Res.* **14**:287 (abst.)

Schroeder, A. C., and Eppig, J. J., 1984, The developmental capacity of mouse oocytes that matured spontaneously in vitro is normal, *Dev. Biol.* **102**:493–497.

Seegar Jones, G., 1983, The use of hMG for ovulation stimulation in patients for in vitro fertilization, *Infertility* **6**:11–29.

Seegar Jones, G., 1984, Update on in vitro fertilization, *Endocrin. Rev.* **5**:62–75.

Shirley, B., Wortham, J. W., Witmyer, J., Condon-Mahony, M., and Fort, G., 1985, Effects of human serum and plasma on development of mouse embryos in culture media, *Fertil. Steril.* **43**:129–134.

Short, R. V., 1979, When a conception fails to become a pregnancy, in: *Maternal Recognition of Pregnancy*, CIBA Foundation Series No. 64, pp. 377–394, Excerpta Medica, Amsterdam.

Smart, Y. C., Cripps, A. W., Clancy, R. L., Roberts, T. K., Lopata, A., and Shutt, D. A., 1981, Detection of an immunosuppressive factor in human preimplantation embryo cultures, *Med. J. Aust.* **1**:78–79.

Smith, B., Porter, R., Ahuja, K., and Craft, I., 1984, Ultrasonic assessment of endometrial changes in stimulated cycles in an in vitro fertilization and embryo transfer program, *J. Vitro Fertil. Embryo Transfer* **1**:233–238.

Soupart, P., and Strong, P. A., 1975, Ultrastructural observations on polyspermic penetration of zona pellucida-free human oocytes inseminated in vitro, *Fertil. Steril.* **26**:523–537.

Steptoe, P. C., and Edwards, R. G., 1978, Birth after the reimplantation of a human embryo, *Lancet* **2**:880–882.

Steptoe, P. C., Edwards, R. G., and Purdy, J. M., 1971, Human blastocysts grown in culture, *Nature (Lond.)* **229**:132–133.

Sundstrom, P., 1982, Interaction between spermatozoa and ovum in vitro, in: *Atlas of Human Reproduction by Scanning Electron Microscopy* (E. S. E. Hafez and P. Kenemans, eds.), pp. 225–230, MTP Press, Lancaster, England.

Sundstrom, P., Nilsson, D., and Liedholm, P., 1981, Cleavage rate and morphology of early human embryos obtained after artificial fertilization and culture, *Acta Obstet. Gynaecol. Scand.* **60**:109–220.

Sundstrom, P., Nilsson, O., and Liedholm, P., 1983, Scanning electron microscopy of human preimplantation endometrium in normal and clomivid/HCG-stimulated cycles, *Fertil. Steril.* **40**:642–647.

Tarlatzis, B. C., Lauter, N., and DeCherney, A. H., 1984, The use of ovarian ultrasonography in monitoring ovulation induction, *J. Vitro Fertil. Embryo Transfer* **1**:226–232.

Temple-Smith, P. D., Southwick, G. J., Yates, C. A., Trounson, A. O., and de Kretser, D. M., 1985, Human pregnancy by in vitro fertilization (IVF) using sperm aspirated from the epididymis, *J. Vitro Fertil. Embryo Transfer* **2:**119–122.

Templeton, A. A., van Look, P., Lumsden, M. A., Angell, R., Aitken, R. J., Duncan, A. W., and Baird, D. T., 1984, The recovery of pre-ovulatory oocytes using a fixed schedule of ovulation induction and follicle aspiration, *Br. J. Obstet. Gynaecol.* **91:**148–154.

Testart, J., Frydman, R., Feinstein, M. C., Thebault, A., Roger, M., and Scholler, R., 1981, Interpretation of plasma LH assay for the collection of mature oocytes from women: Definition of an LH surge initiating rise (LH S.I.R.), *Fertil. Steril.* **30:**50–54.

Testart, J., Frydman, R., and Roger, M., 1982, Seasonal influence on diurnal rhythms in the onset of the plasma luteinizing hormone surge in women, *J. Clin. Endocrinol. Metab.* **55:**374–378.

Testart, J., Lassalle, B., Frydman, R., and Belaisch, J. C., 1983, A study of factors affecting the success of human fertilization in vitro. 2. Influence of semen quality and oocyte maturity on fertilization and cleavage, *Biol. Reprod.* **28:**425–431.

Trapp, M., Baukloh, V., Bohnet, H.-G., and Heeschen, W., 1984, Pollutants in human follicular fluid, *Fertil. Steril.* **42:**146–148.

Trounson, A., 1983*a*, Factors controlling normal embryo development and implantation of human oocytes fertilized in vitro, in: *Fertilization of the Human Egg in Vitro* (H. M. Beier and H. R. Lindner, eds.), pp. 233–246, Springer-Verlag, Berlin.

Trounson, A., 1983*b*, In vitro fertilization, in: *Current Topics in Experimental Endocrinology*, Vol 5: *Fetal Endocrinology and Metabolism* (L. Martini and V. H. T. James eds.), pp. 44–73 Academic Press, New York.

Trounson, A. O., 1983*c*, In vitro fertilization at Monash University, Melbourne, Australia, in: *In Vitro Fertilization and Embryo Transfer* (P. C. Crosignani and B. L. Rubin, eds.), pp. 315–322 Academic Press, London.

Trounson, A., 1984*a*, In vitro fertilization and embryo preservation, in: *In Vitro Fertilization and Embryo Transfer* (A. Trounson and C. Wood, eds.), pp. 111–130, Churchill Livingstone Edinburgh.

Trounson, A., 1984*b*, The preservation and donation of human embryos, in *Recent Progress in Human in Vitro Fertilization* (W. Feichtinger and P. Kemeter, eds.), pp. 345–354, Cofese, Palermo.

Trounson, A. O., and Calabrese, R., 1984, Changes in plasma progesterone concentrations around the time of the LH surge in women superovulated for in vitro fertilization, *J. Clin. Endocrinol. Metab.* **59:**1075–1080.

Trounson, A., and Freemann, L., 1986, Role of cryopreservation of human oocytes and embryos in an IVF program, in: *Progress in Infertility*, 3rd ed. (G. W. Patton, R. Kistner, and J. Behrman, eds.), Little, Brown, Boston (in press).

Trounson, A. O., and Lutjen, J., 1984, In vitro fertilization and embryo transfer in the human, *Biblio Reprod.* **44:**B1–B16.

Trounson, A., and Mohr, L., 1983, Human pregnancy following cyropreservation, thawing and transfer of an eight-call embryo, *Nature (Lond.)* **305:**707–709.

Trounson, A. O., and Moore, N. W., 1974, Effect of progesterone and oestrogen on the survival and development of fertilized sheep ova in the ovariectomized ewe, *Aust. J. Biol. Sci.* **21:**511–517.

Trounson, A., and Sathananthan, A. H., 1984, The application of electron microscopy in the evaluation of two- to four-cell human embryos cultured in vitro for embryo transfer, *J. Vitro Fertil. Embryo Transfer* **1:**153–165.

Trounson, A., and Webb, J., 1984, Fertilization of human oocytes following reinsemination in vitro, *Fertil. Steril.* **41:**816–819.

Trounson, A., and Wood, C., 1984, In vitro fertilization results, 1979–1982, at Monash University, Queen Victoria and Epworth Medical Centres, *J. In Vitro Fertil. Embryo Transfer* **1:**42–47.

Trounson, A. O., Shivers, C. A., McMaster, R , and Lopata, A., 1980, Inhibition of sperm binding and fertilization of human ova by antibody to procine zona pellucida and human sera, *Arch. Androl.* **4:**29–36.

Trounson, A. O., Leeton, J. F., Wood, C., Webb, J., and Wood, J., 1981*a*, Pregnancies in humans by fertilization in vitro and embryo transfer in the controlled ovulatory cycle, *Science* **212:**681–682.

Trounson, A. O., Leeton, J. F., Wood, C., Buttery, B., Webb, J., Wood, J., Jessup, D., McTalbot, J., and

Kovacs, G., 1981*b*, A programme of successful in vitro fertilization and embryo transfer in the controlled ovulatory cycle, in: *Human Reproduction* (K. Semm and L. Metler, eds.), pp. 173–180, Excerpta Medica, Amsterdam.

Trounson, A. O., Mohr, L. R., Wood, C., and Leeton, J. F., 1982*a*, Effect of delayed insemination on in-vitro fertilization, culture and transfer of human embryos, *J. Reprod. Fertil.* **64:**285–294.

Trounson, A. O., Leeton, J. F., and Wood, C., 1982*b*, In vitro fertilization and embryo transfer in the human, in: *Follicular Maturation and Ovulation* (R. Rolland, E. V. van Hall, S. G. Hillier, K. P. McNatty, and J. Schoemaker, eds.), pp. 313–322, Excerpta Medica, Amsterdam.

Trounson, A. O., Wood, C., and Leeton, J., 1982*c*, Freezing of human embryos: An ethical obligation, *Med. J. Aust.* **2:**332–324.

Trounson, A., Burger, H. G., and Kovacs, G. T., 1983*a*, Detection of ovulation for in vitro fertilization, in: *Ovulation: Methods for Its Prediction and Detection* (S. L. Jeffcoate, ed.), pp. 83–102, John Wiley and Sons, Chichester.

Trounson, A., Leeton, J., Besanko, M., Wood, C., and Conti, A., 1983*b*, Pregnancy established in an infertile patient after transfer of a donated embryo fertilised in vitro, *Br. Med. J.* **286:**835–838.

Trounson, A., Hoppen, H.-O., Lutjen, P., Mohr, L. R., Rogers, P. A. W., and Sathananthan, A. H., 1985*a*, In vitro fertilization: the state of the art, in: *Gamete Quality and Fertility Regulation* (R. Rolland, M. J. Heinemann, S. G. Hillier, H. Vemer, eds.), pp. 325–343, Excerpta Medica, Amsterdam.

Trounson, A., Lutjen, P. J., Leeton, J. F., Findlay, J., and Chan, L. K., 1984*b*, Donation of oocytes for the treatment of human infertility, in: *Human in Vitro Fertilization. Present Problems and Prospects* (R. Frydman and J. Testart, eds.), pp. 283–290, Excerpta Medica, Amsterdam.

Tsonis, C. G., Quigg, H., Lee, V. W. K., Levershaw, A., Trounson, A., and Findlay, J. K., 1983, Inhibin in individual ovine follicular in relation to diameter and atresia, *J. Reprod. Fertil.* **67:**83–90.

Uhera, T., and Yanagimachi, R., 1976, Microsurgical injection of spermatozoa into hamster eggs with subsequent transformation of sperm nuclei into male pronuclei, *Biol. Reprod.* **15:**467–470.

Van Blerkom, J., Henry, G. P., and Porreco, R., 1984, Preimplantation human embryonic development from polypronuclear eggs after in vitro fertilization, *Fertil. Steril.* **41:**686–696.

Veeck, L. L., Wortham, J. W. E., Witmyer, J., Sandow, B. A., Acosta, A. A., Garcia, J. E., Jones, G. S., and Jones H. W., 1983, Maturation and fertilization of morphologically immature human oocytes in a program of in vitro fertilization. *Fertil. Steril.* **39:**594–602.

Warnes, G. M., Moor, R. M., and Johnson, M. H., 1977, Changes in protein synthesis during maturation of sheep oocytes in vivo and in vitro, *J. Reprod. Fertil.* **49:**331–345.

Wikland, M., and Hamberger, L., 1984, Ultrasound as a diagnostic and operative tool for in vitro fertilization and embryo replacement (IVF/ER) programs, *J. Vitro Fertil. Embryo Transfer* **1:**213–216.

Wolf, D. P., Byrd, W., Dandekar, P., and Quigley, M. M., 1984, Sperm concentration and the fertilization of human eggs in vitro, *Biol. Reprod.* **31:**837–848.

Wood, C., Leeton, J., McTalbot, J., and Trounson, A. O., 1981, Technique for collecting mature human oocytes for in vitro fertilization, *Br. J. Obstet. Gynaecol.* **88:**756–760.

Wood, C., Downing, B., Trounson, A., and Rogers, P., 1984, Clinical implications of developments in in vitro fertilization, *Br. Med. J.* **289:**978–980.

Wood, C., McMaster, R., Rennie, G., Trounson, A., and Leeton, J., 1985, Factors influencing pregnancy rates following in vitro fertilization and embryo transfer, *Fertil. Steril.* **43:**245–250.

World Health Organization, 1980, *Laboratory Manual for the Examination of Human Semen and Semen–Cervical Mucus Interaction*, Press Concern, Singapore.

Yovich, J. L., and Stanger, J. D., 1984, The limitations of in vitro fertilization from males with severe oligospermia and abnormal sperm morphology, *J. Vitro Fertil. Embryo Transfer* **1:**172–179.

Zeilmaker, G. H., Alberda, A. T., van Gent, I., Rijkmans, C. M. P. M., and Dorgendijk, A. C., 1984, Two pregnancies following transfer of intact frozen-thawed embryos, *Fertil. Steril.* **42:**293–296.

Chapter 6

Embryo Sexing of Farm Animals

W. C. D. HARE

1. Introduction

The obvious purpose in sexing embryos is either simply to have prior knowledge of the sex of the newborn or to have prior knowledge with a view to using this knowledge to influence the sex of the newborn. Certainly, since ancient times, man has been fascinated by sex and how it can be determined before birth, not only for himself, but also for his animals.

In early times, man's interest in knowing the sex of his own unborn was probably based on religious, political or socioeconomic grounds. For his animals, it was undoubtedly on economic grounds. Nowadays, man's reasons for himself may be medical or socioeconomic, but for his animals they are still strictly economic. For example, the severity of some human sex-linked disorders is such that their early diagnosis is desirable, so that, laws permitting, measures may be taken against them (Niermeijer *et al.*, 1976). In countries with a severe overpopulation problem, one way to reduce the birthrate, and thus the population size, would be to increase the ratio of male newborns to female newborns (Cederqvist and Fuchs, 1970). This is happening today in China in a tragic way according to newspaper reports: government policy restricts married couples to one child with strict penalties for those having more than one and, as a result, infanticide of girls is a common practice by couples wanting a son. For the livestock owner, the commercial advantages of controlling the sex of newborns are in obtaining the optimal proportion of males to females in order to (1) benefit from sex-limited traits, such as milk production, and sex-influenced traits, such as body composition and rate of gain; (2) eliminate undesirable recessives; (3) increase selection pressure; and (4) optimize genetic gain (Foote and Miller, 1971). Against these have to be pitted the costs of the technology and the effects of inbreeding.

The chromosome theory of sex determination originated with McClung (1902), who put forward the idea that an accessory chromosome, which he had

W. C. D. HARE • Agriculture Canada, Animal Diseases Research Institute, Nepean, Ontario, Canada K2H 8P9.

observed in half the spermatids of the locust, *Xiphidium*, was male determining. That McClung had been wrong in fact but correct in theory was later established independently by Stevens and Wilson, also working with insects, and reported by them in a series of papers published between 1905 and 1911. It was Wilson (1909) who termed the accessory chromosome, the X chromosome, since earlier it had been labeled X in a diagram by McClung, and a smaller accessory structure observed by him and Stevens, the Y chromosome. In essence, the chromosome theory of sex determination was that, in eukaryotic organisms with separate sexes, sex is determined by genetic regulation from a single chromosome or a group of chromosomes that is represented differently in the two sexes.

Wodsedalak (1913) seems to have been the first to describe a dimorphic chromosome complement in the spermatozoa and somatic cells of a farm animal, namely the pig. However, he got the facts wrong by describing two sets of spermatozoa, one with a pair of accessory chromosomes and the other with none, and somatic cells with a pair of accessory chromosomes in the female that were not present in the male. Subsequently, Wodsedalak described two sets of spermatozoa in the horse (1914) and bull (1920)—one set with one accessory (sex) chromosome and the other with none; the somatic cells of cattle he described as having two sex chromosomes in the female and only one in the male.

Before the factors determining sex and allowing the sex of the conceptus to be known were recognized, the literature contained references to weird and wonderful ways of diagnosing and influencing the sex of the unborn on the basis of superstition and folklore (Parkes, 1971). It is interesting that, in the 75 years since the chromosome theory of sex determination was established, some of the old notions in regard to influencing the sex of the newborn that predated the chromosome theory of sex determination have been the subject of investigation. For example, the old idea that the time of fertilization in relation to estrus affected the sex ratio (Thury, 1863; Dusing, 1884), subscribed to by Pearl and Parshley (1913) and later shown not to be true for cattle (Pearl, 1917), has been reexamined in cattle inseminated early and late in estrus on a number of occasions (Gardner, 1950; Adler and Autrup, 1966; Ballinger, 1970; Foote, 1977; Gebicke-Harter *et al.*, 1977; Deinhardt and Muller, 1978). The old belief that nutrition was a factor in determining the sex of the newborn has been looked at again from the standpoint of mineral intake affecting the sex ratio (Lyster, 1972; Stolkowski, 1977). Even in the age of microchip technology and space exploration, there are those who believe that the points of the compass to which a cow stands during insemination will influence the sex of her calf (Betteridge *et al.*, 1981).

Knowing what we know today about sex determination, there are three main approaches to determining the sex of the embryo with a view to influencing the sex of the newborn. The cheapest and most effective way of doing so, if it could be done, would be to separate semen into X- and Y-chromosome-bearing sperm fractions, so that the sex of the embryo would be determined at the time of fertilization.

The least effective way of determining the sex of an animal prior to birth, while something can still be done about it (i.e., abortion), is to carry out hormone assays on fetal fluids, or sex chromosomal analyses or Y-chromosome-specific DNA probes on cells obtained from the fluids, and then to abort the animal if the sex is not the one desired.

The third and, at the moment, the most practical way of doing it is to sex embryos in the early stages of development in combination with embryo transfer and to transfer only embryos of the desired sex.

Other theoretical approaches to controlling the sex of farm animals are (1) the production of homozygous diploids, (2) oocyte activation to produce parthenotes, and (3) the injection of sperm of known sex into ova.

This chapter reviews briefly the history and present day status of predetermining embryonic sex by semen separation into X- and Y-chromosome-bearing sperm fractions and approaches taken to the diagnosis of fetal (embryonic) sex from fetal fluid cells obtained by centesis; Described at greater length are methods that are or might be used to diagnose the sex of embryos in conjunction with embryo transfer.

2. Prefertilization Sexing

In 1925, Lush reported negative results from attempts to separate the X-chromosome-bearing from the Y-chromosome-bearing spermatozoa in the rabbit. Since then innumerable attempts have been made to separate the two types of spermatozoa based on possible differences in their weight, density, size, motility, or surface change. Other attempts to alter the sex ratio by differentially inactivating either the X- or the Y-chromosome-bearing spermatozoa have involved exposing semen to hormones, altered pH, or altered atmospheric pressure and the use of antibodies to H-Y antigen (for review, see Benjamin, 1979; Ericsson and Glass, 1982; Foote, 1982). None of the attempts in the past to separate X- and Y-chromosome-bearing spermatozoa has been consistently successful, although subsequent to the above reviews fresh claims of having successfully separated human X- and Y-bearing-spermatozoa based on differences in surface charge (Kaneko *et al.*, 1984) and skewed the sex ratio in rabbits by the intravaginal administration of H-Y antiserum (Zavos, 1983) have been made. On the other hand, Hoppe and Koo (1984) found that there is no haploid expression of H-Y antigen by mouse epididymal sperm and that X- and Y-bearing spermatozoa may share the same surface antigen due to their association with testicular constituents.

However, according to Meistrich (1982), the physical techniques available have not been fully exploited; a more informed approach may produce satisfactory and consistent results. Also, if some form of haploid expression occurs, it may be possible to develop procedures that will differentially inactivate one or other type of spermatozoon by using monospecific probes in the form of monoclonal antibodies against surface constituents of the sperm (Bellvé, 1982). It has already been reported (Hedge *et al.*, 1981) that phytohemagglutinin (PHA) binds over the equatorial plane of human sperm with an F body, but not over that of those without. This was interpreted as evidence for haploid expression in X- and Y-chromosome-bearing spermatozoa, although, in keeping with the view expressed by Bellvé (1982), it is difficult to believe that a probe with a broad specificity for glycoproteins could distinguish between X- and Y-chromosome-bearing spermatozoa.

One of the difficulties associated with this work is the evaluation of the success of the separation. Until a few years ago, the only available measure was the sex ratio of progeny. However in man, separation attempts can now be evaluated without resorting to progeny sex-ratio determinations as a result of the following findings. The Y chromosome fluoresces brightly in metaphase spreads with quinacrine mustard (Zech, 1969) and quinacrine dihydrochloride (Vosa, 1970), and in Y-bearing spermatozoa which are similarly stained, it is seen as a bright dot, the F body (Barlow and Vosa, 1970), which is not apparent in X-bearing spermatozoa. It has also been demonstrated (Summer and Robinson, 1976) that there is a phenotypic difference in mass between X- and Y-bearing spermatozoa.

Unfortunately, in farm animals, the Y chromosome does not fluoresce more brightly than the other chromosomes, so there is no basis for distinguishing between X- and Y-chromosome-bearing spermatozoa using fluorescence microscopy and, therefore, no way of verifying the effectiveness of separation techniques using this procedure, despite persistent claims that in the bull the Y-chromosome-bearing spermatozoon can be identified by a fluorescing spot, or B body, (Bhattacharya, 1976; Bhattacharya *et al.*, 1979). Recently, however, it has been reported that flow cytometry can be used to determine rapidly and accurately the proportions of X- and Y-chromosome-bearing spermatozoa in a single insemination dose of semen from bulls, boars, and rams (Garner *et al.*, 1983; Pinkel *et al.*, 1983), but not man (Pinkel, 1983). The potential also exists to examine the chromosomes of spermatozoa of farm animals following entry of their spermatozoa into zona pellucida-free eggs from hamsters or other species (Brackett, 1983), as has been demonstrated with human spermatozoa and zona-free hamster eggs (Rudak *et al.*, 1978; Martin, 1983; Martin *et al.*, 1982, 1983).

3. Preimplantation Sexing

Edwards (1965) first suggested the possibility of controlling sex-linked recessive diseases by sexing human blastocysts following *in vitro* fertilization and before transfer. This was followed by a report from Gardner and Edwards (1968) that the sex of rabbit embryos had been determined successfully by examining the sex chromatin in a few trophoblast cells removed from 5-3/4-day-old blastocysts during embryo transfer. Unfortunately, sexing by sex chromatin cannot be applied to blastocysts of farm animals because, in these species, normal clumping of chromatin in the interphase nucleus makes identification of the sex chromatin body a chancy procedure (Rowson, 1974).

As early as 1966 Rowson and Moor had shown that 13-day sheep blastocysts (i.e., hatched) would develop to full term after removal of some trophoblast. So, when Betteridge *et al.* (1976) showed that day-14 to day-16 cattle embryos could be transferred with acceptable pregnancy rates, it was a small step to sexing these expanded blastocysts during transfer by removing a piece of trophoblast and processing the cells for chromosome analysis using a procedure that had already been developed successfully for whole pig (McFeely, 1966) and cattle (McFeely and Rajakoski, 1968) blastocysts. As far as is known, the first live calf to have had its sex determined when it was a 2-week-old embryo was born on Christmas day 1975 (Mitchell 1977), although it had been reported earlier (Polge and Rowson, 1973) that Cain and Rowson had presexed lambs at the embryo stage.

Initially, the trophoblast cells for analysis were obtained using #5 fine forceps and 3½ inch Vannas curved scissors (Hare *et al.*, 1976). Subsequently, it was possible to work on smaller blastocysts using a home-made biopsy instrument held in a micromanipulator (Hare *et al.*, 1978). Thereafter, the procedure was the same with both techniques (Hare *et al.*, 1978). The pieces of trophoblast were incubated with a mitostatic agent, Colcemid (Ciba Pharmaceutical Products, Inc., Summit, N.J.) (0.025 ×g/ml), for 1 hr. After that they were placed in 0.25% trypsin or 0.05% trypsin–0.02% EDTA to disaggregate the cells, which were then subjected to hypotonic shock (1 part serum to 5 parts distilled water) and fixation (1 part glacial acetic to 3 parts methanol). The cells, suspended in fixative, were dropped onto a wet slide and the alcohol in the fixative ignited by passing the slide rapidly through a flame. The cells, flattened by this procedure, were stained in 10% Giemsa, buffered to pH 6.8, and cleared by passing the slide through acetone, acetone : xylene, and xylene, after which the slide was mounted with a coverglass and synthetic balsam before microscopic examination of the cells.

Essentially the same procedure, namely Colcemid (colchicine) exposure followed by hypotonic treatment and fixation of the cells, has been used by a number of laboratories (Brand *et al.*, 1978; King *et al.*, 1979; L. Kressly, 1979; personal communication; Bouters *et al.*, 1980; Wintenberger-Torres and Popescu, 1980; Ernst *et al.*, 1981; Roschlau *et al.*, 1983), several of which were involved in sexing embryos for commercial transfer. The results, with one or two exceptions, have been similar (Betteridge *et al.*, 1981; Cribiu, 1983; Roschlau *et al.*, 1983), namely that, on average, a two out of three success rate in sexing with a 40% pregnancy rate following transfer of sexed embryos can be expected. One laboratory reported sexing 20 of 21 (95%) embryos, but only three of the 20 embryos transferred resulted in pregnancies (Betteridge *et al.*, 1981). Possibly the high sexing and low pregnancy rates were due to too large a piece of trophoblast having been taken from the blastocyst. Wintenberger-Torres and Popescu (1980) reported obtaining a 45% pregnancy rate when two, three, or four blastocysts (N = 62) were transferred to the horn (N = 22) ipsilateral to the corpus luteum, with breakdown figures of 37% for 8 × 2 blastocysts, 40% for 10 × 3 blastocysts, and 75% for 7× 4 blastocysts transferred. They reasoned that there is a greater probability that one or two sexed embryos will survive when three or four are transferred, but this is an extravagant approach from a commercial standpoint. Also, they did not appear to take into consideration the possibility of a donor factor influencing the pregnancy rate, which was noted by Hare *et al.* (1978).

A major disadvantage of the above procedure from the commercial viewpoint, apart from the low pregnancy rate obtained, is that no one has been successful in freeze-thawing day 12 to day 15 embryos for long-term storage, so transfer of the biopsied embryo has to be delayed for a minimum of 5 hr while the biopsy is being sexed, or the transfer has to be carried out before the sex of the biopsied embryo is known.

Procedures similar to those just described for cattle embryos have been described for whole hatched embryos of the pig (Dolch and Chirsman, 1981; Long and Williams, 1982) and sheep (Long, 1977); undoubtedly they could be modified for sexing and transfer. Murer-Orlando *et al.* (1982) described a culture method for obtaining chromosome preparations from horse embryos which could probably be adapted for use in combination with embryo transfer. Because of the difficulty of identifying the Y chromosome in these species, especially the horse and pig, it would probably be necessary to treat the chromosomes before staining in order to demonstrate heterochromatin (C bands) (Darlington and La Cour, 1976).

In 1978, a report by Moustafa *et al.* held out hope for overcoming some of these problems. They described being able to sex 59% of day 6 and day 7 embryos processed and having had a 100% pregnancy rate with the surgical transfer of five sexed embryos.* The procedure involved the use of micromanipulators with holding and suction micropipettes to remove 7–10 blastomeres from morula and early blastocyst stage embryos. The blastomeres were then processed for chromosome display, after 2–4 hr incubation in Colcemid at 37°C in much the same way as the trophoblast cells of hatched embryos, described above. Although the overall time required for the procedure and the sexing success rate were similar to that achieved with older embryos, the known tolerance to cryopreservation of day 6 and day 7 embryos appeared to make it preferable. Unfortunately attempts to repeat this work have yielded a much lower (30–33%) sexing success rate using short term exposure to Colcemid (4–6 hr) followed by processing (Singh and Hare, 1980), due mainly to a much lower average mitotic index in the embryos examined by them than in the embryos studied by Moustafa and co-workers. Similar low mitotic indices were obtained by Rottman (1981) for rabbit morulae in spite of exposure of the entire embryo to Colcemid for 2–4 hr before removal of a few blastomeres. Subsequently, Dankowski *et al.* (1982) reported that they were able to karyotype 67% of rabbit morula-stage embryos by removing 5–10 blastomeres and culturing them for 24 hr with Colcemid added to the culture medium for the last 4 hr. The possibility of a donor factor being responsible for the discrepancy in cattle seemed possible, but the matter has not been pursued. The mitotic

*This laboratory now reports (J. Hahn, personal communication, 1983) a 70% success rate for sexing day 6 and 7 embryos by chromosomal analysis.

index might be increased by culturing the blastomeres for an increased time in the presence of colcemid. Using entire day 6 and day 7 bovine embryos, King *et al.* (1979) were able to increase an average mitotic index of 7.4% for nine embryos cultured for 20 hr, followed by exposure to colchicine for 2–4 hr to 17.1% for six embryos exposed directly to colchicine for 8½ to 9½ hr. Popescu *et al.* (1982) have been successful in sexing day 8 to day 10 bovine and porcine embryos by taking a few trophoblast cells and culturing them for 5–6 days at 37°C on collagen-coated coverslips in Leighton tubes before processing them for chromosomal analysis. These workers are of the opinion that the system could be adapted for culturing blastomeres from day 7 embroys. However, one of the problems encountered with this length of time in culture is contamination (Popescu and Cribiu, 1982).

Even if a procedure could be developed to regularly obtain a sexing rate of 60% or better for bovine morulae and early blastocysts and it were shown that puncturing the zona pellucida did not affect the success of freeze-thawing too adversely, the method is still hampered by the time it takes to diagnose the sex and the limited number of embryos that can be handled over a given period of time. The time factor obviously does not apply if the biopsied embryo is to be frozen. However, experience with freeze-thawing of day 6 and day 7 bovine embryos with punctured zonae pellucidae (Kanagawa *et al.*, 1979) or with zonae damaged by halving or quartering (Lehn-Jensen and Willadsen, 1983; Lehn-Jensen, 1984; Heyman and Chesné, 1984) indicates that there are problems of embryo survival to be overcome.

The technique of splitting morulae and early blastocysts to produce monozygotic twins obviously lends itself to diagnosing their sex (Popescu and Cribiu, 1982; Williams *et al.*, 1983). Theoretically, it should be easier to sex half of a morula than a few blastomeres removed from a morula. However, in practice so far, this has not proven to be so. Popescu and Cribiu (1982) reported a 26.9% success rate in sexing 26 half demi-day 8 bovine embryos after culturing the embryos for 24–48 hr at 37°C, compared with a 44.2% success rate for 43 whole day 7 embryos at the same stage of development cultured in the same way. L. Picard, W. A. King, and K. J. Betteridge (unpublished data, 1983) obtained a 53.3% success rate in sexing 30 entire day 6 to day 7 bovine embryos after culture for 24 hr at 37°C, compared with a 42.8% success rate for the 35 better halves from 35 bisected day 6 to day 7 embryos and a 5.2% success rate for the 35 poorer halves from the same 35 bisected embryos. In subsequent experiments, Picard *et al.* (1984) were able to improve the success rate significantly for sexing poorer demi-embryos to 60% by reducing the total culture time to 4 hr with Colcemid.

The justification for using one of the halves for sexing as opposed to transferring it, or freeze-thawing it (Lehn-Jensen and Willadsen, 1983) for future transfer, obviously depends on the economics of the situation. Alternatively, it might be possible to maintain one-half in culture long enough to take a few blastomeres from it for sex-chromosome identification before freezing and transferring it later.

The techniques developed by Willadsen (1979, 1980, 1981), by Willadsen and Polge (1979), and by Willadsen *et al.* (1980) for producing identical animals from 2-, 4-, 6-, and 8-cell embryos provide opportunities for sexing either in the early stages of development using either the procedure developed by Long and Williams (1978) or that of King *et al.* (1979), or later when the embryo has developed *in vitro* or in the ligated oviduct of a rabbit to the morula or blastocyst stage. It is obviously a much more economical proposition to use one of three or four embryos for sexing then one of two.

In 1976, Krco and Goldberg reported that when 8-cell mouse embryos were exposed to complement (guinea pig serum) plus mouse antiserum to H-Y antigen* approximately 50% of the embryos were lysed and that lysis was prevented by absorption of the antiserum with male but not female mouse spleen cells. It was inferred, but not proven, that the male embryos were affected. Nevertheless, the fact that H-Y antigen was apprently being expressed by mouse embryos at the 8-cell stage immediately suggested that the expression might be used to sex cattle or other farm animal embryos. The test systems for H-Y antigen available at that time were based on the detection of H-Y antigen on test cells by using them to absorb anti-H-Y-sera developed in inbred strains of mice or rats and then measuring any reduction in the activity of the serum using sperm cytotoxicity or various hemadsorption tests. These procedures were obviously unsuitable for sexing embryos, as was the embryo lysis procedure unless only female embryos were desired.

Early attempts to develop an indirect immunofluorescent test for detecting H-Y antigen on cattle embryos were unsuccessful due to nonspecific staining. A certain percentage of embryos showed increased fluorescence, but this did not correlate with their chromosomal sex (E. L. Singh and W. C. D. Hare, unpublished data, 1977). Other laboratories were also investigating the possibility of sexing early cattle embryos by the detection of H-Y antigen, but nothing of substance emerged; the question arose as to whether H-Y antigen is expressed by bovine embryonic cells at this stage of development.

In the mouse there were continuing reports indicating the expression of H-Y antigen in 8-cell mouse embryos. Epstein *et al.* (1980) reported that one-half the number of 8-cell mouse embryos exposed to complement (Guinea pig serum) plus mouse anti-H-Y-serum were killed or retarded or contained dead cells. The unaffected embryos were karyotyped and found, on average, to be 92% female. However, Epstein *et al.* (1980) also drew attention to one of the

*Recent findings (Melvold *et al.*, 1977; Engel *et al.*, 1981; Simpson *et al.*, 1982) indicate that, at least in rodents, the molecule originally identified (Eichwald and Silmser, 1955) and later named (Billingham and Silvers, 1960) the H-Y (tranplantation) antigen is not the same as the molecule reacting with antiserum produced against male cells inoculated into isogenic mice and rats. This latter molecule has been named the serologically detectable male (SDM) antigen (Silvers *et al.*, 1982) or, alternatively, the serologically defined heterogametic sex-associated (SDHS) antigen (Müller, 1982). However, it has been shown that monoclonal antibody to SDM blocks the testis-inducing rule of H-Y antigen (Müller and Urban, 1981), and there is also a report that SDM antigen is similar to, if not the same as, the antigen that stimulates the cell-mediated-cytoxicity response (Koo and Varano, 1981). Although agreeing that both the H-Y and the SDM antigens are strongly male associated, Silvers *et al.* (1982) are of the opinion that there is sufficient evidence to conclude that neither is the primary cause of gonadal determination. (For a review of role of H-Y antigen in gonadal determination see Wachtel, 1983.)

problems with mouse antisera to H-Y antigen, namely that they have low titers and are unstable, particularly after freezing. Under the best conditions the antisera would kill all or nearly all male embryos, but under less favorable conditions some male embryos were not affected. Elsworthy and Frels were reported by Chapman (1982) to have had similar experiences and Chapman concluded that "more work is required to make a reliable antibody against H-Y antigen which will be effective for sexing early embryos."

The development of monoclonal H-Y antigen-specific antibody (Koo and Varano, 1981; Koo *et al.*, 1981) was a significant step toward having a reliable antibody against H-Y antigen. Likewise the development of detection systems for H-Y antigen by fluorescence microscopy (Galbraith *et al.*, 1978) and a peroxidase–antiperoxidase method (Müller and Bross, 1979) provided the possibility for evaluating either whole embryos or possibly a few blastomeres removed microsurgically from an embryo.

White *et al.* (1982) reported exposing 8- to 16-cell mouse embryos to anti-H-Y-serum and normal guinea pig serum followed by transfer to recipient mice; 86.2% of those born were female. In subsequent experiments, White *et al.* (1983) compared the effectiveness of the cytolytic assay on 8- to 16-cell embryos with an indirect fluorescein-isothiocyanate (FITC)-labeled second antibody technique on morula and blastocysts using a mouse IgM monoclonal antibody to H-Y antigen in both systems. Transfer of embryos unaffected by exposure to monoclonal antibody and normal guinea pig serum resulted in 81% of 53 pups born being female. Transfer of embryos from the fluorescing group resulted in 78% of 23 being male and from the nonfluorescing group in 83% of 23 being female. Survival to birth of transferred embryos unaffected by the cytolytic assay was 20% and of transferred embryos following exposure to FITC-labeled antibody assay 18%—somewhat lower than that expected for cultured embryos. In 1984, White *et al.* reported that when H-Y antiserum prepared in mice was used in an indirect immunofluorescent assay with goat anti-mouse IgG antibody, 52% of 155 bovine embryos (16-cell to early blastocyst stages) displayed cell-specific fluorescence. Of 39 fluorescing embryos in which the chromosomes were analyzable, 85% were male and 18% were female, and of 32 nonfluorescing embryos, 3% were male and 97% were female. In another study (Utsumi *et al.*, 1984), caprine and bovine morula-stage embryos were incubated with H-Y antiserum prepared in rats for 12–24 hr and then evaluated for development. Of nine caprine embryos, four developed further to the blastocyst stage, and five remained as morulae. Of eight bovine embryos, four developed to the blastocyst stage and four remained as morulae. Utsumi and co-workers extrapolated from similar studies with rat embryos, where the embryos were subsequently transferred with 80% of the blastocysts developing into females and 90% of the morulae developing into males, that rat H-Y antiserum can be used to distinguish male from female embryos in goats and cattle.

It would appear that, in the mouse, detection of H-Y antigen using the cytolytic assay on 8- to 16-cell embryos or the FITC-labeled antibody assay on morula and early blastocyst stage embryos gives a 4/5 chance of determining the sex accurately. Even if it is possible to transfer the technology from mouse to farm animal embryos, this degree of accuracy may be adequate for detecting an alteration in sex ratios, but it is questionable as to whether it is economically viable from a commercial standpoint.

In January 1983, a news report from Washington, D.C., appeared in Nature (Shapley, 1983) stating that a biotechnology company in the United States had applied for a patent on a method of sexing day 6 cattle embryos using a monoclonal antibody to H-Y antigen, for which it claimed a 93% accuracy. Taken at face value, this is a significant step toward having frozen sexed embryos. The problem is in obtaining for evaluation the scientific data on which the claim has been made. There is also the question as to whether the results are repeatable. Wachtel (1984) has also reported results from preliminary experiments in sexing day 6 to day 7 bovine embryos using a mouse monoclonal antibody to H-Y antigen in a fluorescent system with an FITC-labeled secondary antibody. Eight sexed embryos were transferred to recipients: seven embryos survived to term, six (85.7%) of which had been sexed correctly; the seventh embryo was a false positive.

The possibility of sexing embryos by measuring metabolic activity has been advanced by Rieger (1984) based on the double complement of X chromosomes in female embryos being reflected in the cellular concentration and activity of certain X-linked enzymes prior to X inactivation. The feasibility of such an approach would be dependent on (1) sufficient time for embryonic enzyme production with two functional X chromosomes to be measured between the depletion of the maternal enzymes initially present in ovum and early cleavage stages of the zygote and X inactivation; (2) a test sufficiently sensitive and specific to be carried out on single embryos; and (3) the viability of the embryos not being prejudiced during the *in vitro* testing period.

Another possible approach to sexing embryos, which will undoubtedly be investigated, is with male-specific DNA probes (Gosden *et al.*, 1982). The technique would have to be done on split embryos or on blastomeres removed from embryos because the probing process is used on fixed cells. Sex differences in mouse DNA have been demonstrated (Jones and Singh, 1981) using a DNA probe derived from the W chromosome of the Indian banded krait snake using the Southern blot technique (Southern, 1975). It has also been shown autoradiographically after labeling this DNA probe with [H^3]thymidine that it hybridizes with the mouse Y chromosome (Singh and Jones, 1982). Thus it may be possible to distinguish a difference between X- and Y-chromosome-bearing embryonic cells using male-specific DNA probes. The time factor involved using autoradiography (2–8 weeks) would be a disadvantage unless the remaining part of the embryo could be cryopreserved satisfactorily.

4. Postimplantation Sexing

An alternative to sexing preimplantation embryos and, perhaps, a last opportunity to influence the sex of the unborn is to sex postimplantation embryos using fetal fluids for hormonal assays or cells present in fetal fluids for chromosomal analysis or hybridization with Y-chromosome-specific DNA probes. The procedure is useful only in the large primiparous animals; to be of practical value to the farmer, the diagnosis has to be made as early as possible in gestation and not later than the time when an unwanted conceptus can be aborted without endangering the dam.

In cattle, the most favorable time to obtain fetal fluids is between 70 and 100 days gestation. The maneuverability of the uterus (Eaglesome and Mitchell, 1977) and the volume of fluids (Arthur, 1957) at this stage combine to make nonsurgical approaches practical and relatively save for the fetus; approx 3–4 weeks remain for tests to be carried out before the safe period for the induction of abortion, i.e., up to 120 days gestation (Dawson, 1974), has expired. Transvaginal (Bongso and Basrur, 1975) and transsacrosciatic (Eaglesome and Mitchell, 1977) approaches have been described, with the latter having the advantage of being applicable to heifers and cows, whereas the former is not always possible in heifers.

A testosterone value of ⩾320 pg/ml in allantoic fluid at 100 days gestation is indicative of a male and a value of ⩽240 pg/ml of a female (Basrur and Bongso, 1980). The main problem is being sure that the aspirated fetal fluids are allantoic rather than amniotic. Color is no guide at this stage, and the purity of the sample has to be determined on the morphological and staining characteristics of the cells it contains (Bongso and Basrur, 1977).

The primary concern using chromosomal analysis is that contamination by maternal cells will lead to a mistaken diagnosis. Therefore, the first 1–2 ml aspirated fetal fluids should be discarded; several cultures should be set up from each sample to increase the opportunities for clones of fetal cells to become established should maternal cells be present (Singh and Hare, 1977). Blood-stained samples should be viewed with extreme caution; if possible, the sample should be discarded and retaken. Also, when male cells are not observed during chromosomal analysis, numerous cells from each culture and from several passages should be studied to minimize the possibility of misinterpreting maternal cells for female fetal cells. Some workers claim that sufficient cell growth for chromosomal analysis is regularly obtained in 4–7 days, whereas experience gained from other cattle studies and many human studies indicates that a more realistic expectation is 14–18 days (Hare and Singh, 1979).

The application of Y-chromosome-specific DNA probes to fetal sex diagnosis has not yet been described in farm animals. Such probes have been used successfully in man for sex determination in the first trimester using chorion biopsy (Gosden *et al.*, 1984; Vergnaud *et al.*, 1984) and amniotic fluid cells (Lau *et al.*, 1984). The advantages claimed for the procedure are that it is extremely accurate, that several tests can be run on the same sample, and that an answer can be obtained in a few days compared with a few weeks using chromosomal analysis of cultured cells. However, with the development of direct methods for chromosomal preparations of chorionic villi (Ford and Jahnke, 1983; Gregson and Seabright, 1983; Simoni *et al.*, 1983), chromosomal analyses can also be carried out in a matter of days. A disadvantage of using Y-chromosome-specific DNA probes rather than chromosomal analysis is that like H-Y antiserum the DNA probe will not detect aberrant sex chromosome complements or autosomal abnormalities.

Before Y-chromosome-specific DNA probes can be used in animals for sex diagnosis, Y-chromosome-specific DNA sequences will have to be identified, possibly for each animal, cloned, and tagged with either an isotopic or preferably a nonisotopic label before hybridization with the DNA in the samples to be studied.

5. Other Approaches to Sex Selection by Predetermination

Parthenogenesis or reproduction without genetic involvement of the sperm occurs naturally in all vertebrate classes, except mammals, and leads to the production of the homogametic sex. Mammalian oocytes, however, can be activated to develop parthenogenetically by various methods, including oocyte fusion (review Markert and Seidel, 1981); for reasons that are not understood development fails, and the furthest it has reached in the mouse has been the 25-somite stage.

Recent work by Hoppe and Illmensee (1982) appeared to hold out hope that the problem may be solvable. Inner cell mass cells from diploid parthogenetically activated mouse oocytes were transferred to fertilized mouse eggs from which the male and female pronuclei were removed. The eggs were cultured to the morula and early blastocyst stage when some of them were transferred to the reproductive tracts of recipients. Four female offspring were born, one of which later proved to be fertile, indicating to Hoppe and Illmensee that the early postimplantation death of parthenogenetic diploids is due to undetermined factors associated with fertilization and normal morphogenetic processes rather than an aberrant genotype. Diploid homozygosity exists when females are produced solely from either the male or the female parent. Briefly, the procedure involves activating ova by fertilization followed in a few hours by the enucleation of one of the pronuclei by microsurgery. The ova are then exposed in culture through one division to an agent, cytochalasin B, that permits karyokinesis while preventing cytokinesis. This results in two pronuclei. The agent is then removed and the pronuclei break down preparatory to the first cell division. The ova are cultured to the morula or early blastocyst stage before being transferred to the reproductive tracts of recipient mice. The success rate for development to term has been very low: thousands of mouse ova have been treated in this way, but only seven live-born young have been reported. All were female—five from the female pronucleus and two from the male pronucleus (Hoppe and Illmensee, 1977). Unfortunately, results on mice published by Surani and Barton (1983), Mann and Lovell-Badge (1984), and McGrath and Solter (1984) do not support these findings, leading Markert (1984) to suggest that "such homozygotes are currently biologically "impossible."

In vitro fertilization using sexed sperm is obviously a means of selecting sex, but it would only be of potential practical value over artificial insemination if, as a system, it would permit the use of sperm incapable of normal fertilization by microinjecting them into the ovum. For example, if the pretreatment of sperm for sorting into X- and Y-chromosome-bearing populations by flow cytometry could be modified so that inoculation of the sperm into eggs would still lead to normal development (Pinkel, 1983).

6. Discussion

The various approaches that can be taken to knowing the sex of the conceptus before its birth have been reviewed. What has to be considered now is the practicality of these approaches from both the commercial and the research standpoint. On a commerical basis, various factors enter into a determination of practicality, including demand, cost, time, numbers, and accuracy.

The demand for embryos of known sex will depend on the degree to which livestock production has been developed in a particular country or part of the world and the economic benefits that can be derived from controlling the sex of the unborn. The value of being able to exercise this control has been the subject of a number of papers (Foote and Miller, 1971; Cunningham, 1975; Van Vleck and Everett, 1976; Seidel and Amman, 1982; Van Vleck, 1981) and was reviewed recently for the United States by Kiddy (1983). His estimate is that the ability to control sex in dairy cattle could be worth about $180 million per year in the United States, less the cost of the technology. For beef cattle, the value of feedlot animals at the time of slaughter could be increased by approximately $425 million per year, if all the females were replaced by males. In addition, the economic value of increased genetic improvement that could be achieved by controlling sex could be substantial. In sheep and swine, the ability to control the sex would have little impact on genetic improvement, but the increased market value for food would be about $5 million and $70 million, respectively.

Unquestionably, an effective method of separating semen into X- and Y-chromosome-bearing fractions would be the ideal approach from a commercial standpoint. It would be easy to apply, and the costs would likely be reasonable once its development had taken place. It would probably be available for all species and therefore within the reach of most livestock producers. Unfortunately, the prospects of having semen from farm animals separated into male and female sperm fractions in the foreseeable future do not look promising. According to Meistrich (1982),

> Based on the known resolution of separation techniques, fractional differences between X- and Y-sperm, and variability between sperm, it appears that none of the physical separation techniques described herein (velocity, sedimentation, equilibrium density, centrifugation, electrophoresis, isoelectric focusing) can effect as good a separation as can be achieved by flow cell sorting based on DNA fluorescence (Meistrich *et al.*, 1978).

Flow sorting based on DNA fluorescence, however, is destructive to sperm; until a method is developed whereby flow sorting can be applied successfully with preservation of fertility, the procedure does not improve the prospects of having sexed semen. What flow sorting can do is to determine the effectiveness of various separation techniques by determining the relative number of X- and Y-chromosome-bearing spermatozoa in a given sample, thereby avoiding the need for field trials.

There is a possibility that there may be a difference in prehaploid gene expression between X- and Y-chromosome-bearing spermatozoa that could be utilized, for example, by using monoclonal antibodies to inactivate either the X or the Y sperm. But this, too, seems a long way off.

As sexing the preimplantation embryo can only be done in conjunction with embryo transfer, application has been limited from a commercial standpoint. The usefulness is still restricted to cattle, although it is possible that it would be of interest in the horse, if the blood horse industry ever fully accepts the practice of embryo transfer. Since sexing the preimplantation embryo is tied to embryo transfer, demand for it has to be looked at in terms of augmenting the advantages to be gained from embryo transfer.

Applications of embryo transfer to the cattle industry have been documented as follows: (1) to increase the number of progeny from valuable dams and sires, (2) to obtain additional progeny from proven valuable but now infertile females, (3) to induce twinning, and (4) to facilitate the international exchange of genetic material.

With the use of superovulation and embryo transfer, reproduction of a fertile female can be increased from a single calf to 10 calves per year. In the dairy cattle industry, for example, in which 70–90% of female calves must be reared as replacements for animals culled for reasons other than low production, this means that, if the replacements were selected from the top 10% of the herd by amplifying the reproduction of superior cows through embryo transfer, the rate of genetic gain could be increased three- to fourfold (Seidel, 1981). Since females are being looked for in this situation, it would undoubtedly be more economical if all these pregnancies were to yield female calves as a result of embryo sexing rather than only 50% of them. In the beef cattle industry, it may be profitable to establish a herd of the polled mutation of a horned breed rapidly using embryo transfer. Here again it would be more economical if the sex of the newborns could be controlled. There may also be an advantage to being able to select the sex of the progeny of an infertile female. Twinning is induced in beef breeds to increase the efficiency of meat production. Steers are worth more than heifers at slaughter, but it is questionable as to whether any economic advantage to be gained by having all male calves would justify the cost of sex selection.

Knowing the sex of embryos for import/export could be economically advantageous. For example, there is more demand in Europe for bull calves from North America than for heifer calves (J. F. Hasler, personal communication, 1983), whereas female calves would be more valuable if embryos were being exported to underdeveloped countries in order to upgrade indigenous breeds.

For embryo sexing to be combined efficiently with embryo transfer, it should meet as many as possible of the following criteria: (1) Not too many of the embryos should be killed or otherwise lost during the sexing process, (2) the method has to be accurate, (3) the diagnosis should be available within 24 hr if fresh embryos are being transferred, (4) it is an advantage to be able to combine the sexing process with cryopreservation, and (5) the costs should not be so high that they exceed the benefits to be gained.

Viewing the various techniques in light of these considerations, sexing prehatched embryos by the sex chromosomal analysis of a few blastomeres is not very efficient in that, on the average, only one in three embryos will be sexed. This means that, on average with the exception of J. Hahn (personal communication, 1983),* only one in six embryos sexed will be of the desired sex, so, with an estimated pregnancy rate of 50% for transferred embryos, 12 embryos will have to be processed in order to be reasonably sure of obtaining one newborn calf of the desired sex. The diagnosis will take 5–6 hr, and 1.5–2 technicians will be required to handle 12 embryos within that time frame. Add to this the cost of equipment (e.g., micromanipulators, microscope) and supplies, and it can be seen that the costs of sexing prehatched embryos by this means are likely to be high. However, the method is accurate, and it has the potential for being combined with low-temperature preservation of the embryo. Although some preliminary results have been disappointing (Popescu and Cribiu, 1982; L. Picard, unpublished data), sexing halved (demi-) embryos by chromosomal analysis should yield a much higher rate of success in sexing, probably in the 90th percentile; thus, with an estimated pregnancy rate of 50% with the transferred other half, an average of four embryos would be required to guarantee having one newborn calf of the desired sex. The time element and other costs will be similar to those for sexing a few isolated blastomeres.

Sexing prehatched embryos by means of H-Y antigen detection using monoclonal antibodies would presumably be much more cost efficient provided that the embryo loss were not great and that the system were sufficiently accurate. It should be possible to sex many more embryos within a given time period with this method than is possible by chromosomal analysis and the only major equipment outlay would be a fluorescence microscope. Presumably quality control will provide the greatest problem. Mouse immunoglobulins are known to be unstable; in addition, mouse × mouse hybrid myelomas from which all the reported monoclonal antibodies to H-Y antigen have been prepared are also unstable, particulary in the early stages of a fusion (Clark *et al.*, 1983). These problems could be circumvented by the use of the rat rather than the mouse system (Clark *et al.*, 1983), as has been done by Utsumi *et al.* (1984). Monoclonal antibodies are likely to be more powerful than conventional anti-H-Y antibody, which is notoriously weak, but "it has proved extremely difficult to make the high titred, high affinity anti-H-Y monoclonal antibody needed to resolve questions of reproducibility" (Simpson, 1982). The tests are difficult to perform and require experienced personnel with adequate training to carry them out (Müller, 1982). Appropriate positive and negative control systems would have to be available to confirm the accuracy of the test: chromosomal analysis is clearly the definitive control system, the questions that arise are:

*See the footnote on page 200.

How many embryos have to be looked at to determine the accuracy of the H-Y test?, and, To what extent will this affect the cost efficiency of sexing by H-Y antigen detection?

Sexing hatched embryos by chromosomal analysis falls between sexing prehatched embryos by chromosomal analysis of a few blastomeres and chromosomal analysis of demi-embryos in terms of efficiency. Two out of three embryos processed will be sexed, meaning that one out of every three embryos processed will be of the desired sex. With a pregnancy rate of 40–50%, on average only one out of every six embryos processed can be expected to yield a newborn calf of the desired sex. The time required for the process is the same as for chromosomal analysis of prehatched embryos. The cost of equipment is considerably less, since a micromanipulator, although helpful, is not necessary and a fluorescent microscope is not required; a distinct disadvantage is that, to date, hatched embryos have not been successfully freeze-thawed, hence the inability to store sexed hatched embryos.

Postimplantation sexing can be done if required, but demand for it will be limited to special situations. With experience and care, the culture success for chromosomal analysis can be expected to be about 90% and the accuracy of the analysis about 95%. With the testosterone assay, the success rate will depend on the ability of the operator to obtain a pure sample of allantoic fluid, not a particularly easy task using nonsurgical procedures at 100 days gestation. Provided that personnel are experienced and strict precautions for asepsis are taken, the risk to the dam and fetus in withdrawing fetal fluids is estimated at 4% or less (Hare and Singh, 1979).

From a research standpoint, the factors that enter into a determination of practicality are probably reduced to numbers and accuracy, since the cost and time required to provide sexed embryos are likely to be fractional compared with the overall cost and importance of the study. For example, of unique value for research could be the sexing of quarter-embryos when identical multiplets are being produced for physiological, endocrinological, nutritional, or genetic studies.

A recent survey (King, 1984) based on the response of nine embryo transfer companies and 11 animal cytogenetics laboratories revealed that none of the embryo-transfer companies offered chromosomal analysis as a service and that only five laboratories were actively engaged in sexing embryos for research purposes. These numbers would undoubtedly increase substantially if a rapid, accurate, and inexpensive method of sexing embryos were to become available.

Acknowledgments. The author acknowledges with gratitude the help that he received from the many scientists who supplied information on request.

References

Adler, H. C., and Autrup, E. H., 1966, Time of insemination, breeding efficiency and sex ratio in cattle, *Aarsberetn. Inst. Sterlitetsfocsk. K. Vet. Landbohojsk.* 155–160.

Arthur, G. H., 1957, Some notes on the quantities of foetal fluids in ruminants with special reference to Hydrops amnii, *Br. Vet. J.* **113**:17–28.

Ballinger, H. J., 1970, Effect of inseminations carried out early or late in oestrus on the sex ratio of calves born, *Vet. Rec.* **86**:631.

Barlow, P., and Vosa, C. G., 1970, The Y-chromosome in human spermatozoa, *Nature (Lond.)* **226**:961–962.

Basrur, P. K., and Bongso, A. T., 1980, Amniocentesis for prenatal detection of sex and cytogenetic defects in cattle, in: *Current Therapy in Theriogenology* (D. A. Morrow, ed.), pp. 1185–1189, W. B. Saunders, Philadelphia.

Bellvé, A. R., 1982, General discussion on genetic expression, in: *Prospects for Sexing Mammalian Sperm* (R. P. Amman and G. E. Seidel, eds.), p. 140, Colorado Associated University Press, Boulder, Colorado.

Benjamin, B. R., 1979, Selective sex determination in farm animals, *Ind. Vet. Med. J.* **3**:141–158.

Betteridge, K. J., Mitchell, D., Eaglesome, M. D., and Randall, G. C. B., 1976, Embryo transfer in cattle 10–17 days after estrus, *Proc. 8th Int. Congr. Anim. Reprod. A.I. Krakow* **3**:237–240.

Betteridge, K. J., Hare, W. C. D., and Singh, E. L., 1981, Approaches to sex selection in farm animals, in: *New Technologies in Animal Breeding* (B. G. Brackett, G. E. Seidel, and S. M. Seidel, eds.), pp. 109–125, Academic Press, New York/London.

Bhattacharya, B. C., 1976, Predetermination of sex in animal reproduction, *Proc. 8th Int. Cong. Anim. Reprod. A.I. Krakow*, **4**:876–879.

Bhattacharya, B. C., Evans, B. M., and Shome, P., 1979, Semen separation technique monitored with greater accuracy by B-body test, *Int. J. Fertil.* **24**:256–259.

Billingham, R. E., and Silvers, W. K., 1960, Studies on tolerance of the Y chromosome antigen in mice, *J. Immunol.* **85**:14–26.

Bongso, T. A., and Basrur, P. K., 1975, Prenatal diagnosis of sex in cattle by amniocentesis, *Vet. Rec.* **96**:124–127.

Bongso, T. A., and Basrur, P. K., 1977, Cytology of bovine foetal fluids, *Can. J. Comp. Med.* **41**:287–293.

Bouters, R., Moyert, L., and Vanderplassche, M., 1980, The primary sex ratio in superovulated bovine embryos, *Proc. 9th Int. Congr. Anim. Reprod. A.I. Madrid* **3** 459 (abst.).

Brackett, B. G., 1983, A review of bovine fertilization "in vitro", *Theriogenology* **19**:1–15.

Brand, A., Aarts, M. H., Zaayer, D., and Oxender, W. D., 1978, Recovery and transfer of embryos by non-surgical procedures in lactating dairy cattle, in: *Current Topics in Veterinary Medicine* (J. M. Sreenan, ed.), Vol. 1, pp. 281–291, Martinus Nijhoff, The Hague/Boston/London.

Cederqvist, L. L., and Fuchs, F., 1970, Antenatal sex determination, *Clin. Obstet. Gynecol.* **13**:159–177.

Chapman, V. M., 1982, Gene products of sex chromosomes, in: *Prospects for Sexing Mammalian Sperm* (R. P. Amman and G. E. Seidel, eds.), pp. 115–117, Colorado Associated University Press, Boulder, Colorado.

Clark, M., Cobbold, S., Hale, G., and Waldmann, H., 1983, Advantages of rat monoclonal antibodies, *Immunol. Today* **4**:100–101.

Cribiu, E. P., 1983, Cytogénétique et détermination du sexe des embryons bovins, *Rec. Med. Vet.* **159**:95–101.

Cunningham, E. P., 1975, The effect of changing the sex ratio on the efficiency of cattle breeding operations, *Livestock Prod. Sci.* **2**:29–38.

Dankowski, K., Köster, U., Schneider, U., and Hahn, J., 1982, Möglichkeiten der zytogenetischen Analyse und Geschlechtsbestimmung von Kaninchen Morulae an isolierten Blastomeren, *Dtsch. Tierärztl Wochenschr.* **89**:248–249.

Darlington, C. D., and La Cour, L. F., 1976, *The Handling of Chromosomes*, 6th ed., pp. 131–132, Allen and Unwin, London.

Dawson, F. L. M., 1974, Methods for early termination of pregnancy in the cow, *Vet. Rec.* **94**:542–548.

Deinhardt, P., and Muller, E., 1978, Untersuchungen uber den Einfluss des Besamungszeitpunktes auf Geschlechtsverhaltnis und Besamungserfolg in einer Voralpenpraxis, *Zuchthyg.* **18**:82 (abst.).

Dolch, K. M., and Chirsman, C. L., 1981, Cytogenetic analysis of preimplantation blastocysts from prepuberal gilts treated with gonadotropins, *Am. J. Vet. Res.* **42**:344–346.

Dusing, C., 1884, Die Regulierung des Geschlechtsverhaltnisses bei der Vermehrung der Meschen, Tiere und Pflanzen, *Jen. Z. Naturwiss.* **17**:593–940.

Eaglesome, M. D., and Mitchell, D., 1977, Collection and cytogenetics of fetal fluids from heifers during the third month of pregnancy. I. Collection, *Theriogenology* **7**:195–201.

Edwards, R. G., 1965, Maturation in vitro of human ovarian oocytes, *Lancet* **2**:926–929.

Eichwald, E. J., and Silmser, C. R., 1955, Untitled communication, *Transplant. Bull.* **2**:148–149.

Engel, W., Klemme, B., and Ebrecht, A., 1981, Serological evidence for H-Y antigen in XO-female mice, *Hum. Genet.* **57**:68–70.

Epstein, C. J., Smith, S., and Travis, B., 1980, Expression of H-Y antigen in preimplantation mouse embryos, *Tissue Antigens* **15**:63–67.

Ericsson, R. J., and Glass, R. H., 1982, Functional differences between sperm bearing the X- or Y-chromosome, in: *Prospects for Sexing Mammalian Sperm* (R. B. Amman and G. E. Seidel, eds.), pp. 201–211, Colorado Associated University Press, Boulder, Colorado.

Ernst, L. K., Gol'dman, I. L., Sergeev, N. I., and Zhivalev, I. K., 1981, A cytogenetic method for determination of the sex of cattle embryos, in: *Doklady Vsesoyuznoi Ordena Lenina Akademii Sel'skokhozyaistvennykh Nauk Im. V.I. Lenina*, Vol. 8, pp. 28–30, Soviet Agricultural Sciences, Allerton Press.

Foote, R. H., 1977, Sex ratios in dairy cattle under various conditions, *Theriogenology* **8**:349–356.

Foote, R. H., 1982, Functional differences between sperm bearing the X- or Y-chromosome, in: *Prospects for Sexing Mammalian Sperm* (R. P. Amman and G. E. Seidel, eds.), pp. 212–218, Colorado Associated University Press, Boulder, Colorado.

Foote, R. H., and Miller, P., 1971, What might sex ratio control mean in the animal world?, in: *Sex Ratio at Birth—Prospects for Control: A Symposium* (C. A. Kiddy and H. D. Hafs, eds.), pp. 1–9, American Society for Animal Science, Champaign, Illinois.

Ford, J. H., and Jahnke, A. B., 1983, Handling chorionic villi for direct chromosome studies, *Lancet* **2**:1491–1492.

Galbraith, G. M. P., Galbraith, R. M., Faulk, W. P., and Wachtel, S. S., 1978, Detection of H-Y antigen by fluorescence microscopy, *Transplantation* **26**:25–27.

Gardner, K. E., 1950, The sex ratio in calves resulting from artificial insemination, *J. Dairy Sci.* **33**:391 (abst.).

Gardner, R. L., and Edwards, R. G., 1968, Control of the sex ratio at full term in the rabbit by transferring sexed blastocysts, *Nature (Lond.)* **218**:346–348.

Garner, D. L., Gledhill, B. L., Pinkel, D., Lake, S., Stephenson, D., Van Dilla, M. A., and Johnson, L. R., 1983, Quantification of the X- and Y-chromosome-bearing spermatozoa of domestic animals by flow cytometry, *Biol. Reprod.* **28**:312–321.

Gebicke-Harter, P. J., Michelmann, H. W., Bonhoff, A., and Paufler, S., 1977, Untersuchungen zur Beiinflussung des Geschlechtsverhaltnisses beim Rind nach Besamung zu unterschiedlichen Zeiten nach Brunstbeginn, *Zuchthyg.* **12**:113–116.

Gosden, J. R., Mitchell, A. R., Gosden, C. M., Rodeck, C. H., and Morsman, J. M., 1982, Direct vision chorion biopsy and chromosome-specific DNA probes for determination of fetal sex in first-trimester prenatal diagnosis, *Lancet* **2**:1416–1419.

Gosden, J. R., Gosden, C. M., Christie, S., Cooke, H. J., Morsman, J. M., and Rodeck, C. H., 1984, The use of cloned Y chromosome-specific DNA probes for fetal sex determination in first trimester prenatal diagnosis, *Hum. Genet.* **66**:347–351.

Gregson, N. M., and Seabright, M., 1983, Handling chorionic villi for direct chromosome studies, *Lancet* **2**:1491.

Hare, W. C. D., and Singh, E. L., 1979, *Cytogenetics in Animal Reproduction*, p. 46, Commonwealth Agricultural Bureaux, Slough.

Hare, W. C. D., Mitchell, D., Betteridge, K. J., Eaglesome, M. D., and Randall, G. C. B., 1976, Sexing two-week old bovine embryos by chromosomal analysis prior to surgical transfer: Preliminary methods and results, *Theriogenology* **5**:254–253.

Hare, W. C. D., Singh, E. L., Betteridge, K. J., Eaglesome, M. D., Randall, G. C. B., and Mitchell, D., 1978, Embryo sexing with particular reference to cattle, in: *Current Topics in Veterinary Medicine* (J. M. Sreenan, ed.), pp. 441–449, Martinus Nijhoff, The Hague.

Hedge, U. C., Shastry, P. R., and Rao, S. S., 1981, Phytohaemagglutinin as a molecular probe to study the membrane constituents of human X- and Y-bearing spermatozoa, *J Reprod. Immunol.* **2**:351–357.

Heyman, Y., and Chesné, P., 1984, Freezing bovine embryos: survival after cervical transfer of one half, one or two blastocysts frozen in straws, *Theriogenology* **21**:240 (abst.).

Hoppe, P. C., and Illmensee, K., 1977, Microsurgically produced homozygous–diploid uniparental mice, *Proc. Natl. Acad. Sci. (USA)* **74**:5657–5661.

Hoppe, P. C., and Illmensee, K., 1982, Full-term development after transplantation of parthogenetic embryonic nuclei into fertilized mouse eggs, *Proc. Nat. Acad. Sci. USA* **79**:1912–1915.

Hoppe, P. C., and Koo, G. C., 1984, Reacting mouse sperm with monoclonal H-Y antibodies does not influence sex ratio of eggs fertilized *in vitro*, *J. Reprod. Immunol.* **6**:1–9.

Jones, K. W., and Singh, L., 1981, Conserved repeated DNA sequences in vertebrate sex chromosomes, *Hum. Genet.* **58**:46–53.

Kanagawa, H., Frim, J., and Kruuv, J., 1979, The effect of puncturing the zona pellucida on freeze-thaw survival of bovine embryos, *Can. J. Anim. Sci.* **59**:623–626.

Kaneko, S., Oshio, S., Kobayashi, T., Iizuka, R., and Mohri, H., 1984, Human X- Y-bearing sperm differ in cell surface sialic acid content, *Biochem. Biophys. Res. Commun.* **124**:950–955.

Kiddy, C. A., 1983, Impact on animal production systems of technology to control the sex of animals born, in: *Symposium on Control of Mammalian Sex-Ratio at Birth, Annual Meeting of the American Association for the Advancement of Science,* Detroit.

King, W. A., 1984, Sexing embryos by cytological means, *Theriogen* **21**:7–17.

King, W. A., Linares, T., Gustavsson, I., and Bane, A., 1979, A method for preparation of chromosomes from bovine zygotes and blastocysts, *Vet. Sci. Commun.* **3**:51–56.

Koo, G. C., and Varano, A., 1981, Inhibition of H-Y cell-mediated cytolysis by monoclonal H-Y specific antibody, *Immunogenetics* **11**:255–266.

Koo, G. C., Tada, N., Chaganti, R., and Hammerling, U., 1981, Application of monoclonal anti-HY antibody for human H-Y typing, *Hum. Genet.* **57**:64–67.

Krco, C. J., and Goldberg, E. H., 1976, H-Y (male) antigen: Detection on eight-cell mouse embryos, *Science* **193**:1134–1135.

Lau, Y.-F., Huang, J. C., Dozy, A. M., and Kan, V. W., 1984, A rapid screening test for antenatal sex determination, *Lancet* **1**:14–16.

Lehn-Jensen, H., 1984, Deep freezing of cattle embryos, *Proc. 10th Int. Cong. Anim. Reprod. A.I. Urbana-Champaign* **4**:11, 1–11, 12.

Lehn-Jensen, H., and Willadsen, S. M., 1983, Deep-freezing of cow "half" and "quarter" embryos, *Theriogenology* **19**:49–54.

Long, S. E., 1977, Cytogenetic examination of preimplantation blastocysts of ewes mated to rams heterozygous for the Massey I(ti) translocation, *Cytogenet. Cell Genet.* **18**:82–89.

Long, S. E., and Williams, C. V., 1978, Chromosomal abnormalities in ova and zygotes from normal ewes mated to normal rams, *Vet. Rec.* **102**:153.

Long, S. E., and Williams, C. V., 1982, A comparison of the chromosome complement of inner cell mass and trophoblast cells in day-10 pig embryos, *J. Reprod. Fertil.* **66**:645–648.

Lush, J. L., 1925, Possibility of sex control by artificial insemination with centrifuged spermatozoa, *J. Agric. Res.* **30**:893–913.

Lyster, W. R., 1972, The sex ratios of human and sheep births in areas of high mineralization, *Int. J. Environ. Studies* **2**:309–316.

Mann, J. R., and Lovell-Badge, R. H., 1984, Inviability of parthenogenones is determined by pronuclei, not egg cytoplasm, *Nature (Lond.)* **310**:66–67.

Markert, C. L., 1984, Genetic manipulation of mammalian embryos: Current techniques and their potential usefulness in livestock improvement, *Proc. 10th Int. Congr. Anim. Reprod A.I. Urbana-Champaign* **4**:11, 13–11, 19.

Markert, C. L., and Seidel, G. E., 1981, Parthenogenesis, identical twins, and cloning in mammals, in: *New Technologies in Animal Breeding* (B. G. Brackett, G. E. Seidel, and S. M. Seidel, eds.), pp. 181–200, Academic Press, New York/London.

Martin, R. H., Lin, C. C., Balkan, W., and Burns, K., 1982, Direct chromosome analysis of human spermatozoa: Preliminary results from 18 normal men, *Am. J. Hum. Genet.* **34**:459–468.

Martin, R. H., Balkan, W., Burns, K., Rademaker, A. W., Lin, C. C., and Rudd, N. L., 1983, The chromosome constitution of 1000 human spermatozoa, *Hum. Genet.* **63**:305–309.

McClung, C. E., 1902, The accessory chromosome-sex determinant?, *Biol. Bull.* **3**:43–83.

McFeely, R. A., 1966, A direct method for the display of chromosomes from early pig embryos, *J. Reprod. Fertil.* **11**:161–163.

McFeely, R. A., and Rajakoski, E., 1968, Chromosome studies on early embryos of the cow, *Proc. 6th Int. Cong. Anim. Reprod. A.I. Paris* **2**:905–907.

McGrath, J., and Solter, D., 1984, Completion of mouse embryogenesis requires both the maternal and paternal genomes, *Cell* **37**:179–183.

Meistrich, M. L., 1982, Potential and limitations of physical methods for separation of sperm bearing an X- or Y-chromosome, in: *Prospects for Sexing Mammalian Sperm* (R. P. Amman and G. E. Seidel, eds.), pp. 143–168, Colorado Associated University Press, Boulder, Colorado.

Meistrich, M. L., Gohde, W., White, R. A., and Schumann, J., 1978, Resolution of X and Y spermatids by pulse cytophotometry, *Nature (Lond.)* **274**:821–823.

Melvoid, R. W., Koln, H. I., Yerganian, G., and Fawcett, D. W., 1977, Evidence suggesting the existence of two H-Y antigens in the mouse, *Immunogenetics* **5**:33–41.

Mitchell, D., 1977, Sexing of embryos, in: *Embryo Transfer in Farm Animals* (K. J. Betteridge ed.), p. 26, Monograph 16, Agriculture Canada, Ottawa.

Moustafa, L. A., Hahn, J., and Roselius, R., 1978, Versuche zur Geschlechtsbestimmung an Tag 6 und 7 alten Rinderembryonen, *Ber. Munch. Tierarztl. Wochenschr.* **91**:236–238.

Müller, U., 1982, Identification and function of serologically detectable H-Y antigen, *Hum. Genet.* **61**:91–94.

Müller, U., and Bross, K., 1979, A highly sensitive peroxidase-antiperoxidase method for detection of H-Y antigen in cultivated human fibroblasts, *Hum. Genet.* **52**:143–147.

Müller, U., and Urban, E., 1981, Reaggregation of rat gonadal cells in vitro: Experiments on the function of H-Y antigen, *Cytogenet. Cell Genet.* **31**:104–107.

Murer-Orlando, M., Betteridge, K. J., and Richer, C.-L., 1982, Cytogenetic sex determination in cultured cells of pre-attachment horse embryos, *Proc. 5th Eur. Colloq. Cytogenet. Dom. Anim. Milano-Gargnano*, 372–378.

Niermeijer, M. F., Sachs, E. S., Jahodora, M., Tichelaar-Klepper, C., Kleijer, W. J., and Galjaard, H., 1976, Prenatal diagnosis of genetic disorders, *J. Med. Genet.* **13**:182–194.

Parkes, A. S., 1971, Mythology of the human sex ratio, in: *Sex Ratio in Birth—Prospects for Control: A Symposium* (C. A. Kiddy and H. D. Hafs, eds.), pp. 38–42, American Society for Animal Science, Champaign, Illinois.

Pearl, R., 1917, The control of the sex ratio, *Maine Agric. Exp. Sta. Bull.* **261**:130–131.

Pearl, R., and Parshley, H. M., 1913, Data on sex determination in cattle, *Biol. Bull.* **24**:205–225.

Picard, L., King, W. A., Betteridge, K. J., 1984, Cytological studies of bovine half-embryos, *Theriogenology* **21**:252 (abst.).

Pinkel, D., 1983, Identification and separation of male gametes, in: *Symposium on Control of Mammalian Sex-Ratio at Birth, Annual Meeting of the American Association for the Advancement of Science*, Detroit (abst.).

Pinkel, D., Garner, D. L., Gledhill, B. L., Lake, S., Stephenson, D., and Johnson, L. A., 1983, The proportions of X- and Y-chromosome-bearing spermatozoa in purportedly enriched bovine semen, *75th Annu. Meet. Am. Soc. Anim. Sci.* (abst.).

Polge, C., and Rowson, L. E. A., 1973, Recent progress in techniques for increasing reproductive potential in farm animals, in: *Proceedings of the Third World Conference on Animal Production*, pp. 633–643, Sydney University Press.

Popescu, C. P., and Cribiu, E. P., 1982, L'étude cytogénétique de l'embryon bovin, *Proc. Int. Congr. Emb. Trans. Mamm.*, pp. 277–283, Annecy.

Popescu, C. P., Boscher, J., and Cribiu, E. P., 1982, Collagen substrates for cytogenetic studies of domestic animal embryos, *J. Hered.* **73**:147–148.

Rieger, D., 1984, The measurement of metabolic activity as an approach to evaluating viability and diagnosing sex in early embryos, *Theriogenology* **21**:138–149.

Roschlau, K., Schwiderski, H., Kauffold, P., Rommel, P., and Rehbock, F., 1983, Geschlechtsdiagnose an praimplantiven Embryonen als Teilverfahren der "Eitransplantation," *Arch. Tierzucht. Berl.* **26**:209–216.

Rottman, O. J., 1981, Chromosome preparations from single blastomeres after colcemid treatment and removal from rabbit morulae unsuitable for sexing in routine embryo transfer, *Theriogenology* **15**:321–326.

Rowson, L. E. A., 1974, The role of research in animal production, *Vet. Rec.* **95**:276–280.

Rowson, L. E. A., and Moor, R. M., 1966, Development of the sheep conceptus during the first fourteen days, *J. Anat.* **100**:777–785.

Rudak, E., Jacobs, P. A., and Yanagimachi, R., 1978, Direct analysis of the chromosome constitution of the human spermatozoa, *Nature (Lond.)* **274**:911–912.

Seidel, G. E., 1981, Superovulation and embryo transfer in cattle, *Science* **211**:351–358.

Seidel, G. E., and Amman, R. P., 1982, The impact of using sexed semen, in: *Prospects for Sexing Mammalian Sperm* (R. P. Amman and G. E. Seidel, eds.), pp. 1–4, Colorado Associated University Press, Boulder, Colorado.

Shapley, D., 1983, News report, Cattle Breeding: Techniques for sexing embryos now possible, *Nature (Lond.)* **301**:101.

Silvers, W. K., Gasser, D. L., and Eicher, E. M., 1982, H-Y antigen, serological detectable male antigen and sex determination, *Cell* **28**:439–440.

Simoni, G., Brambati, B., Danesino, C., Rosella, F., Terzoli, G. L., Ferrari, M., and Fraccaro, M. 1983, Efficient direct chromosome analyses and enzyme determinations from chorionic villi samples in the first trimester of pregnancy, *Hum. Genet.* **63**:349–357.

Simpson, E., 1982, The role of H-Y as a minor transplantation antigen, *Immunol. Today* **3**:97–106

Simpson, E., McLaren, A., and Chandler, P., 1982, Evidence for two male antigens in mice, *Immunogenetics* **15**:609–614.

Singh, E. L., and Hare, W. C. D., 1977, Collection and cytogenetics of fetal fluids from heifers during the third month of pregnancy. II. Cytogenetic studies, *Theriogenology* **7**:203–214.

Singh, E. L., and Hare, W. C. D., 1980, The feasibility of sexing bovine morula stage embryos prior to embryo transfer, *Theriogenology* **14**:421–427.

Singh, L., and Jones, K. W., 1982, Sex reversal in the mouse (*Mus musculus*) is caused by a recurrent nonreciprocal crossover involving the X and an aberrant Y chromosome, *Cell* **28**:205–216.

Southern, E. M., 1975, Detection of specific sequences among DNA fragments separated by gel electrophoresis, *J. Mol. Biol.* **98**:503–517.

Stolkowski, J., 1977, Le magnesium dans la reproduction animale et chez l'homme, *Rev. Can. Biol.* **36**:135–177.

Sumner, A. T., and Robinson, J. A., 1976, A difference in dry mass between the heads of X- and Y-bearing human spermatozoa, *J. Reprod. Fertil.* **48**:9–15.

Surani, A. M. H., and Barton, S. C., 1983, Development of gynogenetic eggs in the mouse: Implications for parthogenetic embryos, *Science* **222**:1034–1036.

Thury, 1863, *Über das Gesetz der Erzeugung der Geschlechter*, Leipzig.

Utsumi, K., Satoh, E., and Yuhara, M., 1984, Sexing of goat and cow embryos by rat H-Y antibody, *Proc. 10th Int. Congr. Anim. Reprod. A.I. Urbana-Champaign* **2**:234.

Van Vleck, L. D., 1981, Potential genetic impact of artificial insemination, sex selection, embryo transfer, cloning and selfing in dairy cattle, in: *New Technologies in Animal Breeding* (B. G. Brackett, G. E. Seidel, and S. M. Seidel, eds.), pp. 221–242, Academic Press, New York.

Van Vleck, L. D., and Everett, R. W., 1976, Genetic value of sexed semen to produce dairy heifers, *J. Dairy Sci.* **59**:1803–1807.

Vergnaud, G., Kaplan, L., Weissenbach, J., Dumez, Y., Berger, R., Tiollais, P., and Guellaen, G., 1984, Rapid and early determination of sex using tropoblast biopsy specimens and Y chromosome specific DNA probes, *Br. Med. J.* **289**:73–76.

Vosa, G. C., 1970, Heterochromatin recognition with flurochromes, *Chromosoma (Berl.)* **30**:366–372.

Wachtel, S. S., 1983, *H-Y Antigen and the Biology of Sex Determination*, Grune and Stratton, New York.

Wachtel, S. S., 1984, H-Y antigen in the study of sex determination and control of sex ratio, *Theriogenology* **21**:18–28.

White, K. L., Bradbury, M. W., Anderson, G. B., and BonDurant, R. H., 1984, Immunofluorescent detection of a male-specific factor on preimplantation bovine embryos, *Theriogenology* **21**:275 (abst.).

White, K. L., Lindner, G. M., Anderson, G. B., and BonDurant, R. H., 1982, Survival after transfer of "sexed" mouse embryos exposed to H-Y antisera, *Theriogenology* **18**:655–662.

White, K. L., Lindner, G. M., Anderson, G. B., and BonDurant, R. H., 1983, Cytolytic and fluorescent detection of H-Y antigen on preimplantation mouse embryos, *Theriogenology* **19**:701–705.

Willadsen, S. M., 1979, A method for culture of micromanipulated sheep embryos and its use to produce monozygotic twins, *Nature (Lond.)* **277**:298–300.

Willadsen, S. M., 1980, The viability of early cleavage stages containing half the normal number of blastomeres in the sheep, *J. Reprod. Fertil.* **59**:357–362.

Willadsen, S. M., 1981, The developmental capacity of blastomeres from 4- and 8-cell sheep embryos, *J. Embryol. Exp. Morphol.* **65**:165–172.

Willadsen, S. M., and Polge, C. E., 1979, Cited by Willadsen, S. M., 1982, Micromanipulation of embryos of the large domestic species, in: *Mammalian Egg Transfer* (C. E. Adams, ed.), pp. 185–210, CRC Press, Boca Raton, Florida.

Willadsen, S. M., Pashen, R., and Allen, W. R., 1980, cited by Willadsen, S. M., 1982, Micromanipulation of embryos of the large domestic species, in: *Mammalian Egg Transfer* (C. E. Adams, ed.), pp. 185–210, CRC Press, Boca Raton, Florida.

Williams, T. J., Elsden, R. P., and Seidel, G. E., 1983, Bisecting bovine embryos: methods, applications and success rates, *Proc. Annu. Conf. A.I. Emb. Trans. in Beef Cattle, Fort Collins, National Association of Animal Breeders*, pp. 45–51.

Wilson, E. B., 1909, Recent researches on the determination and heredity of sex, *Science* **29**:53–70.

Wintenberger-Torres, S., and Popescu, C. P., 1980, Transfer of cow blastocysts after sexing, *Theriogenology* **14**:309–318.

Wodsedalek, J. E., 1913, Spermatogenesis of the pig with special reference to the accessory chromosomes, *Biol. Bull.* **25**:8–45.

Wodsedalek, J. E., 1914, Spermatogenesis of the horse with special reference to the accessory chromosome and the chromatoid body, *Biol. Bull.* **27**:295–313.

Wodsedalek, J. E., 1920, Studies on the cells of cattle with special reference to spermatogenesis, oögonia and sex-determination, *Biol. Bull.* **38**:290–309.

Zavos, P. M., 1983, Preconception sex determination via intravaginal administration of H-Y antisera in rabbits, *Theriogenology* **20**:235–240.

Zecht, L., 1969, Investigations of metaphase chromosomes with DNA binding fluorochromes, *Exp. Cell Res.* **58**:463 (abst.).

Chapter 7

Cryopreservation of Mammalian Eggs and Embryos

IAN WILMUT

1. Introduction

In the decade since reliable methods of freezing mouse embryos were first described (Whittingham *et al.*, 1972; Wilmut, 1972*a*) methods have been developed for freezing the embryos of a number of other species. Frozen embryos have been used in the treatment of human infertility, in farm animal breeding, and in biological research. During this period, the procedures have also been simplified.

In this review, the present procedures and the possibility of their further improvement are discussed within the framework provided by current concepts in cryobiology. Consideration is also given to the failure to develop successful methods for some species. Finally, the application of methods of preserving embryos is considered.

2. Sensitivity to Temperatures above Freezing

The embryos of some species are killed when cooled in ice. In the pig, embryos at all stages from 8-cell to expanded blastocyst have been cooled in a variety of media. No embryos survived when cooled to 0°C regardless of the rate of cooling or the choice of medium. A few cells in later blastocysts did survive, but not in sufficient proportion to ensure embryo survival (Wilmut, 1972*b*; Polge *et al.*, 1974).

In cattle and sheep, the proportion of embryos surviving cooling to 0°C depends on the stage of development (Wilmut *et al.*, 1975; Trounson *et al.*, 1976). In cattle only 1 of 19 8–16-cell embryos (day 4) continued their development after being cooled to 0°C. By contrast, 7 of 9 morulae collected on day 6 developed after this treatment (Wilmut *et al.*, 1975). Further analysis of this

IAN WILMUT • AFRC Animal Breeding Research Organization, Dryden Laboratory, Roslin Midlothian EH25 9PS, United Kingdom.

effect revealed that many day 5 embryos were killed by cooling to a temperature below 7.5°C (Trounson *et al.*, 1976; Willadsen *et al.*, 1976*b*). In sheep only one of 12 2–4-cell embryos survived when cooled to 0°C, whereas 5 of 7–8-cell to morula stage embryos survived the same treatment (Willadsen *et al.*, 1976*a*). It seems that in both species, change occurs in the sensitivity to cooling as development proceeds. Sheep embryos may become tolerant at an earlier stage than cattle embryos, but very limited information is available. It also seems that individual embryos vary in their sensitivity, as occasional embryos survive treatments that kill the great majority of other embryos.

Pig embryos are killed at a particular temperature; the rate of cooling to that temperature has no effect on survival. Embryos at the 8-cell stage were cooled to a temperature between 20° and 0°C in phosphate-buffered saline (PBS), either over a 2-hr period or by placing the test tubes directly into water at the selected temperature (Table I). The rate of cooling had no effect on survival, but all embryos were killed when cooled to a temperature of 10°C or below (Wilmut, 1972*b*). Embryos cooled only to 15°C developed normally in culture and were able to implant after transfer into synchronous recipients.

This effect is in contrast with the phenomenon of "cold shock," which is observed with spermatozoa of some species (see Walton, 1957) and with other cell types if they are suspended in hypertonic salt solutions (Morris and Farrant, 1973). The extent of damage due to cold shock is determined by the rate of cooling. However, cells given the chance to acclimatize become partially resistant to the effect of further stress (Chang and Walton, 1940; Morris and Farrant, 1973).

The mechanism of embryo death at a particular temperature is not known, but it probably involves the cell membranes. Pig embryos at 20°C were allowed to take up the vital stain neutral red before being placed in 0.2 ml PBS in test tubes placed in ice. However, quickly they were recovered, the embryos were found to have lost the dye. By contrast, embryos remaining at room temperature retained the dye over a long period (I. Wilmut, unpublished observations, 1974). The fact that neutral red left the cells within 1 or 2 min implies that the cell membranes are the primary site of injury, as damage to lysosomes or nuclei would take longer to cause changes in membrane permeability.

Table I. Effect of Cooling to a Temperature between Room Temperature and 0°C on Subsequent Development in Culture of 8-Cell Pig Embryos

	Lowest temperature (°C)			
	5	10	15	20
No. developed	0	0	16	13
No. cultured	20	18	19	16
% developed	0	0	84	81

There is evidence of phase changes in the lipids of cell membranes as the temperature is altered (reviewed by Edidin and Petit, 1977). Phase changes of this type are usually reversible, but it is possible that proteins that are not normally able to interact are brought together and that in the case of temperature-sensitive embryos irreversible interactions take place. While the addition of specific phospholipids protects erythrocytes (Lovelock, 1954) and spermatozoa (Butler and Roberts, 1975) against cold shock, no similar effect has been described for embryos. It may be that protection will be provided only by modification of membrane lipids during culture.

Evidence of sensitivity to temperatures above zero has been observed only in pig, cattle, and sheep embryos, but when developing methods of freezing for a species whose embryos have not previously been frozen it would be prudent to confirm that they tolerate cooling in ice.

There is little information about the effect on embryo survival of varying the rate of cooling to 0°C. One experiment suggested that early cow morulae (days 5 and 6) were more likely to be damaged by rapid cooling (Trounson et *al.*, 1976). By contrast, large numbers of embryos of several species have survived freezing and thawing after cooling directly in ice (around 10°C/min). These differences warrant further investigation in a serach for methods giving maximum survival.

3. Current Concepts in Cryobiology

The involvement and significance of water in biological systems is not always fully appreciated (for reviews, see Dick, 1966; Tait and Franks, 1971). As many of the disruptive effects of freezing are due to changes in the relationship between water and other molecules, it is pertinent to outline briefly the structure and role of water in the cell.

3.1. Water in Cell Structure and Function

Pure water may be considered a continuously interchanging mixture of clusters of structured, icelike water within a mass of less-structured molecules (Frank, 1965). Decrease in temperature results in an increase in the proportion of structured water; localized changes in the form of the water are also produced by the introduction of solute molecules. Hydrophilic moieties have been shown to order water, probably in hexagonal ice form, whereas cages of water (clathrates) form around hydrophobic groups (see Tait and Franks, 1971). An interaction between macromolecules and water places a significant constraint on the water molecules and has a major role in determining the form of the macromolecule (Klotz, 1965). Nuclear magnetic resonance (NMR) analysis showed that water "bound" to proteins had far less activity than did pure water at temperatures above 0°C, but during cooling the bound water did not become completely immobile until a temperature of −60°C (Blears and Danyluk, 1968).

In muscle cells, two major phases of water have been detected by NMR studies (Hazlewood *et al.*, 1969). One phase is probably water tightly bound to macromolecules, while the other represents the bulk phase. The water in the bulk phase has lost significant motional freedom in comparison with that of pure water or water in plasma. As much of the lost freedom was restored when the tissue was denatured by heating, Hazlewood and co-workers concluded that the ordering of the water was brought about by interaction with macromolecules.

The role of water in maintaining cell membrane structure and function is well known (see Hechter, 1965). Membranes provide a limit to the cell, but also form the boundary to organelles such as the nucleus, mitochondria, the Golgi apparatus, and lysosomes. In addition, they provide a framework upon which enzymes and substrates can interact.

All these observations suggest that water in different regions of the cell will vary in structure and that disruption of this structure could have catastrophic effects on the cell. It is also established that changes in temperature alter the interrelationships.

At physiological temperature DMSO was able to diffuse throughout the water of guinea pig smooth muscle. By contrast, when this type of muscle was cooled to −8°C not all the water was available, and it was noted that the proportion lost to the DMSO was very similar to that estimated in other ways to be bound to the muscle protein (Elford, 1970*a*,*b*).

Thus it may be envisaged that a very small proportion of water in the cells is intimately bound to macromolecules and is essential for their normal structure and function. Changes in temperature, formation of ice, and addition of cryoprotective agents may all modify this interrelationship and so cause cell death. Under different circumstances it has been suggested that freezing and thawing kills cells because of the increase in concentration of electrolytes (Lovelock, 1953), the formation of intracellular ice (Mazur, 1970), and osmotic shock (Farrant and Morris, 1973). It may be that a more general understanding will become apparent as more is learned of the role of water in cell functions and the effect on water in cells of the many different methods of cooling and warming.

3.2. Changes during Freezing

The probability of cells surviving when frozen and thawed is greatly dependent on the rate of cooling, the rate of warming, the choice and concentration of cryoprotective agent, and the type of cell. In addition, some or all of these factors may interact. Optimum cooling rate varies from around 0.3°C/min for mammalian embryos in 10% glycerol or DMSO (Whittingham *et al.*, 1972; Wilmut, 1972*a*) to several thousand degrees per minute for human erythrocytes with no cryoprotectant (Rapatz *et al.*, 1968).

The demonstration of intermediate optimum cooling rates implies that at least two mechanisms cause damage (reviewed by Mazur, 1970). It is suggested that cooling at rates above optimum causes death after the formation of intracellular ice. By contrast, cooling more slowly than optimum exposes cells to concentrated solutions for prolonged periods, and it is this which is judged to be damaging. The optimum rate of cooling for a particular type of cell is determined by the ratio of surface area to volume and by the permeability of the cell membrane to water (Mazur, 1970). The changes that occur during cooling at several different rates, will be discussed before their implications for embryo survival are considered (see Fig. 1). When a cell is cooled slowly, ice generally forms first in the extracellular fluid (Mazur, 1970), and this is certainly the case in samples induced to freeze by seeding. As a result, the vapor pressure of the water in the extracellular fluid is lower than that in the cell; i.e., the solutes are more concentrated in the extracellular fluid as water is removed into ice. Equilibrium across the cell membrane is restored by loss of water from the cell

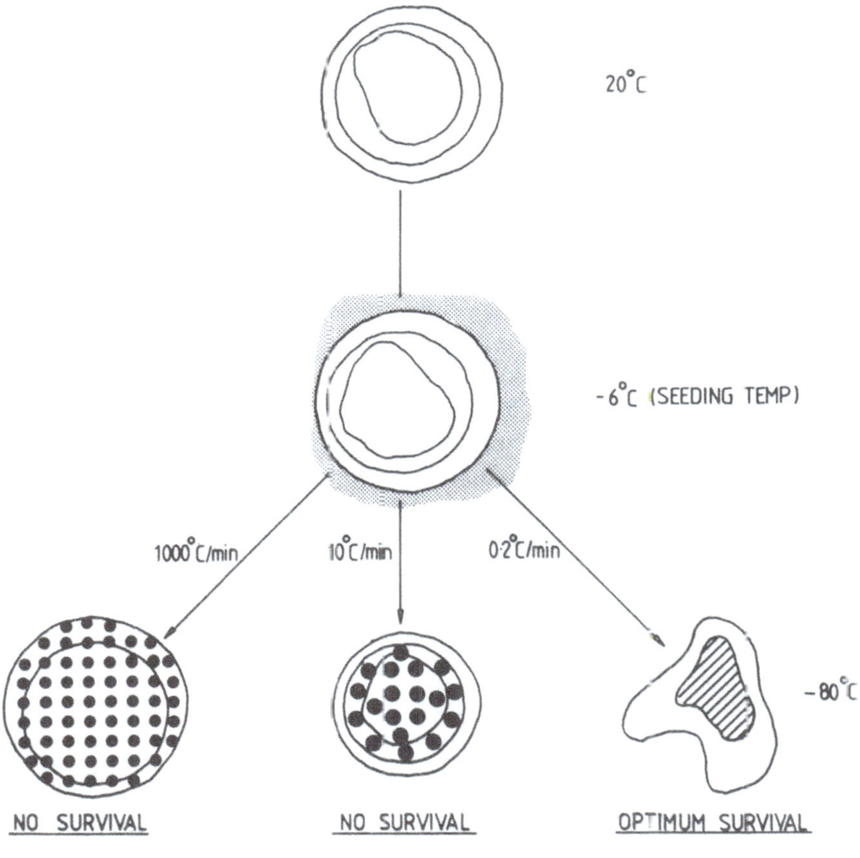

Figure 1. Effect of cooling on mouse 1-cell embryos. At seeding temperature ice is induced to form outside the zona pellucida. If further cooling is at an optimum rate (around 0.2°C/min), no visible ice forms within the embryo, which becomes shrunken. If the cooling rate is 10° or 1000°C, ice forms within the embryo, and at the faster rate in the zona pellucida.

provided that the rate of cooling permits sufficient time. As cooling continues, the cells are exposed to increasing solute concentration, changes in pH, a reduction in the proportion of unfrozen water, and a marked reduction in cell volume. Under these extreme circumstances, some solutes that are normally excluded may be able to enter the cell. There is also the possiblity of physical damage from the extracellular ice crystals. It is not clear which of these many facets of the complex sequence of events is responsible for damage to the cells. There are many hypotheses to account for the death of cells and reviews of the topic are offered by Mazur *et al.* (1970) and Farrant (1980).

If cell suspensions are cooled too rapidly for osmotic equilibration to be maintained, the intracellular water becomes progressively more and more supercooled. Ultimately this leads to the formation of intracellular ice. The amount of intracellular ice and the size of the ice crystals depends upon the rate of cooling (see Fig. 1). After the most rapid cooling, the greatest proportion of intracellular water is retained as ice and the ice crystals are smallest.

The terms rapid and slow cooling do not have absolute values, but vary for each type of cell and with the suspending medium. Those cooling rates found to provide maximum survival are close to the maximum rate at which osmotic equilibrium can be maintained. This rate can vary from less than 1 °C/min to several thousand degrees per minute.

3.3. Effects of Cryoprotective Compounds

A considerable number of compounds have been used to provide protection during freezing and thawing. They are commonly classified as permeating or nonpermeating depending on whether the compound is able to enter cells readily at temperatures above 0°C (although permeation may not be essential for protection). As embryos have been preserved only with low-molecular-weight compounds, the extracellular agents will not be considered. Typically, permeating agents are low-molecular-weight compounds having a high solubility in water at room temperature. As they are used at concentrations of 1–2 M (10–20%; v/v), they must also have a low toxicity for the cell being preserved. In some procedures, the compounds are added in such a way that they reach equal concentrations across the cell membrane, while under other circumstances equilibrium is not achieved. In the latter case, this will have the effect of dehydrating the cells.

The effect of these agents during cooling is to lower the freezing point, typically from −0.5°C to around −4°C. As freezing occurs, they decrease the proportion of water frozen into ice at any particular temperature. As a result, the concentration of electrolyte generated during freezing is much lower in the presence of agent than when the same solution is frozen without additive (Fig. 2); this has two benefits for a cell. First, the maximum concentration of electrolyte is lower in the presence of agent. Second, any particular concentration of electrolyte is reached at a lower temperature. The rate of any chemical effect of concentrated electrolyte on cells is likely to be slower at lower tem-

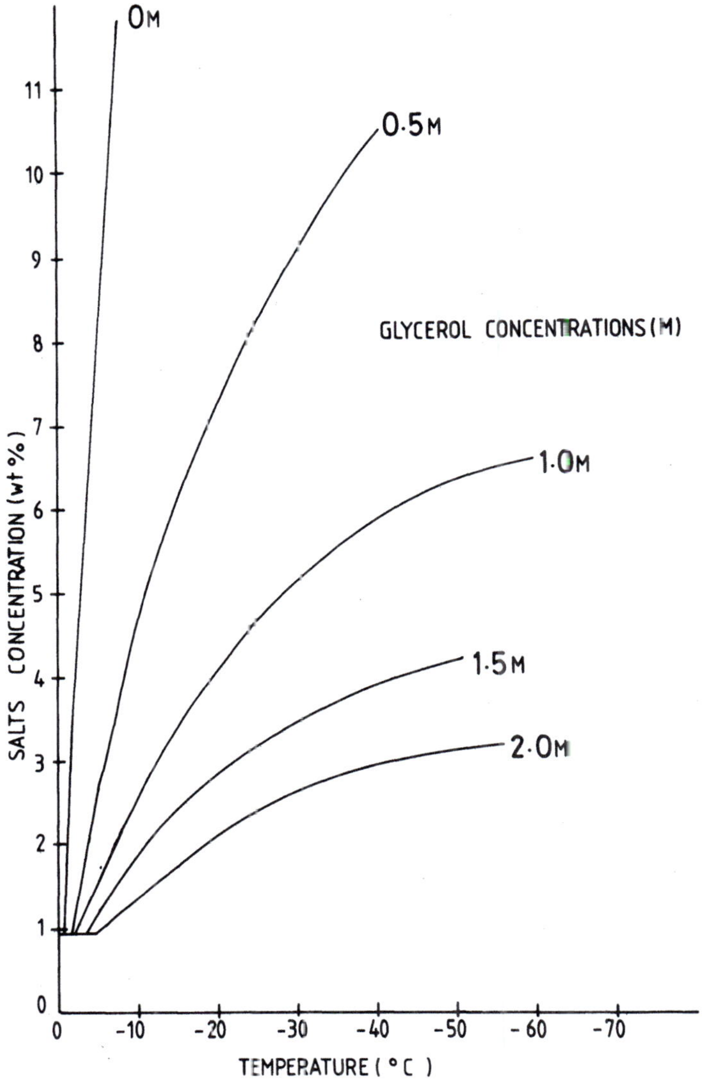

Figure 2. Effect of glycerol concentration on the concentration of salts in phosphate-buffered physiological saline cooled to subzero temperatures.

peratures. Thus, the cells in the presence of agent will be better able to tolerate a particular concentration of electrolyte as it will be created at a lower temperature. By contrast, these agents do not reduce the effect of rapid cooling.

3.4. The Influence of Warming Rate on Survival

The consequences of many of the changes that occur during cooling become apparent only after warming. Similarly, the response to warming is influenced by the rate of cooling and the choice of cryoprotectant, as these factors determine the state of the frozen cell. In particular, rapidly cooled cells are sensitive to the rate of warming (Mazur *et al.*, 1969). When hamster tissue culture cells were cooled at 500°C/min in the presence of 15% PVP, 40% survived if they were warmed at 1100°C/min, whereas none survived if the warming rate was reduced to 29°C/min. It was suggested that during the slower warming of these rapidly cooled cells, small ice crystals were able to grow by the process of migratory recrystallization, forming large crystals that cause cell death, perhaps by physically disrupting cell membranes.

The potential significance of the changes in intracellular electrolyte concentration that may occur during slow cooling was demonstrated by Lovelock (1953). In his classic studies, Lovelock exposed erythrocytes to concentrated salt solution for short periods to mimic the changes that occur during freezing and thawing. He observed that most of the damage occurred during dilution of the hypertonic solutions (Table II). He suggested that, during exposure to the hypertonic solutions, electrolyte entered the cells. As a result, when the cells were returned to isotonic solution, an intolerable amount of water entered the cells and caused their disruption. Thawing returns cells to isotonic solutions from the hypertonic solutions generated by freezing.

More recently, Farrant and colleagues suggested that death after either slow or rapid cooling, may be a result of osmotic shock (Farrant and Morris, 1973; Farrant, 1977). They envisage that when cell membranes are exposed to concentrated electrolyte solutions, as a result of either extracellular or intra-

Table II. Effect on Human Erythrocytes of Freezing and Thawing or Exposure to the Concentration of Sodium Chloride Generated by Freezing[a,b]

Minimum temperature during freezing and thawing (−°C)	Percent hemolysis		
	During freezing and thawing	In concn. NaCl	After dilution of NaCl
2	0	0	0
3	1.4	0	0
4	6.0	0	3.3
5	13	0	11.3
6	26	0	25
7	40	0	38
8	57	0	56
9	76	2	76
10	96	11	89
12	100	13	93
15	100	23	100

[a]Data from Lovelock (1953).
[b]The percentage of cells hemolyzed after freezing and thawing, after transfer to concentrated sodium chloride, and after dilution of the salt solution.

Table III. Effect of Cooling Rate and Warming Rate on the Survival of Mouse Blastocysts in the Presence of 1.5 M DMSO Added at 20°C[a,b]

Warming rate (°C/min)	Cooling rate (°C/min)			
	0.07	0.22	1.2	4.7
1.1	25.0 (16)	22.5 (13)	0 (17)	0 (19)
12	0 (16)	64.7 (17)	0 (19)	0 (18)
360	0 (16)	0 (18)	12.5 (16)	0 (16)

[a]Data from Wilmut (1972*a*).
[b]The percentage that developed in culture is shown and, in parentheses, the number cultured.

cellular ice formation, the membranes are less able to tolerate temperature change or the changes in cell volume that are brought about by dilution.

In the vast majority of cases, optimum survival is obtained by thawing rapidly. By contrast, some very slowly cooled cells survive only if they are warmed slowly (Whittingham *et al*., 1972; Wilmut, 1972*a*; Miller and Mazur, 1976). In the case of embryos the optimum warming rate depends on the rate of cooling (Table III). The events that occur during thawing have been studied in mouse embryos cooled slowly (Leibo *et al*., 1974) or rapidly (Rall *et al*., 1980), although it should be noted that the concentration of DMSO and the method of its addition were different in the two experiments.

The response to rapid cooling has been studied in groups of 8-cell mouse embryos in 1.5 M DMSO cooled at 0.5°C/min to −40°C before being plunged into liquid nitrogen. Maximum survival was obtained when embryos were warmed rapidly by shaking the test tube in a water bath at 37°C. None survived slow warming in an alcohol bath (2°C/min). The range of temperatures over which rapid warming is important was defined by warming the samples slowly to a temperature between −100°C and −40°C before transferring the test tube to a water bath at 37°C. Inspection of the results suggests that survival was similar for embryos warmed rapidly from −196°C or −90°C, whereas survival was reduced when samples were warmed slowly to a temperature of −80°C or higher before being warmed rapidly to 0°C. No embryos survived if they were warmed slowly to −55°C before being warmed rapidly. Cryomicroscopy revealed changes in the intracellular structure at −85°C, −55°C, and −40°C that were judged to reflect the formation of intracellular ice, migratory recrystallization, and melting of the ice, respectively. As embryo death occurred over a range of temperatures, it was not evidently associated with any one of these changes; the authors (Rall *et al*., 1980) concluded that the formation of intracellular ice was innocuous. However, death occurred after the formation of intracellular ice, so that it is not certain that it is innocuous. Any damage arising from the formation of intracellular ice may occur over a period of several minutes, and it would be fascinating to know the fate of embryos held for 15–30 min at −80°C. In the present experiment, the embryos were warmed at 2°C/min, so that it took 15 min for the temperature to increase from −85°C to −55°C.

By contrast, damage to slowly cooled 8-cell mouse embryos was found to occur between −70°C and −20°C (Leibo *et al.*, 1974). Embryos in the presence of 1.0 M DMSO added at 0°C (so that complete permeation could not have occurred before freezing) were cooled at 0.2°C/min to −50°C before being plunged into liquid nitrogen. They were then warmed at 3°C/min to a temperature between −70°C and 0°C before being warmed in a water bath at +35°C. Embryos were able to tolerate rapid warming only if they had been warmed slowly to −20°C. As media containing DMSO begin to thaw at very low temperatures, it was suggested that rapid warming was exposing the embryos to too great an osmotic change.

3.5. Conclusions

The vast majority of the advances in cryobiology were made during empirical studies or by chance. The present description of freezing and thawing merely offers an interpretation of the changes that occur during a successful procedure; it is important to remember that it may be incorrect. The original observations are reliable, whereas the interpretation is incomplete. In developing methods with a different type of cell it may be helpful to adopt an empirical approach with few preconceived ideas.

4. Cryopreservation of Eggs and Embryos

4.1. Introduction

After the discovery of the protection afforded by glycerol to cells that were being frozen and thawed (Polge *et al.*, 1949) a number of papers described attempts to freeze embryos and oocytes (reviewed by Polge, 1977). Oocytes in primary follicles were found to survive when cooled at around 1°C/min in the presence of 10–15% glycerol (Parkes and Smith, 1953), although viability decreased markedly during storage at −70°C (Parrott, 1960). By contrast, only a very small proportion of embryos developed further when frozen and thawed in this way (e.g., Smith, 1952) and no unfertilized eggs survived when frozen at 0.7–0.9°C/min (Sherman and Lin, 1958).

A second era of research was initiated some 20 years later by the demonstration that most mouse embryos survived provided they were cooled and warmed very slowly (Whittingham *et al.*, 1972; Wilmut, 1972*a*). Embryos in the presence of 1.0 or 1.5 M DMSO were cooled at 0.2–0.4°C/min to −70°C or −110°C before transfer directly into liquid nitrogen. Subsequent survival depended on the embryos being warmed at rates between 4° and 25°C/min.

Mouse, rabbit, sheep, and cow embryos were frozen successfully essentially by this method, but it has the disadvantage of being time consuming.

Attempts to reduce the time required for cooling by increasing the rate of cooling were generally unsuccessful. However, Willadsen (1977) found it possible to reduce the time required for cooling by plunging the embryos into liquid nitrogen from a relatively high subzero temperature. The optimum rate of warming depends critically on the temperature from which the embryos are plunged into liquid nitrogen (Table IV). When sheep embryos were cooled slowly to a temperature below −50°C maximum survival was attained if they were warmed slowly. By contrast, if embryos were plunged into liquid nitrogen from a temperature between −30 and −42°C, rapid warming was essential. A similar interaction was observed with mouse embryos (Whittingham *et al.*, 1979). As a result the time required for cooling and warming embryos of some species was greatly reduced.

Practical application of embryo-freezing procedures in cattle breeding was inhibited by the need to remove the cryoprotective agent before transferring the embryos into the recipient animal (Willadsen *et al.*, 1978). Dilution is generally carried out by transferring embryos from one dish to another containing a lower concentration of cryoprotective agent. This is clearly impracticable in a field situation and also demands that the person concerned is able to recognize and manipulate embryos. The entire procedure has been simplified by the use of sucrose as an extracellular compound to reduce the osmotic stress during dilution (Leibo and Mazur, 1978). An extremely efficient procedure incorporates sucrose dissolved in PBS in the plastic insemination straw in which the embryos are frozen and thawed (Leibo *et al.*, 1982*a*,*b*; Renard *et al.*, 1982). After thawing, the tube is shaken to begin dilution by mixing sucrose with the medi-

Table IV. Effect on Survival of Sheep Embryos of Varying the Rate of Warming and the Temperature from Which the Samples Were Transferred to Liquid Nitrogen[a,b]

Transfer temperature (°C)	Slow warming (4°C/min)	Rapid warming (360°/min)
−30	—	2/5 (40)
−36	0/5 (0)	7/10 (70)
−42	2/15 (13)	4/15 (27)
−48	8/20 (40)	7/20 (35)
−54	15/20 (75)	6/20 (30)
−60	12/18 (67)	4/20 (20)

[a]Morulae and blastocysts in the presence of 1.5 M DMSO were cooled at 0.3 or 0.1°C/min to one of the temperatures shown. (See Willadsen, 1977, for details.)

[b]Proportion of embryos surviving with the percentage shown in parentheses.

um in which the embryo is suspended. In this way, transfer can be carried out without the person needing to examine the embryo, and this method is being used commercially (Leibo *et al.*, 1982*a*,*b*; Renard *et al.*, 1982).

An alternative approach to the elimination of post-thaw handling may be provided by the use of agents able to enter and leave cells very rapidly. Mouse embryos frozen and thawed in the presence of methanol and transferred without the methanol being removed gave rise to normal fetuses (Rall *et al.*, 1984).

It can be seen that during the past decade, research has concentrated on simplification of the freezing procedure, rather than searching exclusively for maximum survival. In the future, it may be important to ensure high survival if embryo freezing is to be a commercially useful technique.

The process of freezing and thawing involves several stages, which will be considered in turn, before details of methods for the different species are considered. It must be appreciated that there may be interactions between the different factors. In applying a successful method, the recipe should be followed precisely, as modification of any stage could cause the death of embryos. Mouse embryos cooled slowly to −70°C in the presence of 1.0–1.5 M DMSO would not survive rapid warming, although a majority would survive slow warming (Whittingham *et al.*, 1972; Wilmut, 1972*a*). By contrast, similar embryos plunged into liquid nitrogen from −30°C would only survive if warmed rapidly (Whittingham *et al.*, 1979; Rall *et al.*, 1980). In seeking to develop new methods, it is important to allow for a broad range of conditions; it may be useful to apply optimisation procedures (Maurer and Haseman, 1976; Schneider and Maurer, 1983).

4.2. Choice of Cryoprotective Agent

The vast majority of research with embryos has used low-molecular-weight protective agents that are able to enter cells under certain circumstances. Glycerol and DMSO have been used most frequently and both provide considerable protection for several species. Whereas glycerol seems to provide a broader range of protection than DMSO for mouse and cattle embryos (Kasai *et al.*, 1981; Miyamoto and Ishibashi, 1983*a;* Rall and Polge, 1984; Bilton, 1980; Willadsen, 1980; Bouyssou and Chupin, 1981), a limited study with 4- or 8-cell human embryos suggests that DMSO provides more protection than glycerol (Trounson and Mohr, 1983). A number of other compounds have been studied by Miyamoto and Ishibashi (1977, 1978, 1981). Mouse and rat 8-cell embryos were afforded considerable protection by ethylene, diethylene, triethylene, or polyethylene glycol when the embryos were cooled and warmed slowly (0.5°C/min to −70°C and 15°C/min, respectively). Survival was similar for embryos frozen in the presence of 1.2 M ethylene glycol or 1.5 M DMSO. In general, the survival decreased as the molecular weight of the glycol increased. Over 60% of 8-cell mouse embryos survived when frozen in the presence of 0.6 M erythritol, whereas 76% of similar embryos survived when frozen and thawed rapidly in the presence of 1.5 M DMSO (Miyamoto and Ishibashi,

1981). In a similar study in which mouse embryos were cooled rapidly and warmed very quickly in the presence of 3.0 M methanol, 50% survived (Rall *et al.*, 1984). Comparisons of several different agents are practicable only with laboratory animals because of the requirement for a large number of embryos. It is important, in making comparisons of this type, to use the optimum procedure for each agent.

4.3. Concentration of Agent

The proportion of embryos surviving freezing and thawing may be influenced by the concentration of cryoprotectant. This effect may be seen as a broader range of protection afforded by a higher concentration. When mouse 8-cell embryos were plunged into liquid nitrogen from a temperature between −10° and −80°C and then warmed either rapidly or slowly, survival increased as the concentration of DMSO was increased from 1.0 to 1.5 or 2.0 M (Whittingham, 1981*a*). In other circumstances, the effect is seen as an optimum concentration for high survival. When cryoprotectant was added to 8-cell mouse embryos at 0°C before they were cooled at 0.5°C/min to −70°C and warmed at 15°C/min, the optimum concentration of ethylene, diethylene, triethylene, and propylene glycols was around 1.0 M, whereas the optimum concentration of polyethylene glycol was 0.5 M (Miyamoto and Ishibashi, 1978).

At the highest concentrations employed, some of the compounds may be toxic and this effect sets an upper limit. The concentrations employed most commonly have been around 1–1.5 M, but levels as low as 0.5 M and as high as 3 M have been used with polypropylene glycol and methanol, respectively (Miyamoto and Ishibashi, 1978; Rall *et al.*, 1984).

4.4. Addition of Cryoprotective Agent

Two basic strategies are used in the addition of agent. In one, medium containing agent is added in stages at a temperature of 20°C or above, so that the agent is judged to be able to permeate the cell and restore osmotic equilibrium, whereas in the other strategy the addition is made at 0°C, so that complete permeation before freezing is impossible. The choice may have implications for the method of cooling and warming, and is likely to influence the means by which the agent is removed after thawing.

Addition of medium containing cryoprotectant at 0°C has been used routinely only with mouse embryos (e.g., Whittingham *et al.*, 1972; Leibo and Mazur, 1978; Miyamoto and Ishibashi, 1978). Post-thawing survival of mouse embryos has been excellent after addition of DMSO, (Whittingham *et al.*, 1972) and ethylene glycol (Miyamoto and Ishibashi, 1977, 1978) in one step at 0°C. When adding glycerol or glycols of a greater molecular weight than ethylene glycol at 0°C it was found beneficial to make the addition in steps (Miyamoto and Ishibashi, 1978, 1983*a*).

When freezing rabbit and rat embryos addition of cryoprotectant has sometimes been at 0°C (Bank and Maurer, 1974; Maurer and Haseman, 1976; Miyamoto and Ishibashi, 1977, 1978), but on other occasions at room temperature or 37°C (Whittingham and Adams, 1976; Tsunoda *et al.*, 1981; Wood and Whittingham, 1981).

Embryos of the farm animals have almost always had cryoprotectant added at room temperature in a stepwise manner. This method of addition was developed empirically following the observation that mouse blastocysts were killed when transferred directly into and out of medium containing 2.0 M DMSO at room temperature (Wilmut, 1972*a*). Stepwise addition was applied in the first successful methods with farm animals (Wilmut and Rowson, 1973; Willadsen *et al.*, 1976*a*) and subsequently as a routine with only minor modifications in the number of steps required for the addition. The survival of sheep embryos frozen after addition of DMSO at 6°C was low (Moore and Bilton, 1976, 1977), whereas good survival was obtained when glycerol was added to cattle embryos at 35°C (Bilton, 1980). Further studies in farm animals would be interesting.

It is difficult to make clear generalizations about the relative value of the two approaches in the absence of more comparisons of the two methods of addition, but it is clear that both are effective in some circumstances. However, it is important to consider the method of addition when selecting a method of removal of the agent.

Experiments with embryos have provided evidence to support the conclusion that equilibration of protective agents across cell membranes is not essential for cell survival during freezing and thawing (Sherman, 1963; Whittingham *et al.*, 1972). It seems probable that extracellular agent acts by protecting cell membranes and by altering the composition of extracellular fluid (Mazur, 1970).

4.5. Equilibration Time

If medium containing agent is added at 0°C, there may be marked effects of varying the interval between addition and the initiation of further cooling, i.e., the equilibration time. An extreme observation was made with 8-cell rabbit embryos, when it was found that none survived if cooling was initiated 1 min after the addition of medium containing DMSO, whereas 55% survived when embryos and agent were allowed to equilibrate for 30 min before cooling (Bank and Maurer, 1974). Similar effects have been observed with mouse embryos exposed to DMSO (Whittingham *et al.*, 1972) or glycols (Miyamoto and Ishibashi, 1978).

When the agent is added at a higher temperature, sufficient time is generally allowed during and after the addition for osmotic equilibrium to be fully restored. The extent of equilibration is assessed approximately by watching to confirm that the embryos have recovered their initial volume.

4.6. Cooling above Seeding Temperature

After the addition of cryoprotectant at 0°C, the samples can be transferred directly to a cooling bath at 0°C or at the seeding temperature (Whittingham *et al.*, 1972). If an agent is added at 20°C or above, consideration should be given to the rate of cooling to the seeding temperature. There is evidence that sheep and cattle embryos are more likely to tolerate cooling to temperatures around 0°C if they are cooled slowly rather than transferred abruptly to the lower temperature (Trounson *et al.*, 1976; Willadsen, 1977). All of 22 sheep morulae and blastocysts survived when cooled to −7°C at 1°C/min in the presence of 1.5 M DMSO, whereas only 15 of 21 survived abrupt cooling (Willadsen, 1977). In subsequent studies, Willadsen and co-workers routinely employed a cooling rate of 1°C/min (Willadsen *et al.*, 1978; Willadsen, 1980). Over a period of years, several groups appear to have modified the procedure and now cool the embryos abruptly (Bouyssou and Chupin, 1982; Massip *et al.*, 1982; Renard *et al.*, 1982). No direct study has been made of the effect of the two methods of cooling to the seeding temperature on embryo survival after freezing and thawing.

4.7. Seeding

There is absolute agreement that ice formation must be induced at a temperature just below the freezing point of the suspending medium (Whittingham, 1975; Willadsen, 1977). Embryos in test tubes, ampoules, or plastic straws are cooled in an alcohol bath or in a gas-filled chamber. The temperature of the bath or chamber is lowered either by liquid nitrogen (−196°C) or by solid carbon dioxide (−79°C). If a tube containing salt colution and cryoprotective agent is cooled slowly in such a chamber, ice does not form spontaneously in the sample until its temperature is well below −3°C, the freezing point of the mixture (Fig. 3). As ice forms, the latent heat of crystallization causes the temperature of the sample to rise to the freezing point. When crystallization is complete the sample will be at the freezing point, whereas the chamber that has gone on cooling will be at least 10°C cooler. The sample then cools more rapidly until its temperature approaches that of the chamber. Whereas the aim may be to cool embryos at 0.2–0.5°C/min, during this brief period the actual rate may be 20 times greater (I. Wilmut, unpublished observations, 1972).

In order to prevent this abrupt cooling, the sample is "seeded" (induced to freeze) at a temperature just below freezing point. Seeding can be achieved by placing an ice crystal into the sample (Whittingham *et al.*, 1972; Wilmut, 1972*a*) by tapping the test tube (Wood and Farrant, 1980) or by touching the outside of the container with metal forceps that have been chilled in liquid nitrogen as shown in Figure 4 (Willadsen *et al.*, 1978). The method of seeding is not important, but it is essential that all the samples be frozen before cooling continues.

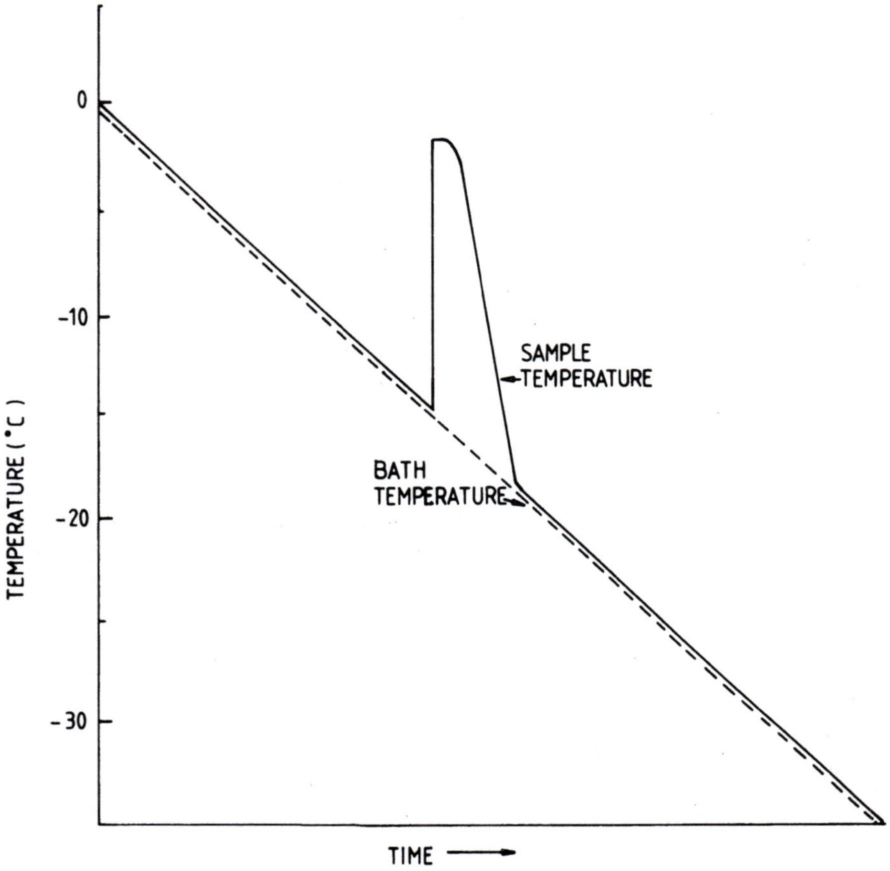

Figure 3. Effect of supercooling on sample temperature when the sample was held in a cooling bath without seeding.

4.8. Cooling Rate

It is well established that the cooling rate critically influences the survival of frozen and thawed cells. The optimum rates of cooling for embryos are very slow in comparison with those required for other types of cell.

Successful procedures can be loosely classified into "slow" and "rapid" methods. In the slow methods, embryos are cooled at a rate of 0.2–1°C/min, to a temperature below −60°C before being plunged into liquid nitrogen. In the rapid methods, embryos are cooled at a similar rate, but only to an intermediate temperature between −20°C and −40°C, before transfer directly into liquid nitrogen. In general, embryos cooled "slowly" require slow warming, whereas embryos cooled "rapidly" depend on rapid warming, but there are several exceptions to this rule.

The cryomicroscope has been used to observe the changes that occur in mouse ova during cooling (Leibo, 1977; Rall *et al.*, 1983). The ova cooled

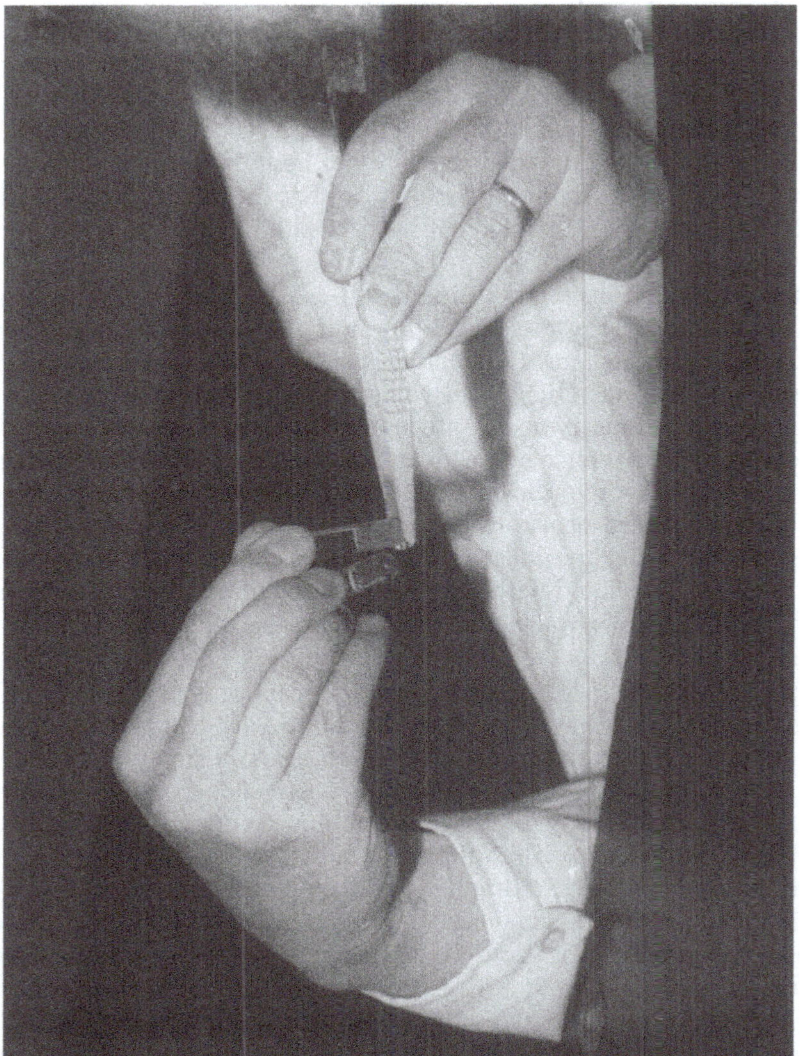

Figure 4. Induction of ice formation seeding by placing cooled forceps against tubes at the seeding temperature. Ice has formed in the right-hand tube.

slowly (less than 1°C/min) did not freeze but lost water and shrank. By contrast, as the rate of cooling was increased to 4°C/min the proportion of cells that froze rose to 100%. The cooling rates at which the ova did not lose water are those that do not yield survival of embryos. The effect of the cryoprotective agents glycerol and DMSO was to lower the temperature at which intracellular ice formed in rapidly cooled cells (Rall *et al.*, 1983). In the absence of agent intracellular ice formed between −10 and −15°C, whereas this change occurred between −38 and −44°C in the presence of 1.0–2.0 M glycerol or DMSO. When embryos are plunged into liquid nitrogen from an intermediate temperature, they contain more water than do embryos cooled slowly to a temperature below −60°C, but it does not form crystals large enough to alter light transmission. These embryos survive provided they are not held or warmed slowly at temperatures above −65°C (Rall *et al.*, 1980). Neither the details of the changes in the cells nor the mechanisms that cause cell death in slowly warmed embryos are known (see Section 3.4).

Attempts have been made to increase the rate of cooling, but they have produced unrepeatable results (I. Wilmut, unpublished data, 1978; Wood and Farrant, 1980; Whittingham, 1981*b*). It is possible to freeze spermatozoa (Luyet and Keane, 1955; Polge, 1957) and some tissue culture cells (Farrant *et al.*, 1978) by a very simple two-step procedure. In these cases, the samples are transferred directly from ice at 0°C to an intermediate temperature that is usually around −25°C, where they are held for a time before being plunged into liquid nitrogen. These observations emphasized that it is not necessary, or perhaps even desirable, to cool at a constant rate. The factors that are critical for cell survival are associated with the degree of dehydration of the cells and the form of the intracellular water when the cells reach the temperature of liquid nitrogen.

It has been suggested that the two-step procedure is particularly effective because the cells are allowed to lose water at a lower temperature, when the harmful effects of the concentrated solutes occur more slowly (Farrant, 1977).

A number of independent attempts have been made to apply this approach to embryos. An initial study established that mouse embryos in the presence of 1.5 M DMSO added at 20°C could tolerate direct transfer from ice to temperatures around −10°C (I. Wilmut, unpublished observations, 1978). By contrast, a reduced proportion of embryos survived transfer to −15°C and very few survived transfer to −20°C (Table V). As a result, all subsequent transfers were made from a seeding bath at −5°C. No holding temperature could be found from which embryos could safely be plunged into liquid nitrogen, although many embryos survived if they were warmed directly from the holding temperature. Embryos were held for periods of 0–30 min. As an alternative approach, embryos were cooled at 2–3°C/min to intermediate temperatures, before being held at the intermediate temperature and finally plunged into liquid nitrogen. Embryos plunged from a temperature around −35°C showed encouraging but extremely unreliable results. In samples drawn from the same pool of embryos survival ranged from 47 to 80%, whereas survival ranged from 20 to 80% for samples frozen on different days (I. Wilmut, unpublished results,

Table V. Effect of Cooling 1-Cell Mouse Embryos in the Presence of 1.5 M DMSO Directly to a Subzero Temperature from an Ice Bath[a]

	Control	Final temperature (°C)			
		−5	−10	−15	−20
No. eggs cultured	149	94	85	83	78
No. experiments	10	4	5	6	5
% developing in culture	62	66	55	31	4

[a]Data from I. Wilmut, unpublished results (1978).

1978). These observations were not published at the time because of their variability, but it has become apparent that other laboratories face a similar problem. The method described by Wood and Farrant (1980) has not found general acceptance and they themselves find the method unreliable (M. J. Wood, personal communication). Whittingham (1981*b*) provided an anecdotal record of another laboratory that found a similar lack of reliability.

It is not clear what factors cause this variation, but it seems to be associated with events within the entire sample. If the variation observed reflects the chance occurrence of intracellular freezing during holding, if the probability of each embryo in the experiment freezing in this way is independent, such gross variation between samples would not be expected. These experiments were carried out with DMSO before it was appreciated that glycerol affords a greater range of protection to mouse embryos (Rall and Polge, 1984); it is also possible that a greater concentration of agent would improve the results.

Further attempts to increase the rate of cooling are being made (e.g., Bouyssou and Chupin, 1981; Miyamoto and Ishibashi, 1983*a*,*b*), and it seems probable that they will become practicable in time. They may offer an additional advantage in not being dependent upon expensive equipment. Constant rates of cooling are generally obtained in apparatus in which electronic circuitry controls the amount of liquid nitrogen that is expelled into an insulated chamber while monitoring the temperature in the chamber (Fig. 5). Such equipment costs several thousand dollars.

4.9. Warming Rate

The choice of warming rate can have a critical influence on whether or not embryos survive the process of thawing. Mammalian embryos have been warmed at rates ranging from 4°C/min, achieved by allowing tubes to warm in an alcohol bath, through 10–20°C/min, achieved by warming a tube in air, to several hundred degrees per minute by warming the tubes in water at 37°C. The rate of warming is also influenced by the volume of the sample and the type of container in which the embryos have been frozen.

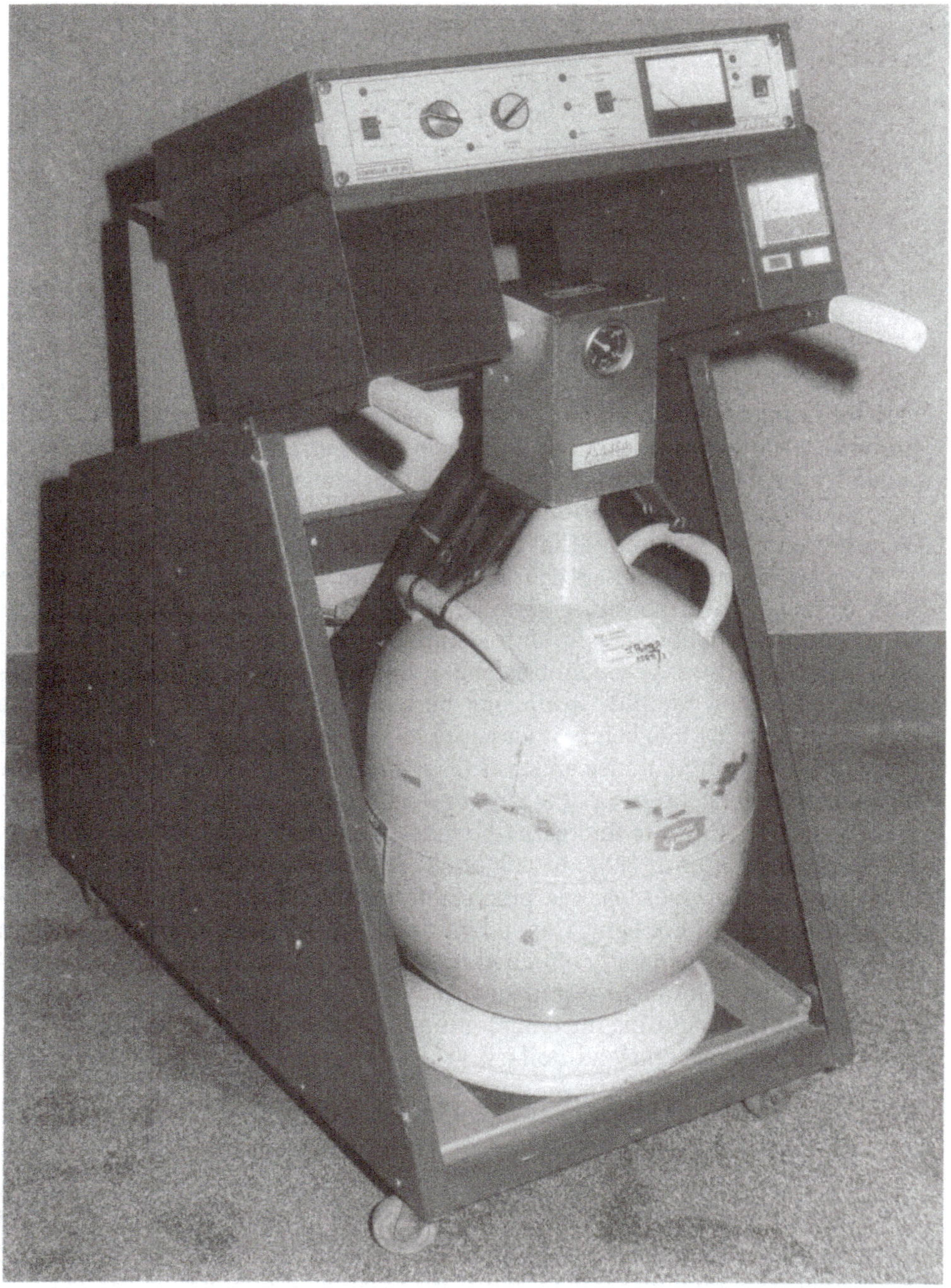

Figure 5. A chamber capable of cooling samples at 0.1 to 100°C/min made by Planer/Products Ltd.

Table VI. Effect of Warming Rate on Survival of 8-Cell Mouse Embryos Cooled at 0.5°C/min to −80°C in the presence of 1.5 M Glycerol[a]

	Warming rate (°C/min)		
	2	25	500
No. of embryos (replicates)	131 (9)	137 (9)	108 (7)
% survival (mean ± SE)	86.0 ±2.5	80.6 ±4.1	75.4 ±4.9

[a]Data from Rall and Polge (1984).

As a generalization, it seems that survival is maximal if embryos that have been cooled "rapidly" are warmed rapidly, and those that have been cooled "slowly" are warmed slowly. However, in some circumstances embryos are much more tolerant of the opposite treatment than the generalization implies (Table VI). When 8-cell mouse embryos were cooled at 0.5°C/min to −80°C after addition of 1.5 M glycerol at room temperature, survival was very similar if they were warmed at 2° or 500°C/min (Rall and Polge, 1984).

Although it has not been defined for many embryo preservation procedures, there will be an optimum rate of warming that also varies with cooling rate (Table III). The existence of such optima is seen most readily when comparing a broad range of rates of cooling and warming (Wilmut, 1972*a*); however, it has also been noted when contrasting two rapid rates of warming (Rall *et al.*, 1984). When 8-cell mouse embryos were cooled at 0.5°C/min to −65°C in the presence of 1.5 M propylene glycol, survival was higher after warming at 500°C/min than 800°C/min or 25°C/min (Table VII).

Greater attention should be given to the selection of warming rate. In order to facilitate very rapid warming, care should be taken in deciding the volume of sample and the material used to contain the sample. Plastic containers conduct

Table VII. Effect of Warming Rate on the Survival of 8-Cell Mouse Embryos Cooled at 0.5°C/min to −65°C in the Presence of 1.5 M Propylene Glycol[a]

	Warming rate (°C/min)			
	2	25	500	800
No. embryos (replicates)	114 (6)	120 (6)	116 (5)	77 (5)
% survival (mean ± SE)	73.6 ±4.4	70.5 ±4.3	83.1 ±3.9	69.5 ±4.2

[a]Data from Rall and Polge (1984).

heat more slowly than do glass containers; thicker walls slow down the rate of heat transfer.

4.10. Removal of Cryoprotectant

If embryos that have been frozen and thawed are removed from medium containing glycerol or DMSO and transferred to medium without agent, they are damaged by the osmotic stress. Only when methanol had been used as the cryoprotective agent have mouse embryos survived direct transfer into recipient mice (Rall *et al.*, 1984).

Two methods have been used to minimize the effect of the stress. In the first method the concentration of agent is reduced in steps (Whittingham *et al.*, 1972; Wilmut, 1972*a*); in the second method an extracellular compound such as sucrose is used to cause embryos to shrink and so reduce the amount of cryoprotectant in the cells before removal of cryoprotectant is completed (Leibo and Mazur, 1978).

The details of the stepwise dilution were initially established by inspired guesswork. There have been few systematic comparisons and the methods vary from one laboratory to another. There have been from three to six steps, and the interval between additions has varied from 45 sec to 10 min (Whittingham *et al.*, 1972; Wilmut, 1972*a*; Willadsen, 1977). There are differences between species in the extent to which they tolerate rapid dilution and the optimum method of dilution may also be influenced by the method of adding the agent and the temperature of dilution.

Mouse embryos at the 8-cell and blastocyst stages are relatively tolerant of rapid dilution and will even survive dilution in two steps at room temperature (Whittingham *et al.*, 1979). If DMSO is added to mouse blastocysts in such a way that complete permeation occurs then dilution must be gradual and at room temperature or above (Wilmut, 1972*a*; Whittingham, 1974). By contrast, if DMSO is added at 0°C, dilution can be at 0°C (Whittingham *et al.*, 1972; Leibo and Mazur, 1978). In these circumstances, it is probable that the cells contain less DMSO. However, even under these circumstances survival is improved by dilution at 20°C or above (Leibo and Mazur, 1978; Schneider and Maurer, 1983).

Sheep and cattle embryos have been frozen and thawed only after addition of the agent in such a way that permeation occurs before cooling begins. Stepwise dilution has almost always been at room temperature (e.g., Wilmut and Rowson, 1973; Willadsen, 1977; Willadsen *et al.*, 1978) or a higher temperature (e.g., Moore and Bilton, 1977; Bilton, 1980).

Rabbit embryos are particularly sensitive to the system of dilution. In some experiments maximum survival was obtained by addition and dilution of DMSO in stages at 35–37°C (Whittingham and Adams, 1976; Tsunoda *et al.*, 1981), while in others DMSO was added at 4°C and hypertonic salt solution was used to remove DMSO from the cell, while maintaining the osmolarity of the medium (Bank and Maurer, 1974; Maurer and Haseman, 1976).

The process of dilution has been greatly simplified by the use of sucrose

(Leibo and Mazur, 1978). When medium containing sucrose is added to embryos they shrink as sucrose does not enter cells readily. If the medium containing sucrose also contains cryoprotectant at the same concentration as that present during freezing and thawing, the embryos shrink in response to the sucrose. When osmotic equilibrium is restored the embryos contain less cryoprotectant, although the concentration is the same as that outside the cells.

This advantage has been exploited in a variety of ways. The diluting medium may contain sucrose but not cryoprotectant (Leibo and Mazur, 1978). In these circumstances three factors are in disequilibrium: (1) the extracellular sucrose will tend to draw water from the cells, (2) the intracellular cryoprotectant will tend to draw water into the cells, and (3) the agent will diffuse from the cells. The effect on the embryos will depend on the concentration of sucrose and the cryoprotective agent and on the permeability of the cell membranes to the agent and to water. If the sucrose concentration is sufficiently high to protect the embryos from the osmotic stress imposed, it may then be necessary to dilute it in steps. This procedure is used with mouse embryos (Leibo and Mazur, 1978).

Alternatively, the medium containing sucrose may also contain the cryoprotectant. The embryo will then be exposed to forces tending to draw water from the cells and, as the cells shrink, to draw the agent from the cells. After equilibration, the embryos may be transferred to medium containing sucrose or medium with no additive. During the final dilution, the cells will swell as water enters the cells more rapidly than the agent is able to leave. However, in the presence of sucrose, the cells may be tolerant of this change as they are smaller than normal at the time of the dilution.

The relative volumes of the sample and the diluent also influence events. The description given so far assumes that the medium added is a much larger volume (×10). A very effective means of removing agent from cattle embryos involves the addition of a similar volume of medium containing sucrose, but no cryoprotectant, before transferring the embryo directly into a recipient cow (Leibo *et al.*, 1982*b*; Renard *et al.*, 1982). This dilution is carried out within the insemination straw in which the embryo was frozen and thawed, and does not involve inspection of the embryo. Stepwise dilution of cattle embryos, using sucrose to protect the embryos from the osmotic stresses, is also effective (Niemann *et al.*, 1982; W. B. Christie, personal communication, 1983). No direct comparison of the two approaches has been made, but there is a suggestion that the simpler method of dilution pays a price in the form of lower survival (see Section 5.6).

4.11. Embryo Stage and Quality

The optimum method of freezing and thawing may vary for the different stages of development (e.g., Whittingham *et al.*, 1972; Schneider and Maurer 1983). In some circumstances, the proportion of embryos that survive when frozen and thawed by present methods is higher with morulae or early blastocysts, than with the very early cleavage stages (Whittingham *et al.*, 1972;

Tsunoda *et al.*, 1981). It is not clear whether this reflects real differences in sensitivity to freezing, or a failure to develop optimal procedures for the early stages.

As development continues through blastulation and hatching from the zona pellucida there are further changes in requirements and survival (Wilmut and Rowson, 1973; Whittingham, 1974; Trounson *et al.*, 1978). In sheep and cattle, maximum survival with rapid cooling and warming is attained with morulae or early blastocysts that are still in the zona (Willadsen *et al.*, 1976; Trounson *et al.*, 1978). Early embryos are damaged by cooling in ice (see earlier discussion), while later stage blastocysts have not been frozen in large numbers by the more recent methods, but were less likely to survive "slow" cooling.

Among those carrying out embryo transfer in cattle, an impression has been formed that the survival of embryos is influenced by "quality." This characteristic is difficult to define or describe but certainly influences the choices made by the operators. Embryos are judged to be of "good quality" if all the cells have been involved in compaction, so that there are no free cells or anuclear fragments in the perivitelline space. The cytoplasm should also be of even density. Embryos with three or four fragments in the perivitelline space, and pale areas of cytoplasm are less likely to survive freezing and thawing (I. Wilmut and M. R. Thomson, unpublished observations, 1982). In order to maximize the yield of calves, such embryos should be transferred immediately, while the better embryos are frozen and thawed (W. B. Christie, personal communication).

5. Embryo Preservation in Practice

While it is possible to outline the general principles that govern embryo preservation it is more difficult to select a procedure and use it satisfactorily. To aid those deciding to use cryopreservation for the first time, the methods that have been applied successfully will be outlined for a number of species. The list is intended to be a guide rather than a complete bibliography.

5.1. Human

There is only one report of cryopreservation of human embryos (Trounson and Mohr, 1983). Although the number of embryos was small it seemed that some 4-cell and 8-cell embryos survived slow or rapid cooling. DMSO seemed to offer greater protection than does glycerol, as it had less effect during addition or removal.

5.2. Mouse

Mouse embryos tolerate a great variety of procedures and thousands of embryos have been frozen and thawed successfully. Embryos at any stage from

2-cell to blastocyst survive the original slow methods of cooling and warming, or the recent rapid procedures (Whittingham *et al.*, 1972, 1979; Wilmut, 1972*a*). They are protected by glycerol, DMSO and ethylene glycol, and methanol (Whittingham *et al.*, 1972; 1979; Rall and Polge, 1984; Rall *et al.*, 1984; Miyamoto and Ishibashi, 1978).

Two cautionary comments may be helpful. First, very few direct comparisons of the different methods have been made so that it may be inappropriate to judge whether one method is more successful than any other. Second there are substantial differences between strains and between females in the yield of embryos and in the proportion of embryos that survive freezing and thawing (Glenister and Lyon, 1981; Yokoyama *et al.*, 1981).

5.3. Rat

In general it seems that embryos of the rat require methods of freezing and thawing that are similar to those described for the mouse (Whittingham, 1975; Miyamoto and Ishibashi, 1977, 1978; Wood and Whittingham, 1981). However, the proportion surviving freezing and thawing is low, and a further loss occurs following transfer of the embryos to recipients (Whittingham, 1975; Wood and Whittingham, 1981). Overall some 14–21% of rat blastocysts survived to become fetuses, compared with 36–54% of mouse embryos (Whittingham *et al.*, 1979; Wood and Whittingham, 1981). This may reflect limitations in the techniques of embryo handling and transfer in the rat, as only 49% of similar untreated embryos developed to fetuses after transfer (Wood and Whittingham, 1981).

5.4. Hamster

No hamsters have been born after being frozen and thawed while at the embryonic stage of development, despite the fact that morphologically normal embryos have been recovered after freezing and thawing (Tsunoda *et al.*, 1981). However, when cooled and warmed slowly in the presence of DMSO, ethylene glycol, or glycerol, up to 89% of embryos were judged to be morphologically normal. It seems probable that the failure to obtain young reflects inadequate methods of culturing and transferring embryos and the fact that relatively few embryos have been studied.

5.5. Rabbit

A number of laboratories have obtained live young after freezing and thawing rabbit embryos (Bank and Maurer, 1974; Whittingham and Adams, 1976; Tsunoda *et al.*, 1981). These embryos seem unusually sensitive to methods of dilution.

Table VIII. Effect of Embryo Quality on Conception Rate after Transfer of Cattle Embryos with or without Freezing and Thawing[a]

Embryo quality	Direct transfer	Frozen and thawed	Proportion surviving freezing and thawing
1	217/276 (78.6)	238/340 (68.8)	87.5
2	43/68 (63.2)	30/66 (45.5)	71.2
3	20/48 (41.7)	3/11 (27.3)	65.5
Overall			74.7

[a]Data from W. B. Christie, personal communication.

5.6. Cattle

Improvements in the methods of cattle embryos are now being made by those carrying out routine transfer of several thousand embryos each year. It is interesting to compare the quest for simplicity (Leibo *et al.*, 1982*a*,*b*; Renard *et al.*, 1982) with a different demand for maximum survival when each selected embryo is very valuable (W. B. Christie, personal communication). Maximum survival has been obtained by adding 10% glycerol at room temperature, cooling to −7°C at 1°C/min seeding, cooling at 0.3°C/min to −30°C, at 0.1°C/min to −33°C before direct transfer to liquid nitrogen. The embryos, were thawed by agitating the ampoules in water at 37°C, before the glycerol was removed in steps. Embryos were considered to be of three qualities and transfer without freezing and thawing yield pregnancy rates of 78.6%, 63.2%, and 41.7%. The effect of freezing and thawing can be estimated by contrasting the proportion of pregnancies after freezing and thawing with that obtained after direct transfer of similar embryos. Considering only the class 1 embryos, 87.5% survived freezing and thawing itself (68.8 of 78.6). Over all groups, 75% of the embryos survived freezing and thawing (W. B. Christie, personal communication) (Table VIII). It is not clear whether this remarkable success reflects the influence of the particular method, great attention to detail or the quality of embryos being obtained from the donors. In view of the commercial interest in embryo transfer in this species, further improvements in technique are likely.

5.7. Sheep and Goat

Sheep embryos survive the original slow cooling or the more recent rapid methods of preservation (Willadsen *et al.*, 1976; Moore and Bilton, 1977; Willadsen, 1977). In the main, they were used as models for those concerned with cattle embryo preservation, and it is probable that improvements in survival could be obtained by refinement of procedures for this species. Similarly, only a small number of goat embryos have been frozen and thawed (e.g., Moore and Bilton, 1977).

6. Application of Embryo Preservation

Human embryos are being frozen as part of treatment involving *in vitro* fertilization (Trounson and Mohr, 1983). When more embryos are obtained than can be transferred back into the patient in the same menstrual cycle, the excess are being frozen. As the number of embryos available for study is low, development of methods may be slow. It is probable that in due time this technique will be used in clinical routines.

In farm animals, cryopreservation is used during international trade to allow embryos to be held during a period of quarantine for the donor animals. Frozen embryos are also used for convenience and to facilitate transfer into a recipient that was not on heat at the same time as the donor (Wilmut and Rowson, 1973). It is well established that maximum conception rate depends on transfer between animals that were on heat at the same time (Rowson and Moor, 1966; Rowson *et al.*, 1969). Unfortunately the yield of embryos may vary from 0 to more than 30 from one donor cow, so that it is difficult to decide on the number of recipients to be made available. Cryopreservation offers a convenient solution to this management problem, but it should be appreciated that it does involve the loss of some embryos.

Cryopreservation is also useful in research. Programs have been established for the banking of mutant stock of laboratory animals (e.g., Glenister and Lyon, 1981; Yokoyama *et al.*, 1981). While such schemes are potentially useful, they are expensive, there is a delay in obtaining the required lines from the store, and it is impracticable to supply very large numbers of any particular line.

The studies of embryo preservation have added greatly to our fundamental understanding of low-temperature biology. Similarly, they have provided methods of clinical, commercial, and scientific value. It is to be expected that cryopreservation of embryos will be more widely used as methods improve.

Note Added in Proof

After this chapter was completed a different approach to the preservation of embryos was described by Rall and Fahy (1985). Their procedure depends upon the ability of concentrated solutions of cryoprotective agents to supercool to very low temperatures during rapid cooling before solidifying without formation of ice. This process of **vitrification** avoids both the formation of ice and the prolonged exposure of cells to concentrated solutes. Almost 90% of mouse embryos at the eight-celled stage survived when cooled and warmed at 2,500°C/min in the presence of a mixture of DMSO, acetamide, propylene glycol and polyethylene glycol (Rall and Fahy, 1985). Application of this procedure to embryos of other species depends upon the development of media that are not toxic to embryos of these species and upon the simplification of procedures of cooling and warming very rapidly in a repeatable manner. Such procedures will probably be quicker and simpler than conventional methods, but it is not clear whether they can achieve better levels of survival.

Acknowledgments. I am grateful to W. B. Christie of Premier Embryos for advice given over a period of several years and for permission to present his unpublished results. I thank Dr. Maureen Wood, who made many helpful comments on the manuscript.

References

Bank, H., and Maurer, R. R., 1974, Survival of frozen rabbit embryos, *Exp. Cell Res.* **89:**188–196.

Bilton, R. J., 1980, Preservation of embryos of the large domestic species, *Proc. 9th Int. Cong. Anim. Reprod. A.I.* **2:**245–253.

Blears, D. J., and Danyluk, S. S., 1968, Proton wide-line nuclear magnetic resonance spectra of hydrated proteins, *Biochim. Biophys. Acta* **154:**17–27.

Bouyssou, B., and Chupin, D., 1981, Two step freezing of cattle blastocysts with dimethysulphoxide (DMSO) of glycerol, *Theriogenology* **17:**159–167.

Butler, W. J., and Roberts, T. K., 1975, Effects of some Phosphatidyl compounds on boar spermatozoa following cold shock or slow cooling, *J. Reprod. Fertil.* **43:**183–187.

Chang, M. C., and Walton, A., 1940, The effect of low temperature and acclimatization on the respiratory activity and survival of ram spermatozoa, *Proc. R. Soc., Lond. B* **129:**517–527.

Dick, D. A. T., 1966, *Cell Water*, Butterworths, Washington, D.C.

Edidin, M., and Petit, V. A., 1977, The effect of temperature on the lateral diffusion of plasma membrane proteins, in: *The Freezing of Mammalian Embryos* (K. Elliott and J. Whelan, eds.), pp. 155–165, CIBA Symposium 52, Elsevier, Amsterdam.

Elford, B. C., 1970*a*, Diffusion and distribution of diethyl sulphoxide in the isolated Guinea-pig taenia coli, *J. Physiol. (Lond.)* **209:**187–208.

Elford, B. C., 1970*b*, Non-solvent water in muscle, *Nature (Lond.)* **227:**282–283.

Farrant, J., 1977, Water transport and cell survival in cryobiological procedures, *Philos. Trans. R. Soc. Lond. B* **278:**191–205.

Farrant, J., 1980, General observations on cell preservations, in: *Low Temperature Preservation in Medicine and Biology* (M. J. Ashwood-Smith and J. Farrant, eds.), Pitman Medical Tunbridge, Wells.

Farrant, J., and Morris, G. J., 1973, Thermal shock and dilution shock as the causes of freezing injury, *Cryobiology* **10:**134–140.

Farrant, J., Lee, H., and Walter, C. A., 1978, Effects of interactions between cooling and rewarming conditions on survival of cells. in: *The Freezing of Mammalian Embryos* (K. Elliott and J. Whelan, eds.), pp. 49–63, CIBA Symposium 52, Elsevier, Amsterdam.

Frank, H. S., 1965, The structure of water, *Fed. Proc. (Suppl.)* **15:**S1–S11.

Glenister, P. H., and Lyon, M. F., 1981, Experience of banking mutant genes, in: *Frozen Storage of Laboratory Animals* (G. H. Zeilmaker, ed.), pp. 157–163, Gustav Fischer, Stuttgart.

Hazlewood, C. F., Nichols, B. L., and Chamberlain, N. F., 1969, Evidence for the existence of a minimum of two phases of ordered water in skeletal muscle, *Nature (Lond.)* **222:** 747–750.

Hechter, D., 1965, Role of water structure in the molecular organisation of cell membranes, *Fed. Proc. (Suppl.)* 15:S91–S102.

Kasai, M., Niwa, K., and Iritani, A., 1981, Effects of various cryoprotective agents on the survival of unfrozen and frozen mouse embryos, *J. Reprod. Fertil.* **63:**175–180.

Klotz, I. M., 1965, Role of water structure in macromolecules, *Fed. Proc. (Suppl.)* **15:**S24–S33.

Leibo, S. P., 1977, Fundamental cryobiology of mouse ova and embryos, (K. Elliott and J. Whelan, eds.), pp. 69–72. Elsevier, Amsterdam.

Leibo, S. P., and Mazur, P., 1978, Methods for the preservation of mammalian embryos by freezing, in: *Methods in Mammalian Reproduction* (J. C. Daniel, Jr., ed.), pp. 179–201, Academic Press, New York.

Leibo, S. P., Mazur, P., and Jackowski, S. C., 1974, Factors affecting survival of mouse embryos during freezing and thawing, *Exp. Cell Res.* **89:**79–88.

Leibo, S. P., West, A. W., Perry, B., and Mills, A. C., 1982*a*, A one-step method for direct nonsurgical transfer of frozen thawed bovine embryos. II. Applied Studies, *Proc. Soc. Cryobiol*, p. 79 (abst.).

Leibo, S. P., West, A. W., and Perry, B. 1982*b*, A one-step method for direct non-surgical transfer of frozen thawed bovine embryos. 1. Basic studies, *Proc. Soc. Cryobiol.*, p. 78 (abst.).

Lovelock, J. E., 1953, The haemolysis of human red blood-cells by freezing and thawing, *Biochim. Biophys. Acta* **10**:414–426.

Lovelock, J. E., 1954, The protective action of neutral solutes against haemolysis by freezing and thawing, *Biochem. J.* **56**:265–270.

Luyet, B., and Keane, J., 1955, A critical temperature range apparently characterised by sensitivity of bull semen to high freezing velocity, *Biodynamics* **7**:281–292.

Massip, A., Van Der Zawlmen, P., Hanzen, C , and Ectors, F., 1982, Fast freezing of cow embryos in French strains with an automatic program, *Theriogenology* **18**:325–332.

Maurer, R. R., and Haseman, J. K., 1976, Freezing morula stage rabbit embryos, *Biol. Reprod.* **14**:251–263.

Mazur, P., 1970, The freezing of biological systems, *Cryobiology* **168**:939–949.

Mazur, P., Farrant, J., Leibo, S. P., and Chu, E. H. Y., 1969, Survival of hamster tissue culture cells after freezing and thawing, *Cryobiology* **6**:1–9.

Mazur, P., Leibo, S. P., Farrant, J., Chu, E. H. Y., Hanna, M. G., and Smith, L. H., 1970, Interactions of cooling rate, warming rate and protective additive on the survival of frozen mammalian cells, in: *The Frozen Cell* (G. E. W. Wolstenholme and M. O'Connor, eds.), pp. 69–85, Churchill Livingstone, London.

Mazur, P., Rall, W. F., and Rigopoulos, N., 1981, Relative contributions of the fraction of unfrozen water and of salt concentrations to the survival of slowly frozen human erythrocytes, *Biophys. J.* **36**:653–675.

Miller, R. H., and Mazur, P., 1976, Survival of frozen-thawed human red cells as a function of cooling and warming velocities, *Cryobiology* **13**:404–414.

Miyamoto, H., and Ishibashi, T., 1977, Survival of frozen-thawed mouse and rat embryos in the presence of ethylene glycol, *J. Reprod. Fertil.* **50**:373–375.

Miyamoto, H., and Ishibashi, T., 1978, The protective action of glycols against freezing damage of mouse and rat embryos, *J. Reprod. Fertil.* **54**:427–432.

Miyamoto, H., and Ishibashi, T., 1981, Survival of mouse embryos after freezing and thawing in the presence of erythritol, *J. Exp. Zool* **216**:337–340.

Miyamoto, H., and Ishibashi, T., 1983*a*, Survival of mouse embryos frozen-thawed slowly or rapidly in the presence of various cryoprotectants, *J. Exp. Zool.* **226**:123–127.

Miyamoto, H., and Ishibashi, T., 1983*b*, Solid CO_2 freezing of mouse embryos, *J. Reprod. Fertil.* **67**:107–111.

Moore, N. W., and Bilton, R. J., 1976, Storage culture and transfer of embryos of domestic animals, *Proc. 8th Int. Congr. Anim. Reprod. A. I. Krakow* **3**:306–308.

Moore, N. W., and Bilton, R. J., 1977, Frozen storage of embryos of farm animals: Progress and implications, in: *The Freezing of Mammalian Embryos* (K. Elliott and J. Whelan, eds.), pp. 203–211, Elsevier, Amsterdam.

Morris, G. J., and Farrant, J., 1973, Effects of cooling rate on thermal shock hemolysis, *Cryobiology* **10**:119–125.

Nieman, H., Sacher, B., Schilling, E., and Smidt, D., 1982, Improvement of survival rates of bovine blastocysts with sucrose for glycerol dilution after a fast freezing and thawing method, *Theriogenology* **17**:102.

Parkes, A. S., and Smith, A. U., 1953, Regeneration of rat ovarian tissue grafted after exposure to low temperatures, *Proc. R. Soc. Lond. B* **140**:455–470.

Parrott, D. M. V., 1960, The fertility of mice with orthotopic ovarian grafts derived from frozen tissue, *J. Reprod. Fertil.* **1**:230–241.

Polge, C., 1957, Low-temperature storage of mammalian spermatozoa, *Proc. R. Soc. B* **147**:498–508.

Polge, C., 1977, The freezing of mammalian embryos: Perspectives and possibilities, in: *The Freezing of Mammalian Embryos* (K. Elliott and J. Whelan, eds., pp. 3–13, CIBA Symposium 52, Elsevier, Amsterdam.

Polge, C., Smith, A. U., and Parkes, A. S., 1949, Revival of spermatozoa after vitrification and dehydration at low temperatures, *Nature (Lond.)* **164**:666.

Polge, C., Wilmut, I., and Rowson, L. E. A., 1974, The low temperature preservation of cow, sheep and pig embryos, *Cryobiology* **11**:560.

Rall, W. F., and Fahy, G. M., 1985, Ice-free cryopreservation of mouse embryos at −196° V by vitrification, *Nature* (*Lond.*) **313**:573–574.

Rall, W. F., and Polge, C. 1984, Effect of warming rate on mouse embryos frozen and thawed in glycerol, *J. Reprod. Fertil.* **70**:285–292.

Rall, W. F., Reid, D. S., and Farrant, J., 1980, Innocuous biological freezing during warming, *Nature (Lond.)* **286**:511–514.

Rall, W. F., Mazur, P., and McGrath, J. J., 1983, Depression of ice-nucleation temperature of rapidly cooled mouse embryos by glycerol and dimethyl sulphoxide, *Biophys, J.* **41**:1–12.

Rall, W. F., Czlonkowska, M., Barton, S. C., and Polge, C., 1984, Cryoprotection of Day 4 mouse embryos by methanol, *J. Reprod. Fertil.* **70**:293–300.

Rapatz, G., Sullivan, J. J., and Luyet, B., 1968, Preservation of erythrocytes in blood containing various cryoprotective agents, frozen at various rates and brought to a give feral temperature, *Cryobiology* **5**:18–25.

Renard, J. P., Heyman, J., and Ozil, J. P., 1982, Congélation de l'embryon bovin. Une nouvelle méthode de décongelation pour le transfert cervical d'embryons conditionnés une seule fois en paillettes, *Ann. Med. Vet.* **126**:23–32.

Rowson, L. E. A., and Moor, R. M., 1966, Embryo transfer in the sheep: The significance of synchronizing oestrus in the donor and recipient animal, *J. Reprod. Fertil.* **11**:207–212.

Rowson, L. E. A., Moor, R. M., and Lawson, R. A. S., 1969, Fertility following egg transfer in the cow; effect of method, medium and synchronisation of oestrus, *J. Reprod. Fertil.* **18**:517–523.

Schneider, U., and Maurer, R. R., 1983, Factors affecting survival of frozen-thawed mouse embryos, *Biol. Reprod.* **29**:121–128.

Sherman, J. K., and Lin, T. P., 1958, Effect of glycerol and low temperature on survival of unfertilised mouse eggs, *Nature (Lond.)* **170**:374–375.

Tait, M. J., and Franks, F., 1971, Water in biological systems, *Nature (Lond.)* **230**:91–94.

Trounson, A. D., and Mohr, L., 1983, Human pregnancy following cryopreservation, thawing and transfer of an eight-cell embryo, *Nature (Lond).* **305**:707–709.

Trounson, A. D., Willadsen, S. M., Rowson, L. E. A., and Newcomb, R., 1976, The storage of cow eggs at room temperature and at low temperatures, *J. Reprod. Fertil.* **46**:173–178.

Trounson, A. D., Shea, B. F., Ollis, G. W., and Jacobson, M. E., 1978, Frozen storage and transfer of bovine embryos, *J. Anim. Sci.* **47**:677–681.

Tsunoda, Y., Soma, T., and Sugie, T., 1981, Frozen storage of rabbit and hamster embryos, in: *Frozen Storage of Laboratory Animals* (G. H. Zeilmaker, ed.), pp. 129–139, Gustav Fischer, Stuttgart.

Walton, A., 1957, Cold shock of spermatozoa, *Proc. R. Soc. B* **147**:508–509.

Whittingham, D. G. 1974, The viability of frozen-thawed mouse blastocysts, *J. Reprod. Fertil.* **37**:159–162.

Whittingham, D. G., 1975, Survival of rat embryos after freezing and thawing, *J. Reprod. Fertil.* **43**:515–518.

Whittingham, D. G., 1981*a*, Sensitivity of mouse embryos to the rate of thawing, in: *Frozen Storage and Laboratory Animals* (G. H. Zeilmaker, ed.), pp. 21–32, Gustav Fischer, Stuttgart.

Whittingham, D. G., 1981*b*, Discussion, in: *Frozen Storage of Laboratory Animals* (G. H. Zeilmaker, ed.), p. 58, Gustav Fischer, Stuttgart.

Whittingham, D. G., and Adams, C. E. 1976, Low temperature preservation of rabbit embryos, *J. Reprod. Fertil.* **47**:269–274.

Whittingham, D. G., Leibo, S. P., and Mazur, P., 1972, Survival of mouse embryos frozen to −196° and −269°C, *Science* **178**:411–414.

Whittingham, D. G., Wood, M., Farrant, J., Lee, H., and Halsley, J. A., 1979, Survival of frozen mouse embryos after rapid thawing from −196°C, *J. Reprod. Fertil.* **56**:11–21.

Willadsen, S. M., 1977, Factors affecting the survival of sheep embryos during deep-freezing and thawing, in: *The Freezing of Mammalian Embryos* (K. Elliott and J. Whelan, eds.), pp. 175–189. CIBA Symposium 52, Elsevier, Amsterdam.

Willadsen, S. M., 1980, Deep freezing of embryos in the large domestic species, *Proc. 9th Int Cong. Anim. Reprod. A.I.* **2**:255–261.

Willadsen, S. M., Polge, C., Rowson, L. E. A., and Moor, R. M., 1976a, Deep freezing of sheep embryos, *J. Reprod. Fertil.* **46**:151–154.

Willadsen, S. M., Trounson, A. O., Rowson, L. E. A., and Newcomb, R., 1976, Low temperature preservation of cow eggs, in: *Egg Transfer in Cattle* (L. E. A. Rowson, ed.), pp. 117–124, Commission of the European Communities, Luxembourg.

Willadsen, S. M., Polge, C., and Rowson, L. E. A., 1978, In vitro storage of cattle embryos, in: *Control of Reproduction in the Cow* (J. R. Sreenan, ed.), pp. 427–436, Martinus Nijhoff, The Hague.

Wilmut, I., 1972a, The effect of cooling rate, warming rate, cryoprotective agent and stage of development on survival of mouse embryos during freezing and thawing, *Life Sci.* **2(2)**:1071–1079.

Wilmut, I., 1972b, The low temperature preservation of mammalian embryos, *J. Reprod. Fertil.* **31**:513–514.

Wilmut, I., and Rowson, L. E. A., 1973, Experiments on the low-temperature preservation of cow embryos, *Vet. Rec.* **92**:686–690.

Wilmut, I., Polge, C., and Rowson, L. E. A., 1975, The effect on cow embryos of cooling to 20, 0 and −196°C, *J. Reprod. Fertil.* **45**:409–411.

Wood, M. J., and Farrant, J., 1980, Preservation of mouse embryos by two-step freezing, *Cryobiology* **17**:178–180.

Wood, M. J., and Whittingham, D. G., 1981, Low temperature storage of rat embryos, in: *Frozen Storage of Laboratory Animals* (G. H. Zeilmaker, ed.), pp. 119–127, Gustav Fischer, Stuttgart.

Yokoyama, M., Wakasuki, N., and Nomura, T., (1981), An attempt to store inbred mouse strains, in: *Frozen Storage of Laboratory Animals* (G. H. Zeilmaker, ed.), pp. 113–117, Gustav Fischer, Stuttgart.

Chapter 8

Embryo Transfer in Animals

FRANK B. KUZAN and GEORGE E. SEIDEL, JR.

1. Background

Almost 100 years have passed since Heape (1891) reported the transfer of two-cell embryos from an Angora rabbit to the oviducts of a pregnant Belgian hare. This experiment showed that the reproductive tract of the recipient had no effect on the genetic composition of the fetus, because transferred embryos resulted in Angora offspring. The basic strategies behind embryo transfer remain the same today: synchrony of reproductive cycles between the recipient and donor, use of appropriate genetic markers, and transfer of preimplantation embryos to the lumen of the reproductive tract.

Embryo transfer procedures have been developed for various species of laboratory animals including the mouse (Bittner and Little, 1937), rat (Nicholas, 1933), and hamster (Blaha, 1964). Furthermore, nonsurgical means of embryo transfer have been developed for the mouse (Beatty, 1951), rat (Vickery *et al.*, 1969), and rabbit (Shah, 1956), but pregnancy rates are somewhat lower than with surgical methods. Although donors are sacrificed for most experiments with laboratory animals, ova can be collected surgically, without killing the donor (Avis and Sawin, 1951).

The development of hormonal treatments to induce a superovulatory response (rat and mouse: Engle, 1927; rabbit: Pincus, 1940; hamster: Greenwald, 1962) has made embryo transfer, embryo transplantation, and superfetation even more useful research tools. We will use the term embryo transfer for placing embryos into the lumen of reproductive tracts and the term embryo transplantation for placing embryos into extrauterine sites or the transfer of postimplantation embryos.

Techniques of embryo transfer have not changed much during the past 30 years and are only considered briefly, with thorough reference to the literature. Experimental applications, commercial applications, and future considerations

FRANK B. KUZAN and GEORGE E. SEIDEL, JR. • Animal Reproduction Laboratory, Colorado State University, Fort Collins, Colorado 80523. *Present address for* F. B. K.: Department of Obstetrics and Gynecology, University of Washington, Seattle, Washington 98195.

are discussed as well. Procedures for laboratory animals are emphasized; detailed information can be found in Chang and Pickworth (1969), Dickmann (1971, 1982), and Adams (1982). For information on methods of embryo transfer in domestic animals, see the reviews by Hafez (1961), Foote and Onuma (1970), Dziuk (1975), Anderson (1977), Betteridge (1977), Murray (1978), Seidel (1979), Sugie *et al.* (1980), Hunter (1980), and Adams (1982). Historical aspects of embryo transfer have been reviewed by Betteridge (1981) and Adams (1982).

2. Procedures

2.1. Donors

2.1.1. Nonsuperovulated Donors

Adult cycling female rodents can be used as donors for embryos. However, to obtain significant numbers of embryos, one must maintain many females, since only about one in four will exhibit estrus on a given day. Donors are placed with males of proven fertility and checked daily for signs of mating. For mice and rats, the presence of a copulatory plug indicates that mating occurred the previous evening. Embryos recovered that day will be at the 1-cell stage. Since copulatory plugs may be expelled, females should be inspected as early as possible each morning. Detection of sperm in vaginal smears confirms that mating occurred the previous evening. Observation of the postovulatory vaginal discharge of the hamster can be used to assess the stage of the reproductive cycle, and females can be placed with males on this basis.

The convention used for dating the age of rodent embryos is to designate the day on which sperm or a copulatory plug is found in the vagina as day 1. Thus, on day 1, 1-cell ova are present in the oviducts, and by day 4 or 5, blastocysts are present in the uterus. In rodents, timing of ovulation usually is controlled by the light–dark cycle, and generally occurs about midway through the dark period.

Rabbits are induced ovulators, so the need arises for either a receptive female for mating, or the intravenous injection of luteinizing hormone (LH), 1–2 mg based on Armour Standard, gonadotropin-releasing hormone (GnRH), 5–10 μg, or human chorionic gonadotropin (hCG), 25–50 IU, to induce ovulation, followed by artificial insemination. Ovulation occurs 10–12 hr after mating or injection.

2.1.2. Superovulation

Donors are usually superovulated to increase the number of ova recovered. An additional advantage of superovulation is that timing of ovulation is intrinsic to the procedure. Gates (1956) found no difference in the developmental capability of 3-day mouse ova recovered from superovulated immature animals compared to those from naturally mated adults. The efficacy of using ova from superovulated animals is reviewed by Fleming (1982).

Either adult or immature females can be superovulated (Mouse: Fowler and Edwards, 1957; Edwards and Fowler, 1960; Lang and Lamond, 1966; McLaren, 1967; Gates, 1971; rat: Austin, 1948, 1950; Rowlands, 1944; Zarrow and Wilson, 1961; Wilson and Zarrow, 1962; rabbit: Pincus, 1940; Parkes, 1942; Warwick *et al.*, 1943; Kennelley and Foote, 1965; and hamster: Greenwald, 1962; Yanagimachi and Chang, 1964). In species such as the rat and mouse, immature animals respond more uniformly to exogenous hormones than do adults (Gates, 1971). Rabbits are not very responsive to gonadotropins before 4 months of age (Kennelley and Foote, 1965), and adult hamsters respond better than immature ones (Yanagimachi and Chang, 1964).

Pregnant mare serum gonadotropin (PMSG) or follicle-stimulating hormone (FSH) is used to increase numbers of mature follicles. In rodents, PMSG usually is the hormone of choice. For mice, 5–10 IU is administered intraperitoneally (IP), followed 44–52 hrs later with 5–10 IU hCG IP, to induce ovulation. Injections usually are given several hours before the onset of the dark period. Depending on the batch of PMSG, age and strain of the donors, and environmental factors, 10–60 ova can be recovered per female. To maximize yields of ova, mice should be superovulated between 3 and 4 weeks of age (Gates, 1971). Rats are given 10–20 IU PMSG IP, followed 48–56 hr later with 10 IU HCG IP. Usually the stage of the reproductive cycle is ignored when timing superovulation in postpubertal mice and rats.

Hamsters are injected with 25 IU PMSG on the morning of the postovulatory discharge, and 3 days later the animals should ovulate; no hCG is required. Ova are at the one-cell stage on the next day (day 4 after PMSG). Immature hamsters can be treated with 25 IU PMSG followed 56 hr later with 10 IU hCG; however, results are variable, and some animals produce no ova.

The treatment of choice for rabbits is twice daily injections of FSH over a 3-day period (Kennelley and Foote, 1965). Rabbits 4–5 months of age are given 0.3 mg FSH (potency based on Armour standard) per injection, and mature rabbits 0.3–0.5 mg FSH per injection. Twelve hr after the last FSH injection, 2.5 mg LH (Armour Standard) or 100 IU hCG is given intravenously to induce ovulation, which occurs 12–14 hr after the injection of LH, slightly later than in unsuperovulated rabbits. Alternatively, 75 IU PMSG can be given as a single injection followed by LH or hCG 3 days later.

Routinely, rabbits are inseminated artificially or mated at the time of LH or hCG administration. Rodents are usually placed with a fertile male overnight, but can also be bred artificially.

2.1.3. Collection of Embryos

Embryos can be collected by killing the donor and removing the reproductive tract or by exposing the reproductive tract surgically. The method of collecting ova depends on their location in the reproductive tract. Therefore, precise knowledge of ovum transport, ovulating hormone injection, and implantation times is required to recover embryos at the desired stage of development (see Table I). Ova are usually obtained by piercing the oviductal ampullae of rodents or by flushing fluid through the lumen of the oviduct or uterus. For *in vivo* collection, fluid is introduced into the reproductive tract and collected through a cannula placed in the uterine lumen or into the fimbriated end of the oviduct.

2.2. Embryo Handling

2.2.1. *In Vitro* Culture

A variety of culture media and body fluids, ranging from simple salt solutions supplemented with bovine serum albumin (BSA) or blood serum to complex tissue culture media, have been employed for the collection and storage of oocytes and embryos prior to transfer. For immediate transfer of embryos to recipients, saline and plasma (Dickmann and DeFeo, 1967) or phosphate-buffered saline (Mantalenakis and Ketchel, 1966) are adequate. For long-term culture of embryos, more complex media are required (Hammond, 1949; Biggers *et al.*, 1962; Brinster, 1963; Hafez, 1965; Whitten and Biggers, 1968; Whittingham and Wales, 1969; Kane and Foote, 1970; Whittingham, 1971; Gwatkin, 1972). Most of these media are bicarbonate buffered, and must be kept under an atmosphere of 5% CO_2 in order to maintain proper pH (see Table II). Frequently media are overlaid with paraffin oil to slow the escape of CO_2 and prevent evaporation.

2.2.2. Equipment for Handling Embryos

Several devices for moving embryos from one place to another are available, e.g., mouth pipettes, pipette bulbs, suction devices, and small syringes (Sugie *et al.*, 1980). Clean, sterile, fire-polished pipettes should be used when

Table I. Expected Location and Stage of Embryos following Ovulation[a]

	Mouse		Rat		Hamster		Rabbit	
Embryo stage	Day	Location	Day	Location	Day	Location	Day	Location
1-cell	1	O	1	O	1	O	½	O
2-cell	2	O	2	O	2	O	1	O
4-cell	Late 2	O	2½–3	O	2½–3	O/U	1½	O
8-cell	3	O	3½	O	3½	U	2	O
Morula	3½	O/U	4	O	3½–4	U	2–3	O/U
Blastocyst	3½–4	U	4–5	U	4–5	U	3–6	U
Implantation	5		7		5–6		7	
Gestation length	20		21		16		31	

[a]O, Oviduct; U, uterus. Data represent days.

Table II. Composition of Media Commonly Used for the Culture of Embryos

Ingredient (mg/liter)	Whitten's[a] (WM)	Brinster's (BMOC-2)[b]	Brinster's (BMOC-3)	Tyrode's	BSM-II[c]
NaCl	4000	5546	5546	8000	6019
KCl	356	356	356	200	356
$CaCl_2 \cdot 2H_2O$	—	189	189	200	251
$MgCl_2 \cdot 6H_2O$	—	—	—	100	—
$MgSO_4 \cdot 7H_2O$	294	294	294	—	294
$NaHCO_3$	2106	2106	2106	1000	2106
$NaH_2PO_4 \cdot H_2O$	—	—	—	50	—
KH_2PO_4	162	162	162	—	162
Glucose	1000	—	1000	1000	1801
(Na) pyruvate	36	28	56	—	—
(Na) lactate	2416	2253	2253	—	—
(Ca) lactate · $5H_2O$	496	—	—	—	—
Bovine serum albumin	4000	1000	1000	5000	15 000
Amino acids, vitamins, trace elements, cofactors	None	None	None	None	Added
Gas atmosphere	5% CO_2/air	5% CO_2/air	5% CO_2/air	Air	5% CO_2/air

[a]Whitten and Biggers (1968).
[b]Brinster (1969).
[c]Kane and Foote (1970).

handling embryos. Embryos are usually collected in small petri dishes or watch glasses and then located with a dissecting microscope, removed from the collection dish, and washed several times in fresh medium to dilute microorganisms and reproductive tract secretions.

2.3. Recipients

2.3.1. Synchronization

The stage of pregnancy or pseudopregnancy of the recipient should be closely synchronized with the developmental stage of the embryos to be transferred. To ensure the greatest degree of success in the transfer of embryos, asynchrony of the reproductive cycles of the donor and recipient should not be greater than 24 hr. Synchrony of recipients and donors is matched on the basis of vaginal smears, observation of copulatory plugs, and so on. Older embryos appear to survive asynchronous transfer better than do younger embryos. Embryos recovered from the oviduct should be returned to the oviduct, and similarly for the uterus. Asynchronous embryo transfer is considered further in Section 3.2.

Rats, mice, and hamsters can be made pseudopregnant by mating to vasectomized males. Such cervical stimulation is the simplest way to ensure proper corpus luteum function. However, prolactin or progesterone can be injected to induce or substitute for pseudopregnancy (Edwards and Austin, 1959; Banik and Ketchel, 1965). In rabbits, recipients can be injected with GnRH, LH or hCG in the same manner as donors, or a vasectomized male can be used.

2.3.2. Embryo Transfer

The process of embryo transfer itself is more of an art form than a technological procedure. Often pregnancy rates increase as experiments progress. Therefore, a large number of transfers should be completed by an individual before embryo transfer is employed in any experimental design. False conclusions can be drawn from experiments in which this caveat is not heeded.

When surgical transfers are performed, the oviducts or uteri are exteriorized either through a midline or flank incision. Ova can be transferred directly into the ostium of the oviduct. In rodents, the ovary and ostium are enveloped by a bursa that may reduce the chance of losing ova into the peritoneal cavity but also increases the difficulty of placing ova into the ostium. For oviduct transfers, a slit should be made in the bursa and the ostium visualized under a dissecting microscope at 4–10X magnification. Under some conditions, cauterization of the blood vessels of the ostium is desirable to control bleeding. A disposable cautery unit for eye surgery works well.

Transfer to the uterus is accomplished by making a small puncture wound through the uterine wall. Embryos are then placed into the uterine lumen using a fire polished pipet. Entry to the uterine lumen may also be gained *via* the cervix. Embryos should be transferred in as small a volume as possible. Care must be taken not to expell air into the oviduct or uterus, and trauma should be avoided as much as possible. Some investigators prefer to use a transfer device like that illustrated in Figure 1.

2.4. Design and Interpretation of Embryo Transfer Experiments

The scientific method consists of the following steps: (1) accumulate related information, (2) formulate a hypothesis, (3) design an experiment to test the hypothesis, (4) carry out the experiment, (5) evaluate the evidence, and (6) accept or reject the hypothesis. When one deviates from this method, there is a good chance of not learning anything new or useful, or, even worse, of misleading oneself or others. Far too often embryo transfer experiments are accumula-

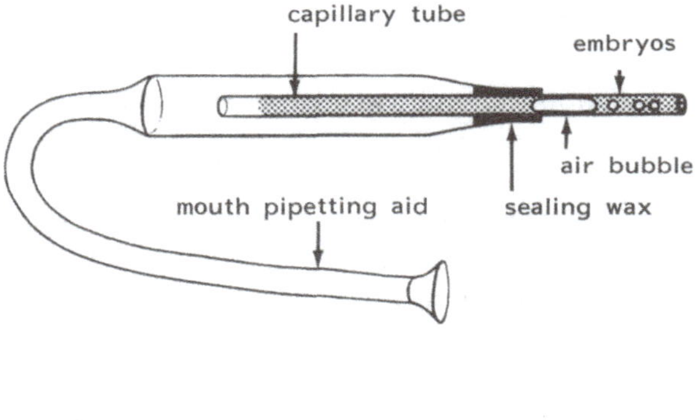

Figure 1. Device for transferring embryos in a small volume of medium.

Table III. *In Vitro* Fertilization Rates

Replicate	Treated[a]	Control[a]
1	3/10 (30)	6/10 (60)
2	20/30 (67)	10/30 (33)
3	4/10 (40)	7/10 (70)
Total	27/50 (54)	23/50 (46)

[a]No. fertilized/no. ova (%).

tions of data rather than well designed and properly analyzed and interpreted experiments.

Here we consider some special problems of designing and analyzing experiments that arise due to the nature of work with embryos; we also consider some common errors. Probably the two most important concepts in experimental design from the statistical standpoint are randomization and replication. Unless experimental units are randomly allocated to treatments and replicated adequately, experiments cannot be interpreted. An experimental unit is defined as the unit to which a treatment is applied; for example, it may be a donor, an embryo, a cumulus clot containing oocytes, a recipient, or a litter.

One common error is the incorrect designation of experimental units. For example, one might randomly apply a treatment to half of the recipients in a litter-bearing species and consider the other half as controls. The response of interest might be birth weight. In most cases, mean birth weight or total birth weight of the litter would be the appropriate response to be analyzed. Use of birth weights of individual offspring would lead to incorrect degrees of freedom, because the experimental unit was the recipient, not the individual offspring. This could lead to the conclusion that there was a statistically significant effect when none existed. A special problem occurs when the experimental unit is sampled repeatedly, for example, taking blood samples from a donor over the course of superovulation. In this case changes in experimental units over time are of interest. Usually, special statistical techniques are required for such data (Gill and Hafs, 1971).

Sometimes it is unclear how to analyze data. Consider the data set shown in Table III. In this experiment, the oocytes available each day were allocated equally to treated and control groups. The overall fertilization rate was 54% for the treated group and 46% for the control. However, if one averages the percentages for each day (30 + 67 + 40)/3 = 46 and (60 + 33 + 70)/3 = 54, one obtains the exact opposite mean response. Most people would analyze the data on the bottom line of the table by chi-square. A more correct approach is to use chi-square procedures that include replicate effects or a least-squares analysis of variance considering the response for each oocyte to be 1 if fertilized and 0 if not. Degrees of freedom would be 1 for treatment, 2 for replicates, 2 for treatment × replicate, and 94 for the error term. The least-squares means (Harvey, 1977) for treated and control groups turn out to be 45.6% and 54.4%, respectively. Fortunately, no matter how one analyzes this particular set of data, the conclusion would be that there was no significant treatment difference, although the least-squares analysis of variance shows a significant treatment by replicate interaction ($p < .05$). This indicates that additional experimentation is required to understand the nature of the treatment effect. Perhaps the number of ova recovered per donor or placed in each experimental drop should be used as a covariate (Snedecor and Cochran, 1977).

Another approach would be to use 10 randomly selected ova per treatment on each day. However, one sometimes does not know how many ova were present until data were recorded (e.g., if one used cumulus clots in which oocytes could not be counted accurately until fertilization occurred). One could, in this case, randomly eliminate ova from each group to equalize the numbers and use chi square, but considerable information would be discarded in some cases.

A very common error in interpretation of embryo transfer experiments arises from data that were not really part of a designed experiment, but rather accumulated over time. The data accumulated initially may have been predominantly from one treatment and those accumulated later from another treatment. Such data cannot be analyzed properly for treatment effects. Skills may improve with time, personnel may change, there may be seasonal effects, chemicals may age, and so on. The problem is one of randomization. As each experimental unit becomes available, it should be allocated to treatment randomly, or at least by some unbiased systematic procedure, such as every second animal is treated, and the others are controls. In most cases, to carry out an experiment properly, one requires a well-thought-out, written protocol with which each of the members of the research team is familiar. In addition to solving the practical problem of accurate communication, such protocols force one to think through allocation of experimental units to treatment.

An extremely useful concept to circumvent the tendency to evaluate this type of accumulated data is the preliminary experiment. During the preliminary experiment one can hone one's skills until they stabilize at an acceptable level of proficiency. Frequently one also can improve the experimental procedure, or determine a feasible experimental design. In many cases, an experiment thought to be well designed becomes (or should become) a preliminary experiment. There is nothing wrong with disregarding data from preliminary experiments, as long as one decides that from a certain point on the experiment is no longer preliminary, and from then on experimental units will be assigned properly to treatments. The starting point must not be decided in retrospect based on examination of the data. One must also design experiments with sufficient replication so that important treatment effects do not go undetected (Snedecor and Cochran, 1967).

3. Experimental Applications

This section consists of a selective literature review to illustrate applications. Thousands of papers could be cited (see Adams, 1982, and references

54–56 cited therein, consisting of compiled bibliographies through 1977). Bibliographies of embryo transfer literature currently are compiled four times per year in the *Embryo Transfer Newsletter* published by the International Embryo Transfer Society.

3.1. Studies of Embryos

3.1.1. Early Embryonic Survival

Studies have been conducted to investigate the effect of transferring varying numbers of embryos to recipients (McLaren and Michie, 1956, 1959; Adams, 1960, 1962). McLaren and Michie (1959) found it difficult to standardize numbers of fetuses resulting from embryo transfer due to embryonic losses and varying ovulation rates. Nevertheless, transferred embryos appeared to inhibit the development of native embryos in utero. Adams (1960) found that fewer 60–90-hr-old embryos survive embryo transfer compared with embryos transferred at later developmental stages.

A most interesting recent finding is that when day 5 and day 7 embryos were transferred to pigs simultaneously, the younger embryos did not survive, whereas survival was satisfactory with day 5 embryos alone (Pope *et al.*, 1982). This may be related to the "runt of the litter" phenomenon. A related study in horses shows that the larger equine embryo wins out when twin ovulations occur (Woods and Ginther, 1983). Elimination of the second embryo appears to explain why twins are rarely born to horses, which is fortunate due to severe problems with twin equine pregnancies. This may be related to the "missing twin" phenomenon frequently found in *Homo sapiens*, which generally occurs at a later embryonic stage. Such mechanisms are the rule rather than the exception in species such as the pronghorn antelope (Wimsatt, 1975). It is very appropriate to use embryo transfer to understand these mechanisms. There are also implications for normality of superovulated embryos in species that regulate numbers of implantations by the demise of excess embryos (Weir, 1971; Wimsatt, 1975). Since approximately 25% of conceptions result in early embryonic death in domestic animals and an even higher proportion in the human, studies concerning this wastage are of high priority. Embryo transfer coupled with innovative approaches should help resolve the causes of early embryonic death.

3.1.2. Chimeras

A chimera or allophenic embryo is formed by mixing cells from two or more embryos. Chimeras provide a means to study many aspects of early embryonic development. Nicholas and Hall (1942) reported transferring "slightly" fused pairs of 1-cell rat embryos to recipients. One offspring resulted, but no attempt was made to confirm chimerism; thus, the result was inconclusive. Tarkowski (1961) and Mintz (1962) developed the technique of mixing 4–8-cell mouse embryos, culturing them *in vitro* to the morula or blastocyst stage, and then transferring them to recipients. Gardner (1978) developed an alternate means of producing chimeras by injecting cells into the blastocoele where they could be incorporated into the inner cell mass of the embryo. With this technique and the use of genetic markers, the fate of individual cells can be followed through development. To date cells from the inner cell mass (Gardner, 1971; Rossant and Lis, 1979; Babinet and Bordenave, 1980), primitive endoderm and ectoderm (Gardner and Papaioannou, 1975; Gardner and Rossant, 1957), extraembryonic ectoderm (Rossant *et al.*, 1978), and teratocarcinoma cells (Mintz and Illmensee, 1975; Papaioannou *et al.*, 1975; Brinster, 1976) have been injected and found to contribute to tissue in the resulting chimeras.

Experimental applications of chimeric or allophenic embryos include studies of the immune system (Gardner and Munro, 1974), parthenogenesis (Stevens *et al.*, 1977), genetic transmittance of reproductive efficiency (Lawrence and Anderson, 1981), activation of the oocyte genome (Balakier, 1979), interspecific individuals (Klein and Markert, 1981; Rossant *et al.*, 1982, 1983; Fehilly *et al.*, 1984), and chimeric cloning (Willadsen and Fehilly, 1983; Seidel, 1983a).

3.1.3. Zona Pellucida and Embryonic Development

Bronson and McLaren (1970) removed the zona pellucida from 8-cell mouse embryos with pronase; no 8-cell embryos developed to term when transferred, but results were confounded due to an impurity in the pronase preparation. Hoppe and Bavister (1983) found that the zona pellucida was not needed for hamster embryos transferred at the morula stage. Brun and Psychoyos (1972) employed acidified medium to remove the zona pellucida from rat embryos; the results after transfer were similar to controls, indicating that the zona was not required for normal embryonic development in the rat. The zona pellucida clearly is required for embryos transferred in precompaction stages. The surrogate zona pellucida plus agar technique of Willadsen (1979) is extremely useful for some studies.

3.1.4. *In Vitro* Culture and Low Temperature Storage

In some experiments, it is necessary to culture embryos prior to transfer (e.g., to make chimeras, expose embryos to chemical teratogens, microsurgical techniques, *in vitro* fertilization). Pre- and postcompaction mouse embryos developing *in vitro* are illustrated in Figure 2. Biggers *et al.* (1965) reported successful transfer of mouse embryos cultured from the 2-cell to blastocyst stage *in vitro*. Previous investigations by Tarkowski (1959) and McLaren and Biggers (1958) resulted in development from the 8-cell to the blastocyst stage. Pincus and Enzmann (1935) successfully transferred 1-cell rabbit embryos after incubation *in vitro* for 8 hr. These techniques are reviewed in the books edited by Daniel (1971, 1978).

Storage of embryos at low temperatures was prompted by the successful low temperature storage of spermatozoa (Chang, 1947). Chang (1947) found that slow cooling (25°C to 0°C in 5 hr) of rabbit embryos was optimal for survival.

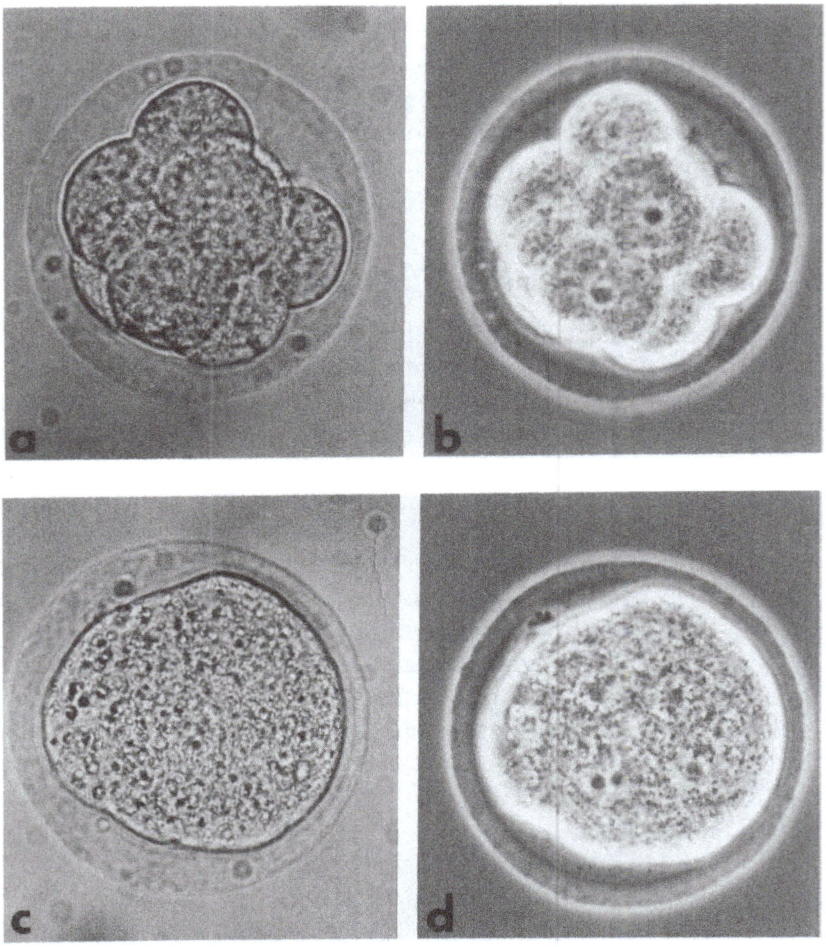

Figure 2. The top mouse embryo (a,b) is starting to compact; the bottom (c,d) is postcompaction. (a,b) Photomicrographs taken with bright-field optics; (b,d) taken with the phase-contrast optics.

Four litters were produced after the transfer of these embryos to eight recipients. Chang also noted that embryos which cleaved *in vitro* were not necessarily viable *in vivo*. Chang (1948*a*,*b*, 1950*a*) conducted a series of experiments to determine the optimal stage of embryonic development for low temperature storage. He found that late rabbit blastocysts could be recovered, stored, and transferred with greater success than younger embryos.

Cryopreservation of embryos is covered in another chapter. This technique coupled with embryo transfer permits preservation of valuable germ plasm, transport of embryos, repopulation of species, experimental studies in cryobiology, and investigation of the effect of cold storage of developmental mechanisms.

3.2. Studies of Maternal Environment

Heape's (1891) experiment showed that there was no effect of maternal environment on the gross appearance of the resulting offspring. However, other basic questions about uterine effects on embryos remain unanswered, such as, why the uterus and embryo need to be in synchrony and why many interspecific transfers fail (Kraemer, 1983).

3.2.1. Donor–Recipient Asynchrony and Site of Transfer

The need for synchrony of developmental stage of the embryo with the recipient oviductal or uterine environment is one of the most interesting facets of embryo transfer. Induced ovulators such as the rabbit are ideal for these studies. Chang (1950*b*) showed that exact synchrony between donor and recipient was optimal for producing pregnancies; 24-hr asynchrony decreased pregnancy rates by 5–10%. Oviductal embryos transferred to the uterus degenerated within 2 days of transfer.

Noyes and Dickmann (1960) transferred 2-, 3-, 4-, or 5-day rat embryos to the uteri of 2-, 3-, 4-, or 5-day recipients. The greatest success was obtained with 4- or 5-day-old embryos transferred to day 5 uteri. Dickmann and Noyes (1960) found that embryos 1 day younger than the uterus were damaged shortly after transfer and were incapable of implantation when removed and retransferred to a synchronous uterus. Noyes and Dickmann (1961) repeated the study with oviductal transfers; although pregnancy rates were somewhat lower than uterine transfers, synchrony was still an important factor in the survival of transferred embryos.

Doyle *et al.* (1963) investigated asynchronous transfer in mice and found synchrony within 24 hr was optimal. Day 4 embryos exhibited the highest pregnancy rate. Wu (1975) transferred 2-cell embryos to uteri of ovariectomized rats. Three to 4 days after transfer, only 25% of the embryos could be recovered, approximately one-third of which had developed into blastocysts.

To investigate oviductal pregnancy in the mouse, blastocysts recovered from the uterus were transferred to the oviducts of recipients (Bronson and Cunnane, 1975). No oviductal pregnancies were observed in mice with ligated oviducts; however, control transfer to the oviduct resulted in a 60% pregnancy rate.

These investigations show that embryos recovered from the oviduct are not suitable for transfer to the uterine environment; however, uterine embryos can survive after transfer to the oviduct, and embryos "older" than the uterus survive transfer better than do embryos "younger" than the uterus. The latter observation does not seem to hold for cattle (Seidel, 1981*a*).

3.2.2. Age and Reproductive Efficiency

The effect of age of recipient and donor on uterine environment was investigated by Blaha (1964). Eight-cell hamster embryos were collected from young

(2½–6 months) and old (14–18 months) animals, and embryos were transferred from young donors to young recipients, young to old, and old to young. Fewer embryos were collected from old animals, and fewer term fetuses resulted from young to old transfers and from old to young transfers (8.3% and 4.5%, respectively) compared with young to young transfers (49.2%). These results suggest that the uterine environment of senescent hamsters is unsuitable for embryonic development and that either oviductal environment or follicular development is impaired in senescent animals.

Maurer and Foote (1971) also transferred embryos from aging (up to 148 weeks) and young (18–30 week) donors to aging and young recipient rabbits. They found a decrease in ovulation rate with increasing age in addition to a decreased capacity of the aged uterus to maintain fetuses. However, the developmental potential of embryos from aged donors was normal if transferred into a young uterus.

Embryo transfer has also been used to determine why very young ewes are subfertile (Quirke and Hanrahan, 1983). In many cases, small, prepubertal animals can reproduce *via* embryo transfer to older, larger animals (Seidel, 1981*b*).

3.2.3. Other Maternal Effects

Several studies deal with the effect of uterine environment on blastocyst formation (Kirby, 1962*b*; Krishnan and Daniel, 1967; Orsini and McLaren, 1967; Murray *et al.*, 1971; Pope and Day, 1972). Bazer and Roberts (1983) recently reviewed biochemical aspects of fetal–maternal interactions. Kirby (1962*b*) found that blastocysts collected from the uterus implanted more successfully at ectopic sites than tube-locked embryos, suggesting that an uterine factor was required for development beyond the blastocyst stage. Krishnan and Daniel (1967) reported that a specific uterine protein (blastokinin or uteroglobin) isolated from rabbit uteri stimulated blastocyst formation and expansion *in vitro*. However, Kane and Foote (1970) showed that BSA was just as effective. Other uterine proteins do, nevertheless, seem to be important (Maurer and Beier, 1976).

Murray *et al.* (1971) reported that pig embryos restricted to the oviduct did not develop beyond the 32-cell or morula stage, while Pope and Day (1972) found that a high percentage of tube-locked pig embryos developed to blastocysts. Only a slight delay in hatching from the zona pellucida was observed in mouse embryos restricted to the oviduct (Orsini and McLaren, 1967). The oviduct of the rabbit has been used successfully to incubate embryos of many species for several days between collection and transfer (Hafez and Sugie, 1963; Adams *et al.*, 1968; Lawson *et al.*, 1972; Boland, 1984). However, interspecific embryo transfer usually does not result in development to term (Averill *et al.* 1955; Briones and Beatty, 1954; Kirby, 1962*a*; Tarkowski, 1962; Hakansson *et al.*, 1978; Adams, 1982; Kraemer, 1983).

Embryo transfer has been used to investigate the effect of X-irradiation on fetal development (Lin, 1980), transuterine migration (Dziuk *et al.*, 1964), and numerous other factors. Using the crosstransfer of embryos between DBA and C57 black mice, Fekete (1947) found that either type of embryo was more likely to produce a viable offspring when transferred to a C57 uterus. These results suggest that the uterine environment of the DBA is less favorable for embryonic development; thus, selection of recipients is a critical consideration.

Bittner and Little (1937) applied both crossfostering and embryo transfer to study breast cancer transmission in mice. Their results indicated that some factor transmitted through milk (now known to be a virus) is the cause of early-onset breast cancer. None of 61 female mice of the high-risk strain obtained by embryo transfer to another strain developed breast cancer.

Embryo transfer can be extremely useful for studying virus–embryo interactions (e.g., Bowen *et al.*, 1983). It appears that embryos with intact zonae pellucidae are protected from most virus infections (Hare, 1983), although there are exceptions (Gwatkin, 1971). In any case, embryo recovery and transfer techniques can be used in such studies. For example, one might wish to study interferon production by preimplantation embryos, or protective effects of interferon. Embryo transfer procedures also can be used to study whether infections are transmitted vertically, transplacentally, or by other routes.

McLaren and Michie (1958) found that the uterine environment influences the number of lumbar vertebrae in offspring produced through embryo transfer, suggesting a uterine effect on extragenetic expression, perhaps *via* space or nutrition. Clearly, it is remarkable how much influence the maternal environment can have on the phenotype of the offspring. Another example is that birth weight of bovine embryo transfer offspring is influenced by breed and parity of recipient (King *et al.*, 1985).

3.3. Separation of Maternal and Fetal Effects

Transfer of embryos between breeds of sheep with differing gestation lengths has been used to determine maternal–fetal interactions in initiating parturition (Kitts *et al.*, 1982). Cytoplasmic factors in the oocyte can also affect embryogenesis. Wagasugi (1974) studied the genetics of early embryonic mortality in DDK mice. By using cross- and backcross matings, he determined that a cytoplasmic factor present in DDK oocytes was not capable of interacting with sperm from other strains of mice. This cytoplasmic defect, contributed by the mother, was the cause of early embryonic losses. An example of a lethal maternal, but not paternal, nuclear defect recently was described by McGrath and Solter (1984).

Runner (1951) attempted to separate intrinsic factors, ascribed to the embryo, from maternal factors associated with prenatal death in mice. The greatest source of prenatal loss, nonpregnant recipients, was not attributable to either maternal or fetal factors. However, when considering large litters, the number of offspring was ascribed mainly to undefined intrinsic factors. On the other hand, small litter size was attributed to maternal effects, i.e., the uterus was not able to support more embryos.

Sometimes detrimental environmental effects are mediated *via* the reproductive tract. For example, embryos recovered from rabbits housed at ambient temperature and transferred to rabbits exposed to 36°C for 6 days (Shah, 1956) did not survive. Results indicated that elevated temperature resulted in a detrimental maternal environment. In an interesting related experiment, Burfening and Ulberg (1968) found that fertilization rates were normal when rabbits were bred with semen incubated at 38°C or 40°C for 3 hr; when embryos were transferred, however, survival was 75% for embryos produced from the 38°C semen but only 53% for those from the 40°C semen.

Yoshinaga and Adams (1967) transferred embryos to cycling rats given progesterone. These workers found that the conceptus exhibited a luteotrophic effect from day 7 to 11; nonfunctional corpora lutea became functional due to the presence of embryos. Dickman and DeFeo (1967) investigated the interaction between nondelayed and delayed embryos and estrogen-sensitized or delayed uteri. They found that dormant blastocysts became active in sensitized uteri, and active blastocysts became dormant in dormant uteri.

3.4. Extrauterine Sites of Implantation

Early investigations of the development of embryos at extrauterine sites were carried out in the rat (Nicholas, 1934) and mouse (Fawcett *et al.*, 1947). Best results were obtained when the oviduct was transected allowing embryos to pass directly into the peritoneal cavity. Embryos injected directly into the peritoneal cavity did not implant (Nicholas, 1934). Jollie (1961) found that local treatment of the peritoneal lining with histamine increased the implantation rate of injected embryos due to induction of a local inflammatory response. Castle (1916) reported embryonic development on the broad ligament in the guinea pig; however, Bland and Donovan (1965) could not repeat his work. Greater success was obtained by transplanting embryos directly into the chamber of the eye (Fawcell *et al.*, 1947; Runner, 1947), or onto the surface of the kidney (Fawcett, 1950; Kirby, 1960, 1962*b*), spleen (Kirby, 1963*a*), or testis (Kirby, 1963b). McLaren and Tarkowski (1963) found that the zona-free mouse blastocysts implanted at a higher rate than zona-intact embryos when injected into the peritoneal cavity, perhaps because the zona pellucida is not shed as readily in extrauterine sites.

Normal embryonic development did not occur in any of these experiments; degeneration of the inner cell mass was common. McLaren and Tarkowski (1963) observed that after the loss of the inner cell mass, an abnormally high proportion of trophoblast cells differentiated into giant cells, suggesting a role for inner cell mass in regulation of placental formation

Bland and Donovan (1965) also found abnormal development when guinea pig embryos were injected into the peritoneal cavity. When implantation occurred, there was no effect on the length of the recipient's estrous cycle.

Kirby (1963*b*) and Billington (1965) found that a small proportion of blastocysts recovered from the uterus formed normal early embryonic structures when transferred to extrauterine sites. Billington *et al.* (1968) then investigated use of oviductal embryos cultured to the blastocyst stage *in vitro*, and then transplanted to the eye, testis, or kidney. A very small number of these embryos developed normally at extrauterine sites, indicating that exposure to the uterine environment was not necessary for embryo development.

In assessing these data, several factors must be considered: strain of animal, incidence of spontaneous teratocarcinoma, incidence of ectopic pregnancy, and method of transplantation. Stevens (1967) found that a large percentage of embryos transplanted to the testes developed into disorganized embryonic tissue. However, the strain of mouse used for the study exhibits a high level of spontaneous teratocarcinoma, which may compromise these data.

3.5. Validation of Other Techniques

Transfer of embryos is commonly used to verify fertilization of oocytes *in vitro* (Noyes, 1952; Chang, 1955, 1959; Brackett and Williams, 1965; Whittingham, 1968; Seitz *et al.*, 1970; Miyamoto and Chang, 1972; Toyoda and Chang, 1974; Seidel *et al.*, 1976; Whittingham, 1977; Thadani, 1982; Brackett *et al.*, 1982). With the advent of microsurgery to embryos (Lin, 1971), the viability of manipulated embryos has been tested by transfer to suitable recipients. Examples include production of homozygous diploid mice (Hoppe and Illmensee, 1977; Modlinski, 1980), injection of sperm into oocytes (Thadani, 1980), derivation of embryos from transplanted nuclei (Illmensee and Burki, 1981; Illmensee and Hoppe, 1981; Hoppe and Illmensee, 1982), microinjection of genes into pronuclei (Palmiter *et al.*, 1982) and bisection of embryos to produce identical twins (Tarkowski, 1959; Moore *et al.*, 1969; Moustafa and Hahn, 1978; Willadsen, 1980; Ozil *et al.*, 1982; Williams *et al.*, 1982). Production of live offspring with appropriate genetic markers *via* embryo transfer was invaluable in these and many other studies.

3.6. Other Applications

Embryo transfer is required to exploit many kinds of genetic engineering. Cloning laboratory animals to produce more uniform experimental units (Seidel, 1983*b*), shortening generation intervals for genetic studies, making litters of all males or all females, and creating clones to be used in assessing environmental effects are just some of the future uses of embryo transfer. These procedures are also necessary to resurrect cryopreserved embryos. More exotic uses of embryo transfer are covered in Section 5.2.

4. Commercial Applications

Many scientists perceive that the commercial implications of their work are best left to others and furthermore that this aspect will have little impact on their research. It appears that such attitudes are outdated, particularly in the case of embryo transfer and related technologies. There is at least some responsibility to see that one's work is used for the benefit of mankind, or at least not misused. There is also the pragmatic consideration of financing one's work. To illustrate interrelationships between basic research and commercial application, a synopsis of the history of the bovine embryo transfer industry will be presented along with related information for other species.

4.1. The Bovine Embryo Transfer Industry

A huge industry, with gross receipts probably in excess of $25 million in North America in 1983, has developed based primarily on amplification of reproductive rates of valuable cattle. The industry has grown exponentially over the past decade; it was described in detail by Seidel and Seidel (1981). The industry developed in response to demands of cattle breeders and entrepreneurs, who recognized the value of amplifying the reproductive rate of cows whose calves were worth thousands of dollars. Initially, available technologies were overextended resulting in damage to the reproductive tracts of some cows due to surgical procedures. However, technology improved rapidly, including the use of prostaglandin $F_{2\alpha}$, nonsurgical methods of embryo recovery, and transfer, and freezing, splitting, and sexing embryos. Current procedures are illustrated in Figures 3 and 4. While considerable information flowed to this industry from universities and government research laboratories, there was also considerable movement of information from the industry to academia, resulting in greatly improved experimental procedures. This exchange of information has been fostered in part by a very active International Embryo Transfer Society. Each new technology contributed to further expansion of the industry, although growth was not without its problems. For example, at times new technologies worked too well, resulting in too many offspring from some animals, which saturated the market. Such "boom and bust" situations were more common in the early days of the industry, but still occur on occasion.

Over the past decade costs of bovine embryo transfer have declined to about one-fourth of what they were if adjusted for inflation. Costs will likely decline even more as more effective delivery systems for this technology evolve. A few statistics may be of interest. It was estimated that about 70,000 embryo transfer pregnancies were established in cattle in North America in 1983 (Seidel, 1984). An average of 6–7 normal embryos is recovered per normal superovulated donor, and this leads to 3–4 calves (Seidel, 1981b). Variation is astounding; in many cases no calves result, and in others 20 or more calves are produced. Superovulation can be repeated at about 2-month intervals. Embryos frequently can be recovered from otherwise infertile cows, and pregnancy rates per embryo transferred are similar to those from normal cows. However, only about one-third as many transferable embryos are recovered per infertile donor.

Probably 20–25% of bovine embryos collected in 1983 were cryopreserved in liquid nitrogen. Under ideal conditions pregnancy rates with frozen embryos

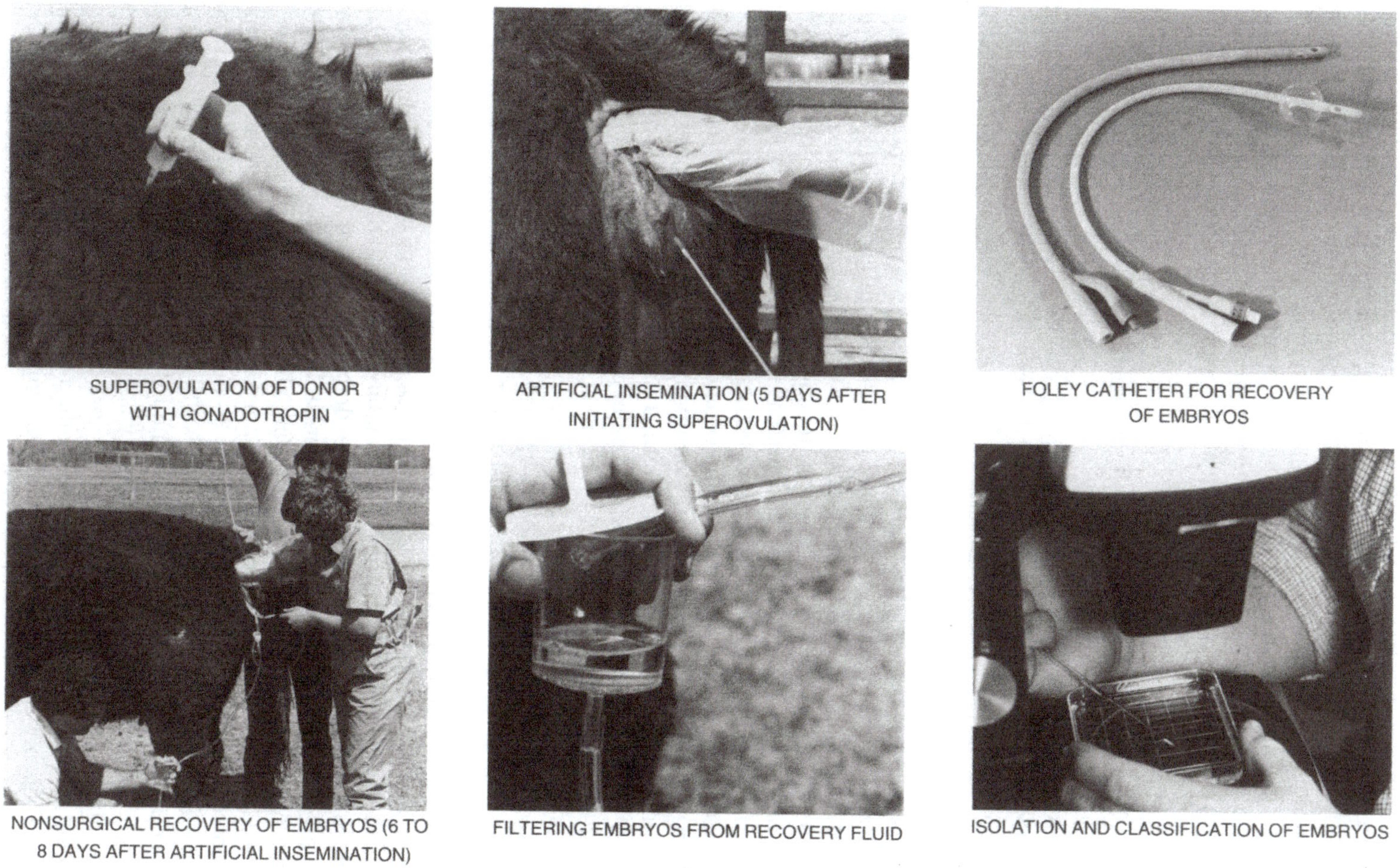

Figure 3. Summary of techniques for obtaining bovine embryos.

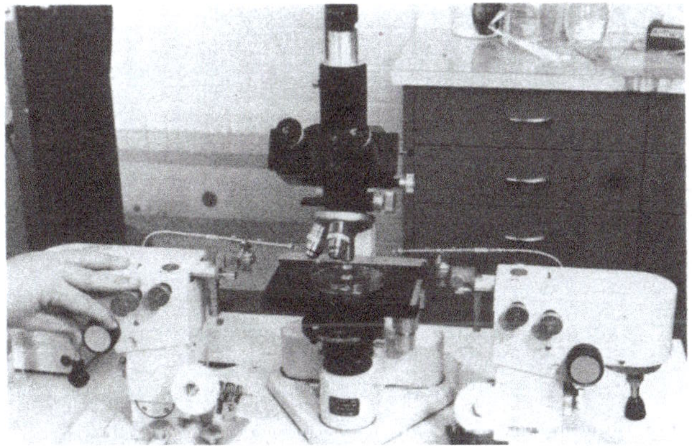

MICROMANIPULATOR FOR MICROSURGERY TO EMBRYOS

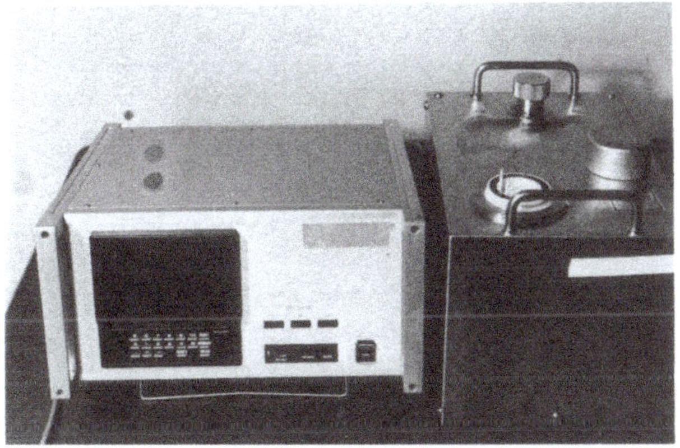

UNIT FOR CRYOPRESERVATION OF EMBRYOS

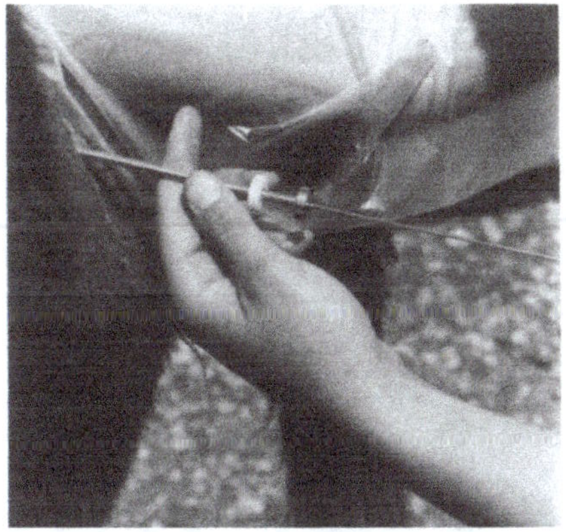

NONSURGICAL TRANSFER OF EMBRYOS TO RECIPIENT

RECTAL PALPATION 1 TO 3 MONTHS AFTER TRANSFER TO DIAGNOSE PREGNANCY

RECIPIENT WITH CALF OF DIFFERENT BREED

Figure 4. Summary of techniques for processing and transferring bovine embryos.

are about three-fourths of those with unfrozen ones, but sometimes are much lower.

There is considerable international movement of bovine embryos, currently on the order of several thousand per year. There are a number of theoretical advantages to moving germplasm internationally *via* embryos rather than *via* live animals. Under some circumstances, the cost is much lower. For example, several thousand bovine embryos can be placed in a 25-kg tank of liquid nitrogen, which can be shipped inexpensively. Another advantage is that animals frequently become debilitated and die prematurely when introduced into radically new environments, especially because of new diseases for which they have no immunological protection. This is less of a problem with embryo transfer because the newborn calf receives appropriate passive immunity from the recipient mother *via* colostrum (first milk), and thus is protected from the microorganisms in the environment. The immune system as well as other physiological and behavioral processes may also evolve more appropriately when young are born in the region in which they will mature. A third reason for replacing animal movements with embryo transfer has to do with humane treatment of animals. At best international transportation of animals is stressful, and it can be traumatic.

Perhaps the overriding reason that movement of embryos will replace movement of animals in many international transactions is prevention of spread of disease. Embryo transfer has been shown to be effective in preventing spread of a number of viral diseases (e.g., Bowen *et al.*, 1983; Hare, 1983; Singh *et al.*, 1983). Since embryos have no respiratory, digestive, circulatory, excretory, etc. systems, pathogens specific to these systems usually are not transmitted. Currently, there are restrictions prohibiting the import and export of animals between many countries, and embryo transfer may be the only approved method of moving germ plasm in some cases. For some situations, it can even be safer than use of frozen semen.

There are also several disadvantages in using embryos rather than live animals. An obvious one is that one loses time in waiting for the animals to be born and mature. Also, in some cases, moving animals is less expensive than embryo transfer procedures.

4.2. Commercial Embryo Transfer in Other Species

The equine embryo transfer industry is much smaller than that for cattle, with only a few hundred pregnancies per year in North America. However, it is growing rapidly. Equine embryo transfer is severely hampered by the difficulty of superovulating mares and restrictions of breed societies. Some societies do not permit registration of embryo transfer foals at all, and others only permit one offspring to be registered per female per year. Because of the latter constraint, most embryo transfer work with horses concerns old, infertile mares, whose reproductive tracts are not capable of maintaining pregnancy, but who still produce normal embryos. Some breeders like the option of selecting from two or three offspring, and register the one that is most desirable, e.g., one that is of the desired sex, has not succumbed to environmental diseases, or has the most desirable color or conformation. The offspring not registered can still be sold as grade animals, sometimes at large profits.

The porcine embryo transfer industry is active but specialized. The major application has been introduction of new genetic material into specific pathogen-free herds (Martin, 1983). Because these procedures are performed under sterile conditions, embryos can be transferred without infecting the recipient herd with pathogens endemic in the donor herds. The second most common use of embryo transfer for swine is for international movement of embryos. This is done with unfrozen embryos because cryopreservation of porcine embryos has not been successful to date.

Commercial embryo transfer is done on a much more limited scale in a number of other species including sheep, goats and rabbits. Possibly some semicommerical work will be done at zoos in the context of rescuing endangered species. Also, a huge industry is likely to evolve in our own species (see Chapter 5, this volume).

5. Future Considerations

5.1. Research Needs

Although embryo transfer is a simple technique, successful use depends on thorough understanding of many aspects of reproductive physiology. For example, if embryos are not normal at the time of transfer, success can hardly be expected. Obtaining large numbers of normal embryos frequently depends on such methods as superovulation, *in vitro* fertilization culture of embryos, and cryopreservation.

One of the biggest problems in most species is superovulation. Superovulated embryos are normal for all practical purposes, and develop into normal young (Seidel, 1981*a*), but the variability in the numbers obtained is phenomenal (Seidel, 1981*b*). It is hardly surprising that superovulation cannot be controlled precisely, as follicular growth and oocyte maturation are poorly understood in the normal, unsuperovulated female. Furthermore, such simple problems as control of gamete and embryo transport are understood only rudimentally, and these probably are modified in the superovulated animal. At present superovulation is a very inefficient method of obtaining oocytes from ovaries. In all likelihood, methods will eventually evolve for removing oocytes from ovaries by the hundreds, and maturing them *in vitro*.

It is clear that additional basic research in many aspects of reproductive physiology will be of great value in improving embryo transfer results. Research is also required on *in vitro* techniques including cryopreservation Sometimes new methodology brings its own set of rules, that cannot be deduced solely from studying natural models. One clear example is artificial insemination (AI) in cattle. Site of semen deposition is different (vagina for the bull, uterus for AI); timing is different (during estrus for the bull, after estrus for AI); number of sperm and volume of inseminate differ (billions of sperm in several milliliters for the bull, tens of millions in <1 ml for AI); and additives differ (none for the bull; antibiotics, buffers, cryoprotectants, and macromolecules for AI). More research is required to modify embryo transfer methodology for optimal success rates as has been done with artificial insemination in cattle. An example along these lines is having the donor and recipient reproductive cycles slightly asynchronous to compensate for retarded embryonic development due to suboptimal *in vitro* culture conditions (Hahn, 1984).

5.2. Unconventional Approaches

One can use embryo transfer to make animals pregnant that are normally sterile. An interesting example is the mule. Mules ovulate in some cases and have reproductive cycles. The ova are infertile due to the impossibility of pairing homologous chromosomes during synapsis in meiosis 1. However, it turns out to be possible to impregnate mules by transfer of horse embryos (Salazar, 1981). Recently, several mules have been made pregnant with donkey or horse embryos to study the endocrinology of PMSG production and parturition (Antczak *et al.*, 1984).

Embryo transfer probably could be used to make males pregnant *via* implantation in the peritoneal cavity. Most of these pregnancies would abort, but occasionally some would probably go to term as rarely occurs in ectopic pregnancies in the female. A concerted effort would likely increase ectopic pregnancy rates as discussed in an earlier section. This cannot be recommended as a routine procedure due to the high incidence of death of both the fetus and mother (or male mother) with ectopic pregnancies. The model would be interesting from the standpoint of inducing lactation in males, among other things.

Another area that has received only marginal attention is *in vitro* implantation. Many workers, notably New (1978) have worked with postimplantation fetuses. But more work on perfusion of gravid uteri *in vitro* or implantation into agar or gelatin embedded with endometrial cells or extracts or various nutrients would be very interesting and might teach us much about implantation and postimplantation physiology.

Much remains to be done on studying various maternal effects, such as having very dissimilar donors and recipients, e.g., Great Dane embryos in Chihuahua recipients or even different species. The latter concept could be useful in rescuing endangered species. Donors and recipients can be made genetically identical by splitting an embryo, freezing one-half, and when the second half matures, making her pregnant with her identical twin sister.

Likely at some future time the genetic donor of the ovum will not be a female, but rather a cell line *in vitro*, possibly even a tumor line. One would simply exchange the genetic information in such cells with that present in a 1-cell embryo. One could go on with even more exotic schemes, such as adding genetic information by means of artificial chromosomes (Murray and Szostak 1983) or even specifying all the genetic information in this way. Eventually it may be possible to have the base sequence of the entire genome stored in a computer connected to a gene machine of some sort that would spew forth the DNA needed for a particular animal in whatever number of chromosomes desired. These chromosomes would be inserted into a 1-cell ovum, which would then proceed to form an embryo and fetus, partly due to intrinsic epigenetic tendencies (Seidel, 1983*a*), but with the genetic characteristics of the artificial genome. Embryo transfer will probably be one of the steps in getting such an animal to term, but perhaps the transfer will be to a petri dish of gelatin.

References

Adams, C. E., 1960, Early embryonic mortality in the rabbit, *J. Reprod. Fertil.* **1**:315–316.

Adams, C. E., 1962, Studies on prenatal mortality in the rabbit, *Oryctolagos cuniculus;* the effect of transferring varying numbers of eggs, *J. Endocrinol.* **24**:471–490.

Adams, C. E., ed., 1982, *Mammalian Egg Transfer*, CRC Press, Boca Raton, Florida.

Adams, C. E., Moor, R. M., and Rowson, L. E. A., 1968, Survival of cow and sheep eggs in the rabbit oviduct, *Proc. VIth Int. Congr. Anim. Reprod. Artif. Insem.* **1**:573–574.

Anderson, G. B., 1977, Fertilization, early development, and embryo transfer, in: *Reproduction in Domestic Animals* (H. H. Cole and P. T. Cupps, eds.), pp. 285–314, Academic Press, New York.

Antczak, D. F., Davies, C. J., and Allen, W. R., 1984, Pregnancy in the mule achieved through embryo transfer, *Proc. Xth Int. Congr. Anim. Reprod. Artif. Insem.* **1**:219 (abst.).

Austin, C. R., 1948, The fragmentation of eggs following induced ovulation in immature rats, *J. Endocrinol.* **6**:104–110.

Austin, C. R., 1950, The fecundity of the immature rat following induced superovulation, *J. Endocrinol.* **6**:293–301.

Averill, R. L. W., Adams, C. E., and Rowson, L. E. A., 1955, Transfer of mammalian ova between species, *Nature (Lond.)* **176**:167–168.

Avis, F. R., and Sawin, P. B., 1951, A surgical technique for the reciprocal transplantation of fertilized eggs in the rabbit, *J. Hered.* **42**:259–260.

Babinet, C., and Bordenave, G. R., 1980, Chimeric rabbits from immunosurgically-prepared inner-cell-mass transplantation, *J. Embryol. Exp. Morphol.* **60**:429–440.

Balakier, H., 1979, Interspecific heterokaryons between oocytes and blastomeres of the mouse and the bank vole, *J. Exp. Zool.* **209**:323–329.

Banik, U. K., and Ketchel, M. M., 1965, Hormonal induction of pseudopregnancy in rats, *J. Reprod. Fertil.* **10**:85–91.

Bazer, F. W., and Roberts, R. M., 1983, Biochemical aspects of conceptus–endometrial interactions, *J. Exp. Zool.* **228**:373–383.

Beatty, R. A., 1951, Transplantation of mouse eggs, *Nature (Lond.)* **168**:995.

Betteridge, K. J., 1977, *Embryo Transfer in Farm Animals*, Monograph 16, Canada Department of Agriculture, Ottawa.

Betteridge, K. J., 1981, An historical look at embryo transfer, *J. Reprod Fertil.* **62**:1–13.

Biggers, J. D., Gwatkin, R. B. L., and Brinster, R. L., 1962, Development of mouse embryos in organ cultures of fallopian tubes in a chemically defined medium, *Nature (Lond.)* **194**:747–749.

Biggers, J. D., Moore, B. D., and Whittingham, D. G., 1965, Development of mouse embryos in vivo after cultivation from two-cell ova to blastocysts in vitro, *Nature (Lond.)* **206**:734–735

Billington, W. D., 1965, The invasiveness of transplanted mouse trophoblast and the influence of immunological factors. *J. Reprod. Fertil.* **10**:343–352.

Billington, W. D., Graham, C. F., and McLaren, A., 1968, Extra-uterine development of mouse blastocysts cultured in vitro from early cleavage stages, *J. Embryol. Exp. Morphol.* **20**:391–400.

Bittner, J. J., and Little, C. C., 1937, The transmission of breast and lung cancer in mice, *J. Hered.* **28**:117–121.

Blaha, G. C., 1964, Effect of age of the donor and recipient on the development of transferred golden hamster ova, *Anat. Rec.* **150**:413–416.

Bland, K. P., and Donovan, B. T., 1965, Experimental ectopic implantation of eggs and early embryos in guinea pigs, *J. Reprod. Fertil.* **10**:189–196.

Boland, M. P., 1984, Use of the rabbit oviduct as a screening tool for the viability of mammalian eggs, *Theriogenology* **21**:126–137.

Bowen, R. A., Howard, T. H., Elsden, R. P., and Seidel, G. E., Jr., 1983, Embryo transfer from cattle infected with Blue Tongue Virus, *Am. J. Vet. Res.* **44**:1625–1628.

Brackett, B. G., Bousquet, D., and Dressel, M. A., 1982, In vitro sperm capacitation and in vitro fertilization with normal development in the rabbit, *J. Androl.* **3**:402–411.

Brackett, B. G., and Williams, W. L., 1965, In vitro fertilization of rabbit ova, *J. Exp. Zool.* **160**:271–282.

Brinster, R. L., 1963, A method for in vitro cultivation of mouse ova from two-cell to blastocyst, *Exp. Cell. Res.* **32**:205–207.

Brinster, R. L., 1969, Mammalian embryo culture, in: *The Mammalian Oviduct* (E. S. E. Hafez and R. J. Blandau, eds.), pp. 419–444, The University of Chicago Press, Chicago.

Brinster, R. L., 1976, Participation of teratocarinoma cells in mouse embryo development, *Cancer Res.* **36**:3412–3414.

Briones, H., and Beatty, R. A., 1954, Interspecific transfers of rodent eggs, *J. Exp. Zool.* **125**:99–118.

Bronson, R., and Cunnane, M., 1975, Transfer of uterine implantation blastocysts to the oviduct in mice, *Fertil. Steril.* **26**:455–459.

Bronson, R. A., and McLaren, A., 1970, Transfer to the mouse oviduct of eggs with and without the zona pellucida, *J. Reprod. Fertil.* **22**:129–137.

Brun, J. L., and Psychoyos, A., 1972, Dissolution of the rat zona pellucida by acidified media and blastocyst viability, *J. Reprod. Fertil.* **30**:489–491.

Burfening, P. J., and Ulberg, L. C., 1968, Embryonic survival subsequent to culture of rabbit spermatozoa at 38° and 40°C, *J. Reprod. Fertil.* **15**:87–92.

Castle, W. E., 1916, Publ. Carnegie Instn. N. 241 cited by Pincus, G., 1965, in: *Preimplantation Stages of Pregnancy* (G. E. W. Wolstenholme and M. O'Connor, eds.), p. 340, Little, Brown and Company, Boston.

Chang, M. C., 1947, Normal development of fertilized rabbit ova stored at low temperature for several days, *Nature (Lond.)* **159**:602.

Chang, M. C., 1948*a*, Transplantation of fertilized rabbit ova: the effect on viability of age, in vitro storage period, and storage temperature, *Nature (Lond.)* **161**:978–979.

Chang, M. C., 1948*b*, Probability of normal development after transplantation of fertilized rabbit ova stored at different temperatures, *Proc. Soc. Exptl. Biol. Med.* **68**:680–683.

Chang, M. C., 1950*a*, Transplantation of rabbit blastocysts at late stage: Probability of normal development and viability at low temperature, *Science* **111**:544–545.

Chang, M. C., 1950*b*, Development and fate of transferred rabbit ova or blastocyst in relation to the ovulation time of recipients, *J. Exp. Zool.* **114**:197–216.

Chang, M. C., 1955, The maturation of rabbit oocytes in culture and their maturation, activation, fertilization and subsequent development in the fallopian tubes, *J. Exp. Zool.* **128**:379–405.

Chang, M. C., 1959, Fertilization of rabbit ova in vitro, *Nature (Lond.)* **184**:466–467.

Chang, M. C., and Pickworth, S., 1969, Egg transfer in the laboratory animal, in: *The Mammalian Oviduct* (E. S. E. Hafez and R. J. Blandau, eds.), pp. 389–405, The University of Chicago Press, Chicago.

Daniel, J. C., Jr., ed., 1971, *Methods in Mammalian Embryology*, W. H. Freeman, San Francisco.

Daniel, J. C., Jr., ed., 1978, *Methods in Mammalian Reproduction*, Academic Press, New York.

Dickmann, Z., 1964, Fertilization and development of rabbit eggs following the removal of the cumulus oophorus, *J. Anat.* **98**:397–402.

Dickmann, Z., 1971, Egg transfer, in: *Methods in Mammalian Embryology*, (J. C. Daniel, Jr., ed.), pp. 131–145, W. H. Freeman, San Francisco.
Dickmann, Z., 1982, Egg transfer in the mouse and rat, in: *Mammalian Egg Transfer* (C. E. Adams ed.), pp. 19–27, CRC Press, Boca Raton, Florida.
Dickmann, Z., and DeFeo, V. J., 1967, The rat blastocyst during normal pregnancy and during delayed implantation, including an observation on the shedding of the zona pellucida, *J Reprod. Fertil.* **13**:3–9.
Dickmann, Z., and Noyes, R. W., 1960, The fate of ova transferred into the uterus of the rat, *J Reprod. Fertil.* **1**:197–212.
Doyle, L. L., Gates, A. H., and Noyes, R. W., 1963, Asynchronous transfer of mouse ova, *Fertil Steril.* **14**:215–225.
Dziuk, P., 1975, Embryo transfer: An experimental tool with practical applications, *Bio. Sci* **25**:102–106.
Dziuk, P. J., Polge, C., and Rowson, L. E., 1964, Intra-uterine migration and mixing of embryos in swine following egg transfer, *J. Anim. Sci.***23**:37–42.
Edwards, R. G., and Austin, C. R., 1959, Induction of oestrus and ovulation in adult rats, *J. Endocrinol.* **18**:VII–VIII.
Edwards, R. G., and Fowler, R. E., 1960, Superovulation treatment of adult mice: Their subsequent natural fertility and response to further treatment, *J. Endocrinol.* **21**:147–154.
Engle, E. T., 1927, Pregnancy following super-ovulation in the mouse, *Proc. Soc. Exp. Biol Med* **25**:84–85.
Fawcett, D. W., 1950, The development of mouse ova under the capsule of the kidney, *Anat. Rec.* **108**:71–91.
Fawcett, D. W., Wislocki, G. B., and Waldo, C. M., 1947, The development of mouse ova in the anterior chamber of the eye and in the abdominal cavity, *Am. J. Anat.* **81**:413–443.
Fehilly, C. B., Willadsen, S. M., and Tucker, E. M., 1984, Interspecific chimerism between sheep and goat, *Nature (Lond.)* **307**:634–636.
Fekete, E., 1947, Differences in the effect of uterine environment upon development in the DBA and C57 black strains of mice, *Anat. Rec.* **98**:409–415.
Fleming, A. D., 1982, Development capability of superovulated ova, in: *In Vitro Fertilization and Embryo Transfer* (E. S. E. Hafez and K. Semm, eds.) pp. 277–286, Alan R. Liss, New York.
Foote, R. H., and Onuma, H., 1970, Superovulation, ovum collection, culture and transfer, a review, *J. Dairy Sci.* **53**:1681–1692.
Fowler, R. E., and Edwards, R. G., 1957, Induction of superovulation and pregnancy in mature mice by gonadotropins, *J. Endocrinol.* **15**: 374–384.
Gardner, R. L., 1971, Manipulations on the blastocyst, in: *Advances in the Biosciences, No 6: Schering Symposium on Intrinsic and Extrinsic Factors in Early Mammalian Development* (G. Raspe, ed.), pp. 279–296, Pergamon Press, Oxford.
Gardner, R. L., 1978, Production of chimeras by injecting cells or tissues into the blastocyst, in: *Methods in Mammalian Reproduction* (J. C. Daniel, Jr., ed), pp. 137–166, Academic Press, New York.
Gardner, R. L., and Munro, A. J., 1974, Successful construction of chimaeric rabbit, *Nature (Lond.)* **250**:146–147.
Gardner, R. L., and Papaioannou, V. E., 1975, Differentiation in the trophectoderm and inner cell mass, in: *Early Development of Mammals* (M. Balls and A. E. Wild, eds.), pp. 107–132, Cambridge University Press, Cambridge.
Gardner, R. L., and Rossant, J., 1975, Determination during embryogenesis, in: *Embryogenesis in Mammals*, pp. 5–25, Ciba Foundation Symposium, Amsterdam.
Gates, A. H., 1956, Viability and developmental capacity of eggs from immature mice treated with gonadotropins, *Nature (Lond.)* **177**:754–755.
Gates, A. H., 1971, Maximizing yield and developmental uniformity of eggs, in: *Methods in Mammalian Embryology* (J. C. Daniel, Jr., ed.), pp. 64–75, W. H. Freeman, San Francisco.
Gill, J. L., and Hafs, H. D., 1971, Analysis of repeated measurements of animals, *J. Anim. Sci.* **33**:331–336.

Greenwald, G. S., 1962, Analysis of superovulation in the adult hamster, *Endocrinol.* **71**:378–389.
Gwatkin, R. B. L., 1971, Studying the effect of viruses on eggs, in: *Methods in Mammalian Embryology* (J. C. Daniel, Jr., ed.), pp. 228–237, W. H. Freeman, San Francisco.
Gwatkin, R. B. L., 1972, Chemically defined media for mammalian eggs and early embryos, *In Vitro* **8**:59–67.
Hafez, E. S. E., 1961, Procedures and problems of manipulation, selection, storage, and transfer of mammalian ova, *Cornell Vet.* **51**:299–333.
Hafez, E. S. E., 1965, Storage media for rabbit ova, *J. Appl. Physiol.* **20**:731–736.
Hafez, E. S. E., and Sugie, T., 1963, Reciprocal transfer of cattle and rabbit embryos, *J. Anim. Sci.* **22**:30–35.
Hahn, J., 1984, The value of laboratory animal models in embryo transfer research, *Theriogenology* **21**: 45–59.
Hakansson, S., Lundkvist, O., and Nilsson, B. O., 1978, Survival of rat blastocysts transplanted into the uterus of hyperimmunized mice during delay of implantation, *Int. J. Fertil.* **23**:148–151.
Hammond, J., 1949, Recovery and culture of tubal mouse ova, *Nature (Lond.)* **163**:28–29.
Hare, W. C. D., 1983, Embryo transfer and disease transmission in farm animals, *Proc. 87th Ann. Mtg. U.S. Animal. Health Assn. Las Vegas,* 303–314.
Harvey, W. R., 1977, *User's Guide for LSML76,* Department of Dairy Science, Ohio State University, Columbus, Ohio.
Heape, W., 1891, Preliminary note on the transplantation and growth of mammalian ova within a uterine foster mother, *Proc. R. Soc. Lond.* **48**:457–458.
Hoppe, P. C., and Illmensee, K., 1977, Microsurgically produced homozygous–diploid uniparental mice, *Proc. Natl. Acad. Sci. USA* **74**:5657–5661.
Hoppe, P. C., and Illmensee, K., 1982, Full-term development after transplantation of parthenogenetic embryonic nuclei into fertilized mouse eggs, *Proc. Natl. Acad. Sci. USA* **79**:1912–1916.
Hoppe, R. W., and Bavister, B. D., 1983, Effect of removing the zona pellucida on development of hamster and bovine embryos in vitro and in vivo, *Theriogenology* **19**:391–404.
Hunter, R. H. F., 1980, *Physiology and Technology of Reproduction in Female Domestic Animals,* Academic Press, New York.
Illmensee, K., and Burki, K., 1981, Nuclear and gene transplantation in the mouse, in: *Developmental Biology Using Purified Genes* (D. D. Brown and C. F. Fox, eds.) pp. 607–619, Academic Press, New York.
Illmensee, K., and Hoppe, P. C., 1981, The potential of transplanted nuclei during mammalian differentiation, *Fortschr. Zool.* **26**:67–74.
Jollie, W. P., 1961, The incidence of experimentally produced abdominal implantations in the rat, *Anat. Rec.* **141**:159–168.
Kane, M. T., and Foote, R. H., 1970, Culture of two- and four-cell rabbit embryos to the expanding blastocyst stage in synthetic media, *Proc. Soc. Exp. Biol. Med.* **133**:921–925.
Kennelley, J. J., and Foote, R. H., 1965, Superovulatory response of pre- and post-pubertal rabbits to commercially available gonadotrophins, *J. Reprod. Fertil.* **9**:177–188.
King, K. K., Seidel, G. E., Jr., and Elsden, R. P., 1985, Bovine embryo transfer pregnancies. I. Abortion rates and characteristics of calves, *J. Anim. Sci.* **61**:747–757.
Kirby, D. R. S., 1960, Development of mouse eggs beneath the kidney capsule, *Nature (Lond.)* **187**:707–708.
Kirby, D. R. S., 1962*a*, Reciprocal transplantation of blastocysts between rats and mice, *Nature (Lond.)* **194**:785–786.
Kirby, D. R. S., 1962*b*, The influence of the uterine environment on the development of mouse eggs, *J. Embryol. Exp. Morphol.* **10**:496–506.
Kirby, D. R. S., 1963*a*, The development of mouse blastocysts transplanted to the spleen, *J. Reprod. Fertil.* **5**:1–12.
Kirby, D. R. S., 1963*b*, The development of mouse blastocysts transplanted to the scrotal and cryptorchid testis, *J. Anat.* **97**:119–130.
Kitts, D. D., Anderson, G. B., Stabenfeldt, G. H., Cupps, P. T., and BonDurant, R. H., 1982, A

preliminary report on the initiation of parturition of ewes carrying fetuses of two different breeds, *Theriogenology* **17**:93 (abst.).
Klein, M. S., and Markert, C. L., 1981, Production of chimeras in the deer mouse *Peromyscus maniculatus bairdi*, *J. Exp. Zool.* **218**:183–193.
Kraemer, D. C., 1983, Intra- and inter-specific embryo transfer, *J. Exp. Zool.* **228**:363–371.
Krishnan, R. S., and Daniel, J. C., 1967, "Blastokinin": inducer and regulator of blastocyst development in the rabbit uterus, *Science* **158**:490–492.
Lang, D. R., and Lamond, D. R., 1966, Some factors affecting the response of the immature mouse to pregnant mare serum gonadotrophin and human chorionic gonadotrophin, *J. Endocrinol.* **34**:41–50.
Lawrence, C. C., and Anderson, G. B., 1981, Reproduction in allophenic mice produced from two lines differing in litter size, *73rd Ann. Proc. Am. Soc. Anim. Sci.* 161–162 (abst.).
Lawson, R.A. S., Rowson, L. E. A., and Adams, C. E., 1972, The development of cow eggs in the rabbit oviduct and their viability after retransfer to heifers, *J. Reprod. Fertil.* **28**:313–315.
Lin, T. P., 1971, Egg micromanipulation, in: *Methods in Mammalian Embryology* (J. C. Daniel, Jr., ed.), pp. 155–171, W. H. Freeman, San Francisco.
Lin, T. P., 1980, Entire litters developed from transferred eggs in whole body X-irradiated female mice, *J. Exp. Zool.* **213**:41–44.
Mantalenakis, S. J., and Ketchel, M. M., 1966, Pseudopregnant recipients for blastocyst transfer in rats, *Int. J. Fertil.* **11**:318–321.
Markert, C. L., and Petters, R. M., 1977, Homozygous mouse embryos produced by microsurgery, *J. Exp. Zool.* **201**:295–302.
Martin, P. A., 1983, Commercial embryo transfer in swine: Who is interested in it and why, *Theriogenology* **19**:43–48.
Maurer, R. R., and Beier, H. M., 1976, Uterine proteins and development in vitro of rabbit preimplantation embryos, *J. Reprod. Fertil.* **48**:33–41.
Maurer, R. R., and Foote, R. H., 1971, Maternal aging and embryonic mortality in the rabbit: I. Repeated superovulation, embryo culture, and transfer, *J. Reprod Fertil.* **25**:329–341.
McGrath, J., and Solter, D., 1984, Maternal Thp lethality in the mouse is a nuclear, not cytoplasmic defect, *Nature (Lond.)* **308**:550–551.
McLaren, A., 1967, Factors affecting the variation in response of mice to gonadotrophic hormones, *J. Endocrinol.* **37**:147–154.
McLaren, A., and Biggers, J. D., 1958, Successful development and birth of mice cultivated in vitro as early embryos, *Nature (Lond.)* **182**:877–878.
McLaren, A., and Michie, D., 1956, Studies on the transfer of fertilized mouse eggs to uterine foster-mothers. I. Factors affecting the implantation and survival of native and transferred eggs, *J. Exp. Biol.* **33**:394–416.
McLaren, A., and Michie, D., 1958, An effect of the uterine environment upon skeletal morphology in the mouse, *Nature (Lond.)* **181**:1147–1148.
McLaren, A., and Michie, D., 1959, Studies on the transfer of fertilized mouse eggs to uterine foster mothers. II. The effect of transferring large numbers of eggs, *J. Exp. Biol.* **36**:40–50.
McLaren, A., and Tarkowski, A. K., 1963, Implantation of mouse eggs in the peritoneal cavity, *J. Reprod. Fertil.* **6**:385–392.
Mintz, B., 1962, Formation of genotypically mosaic mouse embryos, *Am. Zool.* **2**:432 (abst.).
Mintz, B., and Illmensee, K., 1975, Normal genetically mosaic mice produced from malignant teratocarcinoma cells, *Proc. Natl Acad. Sci. USA* **72**:3585–3589.
Miyamoto, H., and Chang, M. C., 1972, Development of mouse eggs fertilized in vitro by epididymal spermatozoa, *J. Reprod. Fertil.* **30**:135–137.
Modklinski, J. A., 1980, Preimplantation development of microsurgically obtained haploid and homozygous diploid mouse embryos and effects of pretreatment with cytochalasin B on enucleated eggs, *J. Embryol. Exp. Morphol.* **60**:153–161.
Moore, N. W., Polge, C., and Rowson, L. E. A., 1969, The survival of single blastomeres of pig eggs transferred to recipient gilts, *Aust. J. Biol. Sci.* **22**:979–982.
Moustafa, L., and Hahn, J., 1978, Experimentelle Erzeugung von identischen Mausezwillingen, *Dtsch. Tierartzl. Wochenschr.* **85**:242–244.

Murray, A. W., and Szostak, J. W., 1983, Construction of artificial chromosomes in yeast, *Nature (Lond.)* **305**:189–193.

Murray, F. A., 1978, Embryo transfer in large domestic animals, in: *Methods in Mammalian Reproduction* (J. C. Daniel, ed.), pp. 285–305, Academic Press, New York.

Murray, F. A., Bazer, F. W., Rundell, J. W., Vincent, C. K., Wallace, H. D., and Warnick, A. C., 1971, Developmental failure of swine embryos restricted to the oviductal environment, *J. Reprod. Fertil.* **24**:445–448.

New, D. A. T., 1978, Whole embryo culture and the study of mammalian embryos during organogenesis, *Biol. Rev.* **53**:81–122.

Nicholas, J. S., 1933, Development of transplanted rat eggs, *Proc. Soc. Exp. Biol. Med.* **30**:1111–1113.

Nicholas, J. S., 1934, Experiments on developing rats: I. Limits of foetal regeneration; behavior of embryonic material in abnormal environments, *Anat. Rec.* **58**:387–413.

Nicholas, J. S., and Hall, B. V., 1942, Experiments on developing rats. II. The development of isolated blastomeres and fused eggs, *J. Exp. Zool.* **90**:441–459.

Noyes, R. W., 1952, Fertilization of follicular ova, *Fertil. Steril.* **3**:1–12.

Noyes, R. W., and Dickmann, Z., 1960, Relationship of ovular age to endometrial development, *J. Reprod. Fertil.* **1**:186–196.

Noyes, R. W., and Dickmann, Z., 1961, Survival of ova transferred into the oviduct of the rat, *Fertil. Steril.* **12**:67–79.

Orsini, M. A., and McLaren, A., 1967, Loss of the zona pellucida in mice, and the effect of tubal ligation and ovariectomy, *J. Reprod. Fertil.* **13**:485–499.

Ozil, J. P., Heyman, Y., and Renard, J. P., 1982, Production of monozygotic twins by micromanipulation and cervical transfer in the cow, *Vet. Rec.* **110**:126–127.

Palmiter, R. D., Brinster, R. L., Hammer, R. E., Trumbauer, M. E., Rosenfeld, M. G., Birberg, N. C., and Evans, R. M., 1982, Dramatic growth of mice that develop from eggs microinjected with metallothionein-growth hormone fusion genes, *Nature (Lond.)* **300**:611–615.

Papaioannou, V. E., McBurney, M. N., Gardner, R. L., and Evans, M. J., 1975, Fate of teratocarcinoma cells injected into early mouse embryos, *Nature (Lond.)* **258**:70–73.

Parkes, A. S., 1942, Induction of superovulation and superfecundation in rabbits, *J. Endocrinol.* **3**:268–279.

Pincus, G., 1940, Superovulation in rabbits, *Anat. Rec.* **77**:1–8.

Pincus, G., and Enzmann, E. V., 1935, The comparative behavior of mammalian eggs in vivo and in vitro. I. The activation of ovarian eggs, *J. Exp. Med.* **62**:665–675.

Pope, C. E., and Day, B. N., 1972, Development of pig embryos following restriction to the ampullar portion of the oviduct, *J. Reprod. Fertil.* **31**:135–138.

Pope, W. F., Maurer, R. R., and Stormshak, F., 1982, Survival of porcine embryos after asynchronous transfer, *Proc. Soc. Exp. Biol. Med.* **171**:178–183.

Quirke, J. F., and Hanrahan, J. P., 1983, Comparison of the survival of fertilized eggs from adult ewes in the uteri of adult ewes and ewe lambs, *J. Reprod. Fertil.* **68**:289–294.

Rossant, J., and Lis, W. T., 1979, The possible dual origin of the extoderm of the chorion in the mouse embryo, *Dev. Biol.* **70**:249–254.

Rossant, J., Gardner, R. L., and Alexandre, H. L., 1978, Investigation of the potency of cells from the postimplantation mouse embryo by blastocyst injection: A preliminary report, *J. Embryol. Exp. Morphol.* **48**:239–247.

Rossant, J., Croy, B. A., Chapman, V. M., Siracusa, L., and Clark, D. A., 1982, Interspecific chimeras in mammals: A new experimental system, *J. Anim. Sci.* **55**:1241–1248.

Rossant, J., Croy, B. A., Clark, D. A., and Chapman, V. M., 1983, Interspecific hybrids and chimeras in mice, *J. Exp. Zool.* **228**:223–233.

Rowlands, I. W., 1944, The production of ovulation in the immature rat, *J. Endocrinol.* **3**:384–391.

Runner, M. N., 1947, Development of mouse eggs in the anterior chamber of the eye, *Anat. Rec.* **98**:1–17.

Runner, M. N., 1951, Differentiation of intrinsic and maternal factors governing intrauterine survival of mammalian young, *J. Exp. Zool.* **116**:1–20.

Salazar, F., 1981, Foaling of a mule pregnant by non-surgical equine embryo transfer, *Theriogenology* **15**:133 (motion picture title).

Seidel, G. E., Jr., 1979, Applications of embryo preservation and transfer, in: *Beltsville Symposia in Agricultural Research. 3. Animal Reproduction* (H. W. Hawk, ed.), pp. 195–212, Allanheld, Osmun and Co., Montclair, New Jersey.

Seidel, G. E., Jr., 1981*a*, Critical review of bovine embryo transfer procedures, in: *Fertilization and Embryonic Development in Vitro* (L. Mastroianni and J. D. Biggers, eds.), pp. 323–353, Plenum Press, New York.

Seidel, G. E., Jr., 1981*b*, Superovulation and embryo transfer in cattle *Science* **211**:351–358.

Seidel, G. E., Jr., 1983*a*, Mammalian oocytes and preimplantation embryos as methodological components, *Biol. Reprod.* **23**:36–49.

Seidel, G. E., Jr., 1983*b*, Production of genetically identical sets of mammals: cloning?, *J. Exp. Zool.* **228**:347–354.

Seidel, G. E., Jr., 1984, Applications of embryo transfer and related technologies to cattle, *J Dairy Sci.* **67**:2786–2796.

Seidel, G. E., Jr., and Seidel, S. M., 1981, The embryo transfer industry, in: *New Technologies in Animal Breeding* (B. G. Brackett, G. E. Seidel, Jr., and S. M. Seidel, eds.), pp. 41–80, Academic Press, New York.

Seidel, G. E., Jr., Bowen, R. A. and Kane, M. T., 1976, In vitro fertilization, culture, and transfer of rabbit ova, *Fertil. Steril.* **27**:861–870.

Seitz, H. M., Brackett, B. G., and Mastroianni, L., Jr., 1970, In vitro fertilization of ovulated rabbit ova recovered from the ovary, *Biol. Reprod.* **2**:262–267.

Shah, M. K., 1956, Reciprocal egg transplantation to study the embryo–uterine relationship in heat-induced failure of pregnancy in rabbits, *Nature (Lond.)* **177**:1134–1135.

Singh, E. L., Hare, W. C. D., Thomas, F. C., Eaglesome, M. D., and Bielanski, A., 1983, Embryo transfer as a means of controlling transmission of viral infections. IV. Non-transmission of Infectious Bovine Rhinotracheitis/Infectious Pustular Vulvovaginitis Virus following trypsin treatment of exposed embryos, *Theriogenology* **20**:169–176.

Snedecor, G. W., and Cochran, W. G., 1977, *Statistical Methods*, Iowa State University Press, Ames, Iowa.

Stevens, L. C., 1967, The development of teratomas from intratesticular grafts of two-cell mouse eggs, *Anat. Rec.* **157**:328.

Stevens, L. C., Varnum, D. S., and Eicher, E. M., 1977, Viable chimeras produced from normal and parthenogenetic mouse embryos, *Nature (Lond.)* **269**:515–517.

Sugie, T., Seidel, G. E., Jr., and Hafez, E., 1980, Embryo transfer, in: *Reproduction in Farm Animals* (E. S. E. Hafez, ed.), pp. 569–594, Lea and Febiger, Philadelphia.

Tarkowski, A. K., 1959, Experiments on the development of isolated blastomeres of mouse eggs, *Nature (Lond.)*. **184**:1286–1287.

Tarkowski, A. K., 1961, Mouse chimaeras developed from fused eggs, *Nature (Lond.)* **190**:857–860.

Tarkowski, A. K., 1962, Inter-specific transfer of eggs between rat and mouse, *J. Embryol. Exp. Morphol.* **10**:476–495.

Thadani, V. M., 1980, A study of heterospecific sperm-egg interactions in the rat, mouse, and deer mouse using in vitro fertilization and sperm injection, *J. Exp. Zool.* **212**:435–453.

Thadani, V. M., 1982, Mice produced from eggs fertilized in vitro at a very low sperm : egg ratio, *J. Exp. Zool.* **219**:277–283.

Toyoda, Y., and Chang, M. C., 1974, Fertilization of rat eggs in vitro by epididymal spermatozoa and the development of eggs following transfer, *J. Reprod. Fertil* **36**:9–22.

Vickery, B. H., Erickson, G. I., and Bennett, J. P., 1969, Nonsurgical transfer of eggs through the cervix in rats, *Endocrinology* **85**:1202–1203.

Wagasugi, N., 1974, A genetically determined incompatibility system between spermatozoa and eggs leading to embryonic death in mice, *J. Reprod. Fertil.* **41**:85–96.

Warwick, E. J., Murphree, R. L., Casida, L. E., and Meyer, R. K., 1943, Maintenance of the pregnancy induced in the female rabbit following treatment with pituitary gonadotropins, *Anat. Rec.* **87**:279–296.

Weir, B. J., 1971, The reproductive organs of the female plains viscacho, *Lagostomus maximus*, *J. Reprod. Fertil.* **25**:365–373.

Whitten, W. K., and Biggers, J. D., 1968, Complete development in vitro of the pre-implantation stages of the mouse in a simple chemically defined medium, *J. Reprod. Fertil.* **17**:399–401.

Whittingham, D. G., 1968, Fertilization of mouse eggs in vitro, *Nature (Lond.)* **220**:592–593.

Whittingham, D. G., 1971, Culture of mouse ova, *J. Reprod. Fertil. (Suppl.)* **14**:7–21.

Whittingham, D. G., 1977, Fertilization in vitro and development to term of unfertilized mouse oocytes previously stored at −196°C, *J. Reprod. Fertil.* **49**:89–94.

Whittingham, D. G., and Wales, R. G., 1969, Storage of two-cell mouse embryos in vitro, *Aust. J. Biol. Sci.* **22**:1065–1068.

Willadsen, S. M., 1979, A method for culture of micromanipulated sheep embryos and its use to produce monozygotic twins, *Nature (Lond.)* **277**:298–300.

Willadsen, S. M., 1980, The viability of early cleavage stages containing half the normal number of blastomeres in the sheep, *J. Reprod. Fertil.* **59**:357–362.

Willadsen, S. M., and Fehilly, C. B., 1983, The developmental potential and regulatory capacity of blastomeres from 2-, 4-, and 8-cell sheep embryos, in: *Fertilization of the Human Egg in Vitro—Biological Basis and Clinical Application* (H. M. Beier and H. R. Lindner, eds.), pp. 353–357, Springer-Verlag, Berlin.

Williams, T. J., Elsden, R. P., and Seidel, G. E., 1984, Pregnancy rates with bisected bovine embryos, *Theriogenology* **22**:521–531.

Wilson, E. D., and Zarrow, M. X., 1962, Comparison of superovulation in the immature mouse and rat, *J. Reprod. Fertil.* **3**:148–158.

Wimsatt, W. A., 1975, Some comparative aspects of implantation, *Biol. Reprod.* **12**:1–40.

Woods, G. L., and Ginther, O. J., 1983, Recent studies relating to the collection of multiple embryos in mares, *Theriogenology* **19**:101–108.

Wu, J. T., 1975, Development of two-cell eggs transferred into the uteri of ovariectomized rats, *Biol. Reprod.* **13**:211–217.

Yanagimachi, R., and Chang, M. C., 1964, In vitro fertilization of golden hamster ova, *J. Exp. Zool.* **156**:361–376.

Yoshinaga, K., and Adams, C. E., 1967, Luteotrophic activity of the young conceptus in the rat, *J. Reprod. Fertil.* **13**:505–509.

Zarrow, M. X., and Wilson, E. D., 1961, The influence of age on superovulation in the immature rat and mouse, *Endocrinology* **69**:851–855.

Chapter 9

Manipulation of Early Mammalian Development: What Does It Tell Us about Cell Lineages?

MARTIN H. JOHNSON

1. Introduction

The mouse early embryo may be manipulated with ease, single blastomeres being isolated readily and reassembled together with other blastomeres in a relatively controlled manner. In addition to their resilience during such manipulations, the blastomeres have long cell cycles (see Fig. 1), and developmental change occurs slowly in comparison with early embryos of, for example, *Xenopus*, the sea urchin, and *Drosophila*. The mouse embryo is therefore a good developmental system in which to study the processes by which cells are specified in space and time. However, just because the mouse embryo offers peculiar advantages to the investigator, it does not therefore necessarily occupy a unique position in the spectrum of early developmental strategies. Its accessibility, and particularly its relatively slow tempo of development, do give prominance to the regulative aspects of early mouse embryogenesis. As other types of early embryo are analyzed more carefully, however, it becomes clear that the general principles underlying their development are not too dissimilar from those that apply to the mouse (see Johnson and Pratt, 1983, for discussion). The capacity to regulate, the importance of cell interaction, and the role of cytoplasmic inheritance are all features of, for example, *Xenopus* development, just as they are of mouse early embryos (e.g., Gerhart, 1984).

This chapter reviews some of the principal observations on the development of the mouse embryo from egg to blastocyst (Fig. 1). Emphasis will be placed on manipulations at the cellular level. Detailed reviews of the sub-

MARTIN H. JOHNSON • Department of Anatomy, University of Cambridge, Cambridge CB2 3DY, England.

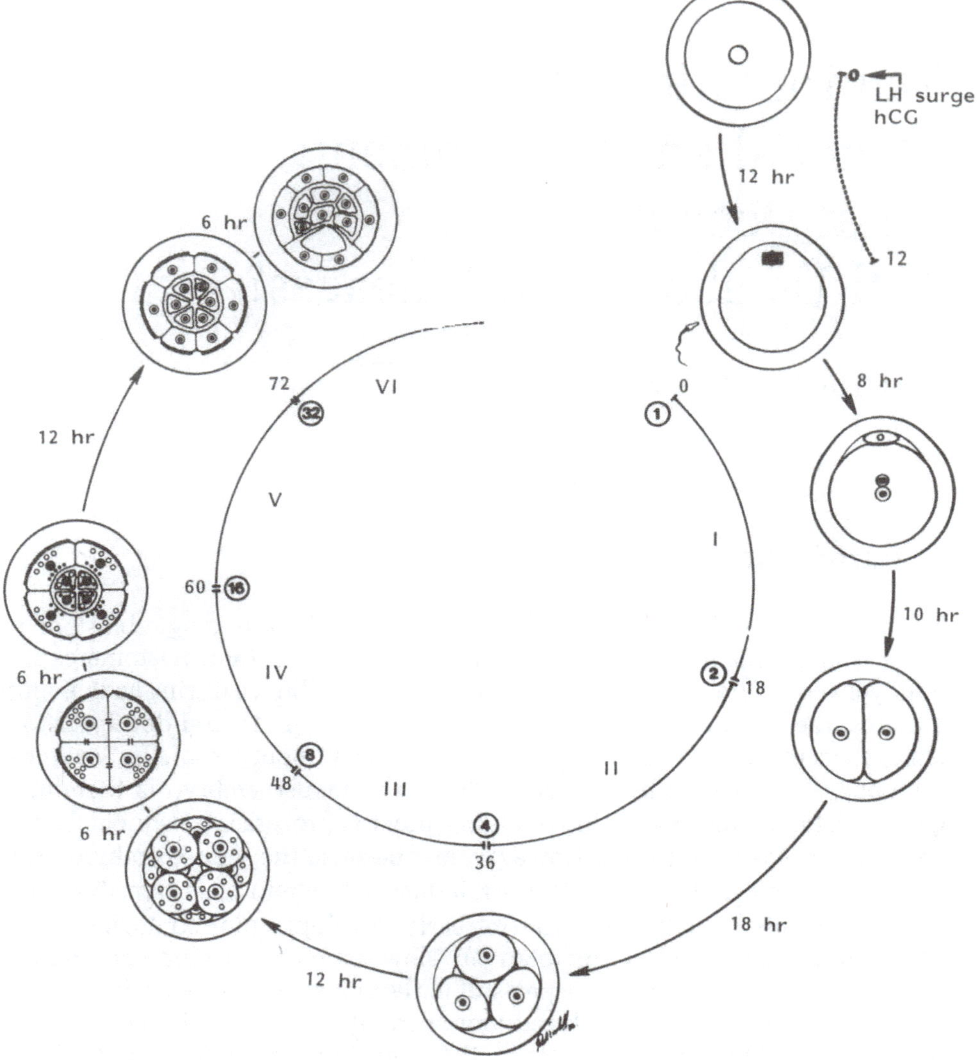

Figure 1. Development of the mouse blastocyst. Development is initiated with the activation of the preovulatory follicle by endogenous luteinizing hormone (LH) or exogenous human chorionic gonadotropin (hCG). The germinal vesicle, containing chromosomes in first meiotic prophase, breaks down and meiosis procedes to second meiotic metaphase over a 12-hr period. During this period the egg is ovulated. Insemination at Oh leads to development through the six developmental cell cycles indicated internally in Roman numerals, cleavage occurring at approximately 18, 36, 48, 60, and 72 hr post insemination. Polarization of blastomeres, intercellular flattening, and junction formation occur first in the IVth cell cycle, distinct inner and outer cell subpopulations are present in the Vth cell cycle and the blastocyst forms in the VIth cell cycle, the blastocoel, inner cell mass and trophectoderm becoming visible.

cellular (Johnson, 1985) and molecular (Johnson *et al.*, 1984) reorganization that occur during blastocyst formation have recently been published.

2. Possibilities and Probabilities

The establishment of cell lineages has occupied a central position in developmental studies. It is clearly important to discover where cells in the adult and in the developing embryo have come from and to define the sequence of phenotypic and developmental changes that marked their emergence. However, such a description is not, as often appears to be the case, an end in itself but rather the necessary beginning that leads onto an analysis of developmental mechanisms. The lineage map of *Caenorhabditis* is complete and precise. But we do not yet know how this developmental precision is achieved.

Results from a range of studies on the mouse early embryo suggest that the lineage of a cell within the blastocyst is related to its earlier relative position within the morula, cells on the inside giving rise to the inner cell mass (ICM) and cells on the outside to trophectoderm (TE) (Mintz, 1965; Wilson *et al.*, 1972; Garner and McLaren, 1974; Graham and Deussen, 1978; Kelly, 1979*a*; Balakier and Pedersen, 1982). Moreover, the prospective fate of a cell, or cell cluster, can be changed by changing its position relative to other cells, suggesting that position and fate are related causally (Hillman *et al.*, 1972; Pedersen and Spindle, 1980; Ziomek *et al.*, 1982*a*; Ziomek and Johnson, 1982). In these studies, the relationship between the relative position of a cell in embryos up to the early blastocyst stage, and the fate of that cell's progeny in the late blastocyst is not absolute and can only be expressed in terms of a probability. Thus, any cell in embryos up to and including the early blastocyst stage has wide developmental possibilities, but a narrower developmental probability the nature of which is conditioned in large part (but not exclusively) by the relative position it occupies in the embryo (Mintz, 1965; Tarkowski and Wroblewska, 1967; Handyside, 1978; Hogan and Tilly, 1978; Spindle, 1978; Johnson, 1979; Kelly, 1979*b*; Rossant and Lis, 1979; Rossant and Vijh, 1980; Ziomek *et al.*, 1982*a*). Such a situation has made the detailed pursuit of lineage relationships a rather frustrating exercise in statistics, in which variation among embryos, strains of mice, culture conditions, analytical techniques, and so forth have all appeared to influence the extent to which any given cell in the embryo expresses its full potential. However, if the simple pursuit of lineage relationships is abandoned in favor of the analysis of the underlying cellular mechanism that mediates recognition of relative position as "inside" or "outside" and thereby results in cell diversification, the mouse early embryo provides an illuminating and experimentally accessible developmental system. Such analyses have revealed that three types of cellular process are adequate to specify a blastocyst and that each depends on clearly definable cell interactions (Johnson, 1985). The evidence for this conclusion, and a reinterpretation of lineage experiments in the light of it, constitute the remainder of this chapter.

3. The Origins of Cell Diversification: Polarization, Cell Flattening, and Cleavage

The earliest stage of mouse embryogenesis at which a consistently recognizable heterogeneity of cell types can be observed is the 16-cell morula. At this stage two populations of cells exist that differ in phenotype, properties, and relative position (Handyside, 1980; Reeve, 1981; Reeve and Ziomek, 1981; Ziomek and Johnson, 1981; Kimber *et al.*, 1982; Randle, 1982; MacQueen and Johnson, 1983). Polar cells are located on the outside of the embryo, nonpolar cells are located on the inside (Fig. 2j,k). How does this heterogeneity arise? Two possible routes to cell diversification at the 16-cell stage have been described.

First, when late 8-cell blastomeres are isolated from compact morulae, cultured *in vitro* and allowed to divide, most generate a 2/16 couplet that is found to be heterogeneous in cell composition, no matter how soon after division it is examined. One cell is larger and polar, the other smaller and nonpolar (Fig. 2h,i). Moreover, with time the polar cell surrounds the nonpolar cell, forming a mini-16-cell "morula" with polar outer and nonpolar, inner cells (Fig. 2l,m,n) (Johnson and Ziomek, 1981*a*; Ziomek and Johnson, 1981; Kimber *et al.*, 1982). Examination of 8-cell blastomeres passing into and through division reveals that the heterogeneity of cell surface phenotype generated in this experiment derives from a heterogeneity in the organization of the cell surface of the 8-cell blastomere itself (Fig. 2b–e) (Ducibella and Anderson, 1975; Handyside, 1980; Ziomek and Johnson, 1980; Johnson and Ziomek, 1981*a*; Reeve and Ziomek, 1981; Izquierdo and Ebensburger, 1982). Thus, late 8-cell blastomeres are highly polarized cells, and this polar organization is sufficiently stable through division that cell diversity can be generated (Fig. 2e–g).

Figure 2. (a–d) Clusters of two or three 8-cell blastomeres stained with FITC-concanavalin A. Note that newly formed 8-cell blastomeres (a) stain evenly, as do 2- and 4-cell blastomeres when examined in the same way (not shown here). By contrast, when 8-cell blastomeres are cultured for several hours, either in the intact embryo or as shown here in smaller clusters (b–d), their surfaces polarize and a pole of ligand-binding forms opposite to the geometric mean of all contact points. The pole so formed can also be visualized by scanning electron microscopy (e), which shows a late 8-cell blastomere entering cytokinesis. Note the pole of microvilli to the left-hand side. As division proceeds this asymmetry of microvilli persists (f,g), generating two cells of dissimilar phenotype—one polar, the other nonpolar (g). (h) A 2/16 couplet generated in this way is shown, in which the cells are stained with FITC-concanavalin A. If the division plane is at right angles to that shown in (e–g), it cuts the pole and generates a 2/16 couplet in which both cells are polar (i), the poles being contiguous at the residuum of the cleavage furrow. The polar and nonpolar 16-cell blastomeres have different properties. In the intact embryo only the polar cells are at the surface, but the nonpolar cells can be exposed by treatment with Ca^{2+}-free medium (j, k—arrows to polar cells, arrowheads to nonpolar cells). If a couplet of a polar plus a nonpolar cell such as that shown in (g) and (h) is cultured, the polar cell engulfs the nonpolar cell (l–n; phase, sectioned and scanned on the electron microscope respectively). Thus, the cells of different phenotype are in different positions, and moreover reinforce that positional difference via their different properties. Magnifications: Intact mouse embryos are approximately 83 μm in diameter, an 8-cell blastomere is approximately 31 μm in diameter and a 16-cell blastomere is approximately 18 μm in diameter.

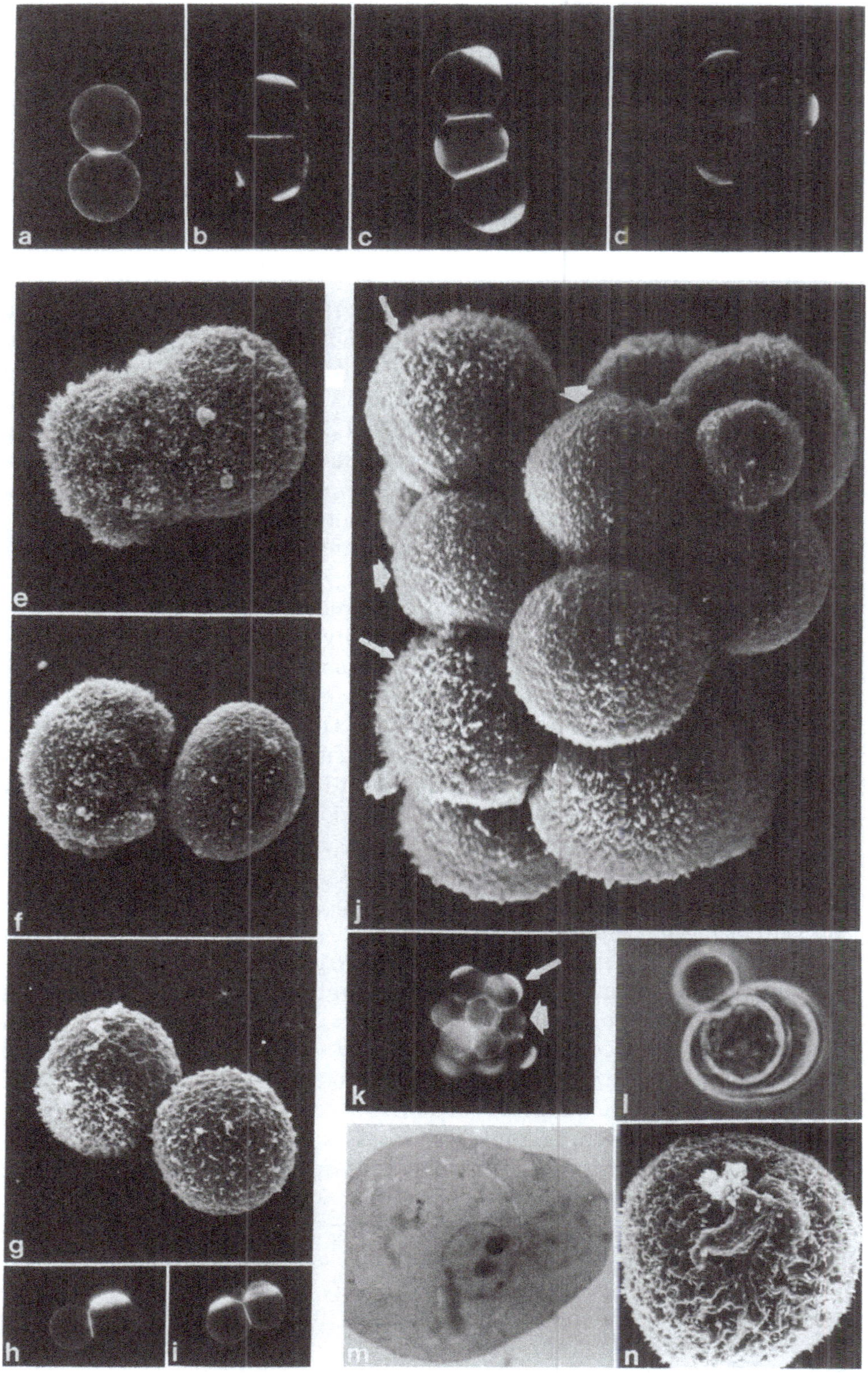
a
b
c
d
e
f
g
h
i
j
k
l
m
n

If the plane of cleavage is approximately perpendicular to the axis of polarity, a $^2/_{16}$ couplet containing a polar and a nonpolar cell will result. The plane of cleavage is so oriented in the majority (7 of 8) of isolated $^1/_8$ blastomeres dividing *in vitro* (Johnson and Ziomek, 1981*a*). *In situ*, however, the cleavage plane appears to be oriented more frequently along the axis of polarity (c. 4 out of 8) generating a greater incidence of $^2/_{16}$ couplets in which both cells are polar (Johnson, 1985). Thus, cell interactions *in situ* may influence the orientation of the cleavage plane and thus the ratio of polar : nonpolar cells generated.

A second possible route to cell diversification occurs not by the process of cleavage of a polarized cell at the 8-cell to 16-cell transition, but rather by events at the 16-cell stage itself. Thus, if the nonpolar, inside cells of the 16-cell morula are isolated and cultured *in vitro* in pairs or in clusters, those that are exposed to an outside position become phenotypically polarized over a period of several hours. Those not so exposed, do not (Ziomek and Johnson, 1981). By this route, cell heterogeneity may be generated within a homogeneous cell population. In contrast, it should be noted here that culture of outer polar 16-cell blastomeres in pairs or in clusters does not result in the detectable loss of polarity in any cell. All cells remain polar throughout the period of the 16-cell stage.

The crucial feature for either route to cell diversification described above is the generation of a relatively stable cell polarity—either *de novo* at the 16-cell stage or at the 8-cell stage before division. It is now clear that in each case the generation of polarity itself depends on cell interaction. If 8-cell and 16-cell nonpolar blastomeres are exposed to no other cell contacts at all or, if they are exposed to contact on all sides either by aggregation in the center of a cluster of cells or by the spreading of one polar cell over the entire surface of the second nonpolar cell, they do not polarize. Only if nonpolar 8-cell or 16-cell blastomeres are aggregated in such a way that they have an asymmetry of cell contacts do they polarize. Moreover they do so with their axis of polarity normal to the geometric mean of the contact points (Fig. 2a–d). This process results in cells that have outward facing apical poles and inward facing basolateral surfaces (Ziomek and Johnson, 1980, 1981; Johnson and Ziomek, 1981*b*).

It can be seen immediately from these simple observations that three types of cellular mechanism are adequate to explain the generation of cell diversity (Fig. 3). First, the process of polarization induced by asymmetry of contact, second the process by which cells flatten on each other to maximize cell contact thereby leading to reduction of asymmetries of contact, and third the process of differentiative cleavage. Each of these processes involves cell interactions that thereby determine, respectively, the axis of polarization, the regulation of developmental potential by suppressing/permitting polarization and the orientation of the cleavage plane. Study of the cell biology underlying each of these processes is now well advanced and has been reviewed in detail by Johnson (1985).

Thus far only the generation of cell diversity in the 16-cell morula has been considered, but, developmental potential exceeds prospective fate up to at least the 32-cell blastocyst. It appears that the same three mechanisms described above may be adequate to explain the continuing regulation within and be-

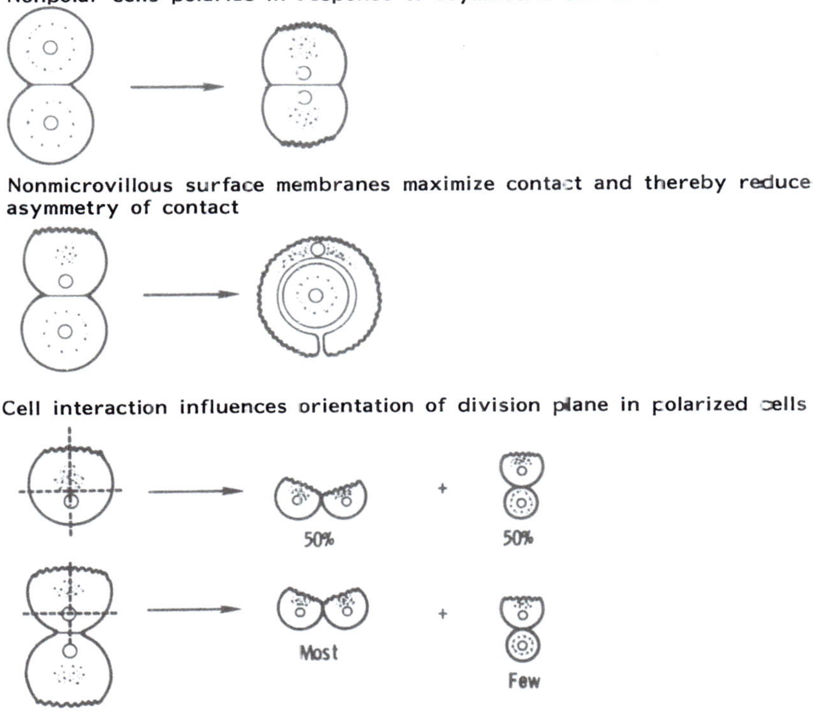

Figure 3. Schematic summary of principal cellular mechanisms involved in generating cell diversity in the blastocyst. (1) Nonpolar cells at the 8-, 16-, and 32-cell stages polarize in response to asymmetric cell contact, their nuclei migrate basally, endosomes aggregate apically and a surface pole develops opposite to the point of contact. (2) Polarization is inhibited if the nonpolar cell (in this case a 1/16 cell) is totally surrounded either by many cells or, as shown here, by a single polar cell enveloping it. In the absence of total cell contact, polarity is generated as shown in (1). (3) The proportion of polar cells dividing conservatively or differentially varies depending upon the contact patterns, as summarized here for polar 1/16 cells.

tween cell lineages, i.e., the extent to which cells that are polar or nonpolar at the 16-cell stage contribute progeny exclusively to the polar (trophectodermal) and nonpolar (ICM) cell populations of the blastocyst. Thus, we have already seen that as long as nonpolar 16-cell blastomeres are surrounded totally by other cells, as they would tend to be by polar cells in the intact embryo, they fail to polarize and would therefore give rise to ICM cells. Conversely, although polar $1/16$ cells have never been observed to "depolarize," they have been shown sometimes to generate one nonpolar cell at the division that yields $2/32$ cells. Again, the frequency with which a nonpolar cell is generated depends on the orientation of the cleavage plane—0 when through the pole and 100% when perpendicular to the axis of polarity. As for the 8-cell to 16-cell transition, the orientation of the cleavage plane is influenced by cell interaction but even more decisively so. Thus, polar $1/16$ cells aggregated with, and enclosing, a nonpolar

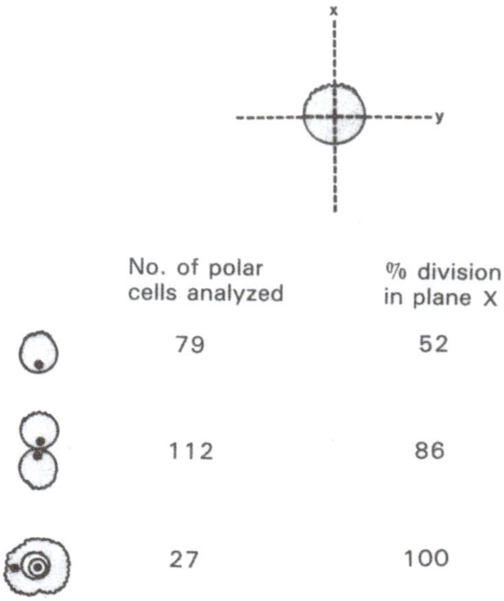

Figure 4. Simplified summary of fate of polar 1/16-cell blastomere during transition to the 2/32-cell stage. The orientation of the cleavage plane, and thus the proportion of the progeny, vary with the cell association of the polar cell. (Data from Johnson and Ziomek, 1983.)

1/16 cell never generate a nonpolar 1/32 cell, polar 1/16 cells alone or aggregated with other polar cells will do so (Fig. 4) (Johnson and Ziomek, 1983). Since *in situ* polar and nonpolar cells are both present, these results argue that *in situ* polar cells will tend to generate mainly more polar cells.

We can thus explain the process by which a blastocyst is generated in terms of precise cellular mechanisms, each of which involves cell interaction, and each of which is susceptible to analysis at the subcellular and molecular level. It is also possible now to reinterpret the many lineage studies undertaken on the early embryo in terms of these mechanisms.

4. The Effect of Spatial Location on Cell Lineage

Classically two extreme types of hypothesis have been proposed to explain the events leading to blastocyst formation. Variations on a *segregation hypothesis* (see Denker, 1976) have proposed a decisive role for the process of differential inheritance of cytocortical or cytoplasmic domains, whereas variations on a *positional hypothesis* (most notably the inside–outside hypothesis of Tarkowski and Wroblewska, 1967) have suggested that cell interaction provides the decisive influence leading to cell diversification. We have suggested that exclusive adherence to either type of hypothesis distorts our perception of the process of cell diversification in the early embryo (Johnson and Pratt, 1983) and have instead attempted to reconcile these more extreme positions by emphasising the importance of both cell interaction and the conservation of defined cytocortical domains during division. Since our conclusions derive from observations on the cell biology of the blastomeres involved, and since these observations indicate that the process of cell polarization is fundamental, both in terms of its dependence on cell interaction and its importance for differential inheritance, we have embodied this compromise proposal in a polarization hypothesis (first outlined in Johnson *et al.*, 1981).

Within the experimental and intellectual framework provided by this hypothesis, it is possible to understand how the peripheral or central position of cytocortex and cytoplasm within the intact embryos will affect its subsequent fate. Thus, experiments in which regions of cytoplasm, or specific cells, are marked *in situ*, and the distribution of marked daughter cells is analysed at subsequent developmental stages, reveal that development is coherent and clonal in nature with little evidence of extensive mixing during early development to the blastocyst. Moreover, when inner embryonic regions or cells are marked, the marker distributes primarily among the cells of the ICM, because the cells being enclosed will not polarize. The converse applies to more peripherally located embryonic cytoplasm or cells, which, being subject to asymmetric exposure, polarize and contribute extensively to the trophectoderm (Wilson *et al.*, 1972; Garner and McLaren, 1974; Graham and Deussen, 1978; Kelly, 1979*a;* Balakier and Pedersen, 1982; Gearhart *et al.*, 1982). In neither case, however, is the contribution exclusive as would be expected for a system in which regulation can occur. The most marked deviation from exclusivity has been reported by Balakier and Pedersen (1982) who also undertook the most detailed analysis. They iontophoresed horseradish peroxidase (HRP) into individual cells and analyzed embryos subsequently for the distribution of labeled cells. Injection of an inside or outside cell at the 16-cell stage resulted in a high incidence of positive cells in both ICM and trophectoderm at the blastocyst stage. From this observation they concluded that regulation (crossing of lineage) occurred on a large scale. It is unfortunate that cleavage in the mouse embryo is incomplete, a tenuous channel persisting through the remnant of the cleavage furrow, the so-called midbody. Such channels can persist for up to at least two mitotic cycles, and HRP can pass through them at or shortly after iontophoretic injection to label up to four blastomeres rather than the ideal single cell. Thus, the relatively high traffic across putative lineages at the 16-cell to 32-cell stage that the results from this study suggested, may in fact reflect a more extensive primary distribution of HRP at injection than was thought originally (Goodall and Johnson, 1984). In addition, it is clear that the injection procedure itself influences either cell viability or the capacity of cells to divide, and so further complicates interpretation of the data in terms of lineage analysis.

In addition to the analysis of cells marked *in situ*, many studies have involved the experimental relocation of a single marked blastomere (or cluster of blastomeres) such as to force it and its progeny primarily into an inside or outside position (see Fig. 5 for strategy). It is easy to see that, when this is done with a 2-cell, 4-cell, or early 8-cell blastomere, the site of relocation will influence whether the blastomere(s) is enclosed or asymmetrically exposed and thus its state of polarization and subsequent fate (Tarkowski and Wroblewska, 1967; Hillman *et al.*, 1972).

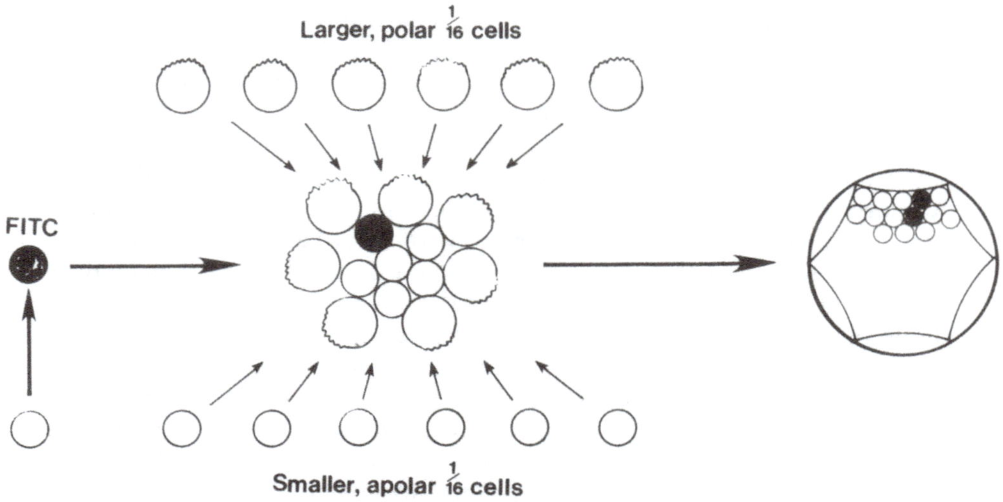

Figure 5. Strategy for marking and relocating single cells from 16-cell embryos. Embryos are disaggregated, cells typed as polar or nonpolar, one is labeled, the cells are reaggregated in defined ratios and spatial arrays, and the aggregate cultured for 18–24 hr. At least two and usually more progeny are recovered and their position and phenotype recorded.

When blastomeres from the late 8-cell stage onward are relocated, the interpretation of results is more complex and depends on both the differentiated state of the cell (polar or nonpolar) at the time of its relocation and the precise experimental protocol used. Several studies have now shown that a polar cell, regardless of the site of its relocation, will always give rise to progeny within the trophectoderm (see below and Randle, 1982; Ziomek and Johnson, 1982; Surani and Handyside, 1983). The appearance of progeny within the inner cell mass as well appears to depend on the position to which the cell was relocated, and on the nature and number of the other cells amongst which it was relocated (Randle, 1982; Ziomek and Johnson, 1982; Surani and Handyside, 1983). Results from a series of experiments from this laboratory are illustrated in Figure 6, and demonstrate three points. First, only one marked and relocated cell, out of 108 cells categorized as polar and so treated, contributed progeny to the ICM only. This result could arise through a low incidence of polarity reversal, through the misassignment of a nonpolar cell as polar at the time of relocation, or through death of the polar progeny of the cell. Whatever the explanation, the result emphasises the stability of the polarized state. Second, when a polar cell is relocated to a site at which it is likely to spread over other cells, a high contribution to trophectoderm only occurs (Fig. 6, rows 1–3). Third, under conditions in which a polar cell is likely to have less surface to flatten on, and therefore is likely to be less attenuated, it will contribute to both trophectoderm and ICM with higher frequency (Fig. 6, rows 4–5). Since the greater the spreading or attenuation of a polar cell, the more likely it is to divide through its pole (see earlier and Fig. 5), these results are entirely to be expected.

Relocation of a marked nonpolar cell, and the analysis of its progeny, is slightly more complicated (see Fig. 7). When a labeled nonpolar cell is relocated internally, it will be totally surrounded; thus it contributes primarily to the ICM only (Fig. 7, line 1). Placement of the non-polar cell on the outside of a compact, intact 16-cell embryo results in an almost exclusive contribution to trophectoderm (Fig. 7, line 5). At such a site of relocation, the cell evidently remains exposed for a sufficiently long period that it polarizes and then behaves like a polar cell yielding mainly trophectodermal progeny. In all other experimental protocols (Fig. 7, lines 2, 3, 4, 6, and 7), the labeled nonpolar cell

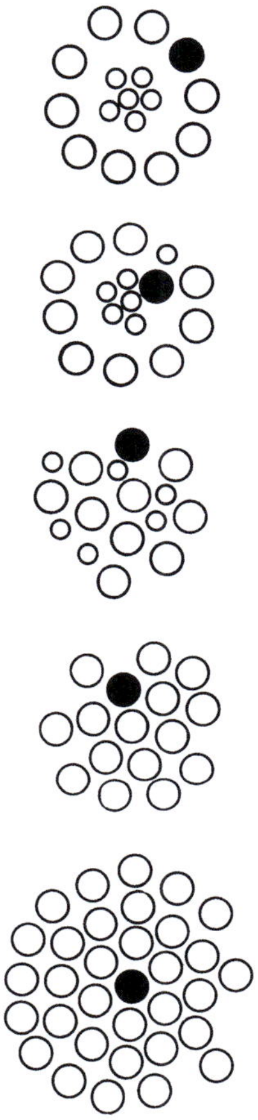

No. of embryos	% blastocysts with 2-4 labelled cells in TE	ICM	TE+ICM
14	93	0	7
18	89	0	11
39	82	0	18
16	75	0	25
21	67	4	29

Figure 6. A single polar 1/16 cell was labeled (large dot) and combined with unlabeled polar (large circles) and nonpolar (small circles) cells in various spatial arrays and ratios (column 1). The numbers of aggregates analyzed is recorded in column 2 and the distribution of the labelled progeny is recorded in columns 3–5. TE, contribution to trophectoderm only; ICM, contribution to inner cell mass only; TE + ICM, contribution to both tissues. (Data from M. H. Johnson, unpublished results, and from Ziomek and Johnson, 1982.)

No. of embryos	% blastocysts with 2-4 labeled cells in		
	TE	ICM	TE+ICM
19	3	84	0
13	54	31	15
32	50	16	34
34	50	32	18
16	94	0	6
20	35	50	15
24	13	71	16

Figure 7. Similar experiment to that illustrated in Figure 6, with two exceptions: A nonpolar cell was labeled (small dot), and in the last three experimental protocols the labeled cell was aggregated not with other single cells but with intact embryos that were compact, 16-cell morulae (line 5), compact 16-cell morulae decompacted in Ca^{2+}-free medium (line 6), and naturally decompacted morulae undergoing division from 8 to 16 cells (line 7).

can contribute progeny to trophectoderm alone, to the ICM alone, or to both tissues. In general, in most cases in which the labeled nonpolar cell is aggregated together with 15 other *isolated blastomeres* (lines 2–4), its progeny contribute heavily to the trophectoderm. In contrast when aggregation is to the outside of an intact embryo that has been decompacted, either artificially by brief exposure to Ca^{2+}-free medium (line 6) or naturally while undergoing the 8-cell to 16-cell transition (line 7), its progeny contribute heavily to the ICM. This result, which is also supported by that of Surani and Handyside (1983) suggests that "sorting" and integration occur more rapidly in decompact embryos than in embryos made from disaggregated/reaggregated blastomeres, giving a nonpolar cell a brief period of asymmetric exposure and thus less opportunity to polarize and thereby contribute progeny across lineages. This conclusion is not surprising, as in decompacted embryos polar and nonpolar cells are already in the optimal spatial relationship to each other and can rapidly integrate and relocate a single new comer. By contrast, disaggregated blastomeres must sort out at reaggregation before assembling an organized morula/blastocyst.

5. The Effect of Division Order on Cell Lineage

The cells of the early embryo do not divide synchronously, indeed the degree of asynchrony among blastomeres within a single embryo increases with each division (Kelly *et al.*, 1978). The cause of the heterogeneity in cell cycle times between blastomeres is not clear, although differences in cell size may be important (MacQueen and Johnson, 1983), and it has also been proposed that the 2-cell blastomere that incorporates the original site of sperm penetration may divide earlier than its companion (Bennet, 1982). Cell division, whatever its timing may be regulated by, is not essential for progress through development, as shown by the effect of cytochalasin D which suppresses cytokinesis but is without significant influence on the tempo or nature of molecular and subcellular differentiation, except for those features that rely on increasing cell number for their expression (Surani *et al.*, 1980; Kimber and Surani, 1981; Pratt *et al.*, 1981; Petzoldt *et al.*, 1983). Thus, cytokinesis is just one manifestation of progress through early developmental cell cycles and is not itself an integral and obligatory component of the temporal control program.

However, cytokinesis clearly is required if cell heterogeneity is to develop. Moreover, it has been proposed that the division order of cells can influence the developmental fate of their progeny. Thus, it has been shown that the progeny of the earliest of the two 2-cell blastomeres in an embryo to divide (1) maintained their temporal advantage over the progeny of the later 2-cell during subsequent divisions, (2) tended to establish more intercellular contacts and thus to contribute a greater proportion of their membrane to the "non-exposed" inner (or deeper) part of the embryo, and (3) tended to contribute disproportionally to the ICM (Graham and Deussen, 1978; Kelly *et al.*, 1978; Graham and Lehtonen, 1979). That these features were indeed consequential on early division has been shown in two types of experiment. First, cells at different points in their progress through the early developmental cell cycles were disaggregated and then reaggregated together in different ways. The division order of the cells in the newly reconstituted embryos exerted the primary influence on both the number of intercellular contacts and the relative contribution to ICM. This latter result was recently supported by Surani and Barton (1984). Second, if a couplet containing two cells that differ in the extent of their passage through the second cell cycle are placed in cytochalasin D (CCD), cytokinesis is inhibited but developmental programs nonetheless unfold. If the CCD is subsequently removed, there is a suggestion that the more advanced cell makes more extensive cell contacts and ends up more completely enclosed by the other cell

(Kimber and Surani, 1981). How might these results be explained in terms of the polarization hypothesis?

The primary route to inside cell, and thus to ICM, formation lies in the differentiative division of polarized 8-cell blastomeres. Since the incidence of differentiative divisions (generating a polar and a nonpolar pair of 16-cell blastomeres) is influenced by the orientation of the cleavage plane, the above result could be obtained if the earliest formed $1/8$ blastomeres had more cell contacts and therefore penetrated more deeply into the embryo having a radically elongated form. The later-formed $1/8$ blastomeres would have less contacts, penetrate less deeply and have a form that was elongated in a relatively tangential direction. It is easy then to evisage that the former might divide across the radial axis of the cell to generate a polar and a nonpolar pair whereas the latter might more readily divide along the radial axis to give two polar cells. In an analysis of cell phenotype in embryos having cell numbers intermediate between 8 and 16, we were unable to detect such a trend for early divisions to be differentiative (Ziomek *et al.*, 1982*b*). However, this explanation awaits a more definitive experimental test before it can be considered disproven.

Similarly at the 16–32-cell transition, the extent to which outer polar blastomeres will divide differentiatively to contribute a nonpolar $1/32$ cell will increase the less extensive the flattening of the outer polar cells. Should there be for any reason a deficiency of inside cells at the 16-cell stage, the earliest polar $1/16$ cells to divide may be more likely to contribute nonpolar cells, which would thereby increasingly extend the later dividing polar $1/16$ cells and so reduce the incidence of their differentiative divisions (Johnson and Ziomek, 1983 and earlier). In this way, "running" adjustments to the inside–outside cell ratio can be made (Johnson *et al.*, 1983; Surani and Barton, 1984).

Recently, it has been conjectured that inside cells, by forcing extensive flattening of polar cells, might not only influence the orientation of their cleavage planes but also delay, or even inhibit, division itself (Surani and Barton, 1984). Such an influence would produce the same result as a reduction in differentiative divisions as just described. It is important to reemphasise, in light of the earlier discussion, that inhibition or delay of cytokinesis does not mean necessarily that nuclear replication and continuing progress through the programme of development are inhibited or delayed. Indeed, since embryos tend to reduce cell contact, and round up, either globally throughout the embryo or locally within it, as cells enter division from 16 to 32 cells, it is possible that as the inner cells divide the outer cells are relieved of their extreme extension and are also permitted to complete their hitherto arrested cytokinesis. In this context it is of interest that smaller, nonpolar $1/16$ cells have been shown to divide later than the larger polar $1/16$ cells when they are isolated *in vitro* (MacQueen and Johnson, 1983), whereas *in situ* the division of inner nonpolar cells has been interpreted as preceding the division of outer polar cells (Barlow *et al.*, 1972; Surani and Barton, 1984). This finding supports a delaying effect of inside cells on outer cells and again stresses that *in situ* (1) reinforcement of lineages established at the 16-cell stage tends to occur, but that (2) if abnormal inside–outside cell ratios occur, regulative adjustment is possible (Johnson *et al.*, 1983).

6. The Effect of Manipulating Polarization, Flattening, or Cleavage on Cell Lineage

If polarization were to be inhibited, no trophectoderm should form and the embryo should consist entirely of ICM cells. Unfortunately, it has proved very difficult to prevent polarization of blastomeres, other than by depriving cells of asymmetry of contact (Ziomek and Johnson, 1980; Johnson and Ziomek, 1981*b*,1983). One agent has been reported to inhibit polarization, namely a monoclonal antibody to a cell surface adhesion molecule.

The monoclonal antibody ECCD-1 identifies a Ca^{2+}-dependent cell–cell adhesion molecule (cadherin or uvomorulin) that plays an important role in the process of intercellular flattening at the 8-cell and 16-cell stages (Hyafil *et al.*, 1980; Ogou *et al.*, 1982). The antibody blocks the intercellular flattening observed at these stages, although it is without effect on subsequent junction formation and does not affect cleavage. Shirayoshi *et al.* (1983) reported that the antibody also inhibits polarization of 8-cell blastomeres, but not the subsequent formation of trophectoderm—the polar cell type of the blastocyst. It has proved difficult to repeat the effect of the antibody on polarization at the 8-cell stage (M. H. Johnson, B. Maro, and M. Takeichi, cited in Johnson, 1985), and we are attempting to determine the reason for this difference in results, and thereby to determine the significance of these observations for lineage studies.

Selective inhibition of intercellular flattening has been achieved. A number of agents give either partial inhibition only or complete inhibition of flattening but also complete or partial inhibition of cleavage (see discussion in Pratt *et al.*, 1982; Johnson, 1985). The results from use of these agents are not therefore easy to interpret but most do suggest that a deficiency of cells in the ICM results. The monoclonal antibody ECCD-1 described above gives excellent inhibition of flattening in both whole embryos, and in small clusters of cells derived from them. Division continues unaffected and blastocystlike structures form that lack, or have a greatly reduced, ICM. If the antibody is removed at varying points during culture from the 4-cell stage to the blastocyst, there is found to be a direct relationship between the length of exposure to antibody and the deficiency of ICM (Shirayoshi *et al.*, 1983; M. H. Johnson and B. Maro, unpublished data). This result would be expected, since the inhibition of flattening at the 16-cell and 32-cell stages would expose inside cells to asymmetry of contact and thereby lead to their polarization and contribution to trophectoderm.

Permanent inhibition of cytokinesis, such as occurs with prolonged exposure to a (18–24 hr) pulse of cytochalasin D, results, as we saw earlier, in no effect on the expression of the temporal programme of development but does inhibit intercellular flattening. On removal of the drug, cells flatten giving a compact 2-cell, 4-cell, or 8-cell embryo. However, although molecular and subcellular aspects of differentiation, including polarization and junction formation occur, cleavage does not. Thus no inner cells are present, all cells polarize, and 2-cell, 4-cell, or 8-cell blastocysts form in which all cells are phenotypically trophectodermal (Surani *et al.*, 1980; Pratt *et al.*, 1981; Kimber and Surani, 1982, Petzoldt *et al.*, 1983). A similar result is achieved by allowing cleavage to procede but instead separating the blastomeres of an embryo into smaller clusters. Such experimental conditions are associated with a reduced incidence of ICM cells (Tarkowski and Wroblewska, 1967).

7. Conclusions

The study of mammalian early development has moved from description *via* experimental manipulation into analysis of cellular mechanisms. The ability to manipulate embryos and their constituent cells has been crucial to this transition. In this chapter, the complex events of early embryogenesis have been reduced to three basic cell processes: (1) induction of a relatively stable cell polarization induced *via* an asymmetry of cell contact, (2) regulation of the degree of asymmetry of contact *via* processes of cell flattening and envelopment, and (3) regulation of cleavage orientation to control the disposition of the components of a polarized cell and thereby the process of cellular diversification. The same manipulative procedures that have led us to this conclusion are already proving to be of value for the reinterpretation of these processes at the subcellular level (Johnson, 1985).

ACKNOWLEDGMENTS. I wish to acknowledge the technical assistant of Gin Flach, Caroline Hunt, Raith Overhill, and Tim Crane. The work reported here was supported by the Medical Research Council and the Cancer Research Campaign.

References

Balakier, H., and Pedersen, R. A., 1982, Allocation of cells to inner cell mass and trophectoderm lineages in preimplantation mouse embryos, *Dev. Biol.* **90**:352–362.

Barlow, P., Owen, D. A. J., and Graham, C. F., 1972, DNA synthesis in the preimplantation mouse embryo, *J. Embryol. Exp. Morphol.* **27**:431–445.

Bennet, J., 1982, Sperm entry point is related to early division of mouse blastomeres, *J. Cell Biol.* **95**:163a.

Denker, H. W., 1976, Formation of the blastocyst: Determination of trophoblast and embryonic knot, *Curr. Top. Pathol.* **62**:59–76.

Ducibella, T., and Anderson, E., 1975, Cell shape and membrane changes in the eight-cell mouse embryo: Prerequisites for morphogenesis of the blastocyst, *Dev. Biol.* **47**:45–58.

Garner, W., and McLaren, A., 1974, Cell distribution in chimaeric mouse embryos before implantation, *J. Embryol. Exp. Morphol.* **32**:495–503.

Gerhart, J., Shaffer, R. M., Mussen, J. M., and Oster-Granite, M. H., 1982, Cell lineage analyses of preimplantation mouse embryos after blastomere injections with horseradish peroxidase, *Pediatr. Res.* **16**:11.

Gerhart, J. C., Vincent, J. P., Scharf, S. R., Black, S. D., Gimlich, R. L., and Danilchik, M., 1984, Localization and induction in early development of *Xenopus*, *Phil. Trans. Roy. Soc. London, Ser. B* **307**:319–330.

Goodall, H., and Johnson, M. H., 1984, The nature of intercellular coupling within the preimplantation mouse embryo, *J. Embryol. Exp. Morphol.* **79**:53–76.

Graham, C. F., and Deussen, Z. A., 1978, Features of cell lineage in preimplantation mouse development, *J. Embryol. Exp. Morphol.* **48**:53–72.

Graham, C. F., and Lehtonen, E., 1979, Formation and consequences of cell patterns in preimplantation mouse development, *J. Embryol. Exp. Morphol.* **49**:277–294.

Handyside, A. H., 1978, Time of commitment of inside cells isolated from preimplantation mouse embryos, *J. Embryol. Exp. Morphol.* **45**:37–53.

Handyside, A. H., 1980, Distribution of antibody- and lectin-binding sites on dissociated blastomeres from mouse morulae: Evidence for polarization at compaction, *J. Embryol. Exp. Morphol.* **60**:99–116.

Hillman, N., Sherman, M. I., and Graham, C. F., 1972, The effect of spatial arrangement on cell determination during mouse development, *J. Embryol. Exp. Morphol.* **28**:263–278.

Hogan, B., and Tilly, R., 1978, *In vitro* development of inner cell masses isolated immunosurgically from mouse blastocysts. I. Inner cell masses from 3.5 day pc blastocysts incubated for 24 hr before immunosurgery, *J. Embryol. Exp. Morphol.* **45**:93–105.

Hyafil, F., Morello, D., Babinet, C., and Jacob, F., 1980, A cell surface glycoprotein involved in the compaction of embryonal carcinoma cells and cleavage stage embryos, *Cell* **21**:927–934.

Izquierdo, L., and Ebensburger, C., 1982, Cell membrane regionalisation in early mouse embryos as demonstrated by 5′-nucleotidase activity, *J. Embryol. Exp. Morphol.* **69**:115–126.

Johnson, M. H., 1979, Molecular differentiation of inside cells and inner cell masses isolated from the preimplantation mouse embryo, *J. Embryol. Exp. Morphol.* **53**:335–344.

Johnson, M. H., 1985, Three types of cell interaction regulate the generation of cell diversity in the mouse blastocyst, in: *The Cell in Contact: Adhesions and Junctions as Morphogenetic Determinants* (G. Edelman and G. P. Thiery, eds.), pp. 27–48, Neurosciences Institute Publications Series, Wiley, New York.

Johnson, M. H., and Pratt, H. P. M., 1983, Cytoplasmic localisations and cell interactions in the formation of the mouse blastocyst, in: *Time, Space and Pattern in Embryonic Development* (W. R. Jeffrey and R. Raff, eds.), pp. 287–312, Alan R. Liss, New York.

Johnson, M. H., and Ziomek, C. A., 1981*a*, The foundation of two distinct cell lineages within the mouse morula, *Cell* **24**:71–80.

Johnson, M. H., and Ziomek, C. A., 1981*b*, Induction of polarity in mouse 8-cell blastomeres: Specificity, geometry, and stability, *J. Cell Biol.* **91**:303–308.

Johnson, M. H., and Ziomek, C. A., 1983, Cell interactions influence the fate of mouse blastomeres undergoing the transition from the 16- to the 32-cell stage, *Dev. Biol.* **95**:211–218.

Johnson, M. H., McConnell, J., and Van Blerkom, J., 1984, Programmed development in the mammalian embryo, *J. Embryol. Exp. Morphol.* **83**(Suppl):197–231.

Johnson, M. H., Pratt, H. P. M., and Handyside, A. H., 1981, The generation and recognition of positional information in the preimplantation mouse embryo, in: *Cellular and Molecular Aspects of Implantation* (S. R. Glasser and D. W. Bullock, eds.), pp. 55–74, Plenum Press, New York.

Johnson, M. H., Ziomek, C. A., Reeve, W. J. D., Pratt, H. P. M., Goodall, H., and Handyside, A. H., 1983, The mosaic organisation of the preimplantation mouse embryo, in: *Ultrastructure of Reproduction* (J. Van Blerkom and P. M. Motta, eds.), pp. 205–217, Martinus Nijhoff, Boston.

Kelly, S. J., 1979*a*, Investigations into the degree of cell mixing that occurs between the 8-cell stage and blastocyst stage of mouse development, *J. Exp. Zool.* **207**:121–130.

Kelly, S. J., 1979*b*, Studies of the developmental potential of 4- and 8-cell stage mouse blastomeres, *J. Exp. Zool.* **200**:365–376.

Kelly, S. J., Mulnard, J. G., and Graham, C. F., 1978, Cell division and cell allocation in early mouse development, *J. Embryol. Exp. Morphol.* **48**:37–51.

Kimber, S. J., and Surani, M. A. H., 1981, Morphogenetic analysis of changing cell associations following release of 2-cell and 4-cell mouse embryos from cleavage arrest, *J. Embryol. Exp. Morphol.* **61**:331–345.

Kimber, S. J., Surani, M. A. H., and Barton, S. C., 1982, Interactions of blastomeres suggest changes in cell surface adhesiveness during the formation of inner cell mass and trophectoderm in the preimplantation mouse embryo, *J. Embryol. Exp. Morphol.* **70**:133–152.

MacQueen, H. A., and Johnson, M. H., 1983, The fifth cell cycle of the mouse embryo is longer for smaller cells than for larger cells, *J. Embryol. Exp. Morphol.* **77**:241–308.

Maro, B., and Pickering, S., 1984, Microtubules influence compaction in preimplantation mouse embryos, *J. Embryol. Exp. Morphol.* **84**:217–232.

Mintz, B., 1965, Experimental genetic mosaicism in the mouse, in: *Preimplantation Stages of Pregnancy*, CIBA Foundation Symposium, pp. 194–209, Churchill Livingstone, London.
Ogou, S-I., Okada, T. S., and Takeichi, M., 1982, Cleavage stage mouse embryos share a common cell adhesion system with teratocarcinoma cells, *Dev. Biol.* **92:**521–528.
Pedersen, R. A., and Spindle, A. I., 1980, Role of the blastocoele microenvironment in early mouse differentiation, *Nature (Lond.)* **284:**550–552.
Petzoldt, U., Burki, K., Illmensee, G. R., and Illmensee, K., 1983, Protein synthesis in mouse embryos with experimentally produced asynchrony between chromosome replication and cell division, *Roux Arch. Dev. Biol.* **192:**138–144.
Pratt, H. P. M., Chakraborty, J., and Surani, M. A. H., 1981, Molecular and morphological differentiation of the mouse blastocyst after manipulation of compaction with Cytochalasin D, *Cell* **26:**279–292.
Pratt, H. P. M., Ziomek, C. A., Reeve, W. J. D., and Johnson, M. H., 1982, Compaction of the mouse embryo: An analysis of its components, *J. Embryol. Exp. Morphol.* **70:**113–132.
Randle, B., 1982, Cosegregation of monoclonal reactivity and cell behaviour in the mouse preimplantation embryo, *J. Embryol. Exp. Morphol.* **70:**261–278.
Reeve, W. J. D., 1981, Cytoplasmic polarity develops at compaction in rat and mouse embryos, *J. Embryol. Exp. Morphol.* **62:**351–367.
Reeve, W. J. D., and Ziomek, C. A., 1981, Distribution of microvilli on dissociated blastomeres from mouse embryos: Evidence for surface polarization at compaction, *J. Embryol. Exp. Morphol.* **62:**339–350.
Rossant, J., and Lis, W. J., 1979, Potential of isolated mouse inner cell masses to form trophectoderm derivatives *in vivo*, *Dev. Biol.* **76:**475–482.
Rossant, J., and Vijh, K. M., 1980, Ability of outside cells from preimplantation mouse embryos to form inner cell mass derivatives, *Dev. Biol.* **76:**475–482.
Shirayoshi, Y., Okada, T. S., and Takeichi, M., 1983, The calcium-dependent cell–cell adhesion system regulates inner cell mass formation and cell surface polarization in early mouse development, *Cell* **35:**631–638.
Spindle, A. I., 1978, Trophoblast regeneration by inner cell masses isolated from cultured mouse embryos, *J. Exp. Zool.* **203:**483–489.
Spindle, A., 1982, Cell allocation in preimplantation mouse chimaeras, *J. Exp. Zool.* **219:**361–367.
Surani, M. A. H., and Barton, S. C., 1984, Spatial distribution of blastomeres is dependent on cell division order and interactions in mouse morulae, *Dev. Biol.* **102:**335–343.
Surani, M. A. H., and Handyside, A. H., 1983, Reassortment of cells according to position in mouse morulae, *J. Exp. Zool.* **225:**505–511.
Surani, M. A. H., Barton, S. C., and Burling, A., 1980, Differentiation of 2-cell and 8-cell mouse embryos arrested by cytoskeletal inhibitors, *Exp. Cell Res.* **125:**274–286.
Tarkowski, A. K., and Wroblewska, J., 1967, Development of blastomeres of mouse egg isolated at the 4- and 8-cell stage, *J. Embryol. Exp. Morph.* **18:**155–180.
Wilson, I. B., Bolton, E., and Cuttler, R. H., 1972, Preimplantation differentiation in the mouse egg as revealed by microinjection of vital markers, *J. Embryol. Exp. Morphol.* **27:**467–479.
Ziomek, C. A., and Johnson, M. H., 1980, Cell surface interaction induces polarisation of mouse 8-cell blastomeres at compaction, *Cell* **21:**935–942.
Ziomek, C. A., and Johnson, M. H., 1981, Properties of polar and apolar cells from the 16-cell mouse morula, *Roux Arch. Dev. Biol.* **190:**287–296.
Ziomek, C. A., and Johnson, M. H., 1982, The roles of phenotype and position in guiding the fate of 16-cell mouse blastomeres, *Dev. Biol.* **91:**440–447.
Ziomek, C. A., Johnson, M. H., and Handyside, A. H., 1982*a*, The developmental potential of mouse 16-cell blastomeres, *J. Exp. Zool.* **221:**345–355.
Ziomek, C. A., Pratt, H. P. M., and Johnson, M. H., 1982*b*, The origins of cell diversity in the early mouse embryo, in: *Functional Integration of Cells in Animal Tissues* (J. D. Pitts and M. E. Finbow, eds.), pp. 149–165, Cambridge University Press, Cambridge.

Chapter 10

Tissue Interactions in Development of Teeth and Related Ectodermal Derivatives

EDWARD J. KOLLAR

1. Introduction

This chapter deals with the tissue interactions in developing skin, teeth, and related ectodermal derivatives. Epithelial–mesenchymal interactions that take place between labile epithelia and mesenchyme are the inductive interactions that occur after primary induction of the nervous system and the axial skeleton. These interactions are responsible for the differentiation of the major organ systems of vertebrate embryos.

Since the earliest days of experimental embryology, the ectodermal derivatives have provided the greatest challenge to understanding the emergence of form and specialized phenotypic expression. The induction of the neural tube and the axial skeleton are the *sine qua non* of vertebrate organization. Unless the ectoderm is specialized on cue from the chordamesoderm, the axiation of vertebrate form is aborted and further development of organized animal form is lost. In later stages, all the developing structures of the skin and oral mucosa similarly depend on the interaction between the ectodermaly derived epithelium and oral mucosa and the underlying neural crest or mesenchymally derived stroma. Similar examples of this general developmental phenomenon can be found in such well studied models as the developing limb (Fallon and Caplan, 1983), or the developing urogenital system (Saxen *et al.*, 1980, 1981; Cunha, 1971; Cunha *et al.*, 1980). The developmental versatility of the developing epithelia that results in the development of such complex derivatives as hair, feathers, scales, and teeth, has made the developing skin and oral mucosa a useful model for studying the development of complex embryonic structures. The questions of induction, cytodifferentiation, morphogenesis, and complex organization of organ patterns have all been addressed, using the skin and oral

EDWARD J. KOLLAR • Department of Oral Biology, School of Dental Medicine, The University of Connecticut Health Center, Farmington, Connecticut 06032.

mucosa as the paradigm. Most of the advances in this area of experimental interest can be traced to several technical and conceptual breakthroughs that can be attributed to the work of Grobstein (1953, 1955), Moscona (1952), and Rawles (1963).

This chapter examines the development of some of the derivatives of the skin and oral mucosa and describes the various manipulations of the mammalian embryo that have generated much of the data that have contributed to the present state of understanding of the development of these structures.

2. Ectodermal Differentiation

Several recent reviews have discussed the development of the skin and oral cavity (Fallon and Caplan, 1983; Sawyer and Fallon, 1983; Pratt and Christensen, 1980; Connelly *et al.*, 1981; Kollar, 1983; Sengel, 1983; Briggaman, 1982).

It has been clear since the early work of McLoughlin (1966), Rawles (1963), and Billingham and Silvers (1968) that the developing integument and oral epithelium have obligatory requirements for histodifferentiation and cytodifferentiation. The connective tissue component (dermis, lamina propria) is necessary to support the continued survival of epithelial cells and it contains inductive information that regionally specifies the morphology and specialized genetic expression in local epithelial derivatives (e.g., hair, tooth, feather, gland).

In mammals, there is a consistently uniform sequence of developmental events associated with the expression of these complex derivatives of the epithelium. A thickened epithelium can be seen in association with a condensation of the connective tissue cells, the papilla. Then, the epithelium begins to invade the underlying stroma always maintaining its relationship to the papilla. Variations in the morphology of invading epithelial follicles are obvious from the very earliest stages; hair follicles are distinctively shaped, as are incisor and molar tooth buds. Once early rudiments are definitively formed the cells of the epithelium (and, in the case of the teeth, the mesenchyme) begin to express specialized function and secrete unique proteins characteristic of each derivative.

In a variation on this pattern that is characteristic of glandular development, instead of connective tissue papillae, that seemingly control epithelial form, the epithelium, branches into the underlying connective tissue is invested by a mesenchymal capsule that surrounds the epithelial rudiment (see Chapter 11 in this volume for details of glandular development.) The epithelial–mesenchymal relationship in glands is apparently different from nonglandular derivatives. Thus, the basic mode of interaction may not be equivalent between glands and hair or teeth. Capsular and papillar mesenchyme may represent two different modes of induction.

The teeth present an especially interesting model. While the early events of development are superficially identical to those in developing hair and vibrissae development, some essential differences soon appear that distinguish these two epithelial rudiments. In hair and vibrissae follicles the papilla seems to have a somewhat passive and supporting role in that it does not express any obvious specialized protein. It is the organization of the keratinocytes and keratin filaments into a hair shaft that characterizes hair and vibrissae follicles. In teeth, on the other hand, the papilla will express increased secretion of type I collagen and thus provide the organic matrix for dentin formation and calcification. Following this deposition of dentin matrix by the dental papilla, epithelial cells begin to specialize into secretory ameloblasts that will begin to secrete two specialized proteins, enamelins and amelogenins, that will constitute the organic matrix of the enamel of the tooth crown.

Thus, in teeth, both interacting tissues of the rudiment will express distinctive genetic information that in the case of the enamel proteins cannot be found until this stage of development has been achieved (Slavkin *et al.*, 1982).

There is a great deal of interest in the role of the extracellular matrix—especially the basement membrane—which often separates the epithelium from the papilla. At early stages of epithelial morphogenesis, basal laminae and associated acellular elements are present separating the two tissues. There is ambiguity and controversy about whether the basal lamina breaks down or is perforated during critical periods of interaction [see Hardy and Goldberg (1983) for a discussion of these events during hair development.] But, in the teeth it has been established (Thesleff *et al.*, 1981) that just after the appearance of the dentin matrix the basal lamina adjacent to the epithelial cells of the enamel organ does degenerate and that event is coincident with the first evidence that the epithelial cells are beginning to differentiate into ameloblasts, the enamel protein secreting cell of the tooth bud.

As you can see from this short summary of the early events of epithelial differentiation in the skin and oral mucosa, all the derivatives have a general sequence in common. Certain details vary, but the overall picture hints at a commonality of mechanism operating in all these experimental models. Furthermore, the variety of expression presented in the developing ectoderm offers even greater experimental advantages. For example, the importance of pattern regulation is obvious in the skin and in the dentition. The arrangement of hair follicles—especially in the vibrissae pattern on the snout—and the arrangement of metameric subunits precisely arrayed in the dentition pique one's curiosity about the control mechanisms that not only initiate the structures but also organizes the units into functional groups.

On a similar level, note that the precision of instruction evokes exquisite structural detail in each of the developing derivatives. Molars and incisors are clearly designed for precise use in mastication. The patterns of enamel and dentin deposition result in two distinctive shapes, each suited to a different function. The cells must respond to precise cues with exacting fidelty in order to reproduce the shape of the structure. Thus, the skin and the dentition have been useful in asking a number of questions that relate to organization above the cellular level.

Fortunately, these models of development have been very amenable to experimental manipulation. What has become clear to me is that such factors as the method used to dissect the component tissue, the method of culture, the use of long-term explanation, the use of xenoplastic combinations, and so forth can dramatically alter the results; the conclusions made in one laboratory versus those of another can vary a great deal as a result of the methods employed. The body of this chapter describes some of my experiences with manipulating the mammalian (and sometimes the avian) embryo.

3. Tissue Isolation Methods

Almost universally, two methods have been employed to separate tissue components in complex structures. The cold trypsin method described by Rawles (1963) using crude trypsin has been most useful in my laboratory. The method is a variant of other trypsin dissociation methods devised to disperse tissues into cell suspensions. The advantage of the cold (4°C) incubation is that the extracellular matrix is digested and the tissues are not completely dissociated; rather, the epithelium can be retrieved as an intact sheet leaving behind an intact stromal mirror image. Naturally, the method does result in the degredation of the basal lamina and possibly other acellular portions of the tissue interface (Kollar, 1983; Osman and Ruch, 1981; Hardy and Goldberg, 1983).

By contrast, the other method using EDTA, a chelator, as a method of tissue separation spares portions of the basement so that at least part of the original tissue interface is left intact for the first hours of the experimental period. How this alters the response of the ensuing experimental confrontations is not certain at this time; but clearly this mode of tissue separation does not entirely destroy the previous developmental history of the tissue. Whether the information stored as the remnant of the basal lamina can affect the result of an experimental tissue interaction merits investigation.

4. *In Vitro* Culture Methods: Cell versus Organ Cultures

There is no question that in the future any tissue or cell type that is now refractory to *in vitro* propagation will submit to technology and will be cultured with great success. A case in point is the recent history of success with epithelial cells. Epithelial cells, especially from the skin now can be cultured *in vitro* with relative ease. Feeder layers, serum-free media, complex formulations of synthetic media, and so forth are all having their impact on the culture of epithelial cells.

The difficulty with cell culture is that (as is true of all cells growing on artificial subtrates in two dimensions) the cells assume what I call the "Falcon phenotype." The imposed restriction that cells must grow in a two-dimensional environment without a biological substrate severly limits the capability of cells *in vitro*. The result reported by Pitelka and her group (Emerman *et al.*, 1979; Burwen and Pitelka, 1980), in which cells are released from the plastic substrate, or those of Greenburg and Hay (1982), in which cells are allowed to grow in three-dimensional collagen gels, demonstrate that the expected behavior of cells can only be surmised when a range of environmental conditions are tested that freely permit a range of cell responses to laboratory conditions.

We have had similar experiences (Kollar and Kerley, 1980; Kollar, 1980). If the tooth bud is separated into the epithelial enamel organ and the dental papilla, these components can be cultured *in vitro* with relative ease. The epithelium assumes the typical cohesive sheet of cells and, by contrast, the papilla cells in isolation *in vitro* appear as any other fibroblastlike cell culture. Neither cell population secreted differentiated cell products *in vitro* typical of the differentiated tissues *in vivo*.

Mixing the cultures, or adding intact epithelial tissue to the fibroblast cultures (or the reciprocal) does not result in any change in the differentiation of the cells. On the other hand, if cell cultures are released from the substrate with enzymes and the cells are pelleted and grafted to an explanation site such as the intraocular anterior chamber or beneath the kidney capsule, these cultured cells will sort out, recapitulate their developmental histories, and participate in tooth development. Development is complete; all dental structures are present as recognizable teeth and the specific protein matrices are present. Thus, cells seemingly dedifferentiated and surviving minimally in culture can re-express their developmental potential and re-express their phenotype.

This is not to say that *in vitro* culture techniques are not useful—or even necessary. We find it essential when tissue combinations are made that the recombinants be passaged in organ culture overnight to ensure that the basement membrane reorganizes and that the two tissues are once again interacting. In addition, transfilter experiments have been a routine procedure for studying the interactions between tissues. Koch (1967) and Thesleff *et al.* (1978) have used this technique especially to demonstrate that the tissues of the developing tooth bud do interact across filters and that it is not cell contact that is important. Rather, it is now clear that the extracellular matrix is the important element in mediating tissue interactions in the developing integument and the oral mucosa (Hay, 1981, 1982, 1983; Kleinman *et al.*, 1981; Martin *et al.*, 1983).

5. *In Vitro* Grafting Methods

An alternative approach to using standard *in vitro* culture methods is to graft tissues and experimental tissue combinations to sites that offer the advantage of a blood supply. Four techniques are standard, but many variations exist.

Grafting to the chorioallantoic membrane (CAM) of embryonated eggs has been used for a wide variety of studies, but it does have the disadvantage that the best results are obtained when avian tissue is being studied. When mammalian tissue is grafted to the CAM, there is a limited amount of time available for growing the grafts before the avian cells begin to reject the graft as foreign.

As an alternative to CAM grafting, explanation to sites in mammalian hosts have been used. Grafts can be placed beneath the kidney capsule (e.g., Cunha *et al.*, 1980; Svajger *et al.*, 1981) or in the anterior chamber of eyes (Kollar, 1972).

In both techniques, the grafts elicit an invasion of the host blood supply in a very short time. The choice of one or the other technique appears to be a matter of personal opinion. The subcapsular grafts do somewhat compress and distort the graft as it grows. Intraocular grafts can be observed at anytime during the grafting period and have the advantage that the host tissue does not interact with the graft. The anterior chamber is lined by endothelium of the cornea and the epithelium of the iris; thus, the only host involvement with the grafted tissue is with the vascular elements.

The fourth alternative is grafting directly to the dorsum of the animal either as surface grafts (Briggaman, 1982) or in the subcutaneous space either as cell suspensions (Lavker and Sun, 1983) or in a variety of chambers (Mackenzie and Fusenig, 1983). The availability of athymic nude mice has provided an added advantage because tissues of diverse species origin can be grafted without rejection. Thus, long-term grafting can be achieved for xenoplastic combinations.

The advantages of these grafting techniques over routine *in vitro* methods are that optimal differentiation is achieved and, in the case of tissue recombination, negative responses seen *in vitro* are often positive when given the advantage of optimal nutritional conditions. For example, the development of tooth buds *in vitro* does not result in enamel deposition unless very exacting media requirements are met and even then only minimal quantities of protein are deposited. By contrast, tooth buds grown in the anterior chamber develop full thicknesses of enamel protein, which then calcifies as true enamel. Calcification *in vitro* has not been possible.

In addition to the optimal expression of the differentiated state, the intraocular site has proved favorable for growing experimental tissue combinations and for permitting them to express new developmental capabilities. For example, as you will see in the next section, attempts to investigate the potential interaction between avian and mouse oral tissues *in vitro* (Cummings *et al.*, 1981; Ruch and Karcher-Djuricic, 1975) have failed, as did our attempts *in vitro*. Yet, the results were positive when the tissue combinations were grown as grafts to the CAM or intraocularly in nude mice (Kollar, 1972; Kollar and Fisher, 1980).

6. Xenoplastic Tissue Combinations

The development of the integument and oral cavity in avian (Sengel, 1976) and mammalian (Kollar, 1972, 1980, 1983) species has naturally raised questions about the nature of the signals that specify such diverse vertebrate structures as feathers, hair, and teeth.

Moscona (1952, 1960) was the first to demonstrate that dissociated cells reaggregate not according to species specificities but according to tissue specificities; epithelial cells from avian and mammalian skin comingle into a chimeric epidermis. Garber *et al.* (1968) found that mouse epithelial cells could be found in feather ridges of feather follicles in CAM grafts of co-mingled chick and mouse skin cell suspensions. However, the striking finding of Coulombre and Coulombre (1971) that embryonic mouse dermis from the dorsum could induce the embryonic chick corneal epithelium to form feathers is one of the best examples of the importance of using xenoplastic tissue combinations. The result of that experiment reconfirmed the notion that inductive influences also are not species specific but rather inductive signals cross species and class phylogenetic lines.

Our experience with using oral tissue from chick and mouse embryos mentioned above is another example. Fell and Jacobson (1941) noted in a report of their studies of the developing chick mandible that a structure very much like a dental lamina (the first thickening of the epithelium before it grows into the stroma in association with dental papilla) was present in the five day old chick embryo. This intriguing observations suggested that some of the capabilities for making teeth were present in extant birds. We combined the embryonic 5-day-old chick epithelium with dental papillae from mouse embryos. When cultured *in vitro* as organ cultures, the results were negatives. But, if the grafts were placed on the CAM (Kollar, 1972) or in the anterior chamber (Kollar and Fisher, 1980), dental structures did form, and even enamel was deposited.

7. Mutants as Experimental Tools

Despite the obvious usefulness of animal mutations for studying abnormal development, mutants that seriously affect the skin and oral mucosa have not been readily available. Alterations of coat color or tooth size and shape have been identified, but serious developmental aberrations of the epithelium or the dentition have been reported less frequently. However, two mutations have been studied that do exhibit serious defects in these important structures. Pupoid fetus (*pf/pf*) and repeated epilation (*Er/Er*) both exhibit hyperplasia of the skin and alterations of the development of the oral mucosa.

Our preliminary work has concentrated on the integument of *pf/pf*. When mutant pups of 13 or 14 days gestation are contrasted to their normal littermates, the skin is variably thickened and consists of normal areas that range from a few cells in thickness to areas that are obviously hyperplastic. By 16 days of development, the differences between *pf/pf* and normal skin are easily visible and indicate that patches of epithelial cells are degenerating; the normally smooth epidermal–dermal junction is not maintained and there are indications that the dermal cells are exfoliating onto the epidermal surface through breaks in the basement membrane.

The skin is grossly hyperplastic in *pf/pf* animals at birth. The orderly cellular stratification of the skin is lost and the epidermis is obviously thickened. Normal littermates express the usual stratified granular and keratinized layers as well as hair follicles; all these structures are absent in *pf/pf* mutants. Pockets of dermal cells within the epithelium are diagnostic of the hyperplasia in *pf/pf* mutants. Because the intraepithelial pockets of dermal tissue contain fibroblastlike cells, nerve bundles, and vascular elements, the skin is visibly altered. These cellular intrusions are not hemorraghic and are apparently orderly extensions of the dermis into the epidermis *via* perforations in the basal lamina—a sort of repeated wound healing. Dermal cells exfoliating onto the skin surface are recognized by the outermost epidermal cells as true dermis. Consequently, the epidermal cells grow around and over the dermal cells, as in wound healing. The dermal components are surrounded by epidermal cells that now reorganize themselves as basal cells; a well defined basal lamina and hemidemosomes can be seen surrounding the invading dermal cells. The result of this invasion by the dermis is to completely confuse the normal stratification patterns of the epidermis (Fisher and Kollar, 1984*a*).

While we have not examined *Er*/*Er* mutant skin detail, the development of the integument has been thoroughly described by Holbrook *et al.* (1982). In this mutation, which appears to be a totally different mutation, the superficial resemblance to *pf*/*pf* is that the skin is hyperplastic and that the thickness of the skin is variable from site to site. But, there is little else in common with *pf*/*pf*. The invasion of dermal tissue into the epidermis is lacking in *Er*/*Er*, and the layers of the epidermis are less abnormally stratified.

Two proteins distinguish the differentiation of epithelial cells of the epidermis and the oral mucosa; these are keratin and filaggrin. Keratin is composed of a family of 10-nm filaments whose structure is increasingly well known (Steinart *et al.*, 1976) and for which a series of antibodies is available. They are water insoluble, 40,000–70,000-M_r proteins related to the intermediate filaments. The second protein, filaggrin, is found in the keratohyalin granules of the integument and the oral mucosa (Matoltsy, 1975; Dale *et al.*, 1978, 1981; Steinart *et al.*, 1981) and is a 27,000-M_r protein that organizes isolated keratin filaments into complexes that mimic the structure of mature keratin as it appears *in vivo*. Antibodies to this protein interact with filaggrin and decorate keratohyalin granules *in situ*.

Variations in the keratohyalin granules and the protein filaggrin are the most obvious differences that distinguish the two mutants. Keratohyalin granules are virtually absent in *pf*/*pf*, and only on the rarest occasion can small foci be seen. On the other hand, in *Er*/*Er* mutants, keratohyalin granules are sometimes present but they are minute and diffuse (Fisher and Kollar, 1984*b*). While these variations in keratohyalin granule formation are convincing, the biochemical variation between the two mutants is even more dramatic.

The 27,000-M_r band that is identifiable in immunoblots as filaggrin is absent in SDS-PAGE gels of skin extracts of *pf*/*pf* indicating the absence of filaggrin in *pf*/*pf*. The precursor bands are absent as well. By contrast, in *Er*/*Er* as described by Holbrook *et al.* (1982), the 26,500-M_r band of their gels is absent, but the precursors are present. This is in contrast to *pf*/*pf*, where the precursors as well as the mature protein band are absent. This is demonstrated in their pictures of the immunofluorescent staining using antifilaggrin, which decorates a broad layer of cells throughout the epidermis in *Er*/*Er* tissue, indicating the presence of the gene product. Antifilaggrin staining of our normal tissue indicates that the filaggrin is confined to the outermost epidermal layers, while in the *pf/pf* mutant tissue the staining is absent. Thus, the participation of keratohyalin granules and the protein filaggrin in skin differ in the two mutants is considerably different.

The invasion of the epidermis by the dermis is such an unusual finding that we are examining the nature of the dermal basal lamina or basement membrane defect that permits the invasion to take place (Fischer and Kollar, 1985). We know that the dermal pockets that invade the epidermis are normal with respect to many of the usual constituents of the dermis and the basal lamina. Immunofluorescent staining of fibronectin, collagen IV, and laminin molecules indicates that these components are present. Fibronectin is present in these dermal pockets and can be seen to be contiguous with the dermal components and present to the same degree of brightness in the sections labeled with antifibronectin (Fisher and Kollar, 1984*b*).

Laminin in the epidermal basal lamina and surrounding the vessels and nerves is distributed normally and can be seen to be contiguous with the epidermal basal lamina around the aberrant dermal pockets. Even though the appearance of dermal cells on the surface of the developing epithelium (and later as incorporated into dermal pockets in the hyperplastic epithelium) is an unusual finding, the dermal elements and their relationship to the epithelial cells seem quite normal with respect to the basal lamina.

We do not know how the initial breakdown of the basal lamina occurs. The dermal cells do not behave in the same way in the mutant as they do in normal skin. In mutant skin, the cellular density of the dermis is markedly higher. In mutant tissue, cell density is 1.5–2.0 times higher than the number of dermal cells in the area immediately adjacent to the basal lamina as compared with normal tissue. Mutant dermal cells pallisade next to the basal lamina, suggesting that the normal epidermal–dermal interaction has been upset. However, there is no firm indication of causality in this kind of cellular behavior. It is tempting to think that basal lamina dissolution is due to dermal cell aggregation, but it could be quite the opposite case. The lamina might break down because of an epidermal defect or deficiency thereby permitting the dermal cells to migrate into the breach and to exfoliate onto the surface. *In vitro* and transfilter studies may help resolve these questions in the future.

Normal epidermis recombined with normal dermis (*+E/+D*) after the tissues have been separated with trypsin and then grown in the anterior chamber of the eyes of adult normal hosts maintain the usual pattern of stratification and keratinize with a prominent granular layer beneath the stratum corneum as expected in these control cultures. Experimental intraocular grafts of recombined normal epidermis and *pf/pf* dermis (*+E/pfD*) do not express abnormal development. Instead, the tissue in the grafts is normal and indistinguishable from the controls. We conclude that the defect is not in the dermis.

The reciprocal combinations of *pf/pf* epidermis and normal epidermis (*pfE/+D*) were similarly in intraocular grafts. These recombinations are also normal, indicating that the defect is not in the epidermis either. Surprisingly, if the intact *pf/pf* skin is grown as a graft to the intraocular site or under the kidney capsule, it reverts to a normal histotypic pattern of stratification and keratinization. The defect is not localized in either of the tissues, and the data suggest either a systemic, metabolic, or epigenetic defect (Fisher *et al.*, 1983).

In thin sections that enhance the visibility of the keratohyalin granules and illustrate the ultrastructural organization, the normal appearance of the mutant tissue in an ectopic site is more clearly seen. By all histological criteria the tissue is morphologically normal. But, the most convincing argument for the reorganization of *pf/pf* tissue in the ectopic explanation site is the reappearance of filaggrin, the protein related to keratohyalin granules, in its normal configuration in the stratum granulosum and seen as keratohyalin granules (Fisher *et al.*, 1984).

The keratins of *pf/pf* have been examined with electron microscopy and immunofluorescent staining. The dramatic upset of the stratification pattern and the absence of the filaggrin in the *pf/pf* mutant tissue, notwithstanding, intermediate filaments are present in a very aberrant stratum corneum in the upper most layers of the skin of newborn animals. Staining with polyclonal antibodies (supplied by Dr. T.-T. Sun) indicates that the filaments seen ultrastructurally are keratins as the entire thickness of the epidermis is positive for the antibody–keratin interaction. Thus it is the organization of the filaments, and not their synthesis, that is affected by the mutation. It may be that the loss of normal stratification patterns leads to the abnormal arrangement of the filaments.

Abnormal development of the oral cavity has been examined superficially in *pf/pf* mutants. The oral cavity is fused along the entire length. The tongue is grooved and fused to the vertical palatine shelves. The teeth are present, but they are aberrant. The upper and lower molars are fused, and the stellate reticulum is confluent from one molar to the other. The oldest (18-day) specimens examined suggest that hard tissue deposition may not be normal. The inner dental epithelium is hyperplastic, and there is not any indication of odontoblast differentation which should be present at this stage. Only a thin layer of dentin is present in the incisor, and there is no indication of incipient ameloblast differentiation.

Pupoid fetus and repeated epilation mutants have well-described genetic histories (Guenet *et al.*, 1979; Meredith, 1954; Green, 1981). They are close to each other on the same chromosome (Green, 1981), and they express similar pleiotropic expressions of their mutations. But, the mutations are quite distinct. Breeding patterns, segregation of phenotypes, different heterozygote expression, and unequivocal differences in the defects in the integument make these two mutations related but quite distinct.

It is not understood how these two mutations that affect the mouth as well as the skin, the genetic history and chromosomal location of which are known in some detail and whose biochemical phenotypes are well characterized are expressed. While a simple defect in a structural or enzymatic gene is an attractively simple hypothesis (especially since filaggrin seems to be absent in pupoid fetus), the pleiotropic effects of the defect and the somewhat different modes of expression in the oral cavities of the two mutants suggest that a more complicated pathway exists between the genome and the abnormal structures.

Other possibilities include malfunctioning regulatory genes, metabolic defects, hormonal receptor defects, and inappropriate extracellular matrix signals. All are equally plausible mediators of the abnormal development at this stage of the analysis. In pupoid fetus, that the integumental defects (both morphological and biochemical) are restored to normal function when the tissue is grown in a normal adult environment suggests that the defect is mediated by some systemic influence on tissue organization and that the biochemical defects are perhaps secondary to the dysmorphology.

Recently, a new approach to altering mammalian development has been reported by Jaenisch and co-workers (Schnieke *et al.*, 1983). The Moloney leukaemia virus, a retrovirus, has been inserted into the genome of developing mouse embryos. The retrovirus was injected *intra utero* and was apparently incorporated into the germ cell line of the developing 8-day postimplantation embryos because the subsequent progeny of the infected litter produced offspring that incorporated the virus. During the breeding of these animals for an analysis of whether the animal had incorporated the virus into one or both alleles, it became apparent from the phenotypic ratios that homozygotes were lethal. Using the viruses, then, as genetic insertion mutagens, a developmentally lethal mutation had been achieved. But, the most important finding was that homozygous embryos develop normally until day 11–12 of gestation and then become arrested; death occurs by day 13 or 14.

The location and nature of the genetic defect were examined in some detail. Schnieke *et al.* (1983) were able to show that the virus is inserted into the 5′ end of the 1(I) collagen gene. In heterozygotes and in normal littermates this gene begins to be expressed at about day 9 of gestation and is fully expressed by day 12. Apparently, in the infected animals carrying the inserted virus this gene is not expressed. The mesodermal cells of the mutant animals were grown in cell cultures and were found to be viable and transformable. Growth was normal as cell cultures, but the gene was not expressed.

While this approach uses the most advanced of biological techniques, it is not remarkably different from traditional strategies for producing experimental abnormalities. The unquestionable advantage of the model is that the technique can probe the innermost details of genetic expression in mutants that express developmental defects.

The questions of why this important gene is not being expressed in the early embryo before day 12 and what regulates its activation will be of great interest to those of us working with epithelial–mesenchymal interactions. By day 12 of gestation, the important maneuvers of proliferation, migration and mechanical folding has produced an embryo now poised for organogenesis. The results suggest that an ECM much different from later stages after day 12 is responsible for these early events. At day 12, the teeth and vibrissae will begin to develop; slowly, with increasing age, more and more of the major organ systems will be initiated. All these interactions depend on the ECM and its relationship between the developing epithelium and the underlying mesenchyme.

It has always been difficult to grow and maintain presumptive organ rudiments from embryos younger than 12 days gestation *in vitro*. Supposedly, when younger tissue is removed, the gene for 1(I) collagen is not being expressed even though the cells appear normal in organ culture. Thus, our attempts to investigate the early stages of tooth initiation may have failed because this important gene product is not being produced and the requisite tissue interactions are aborted. Future work should be addressed to this issue since the early stages of organ rudiment formation may be the most informative (Kollar 1980, 1983).

One could not hope for a better demonstration than in the example cited above for the relationship of the timing of gene expression to harmonious development. The issue of what regulates the turning on of this gene in mesodermally derived cells *in situ* is the greater issue, since all of organogenesis is dependent on this crucial event.

8. The Nature of Epithelial–Mesenchymal Inductive Interactions

It is over 30 years since Grobstein (1955) had first pointed out the importance and far-reaching consequences of epithelial–mesenchymal interactions and the importance of the extracellular matrix. In the intervening years, we have honed the original concepts into a number of guidelines that help frame our thinking about inductive tissue interactions.

The integument and oral mucosa provide the most stringent set of requirements for proper development to take place. Locally the mesenchyme is responsible for regional specificity. A labile epithelium of ectodermal origin will respond to the command of an inducing mesenchymal tissue and will develop in accordance with the origin of the mesenchyme cells. Thus, vibrissae mesenchyme induces hair follicles in almost any reactively labile epithelium, and tooth mesenchyme will induce teeth in ectodermally derived epithelium from a wide variety of sources in the embryo and phylogenetic histories. This type of interaction has been called instructive induction. The specificity resides in the inducer, and the responding epithelium always conforms to the inducing signal by responding with specific morphogenesis and genetic expression for specialized molecules.

In contrast to instructive induction, another less demanding interaction has been described. These are the permissive interactins that characterize a wide variety of tissue interactions in the developing gut and urogenital system. For example, a wide variety of tissues and even acellular and cellular extracts will support pancreatic development. If a very demanding and restrictive inductive interaction does take place at some early stage of development, it has not been discovered because of the technical difficulties of working with early stages of mammalian embryos. For example, tubules are induced by the ureteric bud but this tissue can be substituted by the similarly branching epithelium of the salivary gland. These permissive interactions have been reviewed by Saxen *et al.* (1980) and Wessells (1977).

Similarly, less complicated tissues (e.g., single-cell systems) that do not depend on restrictive tissue interaction such as cartilage, bone, and pigment cells are very accommodating in their requirements for differentiation and, thus, their inductive interactions are extremely permissive. Any number of tissue combinations or experimental conditions will elicit differentiated phenotypic expression in these cells.

The signal and mode of signal transmission from one tissue to the other and the mechanism of signal transduction and interpretation are not known in instructive tissue interactions. How the genome is turned on for specific gene transcription in these interactions remains the major unsolved problem in biology.

Embryonic induction as illustrated in the skin and oral tissues is clearly distinct from enzyme induction or hormone activation. In the latter two modes of gene activation, the signal is unique to each interaction, and the response is transient; when the signal is withdrawn, the cells revert to their former state. By contrast, in embryonic induction, the signal seems to be general; mouse dorsal dermis induces feathers in chick cornea, and mouse dental papillae induce teeth in chick epithelium. In addition, the signal imparts long-term changes in cells that even if deprived of their ability to express differentiate states will revert to their former inductive and responsive states. That is why dedifferentiated cultured cells can re-form teeth given a suitable environment and why regeneration is possible from blastema cells.

There is no doubt that the extracellular matrix is an important element in this puzzle. Any perturbation of the basal lamina or associated portions of ECM will abort development. The reviews of Hay (1981, 1982, 1983) will support this notion, as will our work on developing teeth (Kollar, 1980, 1983). The exciting and innovative work on Schnieke *et al.* (1983) underscores the universal importance of type I collagen in the epithelial–mesenchymal interactions that are responsible for organizing all the organ systems that maintain the embryo after the development of the brain axial skeleton and cardiovascular system and of the nervous system, axial skeleton, and cardiovascular systems.

Despite the importance and interest in this subject (see Kollar, 1983), a more intriguing question may be the mechanism of setting up the individual anlage. It is obvious that the sites of the structures and the complex distribution of their final patterns can be discerned days before the developing anlage can be dissected and manipulated. The distribution of vibrissae follicles and teeth are probably the best example of the influence of early events on the initiation and pattern formation.

Svajger *et al.* (1981) provided an interesting hypothesis that results from their exquisite studies on transplanting early mouse embryo tissues under the kidney capsule. It is astounding that this very early ectoderm (which does give rise to a mesenchymal cell population) can develop into the array of differentiated and organized structures. The outcome of any experiment seems to be random and ostensibly depends on which portions of the ectoderm survive the experimental trauma.

Nonetheless, one is led to the speculation that it is the epithelium that transmits organizing signals to the mesenchyme, which then specifies it as the inducing mesenchyme that can subsequently, with appropriate experimental manipulation, induce nascent epithelium to undergo type-specific differentiation.

One might speculate further that the dental lamina present in the mandibular and maxillary arches may be interacting with the underlying neural crest cells to determine the sites of tooth placement, perhaps even endowing the mesenchyme with their inductive capabilities. These notions need testing. The implication that the mammalian ectoderm is a mosaic of developmental signals is raised by these studies.

9. Summary

A number of traditional techniques have been used to examine epithelial–mesenchymal interactions; they remain to this day the only means to examine the problems of tissue interactions. Problems at the tissue level of organization require that the approaches to these problems ask questions and use techniques that address the issues of tissue organization and cellular morphology in the context of three-dimensional organization.

Manipulating the embryo is exactly that—manipulating a three-dimensional organized organism that must express its differentiated functional organs and organ systems. When the parallel and correlative studies of cell biological function achieve new insights, they will have to be tested in terms of the cellular response at the tissue level of organization.

ACKNOWLEDGMENTS. My thanks are due to C. Fisher, who participated in the most recent experiments, to V. Shockley, for typing the manuscript, and to the National Institute of Dental Research for grant no. DE 06298.

References

Billingham, R. E., and Silvers, W. K., 1968, Dermo–epidermal interactions and epithelial specificity, in: *Epithelial–Mesenchymal Interactions* (R. Fleischmajer and R. E. Billingham, eds.), pp. 252–263, Williams and Wilkins, Baltimore.

Briggaman, R. A., 1982, Epidermal–dermal interaction in adult skin, *J. Invest. Dermatol.* **79:**21s–24s.

Burwen, S. J., and Pitelka, D. R., 1980, Secretory function of lactating mouse mammary epithelial cells culture on collagen gels, *Exp. Cell Res.* **126:**249–262.

Connelly, T. G., Brinkley, L., and Carlson, B., (eds.), 1981, *Morphogenesis and Pattern Formation,* Raven Press, New York.

Coulombre, J. L., and Coulombre, A. J., 1971, Metaplastic induction of scales and feathers in the corneal anterior epithelium of the chick embryo, *Dev. Biol.* **25:**464–478.

Cummings, E. G., Bringas, P., Grodin, M. S., and Slavkin, H. C., 1981, Epithelial-directed mesenchyme differentiation *in vitro.* A model of murine odontoblast differentiation mediated by quail epithelia, *Differentiation* **20:**1–9.

Cunha, G. R., 1971, Support of normal salivary gland morphogenesis by mesenchyme derived from accessory sexual glands of embryonic mice, *Anat. Rec.* **173:**205–212.

Cunha, G. R., Lung, B., and Reese, B., 1980, Glandular epithelium induction by embryonic mesenchyme in adult bladder epithelium of BALB/c mice, *Invest. Urol.* **17:**302–304.

Dale, B. A., Holbrook, K. A., and Steinart, P. M., 1978, Assembly of stratum corneum basic protein and keratin filaments in macrofibrils, Nature **276:**729–731.

Dale, B. A., Vadlamndi, B., Delap, L. W., and Bernstein, I. A., 1981, Similarities between stratum corneum basic proteins and histidine-rich protein II from newborn rat epidermis, *Biochim. Biophys. Acta* **668:**98–106.

Emerman, J. T., Burwen, S. A., and Pitelka, D. R., 1979, Substrate properties influencing ultrastructural differentiation of mammary epithelial cells in culture, *Tissue Cell* **11:**109–119.

Fallon, J. A., and Caplan, A. I., (eds.), 1983, *Limb Development and Regeneration, Part A*, Alan R. Liss, New York.

Fell, G. B., and Jacobson, W., 1941, The development mechanics and potencies of the undifferentiated mesenchyme of the mandible, *Q. J. Microsc. Sci. N.S.* **82:**563–586.

Fisher, C., and Kollar, E. J., 1984*a*, The abnormal development of the skin of pupoid fetus (pf/pf) mutant mice. I. Light and electronmicroscopic observations, *J. Embryol. Exp. Morphol.* (in press).

Fisher, C., and Kollar, E. J., 1984*b*, The abnormal development of the skin of pupoid fetus (pf/pf) mutant mice. II. Immunofluorescent labelling studies, *J. Embryol. Exp. Morphol.* (in press).

Fischer, C., and Kollar, E. J., 1985, Abnormal skin development in pupoid fetus (pf/pf) mutant mice, *J. Embryol. Exp. Morph.* **87:**47–64.

Fischer, C., Dale, B. A., and Kollar, E. J., 1984, Abnormal keratinization in the pupoid fetus (pf/pf) mutant epidermis, *Dev. Biol.* **102:**290–299.

Garber, B., Kollar, E. J., and Moscona, A. A., 1968, Aggregation *in vivo* of dissociated cells. II Effect ot state of differentiation of cells on feather development in hybrid aggregates of embryonic mouse and chick cells, *J. Exp. Zool.* **168:**452–472.

Green, M. C., (ed.), 1981, *Genetic Variants and Strains of the Laboratory Mouse*, Gustav Fischer Verlag, Stuttgart.

Greenburg, G., and Hay, E. D., 1982, Epithelia suspended in collagen gels can lose polarity and express characteristics of migrating mesenchymal cells, *J. Cell Biol.* **95:**333–339.

Gobstein, C., 1953, Epithelio-mesenchymal specificity in the morphogenesis of mouse submandibular rudiments *in vitro*, *J. Exp. Zool.* **124:**383–414.

Grobstein, C., 1955, Inductive tissue interactions in development, *Adv. Cancer Res.* **4:**187–189.

Guenet, J. L., Salzgeber, B., and Tassin, M. T., 1979, Repeated epilation: A genetic epidermal syndrome in mice, *Heredity* **70:**90–94.

Hardy, M. H., and Goldberg, E. A., 1983, Morphological changes in the basement membrane during some tissue interactions in the integument, *Can. J. Biochem. Cell Biol.* **61:**957–966.

Hay, E. D., 1981, Collagen and embryonic development, in: *Cell Biology of Extracellular Matrix* (E. D. Hay, ed.), pp. 374–409, Plenum Press, New York.

Hay, E. D., 1982, Interaction of embryonic cell surface and cytoskeleton with extracellular matrix, *Amer. J. Anat.* **165:**1–12.

Hay, E. D., 1983, Cell and extracellular matrix: Their organization and mutual dependence, *Mod. Cell Biol.* **2:**509–548.

Holbrook, K. A., Dale, B. A., and Brown, K. S., 1982, Abnormal epidermal keratinization in the repeated epilation mutant mouse, *J. Cell Biol.* **92:**387–397.

Kleinman, H. K., Klebe, R. J., Martin, G. R., 1981, Role of collagenous matrices in the adhesion and growth of cells, *J. Cell Biol.* **88:**473–485.

Koch, W. E., 1967, *In vitro* differentiation of tooth rudiments of embryonic mice. I. Transfilter interaction of embryonic incisor tissues, *J. Exp. Zool.* **165:**155–169.

Kollar, E. J., 1972, The development of the integument: Spatial, temporal and phylogenetic factors, *Am. Zool.* **12:**125–135.

Kollar, E. J., 1980, Tooth development and dental patterning, in: *Morphogenesis and Pattern Formation* (T. G. Connelly, L. L. Brinkley, and B. M. Carlson, eds.), Raven Press, New York.

Kollar, E. J., 1983, Epithelio–mesenchymal interactions in the mammalian integument: Tooth development as a model for instructive induction, in: *Epithelio-Mesenchymal Interactions in Development* (R. H. Sawyer and J. F. Fallon, eds.), Praeger Publishers, New York.

Kollar, E. J., and Fisher, 1980, Tooth induction in chick epithelium: Expression of quiescent genes for enamel synthesis, *J. Science* **207:**993–995.

Kollar, E. J., and Kerley, M. A., 1980, Odontogeneic interactions between isolated enamel organ epithelium and dental papilla cells, *Int. J. Skeletal Res.* **6:**163–170.

Lavker, R. M., and Sun, T-T., 1983, Rapid modulation of keratinocyte differentiation by the external environment, *J. Invest. Dermatol.* **80:**228–237.

Mackenzie, I. C., and Fusenig, N. E., 1983, Regeneration of organized epithelial structure, *J. Invest. Dermatol.* **81:**189s–194s.

Martin, G. R., Rohrbach, D. H., Terranova, V. P., and Liotta, L. A., 1983, Structure, function, and pathology of basement membranes, in: *Connected Tissue Diseases* (B. Wagner, R. Fleishmajor, and N. Kaufman, eds.), pp. 16–30, Williams and Wilkins, Baltimore.

Matoltsy, A. G., 1975, Desmosomes, filaments and keratohyaline granules: Their role in the stabilization and keratinization of the epidermis, *J. Invest. Derm.* **65:**127–142.

McLoughlin, C. B., 1966, The importance of mesenchymal factors in the differentiation of chick epidermis. II. Modification of epidermal differentiation by contact with different types of mesenchyme, *J. Embryol. Exptl. Morphol.* **9:**385.

Meredith, R., 1954, Private communication, *Mouse Newsl.* **31:**25.

Moscona, A. A., 1952, Cell suspensions from organ rudiments of chick embryos, *Exp. Cell Res.* **3:**535–539.

Moscona, A. A., 1960, Patterns and mechanisms of tissue reconstruction from dissociated cells, in: *Developing Cell Systems and Their Control Systems* (D. Rudrick, ed.), *Eighteenth Growth Symposium*, pp. 45–70, Ronald Press, New York.

Osman, H. M., and Ruch, J. V., 1981, Behavior of odontoblasts and basal lamina of trypsin or EDTA-isolated mouse dental papillae in short-term culture, *J. Dent. Res.* **60:**1015–1027.

Pratt, R. M., and Christensen, R. L., 1980, *Current Research Trends in Prenatal Craniofacial Development*, Elsevier/North Holland, New York.

Rawles, M. E., 1963, Tissue interactions in scale and feather development as studied in dermal–epidermal recombinations, *J. Embryol. Exp. Morphol.* **11:**765–789.

Ruch, J. V., and Karcher-Djuricic, V., 1975, On odonogenic tissue interactions, in: *Extracellular Matrix Influences on Gene Expression* (H. C. Slavkin and R. C. Greulich, eds.), pp. 549–554, Academic Press, New York.

Sawyer, R. H., and Fallon, J. F., (eds.), *Epithelio–Mesenchymal Interactions in Development*, Praeger Publishers, New York.

Saxen, L., Ekblom, P., and Thesleff, I., 1980, Mechanisms of morphogenetic cell interactions, in: *Development in Mammals, Vol. 4* (M. H. Johnson, ed.), pp. 1161–1201, Elsevier/North Holland, New York.

Saxen, L., Ekblom, P., and Lehtonen, E., 1981, The kidney as a model system for determination and differentiation, in: *The Biology of Normal Human Growth* (M. Ritzen, A. Aperia, K. Hall, A. Larsson, A. Zetterberg, and R. Zetterstrom, eds.), pp. 117–127, Raven Press, New York.

Schnieke, A., Harbers, K., and Jaenish, E., 1983, Embryonic lethal mutation in mice induced by retro virus insertion into the 1 (I) collagen gene, *Nature* **304:**315–320.

Sengel, P., 1976, *Morphogenesis of Skin*, Cambridge University Press, Cambridge.

Sengel, P., 1983, Epidermal dermal interactions during formation of skin and cutaneous appendages, in: *Biochemistry and Physiology of the Skin* (L. A. Goldsmith, ed.), pp. 102–131, Oxford University Press, New York.

Slavkin, H. C., Zeichner-David, M., MacDougall, M., Bringas, P., Jr., Bessem, C., and Honig, L. S., 1982, Antibodies to murine anmelogenins: Localization of enamel proteins during tooth organ development *in vitro*, *Differentiation* **23:**73–82.

Svajger, A., Levak-Svajger, B., Kostovic-Knezevic, L., and Bradamante, Z., 1981, Morphogenetic behaviour of the rat embryonic ectoderm as a renal homograft, *J. Embryol. Exp. Morphol.* **65:**243–267.

Steinart, P. M., Idler, W. W., and Zimmerman, S. B., 1976, Self assembly of bovine epidermal keration filaments *in vitro, J. Mol. Biol.* **108:**547–567.

Steinart, P. M., Cantieri, J. S., Teller, D. C., Lonsdale-Eccles, J. D., and Dale, B. A., 1981, Characterization of a class of cationic proteins that specifically interact with intermediate filaments, *Proc. Natl. Acad. Sci. USA* **78:**4097–4101.

Thesleff, I., Lehtonen, E., and Saxen, L., 1978, Basement membrane formation in transfilter tooth cultures and its relation to ondotoblast differentiation *Differentiation* **10:**71–79.

Thesleff, I., Barrach, H. J., Foidart, J. M., Vaheri, A., Pratt, R. M., and Martin, G. R., 1981, Changes in the distribution of type IV collagen, laminin, proteoglycan and fibronectin during mouse tooth development, *Dev. Biol.* **81:**182–192.

Wessels, N. K., 1977, *Tissue Interactions and Development*, Benjamin/Cummings, Menlo Park, Calif.

Chapter 11

Tissue Combination and Organ Culture Studies in the Development of the Embryonic Mammary Gland

KLAUS KRATOCHWIL

1. Introduction

The mammary gland is a typical representative of the epithelial–mesenchymal organs; and since the pioneering studies of Grobstein and associates we know of the great importance of tissue interaction in the development of these organs. Two special features distinguish the mammary gland from otherwise comparable organs. First, while the development of most epithelial–mesenchymal organs is more or less completed during the embryonic or fetal period, the adult mammary gland still undergoes major developmental changes, suggesting that morphogenetic interaction, more amenable to study in the embryo, may equally take place between epithelium and stroma in the adult organ. Second, unlike the situation found in most other glands and glandlike organs, growth and morphogenesis of the mammary gland are clearly regulated by hormones. Because of the established role of epithelial–mesenchymal interaction in the morphogenesis of other organs, one may speculate about a possible interplay between hormone action and tissue interaction in the mammary gland.

This brief chapter concentrates on the experimental analysis of developmental tissue interaction in the *embryonic* gland only, with particular emphasis on the mutual dependence of hormone action and tissue interaction, as revealed in the mouse mammary rudiment.

2. Formation of Mammary Buds

The epithelial rudiments of the mammary gland are formed by the epidermis at a relatively advanced stage of embryonic development, but still before

KLAUS KRATOCHWIL • Institute of Molecular Biology, Austrian Academy of Sciences, A-5020 Salzburg, Austria.

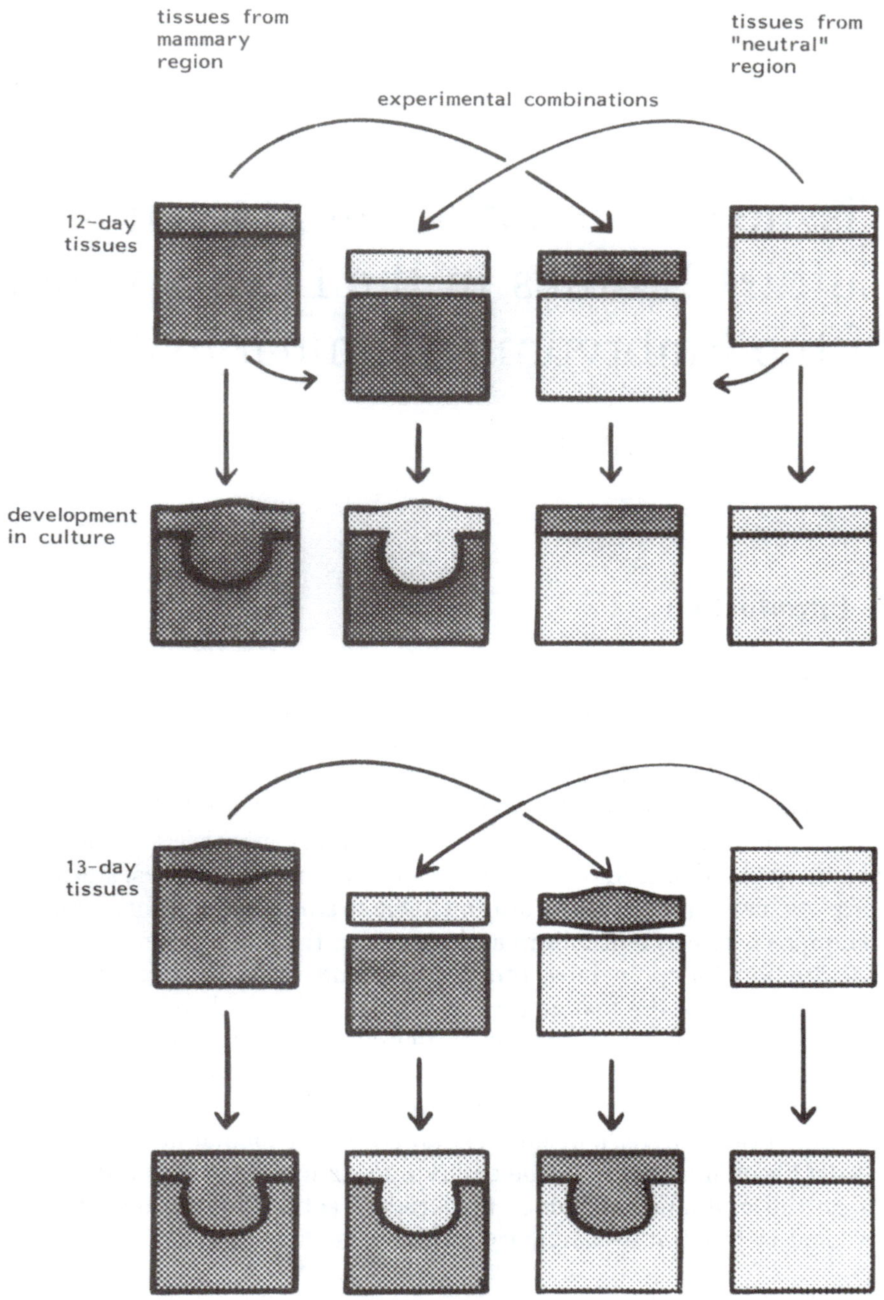

Figure 1. Schematic representation of Propper's (1968) experiments defining the roles of mesenchyme and of epithelium in initiating mammary development. Heavy shading: Epidermis and underlying mesenchyme of the (future) mammary region; light shading: the same tissues from a "neutral," nonmammary region (head or neck). For experimental tissue combinations, epidermis and mesenchyme are first separated after incubation in a 0.5% solution of crude trypsin (5–20 min at 4°C; trypsin action is then stopped by the addition of serum or—as in these experiments—of embryo extract). After reciprocal combination, as indicated in the scheme, the tissues were cultured for 3–6 days on 0.5% agar enriched with chick embryo extract, using the medium of Wolff and Haffen (1952). When the experiment was done on day 12, only the combinations containing

other skin structures (vibrissae, hair) become apparent. In most mammals, formation of the individual gland buds is preceded by the development of a milkline (mammary crest, or *Milchleiste*), a band of thickened epidermis stretching from the axillary to the inguinal region. This milk line is subsequently reduced in length, or its fragments, giving rise to the definitive epithelial mammary buds. In the mouse embryo, however, there is no evidence for a milkline, and the five pairs of the gland form as individual buds (only buds 2 and 3, and 4 and 5, respectively, may be linked for a short period by thickened epidermis).

From the experiments of Propper (1968) in the rabbit embryo it appears that mammary gland development is initiated in the mesenchyme (Fig. 1). In the rabbit, the mammary bands form at 13 days gestation. When 12-day epidermis from the prospective mammary region was combined with "neutral" mesoderm, no glands developed. On the other hand, if the mesoderm was taken from the mammary region of 12-day embryos and associated with 12-day "neutral" epidermis, the combination gave rise to mammary buds. From day 13 onward, epidermis of the mammary region develops mammary bands and buds, also in association with "neutral" mesenchyme. Propper (1968) concludes that epithelial mammary development is induced in the epidermis by the underlying mesenchyme between days 12 and 13; after this event, the epithelium is capable of proceeding with mammary development even in association with nonmammary mesenchyme.

There are indications that the milk line, or the individual mammary buds of the mouse, do not arise by locally elevated proliferative activity in the epidermis (Balinsky, 1949, also our own unpublished results). It has therefore been postulated that the band forms by migration of epidermal cells (Balinsky, 1949), and the characteristic shape of epidermal cells seen in scanning electron micrographs of the rabbit milk line was believed to support this hypothesis (Propper, 1978). From the tissue combination experiments cited above, it follows that this epidermal streaming is in some way caused by the underlying mesenchyme. This may explain the puzzling results obtained by Propper (1969, 1972), who placed epidermis of chick or duck embryos on mammary mesenchyme of the rabbit. In some of these associations, the bird epithelium formed structures reminiscent of mammary buds and even primary sprouts. (Unfortunately, no quail epidermis was used, which would have allowed unquestionable identification of the species origin of the epithelium.) Even if we

---

mesenchyme of the mammary region developed mammary buds in culture (irrespective of the source of the epithelium), while 12-day epithelium of the mammary region could not form mammary structures when placed on "neutral" mesenchyme. Apparently, the mesenchyme is the first tissue to possess the cues for mammary development; it is also capable of imposing its program on neutral epidermis ("induction"). When the same experiment was done on day 13, the epidermis of the mammary region has become independent of regionally specific mesenchyme and undergoes mammary development in association with "neutral" mesenchyme, as well. "Induction" of the epithelium is thus thought to have occurred between days 12 and 13. Nevertheless, "mammary" mesenchyme of 13-day and even of 14-day (not shown) embryos remains capable of inducing neutral epidermis for mammary development.

are not ready to accept avian mammary glands, the experiment can be considered to indicate that the mesenchyme of the prospective mammary region causes epidermal movement, eventually resulting in epithelial ingrowth, by "signals" that are understood across class barriers, as are other inductive signals.

The limited experimental evidence thus suggests that formation of the mammary primordia is induced by the mesenchyme—a situation comparable to the induction of other skin structures, such as feathers and scales in birds, or hair in mammals (for a review, see Sengel, 1976). As in all other inductive systems, we do not know the type of "signal" that causes small groups of epidermal cells to enter a developmental pathway different from their neighbors. One may speculate that this (mesenchyme-induced) epithelial movement to, and compaction in, the milkline, accompanied by changes in cell shape and association, could constitute the trigger for the developmental switch. After this event, at least, mammary development is no longer dependent on association with mammary mesenchyme.

3. Resting Phase and Early Morphogenesis

Once the mammary buds are formed, they go through a prolonged period of conspicuous morphogenetic inactivity—a feature distinguishing the mammary gland from all other epithelial–mesenchymal organs. In the mouse, the buds form in the night between day 11 and day 12, but outgrowth of the primary sprout does not occur before late on day 16. The first half of this "resting phase" is characterized by low mitotic activity (Balinsky, 1950) and a 30-hr period of almost no DNA synthesis by cells of the epithelial bud (unpublished). DNA synthesis is resumed on day 14 and remains high throughout the rest of the fetal period. At term (day 19), each gland is represented by a modestly branched system of slim ducts without any alveoli. Although the early morphogenesis of the gland differs slightly between species, the "resting phase" seems to be a common feature.

Since growth and development of the (adult) mammary gland is under strict hormonal control, it was assumed that this transition from resting phase to actively branching morphogenesis reflects changes in the fetal hormonal environment. This idea had to be dismissed after finding that explanted mammary rudiments followed precisely the same temporal pattern in organ culture (e.g., explanted 12-day glands started to branch after 5 days *in vitro,* 14-day glands after only 3 days, and so on; Kratochwil, 1969). Apparently, the gland follows an intrinsic developmental program; by experimental combination of epithelia of 14–16-day glands with 12-day mesenchyme ("heterochronic" combinations), it was found that the "clock" for outgrowth resides in the epithelium (Kratochwil, 1969).

At outgrowth, the primary sprout breaks through the few layers of mesenchymal cells that had arranged themselves around the bud during the resting phase, and invades deeper mesenchyme that will later form the fat pad of the gland. Experiments in a number of epithelial–mesenchymal organs had shown before that interaction with the surrounding mesenchyme is essential for epithelial morphogenesis, and this tissue interdependence was most impressingly demonstrated in the mammary gland. The typical ductal morphology, with "monopodial" and widely spaced branching, is altered when the epithelium is forced to grow in mesenchyme from other sources. When 17-day epithelium (the stage of the first branch of the outgrowing sprout) was associated with the nonadipose, "fibroblastic" mesenchyme that surrounds the mammary bud of the resting phase, branching occurred much more frequently with ducts more closely packed, leading to a morphology called "ductal hyperplasia" (Sakakura *et al.*, 1982). The most dramatic effect, however, was seen when fetal mammary epithelium was combined with mesenchyme of the 13-day salivary gland: its morphogenesis became salivarylike with terminal adenomeres and a dichotomous branching pattern (Kratochwil, 1969) (see Fig. 2). Interestingly, altered morphogenesis in either "mammary" (i.e., non-fat pad) or salivary mesenchyme does not affect cytodifferentiation of the epithelium Such experimental tissue combinations grafted under the kidney capsule of adult syngeneic females were found to produce milk when the host had given birth and had begun to lactate (Sakakura *et al.*, 1976, 1982).

One of the pecularities of mammary gland development is that it is not completed within the fetal period. In a multiparous female, the adult gland undergoes several cycles of ductal outgrowth and branching, alveolar morphogenesis, and finally extensive degeneration again. Although this development is controlled by reproductive hormones, it was not unreasonable to assume that the morphogenetic behavior of the adult gland epithelium is equally or similarily influenced by the mesenchymal (stromal) component. It has, in fact, been shown that the mammary gland fat pad is an ideal substrate for transplants for adult gland epithelium or of hyperplastic alveolar nodules (DeOme *et al.*, 1959). Even dispersed mammary epithelial cells, after being grown as monolayer *in vitro*, reorganized to histotypical ductal structures when reimplanted into cleared fat pads (Daniel and DeOme, 1965). The fat pad also permits development of implanted fatal mammary epithelia but not of epithelia from other organ rudiments (Sakakura *et al.*, 1979*a*).

On the other hand, adult mammary epithelium also retains its capacity to respond to mesenchymal stroma other than the fat pad (Sakakura *et al.*, 1979*b*). Pieces of fetal mammary or of salivary mesenchyme grafted into the fat pad were invaded by the host's gland epithelium. Either mesenchyme appeared to stimulate epithelial proliferation and caused the gland to branch more densely and less regularly than within adult stroma. In salivary mesenchyme, epithelial branching was again salivarylike and extremely dense, so that the epithelium eventually filled the entire mesenchymal implant, assuming the appearance of a hyperplastic alveolar nodule. It was subsequently observed that mammary cancer in response to virus or chemical induction arose at high incidence from these "duct–alveolus nodules" (Sakakura *et al.*, 1981).

The adult mammary gland does not respond to just any kind of fetal mesenchyme, as pulmonary, pancreatic, or metanephric mesenchymal implants

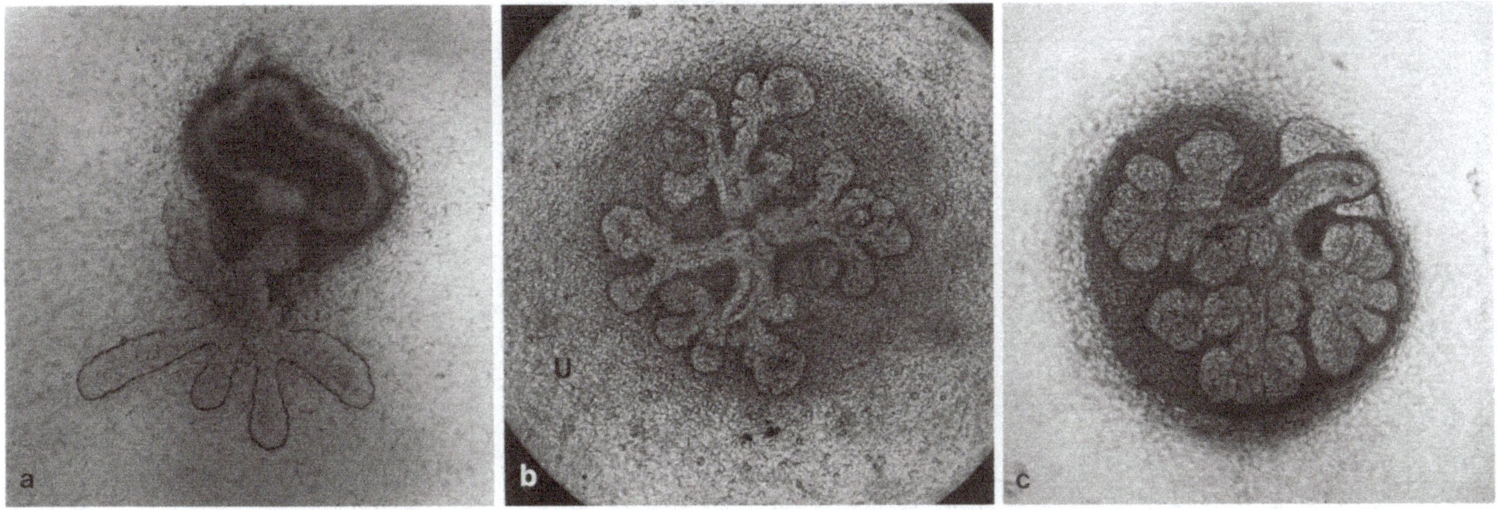

Figure 2. Branching morphogenesis of mammary epithelium is strongly influenced by its associated mesenchyme. (a) Normal outgrowth of the gland explanted with its own mesenchyme. Living culture of a 13-day rudiment after 6 days *in vitro*. (b) Morphology of mammary epithelium in experimental association with mesenchyme of the submandibular salivary gland (6 days in culture). Note that the branching pattern of the epithelium resembles that of a salivary gland (c), especially in the cleavage of terminal bulbs reminis- cent of salivary gland adenomeres. (c) Explanted 13-day salivary gland after 2 days *in vitro*. (From Kratochwil, 1969.)

were not invaded. Also, during lactation, mammary epithelium does not grow into mesenchymal implants (Sakakura *et al.*, 1979*b*), indicating that its morphogenesis is governed both by hormones and by epithelial–stromal interaction.

4. Hormone Response and Hormone Responsiveness

The mammary gland is a classic target organ for various hormones controlling its growth, morphogenesis, and cytodifferentiation. Since the experiments just discussed, as well as experience from other organ systems, had demonstrated the important role of epithelial–mesenchymal interaction, at least for growth and morphogenesis, it seemed worth investigating the interplay between hormone action and tissue interaction. Some insight has been obtained from the androgen response of the mammary gland in mouse embryos.

4.1. The Androgen Response of the Embryonic Mouse Mammary Rudiment

Unlike the situation found in most mammals, mammary gland development in the mouse shows a characteristic sexual dimorphism already during fetal life. Initially, gland buds from in both sexes (between day 11 and 12), and their development is identical until late on day 13. On day 14, however, the epithelial gland buds of male fetuses become surrounded by a conspicuous condensation of mesenchymal cells (Figs. 3 and 4a). A few hours later the "stalk" of the gland is stretched and finally ruptures, thereby separating the mammary epithelium proper from the epidermis. During this process, most or all cells of the gland epithelium become necrotic; whatever may remain of the anlage forms a rudimentary duct system lacking an outlet duct and a nipple (Raynaud and Raynaud, 1953). No such destruction is seen in gland rudiments of female littermates whose mammary buds persist and begin sprouting into the future fat pad tissue late on day 16.

It had been shown that the mammary rudiments of female fetuses are destroyed by the action of testicular hormones. Injection of testosterone into pregnant animals caused male-type regression of the glands in female fetuses (Raynaud, 1947) and conversely, antiandrogenic compounds (Neumann and Elger, 1966) or destruction of the gonads *in utero* (Raynaud and Frilley, 1947) permitted mammary gland survival in male fetuses. It proved of great advantage for further experimentation, that all the processes taking place in the male gland could be reproduced *in vitro* in explanted mammary rudiments by exposing them to low, physiological (1 nM) concentrations of testosterone. Glands taken from either male or female 13-day embryos were equally susceptible to the hormone and they were equally capable of female-type (i.e., undisturbed) development in its absence (Kratochwil, 1971). These experiments established that testosterone acts directly on the mammary gland and not *via* the pituitary

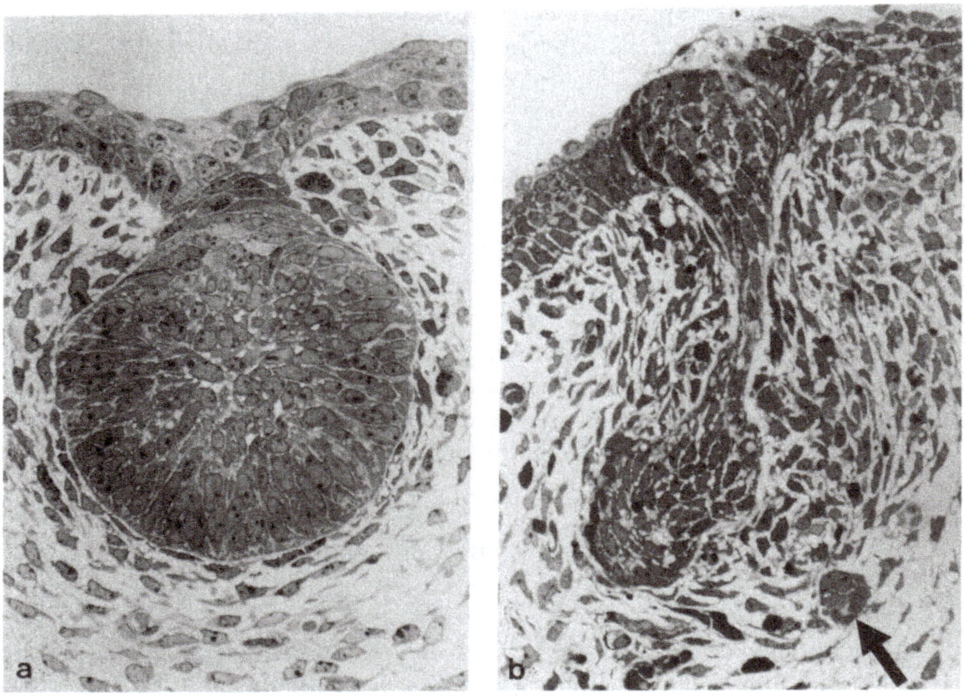

Figure 3. Histological sections of 14-day mammary rudiments of a female (a) and a male (b) embryo. In the male, the gland is at an early stage of the androgen-induced destruction. The epithelial bud has become surrounded by dense mesenchyme and is still connected to the epidermis by a narrow "stalk." At the lower right (arrow) the section passes again through the mammary epithelium, indicating its grossly distorted shape. A few hours later, the lower portion of the gland epithelium will break free from the epidermis and become necrotic.

as had been suggested before. They also showed that testosterone is the only agent responsible for the sexual dimorphism (and not, for instance, a specific "female" hormonal environment enabling the gland's further development) and that neither the genetic sex type of the tissue nor its preexposure to a male or female environment up to day 13 had any influence on the gland's developmental potential or its responsiveness to the hormone.

4.2. Development of Androgen Responsiveness

Destruction of the gland anlage in male fetuses occurs in the second half of day 14, about 40–48 hr after differentiation of the morphological testis. The time of the reaction *in vivo* could therefore reflect either the onset of testosterone release by the gonad or the acquisition of hormone responsiveness by the mammary gland. This question could conveniently be answered by exposing pre-day 14 gland rudiments to testosterone *in vitro*. We found that although the hormone eventually causes destruction of the mammary buds it does not affect their initial development *in vitro* (Kratochwil, 1977). In explanted 11-day skin, mammary anlagen are formed even in the presence of high levels of testosterone, the buds persist for almost 2 days and only then, at a stage corresponding to the 14-day embryo, are they rapidly destroyed. By delaying the addition of testosterone to cultured 11-day explants we could determine that the gland does not become responsive to the hormone before the second half of day 13 (Fig. 4b). Accordingly, explants of 11-day skin and of 12-day or 13-day glands respond to the hormone after different exposure times but at the same developmental stage. Unexpectedly, glands taken from 15-day or 16-day (female) embryos are no longer affected by testosterone (Fig. 4b); we have therefore concluded that the mammary gland goes through a relatively short testosterone-responsive "window" (Kratochwil, 1977).

The fact that mammary buds that had formed *in vitro* from 11-day (and even from 10-day) skin explants were eventually destroyed by testosterone showed that the development of androgen responsiveness had occurred in culture and independently of the embryonic environment. Similarly, 11-day glands kept *in vitro* for 4 or more days before addition of the hormone were as unresponsive as glands taken from 15-day or 16-day embryos (Fig. 4b). Since both acquisition and loss of androgen responsiveness occurred in explanted rudiments, the hormone-responsive "window" must reflect an entirely intrinsic developmental program of the gland itself.

We later found that the appearance and development of androgen receptors—a necessary though not the only prerequisite for androgen responsiveness—correlates well with the acquisition of androgen responsiveness (Fig. 4c). The first receptors are demonstrable in 12-day buds; their number increases about 20-fold to the responsive stage (day 14). Then, when the gland becomes insensitive to testosterone again (on day 15), however, receptors are still present in the tissue and at even higher numbers than during the responsive window (Wasner *et al.*, 1983). These findings suggest that the question to the development of androgen responsiveness can be formulated as the question of the control of androgen receptor development, whereas loss of responsiveness cannot be explained at the receptor level.

4.3. Identification of the Target Cells for Testosterone

Although testosterone visibly affects both tissues of the mammary rudiment (the epithelium becoming necrotic, the mesenchyme forming the characteristic condensation), we considered the possibility that the hormone acted directly on only one of them and that the effect on the other was mediated by tissue interaction. The availability of the androgen-insensitive *Tfm*-mutant (Lyon and Hawkes, 1970) enabled us to prepare chimaeric epithelial–mesenchymal tissue combinations in which only one tissue was of mutant origin and therefore insensitive to testosterone. These combination explants were

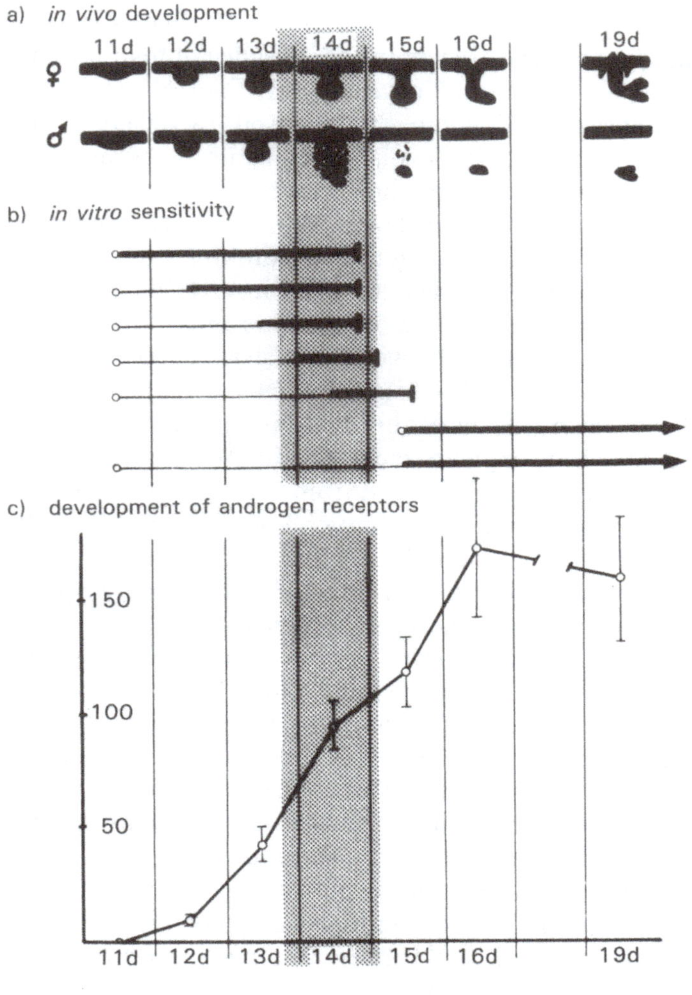

Figure 4. Schematic representation of the experiments establishing the androgen-responsive "window" in the embryonic development of the mammary gland, correlating with the development of androgen receptors. (a) Morphological stages of mammary gland development in female and male fetuses. Sexual dimorphism becomes apparent on day 14, about 36–48 hr after morphological differentiation of the gonad (testis). (b) Development of explanted 11-day rudiments exposed to testosterone *in vitro*. *Thin line:* culture in the absence; *heavy line:* culture in the presence of the hormone. When testosterone was present from the beginning (first line), formation and early development of the gland buds was not affected. After 3 days *in vitro*, however, at a stage corresponding to a 14-day embryo, the gland rudiments were destroyed. The time of hormone addition during the first 48 hr *in vitro* was found to be irrelevant for the time of the response (destruction), indicating that the tissues have not yet become responsive. The first groups of explants showing delayed reaction were those having received the hormone at a stage corresponding to late day 13 (4th and 5th lines), indicating the onset of the sensitive period. Explants of 15-day (female) glands were no longer affected by testosterone (6th line) showing that the organ has again become insensitive to the hormone. The short period of the testosterone-responsive "window," as determined by these experiments, is indicated by the shaded area. Last line: Not only glands from 15-day embryos, but also explants of earlier stages cultured for the corresponding period *in vitro*, lose their sen-

exposed to testosterone *in vitro*, and the results clearly showed that association of mutant epithelium with normal mesenchyme were responsive to the hormones whereas no reaction was seen in reciprocal combinations (Kratochwil and Schwartz, 1976; Drews and Drews, 1977) (see Fig. 5a). It was thus established that the hormone sensitivity of these experimental tissue combinations resides in the mesenchyme alone and, therefore, that this tissue is the target for testosterone. This system also provided the first evidence for hormone action *via* tissue interaction, as it shows that necrosis of mammary epithelium in male fetuses does not reflect (direct) hormone-induced cell death but rather hormone-induced cell killing by the surrounding mesenchyme.

Although the mesenchyme of the mammary gland was identified as the sole target tissue for testosterone, it was soon found that it does not respond to the hormone (i.e., it does not condense) in the absence of mammary epithelium. Even when the latter was replaced by epithelia of other organ rudiments, such as the salivary gland, the pancreas, or the lung, neither mesenchymal condensation nor epithelial necrosis was observed. When mammary and other epithelia were placed on the same piece of mesenchyme, testosterone caused mesenchymal condensations strictly confined to the periphery of mammary epithelium, even when the two epithelia of different origin had fused. By contrast, mammary epithelia of other species (rat, rabbit) became surrounded by mesenchymal condensation and were eventually destroyed (Dürnberger and Kratochwil, 1980). It should be noted that the embryonic rabbit mammary gland does not show any response to testosterone; nevertheless, its epithelium is "recognized" by mouse mammary mesenchyme.

The testosterone-stimulated cells of mammary mesenchyme thus show two interesting properties: (1) they are cytotoxic (destroying mammary epithelium) and (2) they recognize their target tissue. Neither the adjacent epidermis nor any other epithelium in experimental combinations is attacked. These properties are characteristics of immunocompetent cells; therefore we next asked the question as to the origin of these mesenchymal cells. Two alternatives were considered: (1) the mesenchymal condensation could form by migration of specialized cells from distant sites that become attracted by the mammary bud. Alternatively, the mesenchymal cells lining the mammary bud are somehow influenced ("induced") by the epithelium, thereby acquiring the capacity to respond to the hormone. This question could be answered again by tissue combinations utilizing the *Tfm* mutant; this time, gland epithelia carry-

---

sitivity to testosterone, showing that both acquisition (top lines) and loss of androgen responsiveness occur *in vitro*, apparently governed by an intrinsic developmental program of the gland itself. (Data from Kratochwil, 1977.) (c) Development of specific [^{3}H]testosterone-binding (with the characteristics of an androgen receptor) in the mammary gland throughout the embryonic and fetal period. Freshly dissected glands were incubated with the ligand *in vitro* at a low (0.6 nM) and high (10 nM) concentration. The number of binding sites was obtained by calibrating the graph with the value obtained from Scatchard plots of 14-day rudiments. Note that the number of testosterone-binding sites increases markedly with the acquisition of testosterone responsiveness, whereas the loss of hormone sensitivity at day 15 is not at all correlated with a loss or even decrease of receptors. (Data from Wasner *et al.*, 1983.)

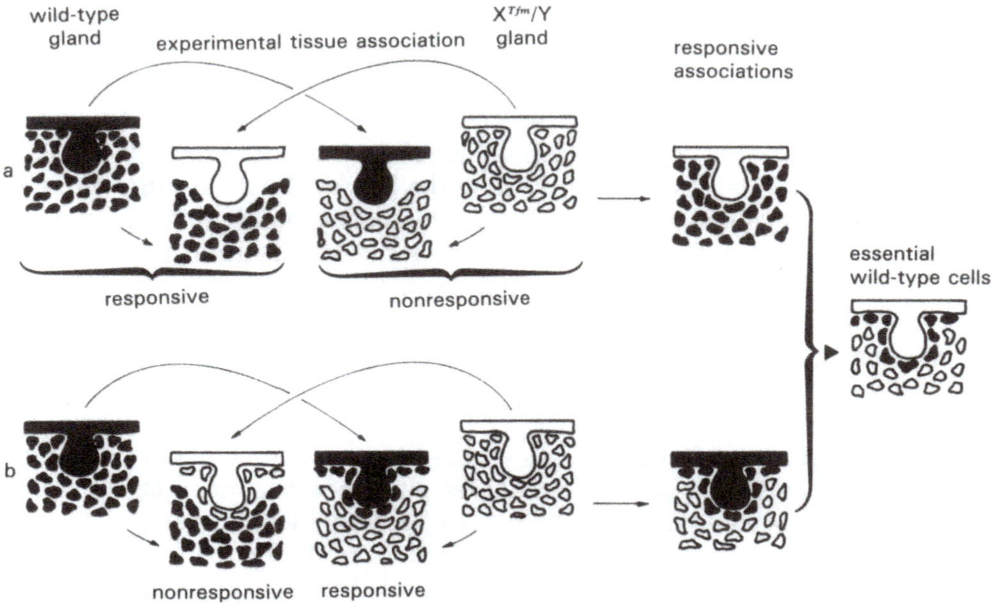

Figure 5. Schematic representation of the tissue combination experiments that led to the identification of the target cells for testosterone in the mammary gland. Use was made of an X-linked mutation *Tfm* (testicular feminization) that renders affected hemizygous (X^{Tfm}/Y) carriers entirely insensitive to testosterone (tissues shown in white). Wild-type (with respect to the *Tfm* locus) tissues are shown in black. (a) In the first type of combination experiments, epithelium and mesenchyme were cleanly separated with the aid of trypsin (as in Fig. 1) and then recombined reciprocally. Subsequently cultured in medium with 10^{-7} M testosterone, only combinations containing wild-type mesenchyme were found responsive to the hormone, whereas the genotype of the epithelium was of no relevance. (b) For the second series of combination experiments, the epidermis with the mammary buds was mechanically "peeled" from the underlying mesenchyme (without applying trypsin) so that a few mesenchymal cells remained adhering to the epithelium. After reciprocal combination, the adepithelial mesenchymal cells differed in their genotype from the rest (much larger than shown in the scheme). The results showed that these few cells are sufficient for the reaction and that no mesenchymal cells migrate from distant sites to the gland bud. (Right) The responsive associations of the two experiments rule out the epithelium (a) and the more distant mesenchyme (b) as possible target cells for the hormone, leaving the adepithelial mesenchymal cells as the only population responsible for the hormone reaction (shown is a situation that in fact cannot be created experimentally). Compare this scheme with Figure 6. (Data for experiment (a) from Kratochwil and Schwartz, 1976, and for (b) from Dürnberger and Kratochwil, 1980.)

ing a few adhering mesenchymal cells were associated with a large piece of mesenchyme of the opposite genotype. The mesenchyme of these explants was thus heterogeneous with respect to its sensitivity to testosterone, specifically, the cells closest to the epithelium differed from the rest of the more distant mesenchyme. (The previous experiments had already shown that the genotype of the epithelium was of no relevance for the response.) After testosterone application, it was found that the response of these chimeric explants depended entirely and exclusively on the few mesenchymal cells at the tissue interface (Fig. 5b). Since the same result was obtained when 12-day tissues were combined we could conclude that all the mesenchymal cells required for the reaction on day 14 were already situated at the epithelial periphery as early as day 12 (Dürnberger and Kratochwil, 1980). The same conclusion was drawn from the histological appearance of the response of explanted glands taken from X^{Tfm}/X^{+} heterozygous animals. Because of random but clonally inherited X-chromosome inactivation (Lyon, 1961), their mesenchyme—as any other tissue—is composed of androgen-sensitive and androgen-insensitive cell clones. Oriented migration to the mammary bud in response to testosterone would have resulted in local sorting out of these clones and in a complete, though perhaps thinner mesenchymal envelope around the bud. Instead, the patchy condensation seen near the gland epithelium argued against cell migration and for local compaction of mesenchymal cells in the immediate vicinity of the gland (Dürnberger and Kratochwil, 1980).

4.4. Induction of Androgen Receptor Formation by Mammary Epithelium

The experiments discussed in Section 4.3 suggested that mammary epithelium plays an essential role in the mesenchyme's response to testosterone, and not by attracting but rather by influencing mesenchymal cells. In a series of combinations involving tissues from 10–14-day embryos it was also found that the association between mammary epithelium and mesenchyme must be done on or before day 12 to permit a mesenchymal response on day 14. Later association, e.g., on day 13 or 14, still enabled typical development of the gland, but the explants were refractory to testosterone (unpublished). It has thus become unlikely that the role of the epithelium is merely to provide a target for the mesenchymal "attack"; instead, the result suggested that the presence of mammary epithelium is required for the development of mesenchymal testosterone responsiveness between days 12 and 14.

The role played by mammary epithelium has finally become clear through [^{3}H]testosterone autoradiographs (Heuberger *et al.*, 1982). First, it was seen that only mesenchymal cells, but not the epithelium, retained the labeled hormone, corroborating our earlier conclusion drawn from the result of combinations between *Tfm*-mutant and wild-type tissues. Labeling of the mesenchyme, however, was not uniform but restricted to a few layers of cells surrounding each epithelial gland bud (Fig. 6). Estimates indicate that only 3000 mesenchymal cells in each 14-day gland rudiment possess high-affinity binding sites for [^{3}H]testosterone. The earliest of these cells were seen in sections of 12-day glands and were situated immediately at the epithelial surface; testosterone-binding cells persist at the same location at least until birth.

The question of whether mammary epithelium in fact induces neighboring mesenchymal cells to form androgen receptors was finally tested directly in experimental tissue combinations. Trypsin-isolated gland epithelia of 12-day

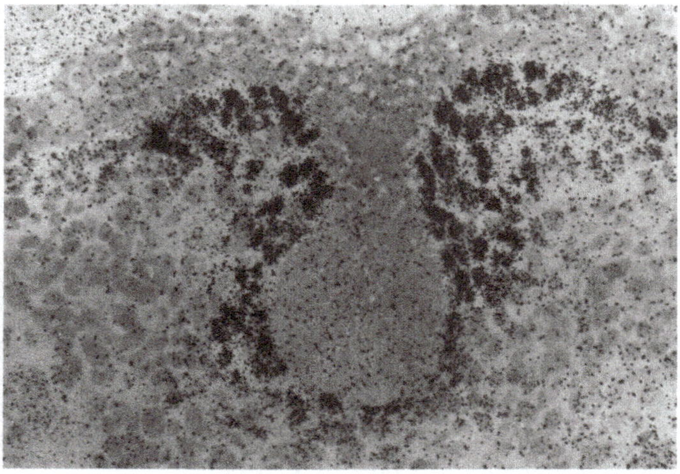

Figure 6. Distribution of [^{3}H]dihydrotestosterone-binding cells in an autoradiograph of a mammary gland from a 14-day female embryo. Immediately after dissection, the living tissues were incubated for 1 hr at 37°C in 0.6 nM [^{3}H]dihydrotestosterone in medium. After washing, the tissues were frozen, sectioned, and placed on emulsion-coated slides in the dark (thaw-mount autoradiographic method of Stumpf and Sar, 1975.) After an exposure time of 7 months, the sections were developed and stained. Note that hormone binding is restricted to the mesenchymal cells immediately adjacent to the mammary bud—the same cells that had been identified as the target cells for the hormone in the tissue combination experiments shown in Figure 5. (From Heuberger *et al.*, 1982.)

embryos (without any adhering mesenchymal cells) were placed on 12-day mesenchyme of the mammary region and subsequently grown *in vitro* for 3 days. During culture, the mesenchyme completely surrounds the mammary epithelia. Autoradiographs of these combination explants showed that each piece of mammary epithelium was surrounded by a halo of [^{3}H]testosterone-retaining mesenchymal cells, the rest of the mesenchyme remaining free of label (Fig. 7). By contrast, epithelia from other sources, such as salivary gland or pancreas, did not cause the appearance of [^{3}H]testosterone-binding sites in mammary mesenchyme. The induction of receptors thus showed the same organ specificity as was previously seen when the hormone sensitivity of mammary mesenchyme was tested in combination with various epithelia. It is therefore reasonable to conclude that mammary epithelium governs the development of mesenchymal androgen responsiveness by inducing the formation of androgen receptors.

4.5. Interrelationship between Hormone Action and Tissue Interaction

The experiments on the mammary rudiment have established the requirement for interaction between epithelium and mesenchyme in at least two steps

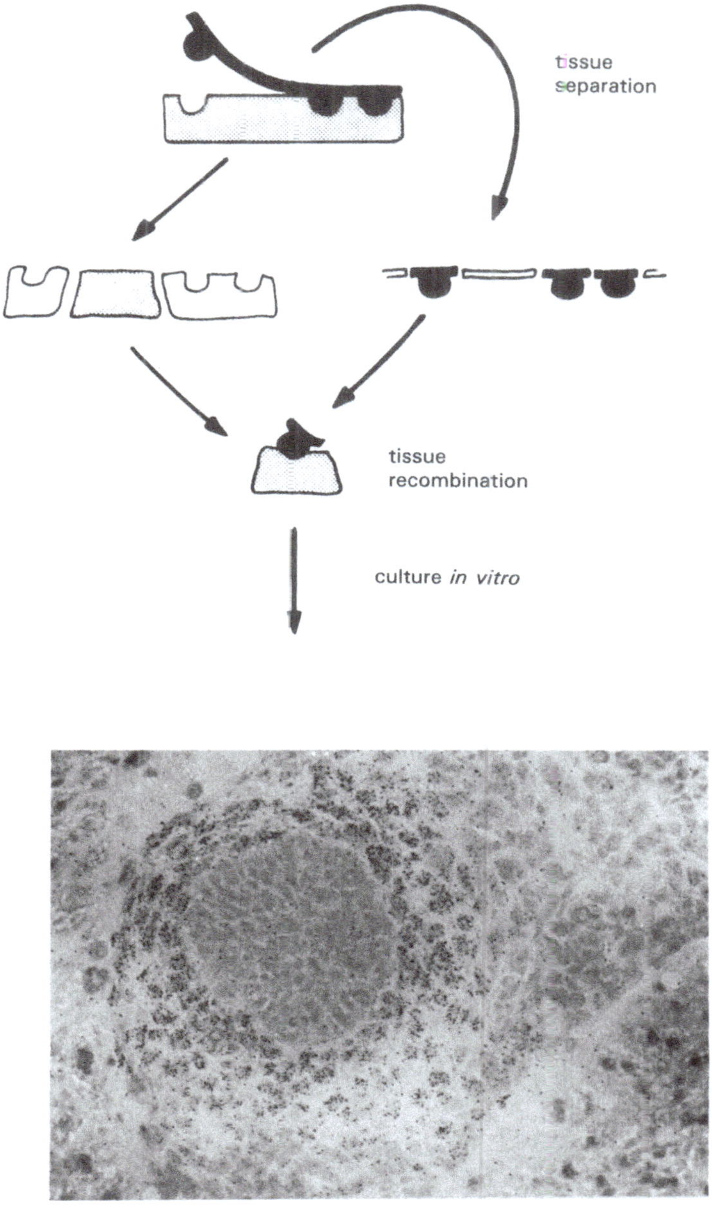

Figure 7. Epithelial–mesenchymal combinations revealing that mammary epithelium in fact induces the formation of androgen receptors in surrounding mesenchymal cells. At day 12, trypsin-isolated (i.e., mesenchyme-free) mammary epithelia were associated with mesenchyme of the mammary region, which, however, had not been in contact with a mammary bud before. After 3 days in culture, autoradiographs were prepared as in Figure 6. Each epithelium is surrounded by hormone-binding mesenchymal cells. Experiments measuring the appearance of [^{3}H]testosterone-binding sites by liquid scintillation counting established that the epithelia in fact cause formation of new binding sites in the explants and not perhaps migration of preexisting, receptor-containing mesenchymal cells to the epithelium.

in the androgen response of the gland (Fig. 8). The first interaction takes place when the early mammary bud causes development of testosterone responsiveness in the surrounding mesenchyme by inducing it to form androgen receptors. Then, during the androgen response on day 14, this mesenchyme mediates testosterone action on the gland by effecting the partial or complete destruction of the epithelium. The study of a rather specialized process, occurring in a few mammalian species only, has thus revealed the close interrelation between hormone action and cell interaction. On the one hand, hormone action depends on the interaction between cells and tissues, while on the other, it influences it.

Presumably similar situations will be found in other systems. Cunha and Lung (1978) have already shown—by the same experimental approach, using *Tfm* and wild-type tissue combinations—that the effect of testosterone on the development of prostatic epithelium is equally mediated by the mesenchyme. Therefore, this type of indirect hormone action is not limited to instances of tissue destruction. The prostate is a particularly interesting example illustrating the interplay between hormone action and tissue interaction. The mes-

Figure 8. Schematic illustration of our current concept of epithelial–mesenchymal tissue interaction involved in the development of androgen responsiveness and in the androgen response of the embryonic mammary gland. At an early stage, apparently beginning with the formation of the primordial bud, mammary epithelium acts on neighboring mesenchyme cells and causes them to form androgen receptors. Since the epithelial "signal" is of relatively short range, cells with receptors are only found in the vicinity of the gland bud. Later, at 14 days, testosterone acts on this mesenchymal subpopulation which then mediates indirectly the effect of the hormone on the epithelium of the gland, causing its destruction by a mechanism that remains unknown.

enchyme is the target for testosterone only during fetal morphogenesis, whereas at later stages the hormone acts directly on the epithelium (Cunha *et al.*, 1980). Apparently, when testosterone controls organogenesis it acts on the tissue that plays the dominant role in morphogenetic epithelial–mesenchymal interaction.

The regression of the Mullerian ducts may possibly constitute another example of hormone-induced and mesenchyme-mediated destruction of an epithelium. Although the target tissue for the Mullerian-inhibiting substance (a glycoprotein hormone) has not yet been identified, the striking similarity of the morphological processes described in Mullerian duct regression (Trelstad *et al.*, 1982) and in mammary gland destruction (P. Schwartz and K. Kratochwil, in preparation) strongly suggests a similar mechanism of action.

The regulation of receptor development by tissue interaction may be the more important aspect, as it determines the specific hormone sensitivity of an organ. Hormone responsiveness is not only required for the proper function of several organs, it also permits control of growth and cell proliferation. So far, only the estrogen receptor of the embryonic mammary gland has definitely been shown to be inducible in the same way as its androgen receptor (in preparation), but steroid autoradiographs of other embryonic organs are highly suggestive (Stumpf *et al.*, 1980; Gasc and Stumpf, 1981*a*,*b*). If dependence of hormone receptor development on specific tissue interaction would prove to be a more general phenomenon (like the development of other differentiative functions), one may speculate further that continued receptor synthesis—and thereby maintenance of responsiveness—may equally depend on adequate tissue association and interaction. This would then obviously be relevant for the responsive status of hormone-sensitive (epithelial) tumors, and especially of their metastases, which in fact represent inappropriate epithelial–stromal associations.

References

Balinsky, B. J., 1949/50, On the developmental processes in mammary glands and other epidermal structures, *Trans. R. Soc. Edinburgh* **62**:1–31.

Balinsky, B. J., 1950, On the prenatal growth of the mammary gland rudiment of the mouse, *J. Anat.* **84**:227–235.

Cunha, G. R., and Lung, B., 1978, The possible influence of temporal factors in androgenic responsiveness of urogenital tissue recombinants from wild-type and androgen-insensitive (Tfm) mice, *J. Exp. Zool.* **205**:181–194.

Cunha, G. R., Reese, B. A., and Sekkingstad, M., 1980, Induction of nuclear androgen-binding sites in epithelium of the embryonic urinary bladder by mesenchyme of the urogenital sinus of embryonic mice, *Endocrinology* **107**:1767–1770.

Daniel, C. W., and DeOme, K. B., 1965, Growth of mouse mammary glands in vivo after monolayer culture, *Science* **149**:634–636.

DeOme, K. B., Faulkin, L. J., Jr., Bern, H. A., and Blair, P. B., 1959, Development of mammary tumors from hyperplastic alveolar nodules transplanted into gland-free mammary fat pads of female C3H mice, *Cancer Res.* **19**:515–520.

Drews, U., and Drews, U., 1977, Regression of mouse mammary gland anlagen in recombinants of Tfm and wild-type tissues: Testosterone acts via the mesenchyme, *Cell* **10**:401–404.

Dürnberger, H., and Kratochwil, K. 1980, Specificity of tissue interaction and origin of mesenchymal cells in the androgen response of the embryonic mammary gland, *Cell* **19:**465–471.

Gasc, J.-M., and Stumpf, W. E., 1981*a*, Sexual differentiation of the urogenital tract in the chicken embryo: Autoradiographic localization of sex-steroid target cells during development, *J. Embryol. Exp. Morph.* **63:**207–223.

Gasc, J.-M., and Stumpf, W. E., 1981*b*, The bursa of Fabricius of the chicken embryo: localization and ontogenic evolution of sex-steroid target cells, *J. Embryol. Exp. Morph.* **63:**225–231.

Heuberger, B., Fitzka, I., Wasner, G., and Kratochwil, K., 1982, Induction of androgen receptor formation by epithelium–mesenchyme interaction in embryonic mouse mammary gland, *Proc. Natl. Acad. Sci. USA* **79:**2957–2961.

Kratochwil, K., 1969, Organ specificity in mesenchymal induction demonstrated in the embryonic development of the mammary gland of the mouse, *Dev. Biol.* **20:**46–71.

Kratochwil, K., 1971, In vitro analysis of the hormonal basis for the sexual dimorphism in the embryonic development of the mouse mammary gland, *J. Embryol. Exp. Morphol.* **25:**141–153.

Kratochwil, K., 1977, Development and loss of androgen responsiveness in the embryonic rudiment of the mouse mammary gland, *Dev. Biol.* **61:**358–365.

Kratochwil, K., and Schwartz, P. 1976, Tissue interaction in androgen response of embryonic mammary rudiment of mouse: Identification of target tissue of testosterone, *Proc. Natl. Acad. Sci. USA* **73:**4041–4044.

Lyon, M. F., 1961, Gene action in the X-chromosome of the mouse (*mus musculus* L.), *Nature (Lond.)* **190:**372–373.

Lyon, M. F., and Hawkes, S. G., 1970, X-linked gene for testicular feminization in the mouse, *Nature (Lond.)* **227:**1217–1219.

Neumann, F., and Elger, W., 1966, The effect on the anti-androgen 1,2α-methylene-6-chloro-$\Delta^{4,6}$-pregnadiene-17α-o1,-3,20-dione-17α-acetate (cyproterone acetate) on the development of the mammary glands of male foetal rats, *J. Endocrinol.* **36:**347–352.

Propper, A., 1968, Relations épidermo-mésodermiques dans la différenciation de l'ébauche mammaire d'embryon de lapin, *Ann. Embryol. Morphol.* **1:**151–160.

Propper, A., 1969, Compétence de l'épiderme embryonnaire d'oiseau vis-à-vis de l'inducteur mammaire mésenchymateux, *C.R. Acad. Sci.* **268:**1423–1426.

Propper, A., 1972, Rôte du mésenchyme dans la différenciation de la glande mammaire chez l'embryon de lapin, *Bull. Soc. Zool. Fr.* **97:**505–512.

Propper, A., 1978, Wandering epithelial cells in the rabbit embryo milk line. A preliminary scanning electron microscope study, *Dev. Biol.* **67:**225–231.

Raynaud, A., 1947, Effet des injections d'hormones sexuelles à la souris gravide sur le développement des ébauches de la glande mammaire des embryons. I. Action des substances androgènes, *Ann. Endocrinol.* **8:**248–253.

Raynaud, A. and Frilley, M. 1947, Etat de développement des ébauches mammaires et du cordon vaginal chez les foetus mâles et femelles de souris, dont les ébauches des glandes génitales ont été détruites par une irradiation au moyen des rayons X à l'âge de treize jours, *C.R. Acad. Sci.* **225:**1380–1382.

Raynaud, A., and Raynaud, J., 1953, Les principales étapes de la séparation, d'avec l'épiderme, des ébauches mammaires des foetus mâles de souris; recherches sur les processus de la rupture de la tige du bourgeon mammaire, *C.R. Soc. Biol.* **147:**1872–1876.

Sakakura, T., Nishizuka, Y., and Dawe, C. J., 1976, Mesenchyme-dependent morphogenesis and epithelium-specific cytodifferentiation in mouse mammary gland, *Science* **194:**1439–1441.

Sakakura, T., Nishizuka, Y., and Dawe, C. J., 1979*a*, Capacity of mammary fat pads of adult C3H/HeMs mice to interact morphogenetically with fetal mammary epithelium, *J. Natl. Cancer Inst.* **63:**733–736.

Sakakura, T., Sakagami, Y., and Nishizuka, Y., 1979*b*, Persistence of responsiveness of adult mouse mammary gland to induction by embryonic mesenchyme, *Dev. Biol.* **72:**201–210.

Sakakura, T., Sakagami, Y., and Nishizuka, Y., 1981, Accelerated mammary cancer development by fetal salivary mesenchyme isografted to adult mouse mammary epithelium, *J. Natl. Cancer Inst.* **66:**953–959.

Sakakura, T., Sakagami, Y., and Nishizuka, Y., 1982, Dual origin of mesenchymal tissues participating in mouse mammary gland embryogenesis, *Dev. Biol.* **91**:202–207.

Sengel, P. 1976, *Morphogenesis of Skin*, Cambridge University Press, Cambridge.

Stumpf, W. E., and Sar, M. 1975, Autoradiographic techniques for localizing steroid hormones, *Methods Enzymol.* **36**:135–156.

Stumpf, W. E., Narbaitz, R., and Sar, M., 1980, Estrogen receptors in the fetal mouse, *J. Steroid Biochem.* **12**:55–64.

Trelstad, R. L., Hayashi, A., Hayashi, K., and Donahoe, P. K., 1982, The epithelial–mesenchymal interface of the male rat Mullerian duct: Loss of basement membrane integrity and ductal regression, *Dev. Biol.* **92**:27–40.

Wasner, G., Hennermann, I., and Kratochwil, K., 1983, Ontogeny of mesenchymal androgen receptors in the embryonic mouse mammary gland, *Endocrinology* **113**:1771–1780.

Wolff, E., and Haffen, K., 1952, Sur une méthode de culture d'organes embryonnaires in vitro, *Texas Rep. Biol. Med.* **10**:463–472.

Chapter 12

Immunogenetic Approaches to the Analysis of Mammalian Development

SUSAN HEYNER

1. Introduction

The mammalian embryo begins life as a free-living organism and, after approximately 4½ days, implants into the endometrial surface of the uterus (Fig. 1). During this free-living period, two cell populations emerge; the inner cell mass, which will give rise to the embryo, and the trophectoderm, which will give rise to the placenta.

These two tissues exhibit differences in structure, biochemistry, membrane specializations, and function. The cues for this initial differentiation are not known, although a number of studies support the idea that cell position is of paramount importance in determining the initial differentiation of the mouse embryo into trophoblast and inner cell mass (ICM). Blastomeres retaining an "outside" position contribute to trophectoderm, while those on the "inside" are destined to form the ICM (Hillman *et al.*, 1972).

Whatever the mechanisms underlying such early differentiation, they must account for the extensive regulatory capacity of the cells of the cleavage stage embryo, the morula, and even the early blastocyst. A popular hypothesis is that molecules on the surface of the cell membrane may be "recognition" molecules that act as signals for differential development and resultant cell lineages. This hypothesis is implicit in many of the studies of the past two decades, but so far the "signal" molecule has proved elusive (see Chapter 2).

One approach to the study of cell surface membrane structures has been to raise antisera, elicited by a variety of immunogens, and to study the cross reactivity of these reagents with surface molecules on developmental stages. The advantages of this approach include the very powerful specificity of such immunological probes, particularly evident with the advent of monoclonal antibody techniques.

SUSAN HEYNER • Philadelphia College of Pharmacy and Science, Philadelphia, Pennsylvania 19104. *Present address:* Department of Obstetrics and Gynecology, Albert Einstein Medical Center, Northern Division, Philadelphia, Pennsylvania 19141.

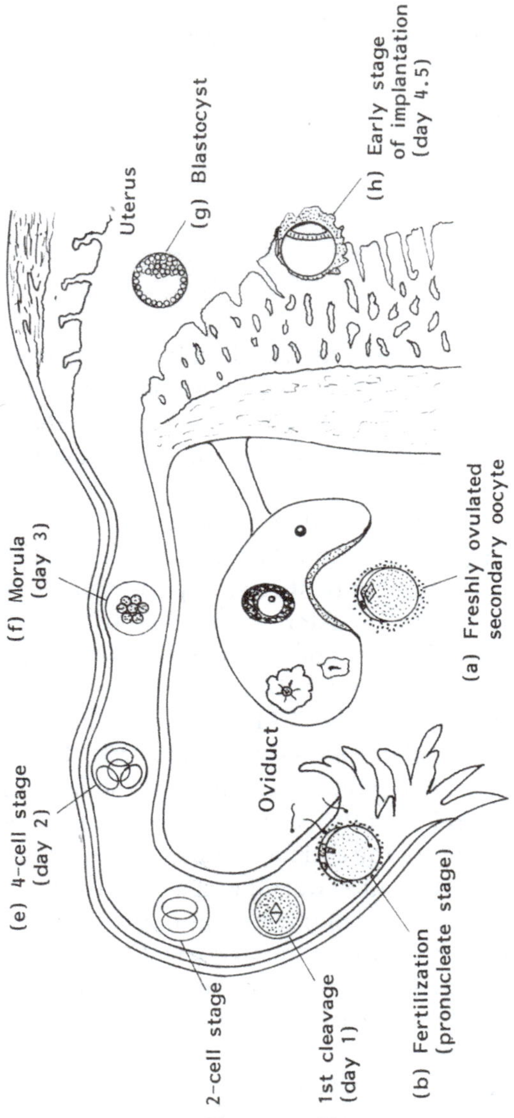

Figure 1. Ovulation and preimplantation stages in the mouse.

Another approach to the study of cell membrane regionalization is the study of enzyme activity in early mammalian embryos; recent studies by Izquierdo and Ebensperger (1982) and by Vorbrodt *et al.* (1977) indicate that a gradual membrane regionalization unfolds, revealing a pattern that may be related to morphogenesis, and in particular, the localization of zonular tight junctions around the peripheral blastomeres of the morula (Izquierdo, 1977; Izquierdo *et al.*, 1980). Other studies have shown that by the 8-cell stage, the blastomeres are polarized with clearly defined apical and basolateral cell features. This polarization can be observed at several levels of organization; thus, short microvilli are restricted to the apical pole, while elsewhere, longer and thicker microvilli are more sparsely distributed, and only persist as long as the basolateral surfaces of adjacent cells remain in contact (Reeve and Ziomek, 1981). The differential binding of a number of ligands to the apical surface of the late 8-cell blastomere may therefore be explained by the concentration of a large amount of membrane at one surface of the cell (Handyside, 1980; Ziomek and Johnson, 1980).

The processes of polarization described above can be seen most dramatically when individual blastomeres are isolated; however, studies on intact embryos have shown that the embryo as a whole assumes a regional organization and becomes a regional mosaic in terms of membrane differentiations (Reeve, 1981).

Studies in which 8-cell blastomeres have been cultured separately show that polarization is limited. On the other hand, if such cells are aggregated with a second cell, polarization ensues with the axis of polarity apparently determined by the point of intercellular contact (Ziomek and Johnson, 1980).

The regionalization of the early mouse embryo, as well as its differentiation into separate lineages can be explained by membrane-situated molecules. This chapter reviews the possible role of the histocompatibility antigens.

2. Histocompatibility Antigens

Histocompatibility genes are widespread throughout the animal kingdom. A number of invertebrate phyla show a range of histocompatibility reactions, while true allograft reactions, mediated by cells capable of specific recognition and development of immunological memory have been described in two phyla of advanced invertebrates, the Echinodermata and Annelida (Hildemann and Dix, 1972; Valembois, 1974). Fundamental expressions of immune competence are found throughout the chordates, although the capacity to effect an acute rejection of tissue grafts is found only in four groups; certain Teleosts, a number of Anurans, birds, and mammals (Cohen and Collins, 1977; Klein, 1977). The idea that histocompatibility molecules are related to graft rejection is so widespread that for many years histocompatibility genes were defined as genes that govern the acceptance or rejection of tissue grafts (Snell, 1971) and their products were referred to as transplantation antigens.

In most mammalian species that have been examined critically, there is a

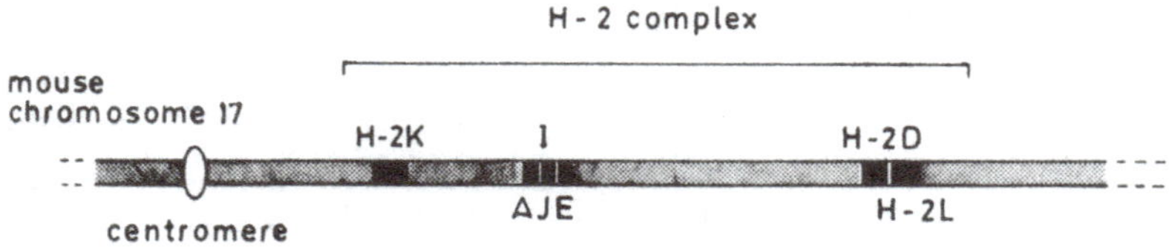

Figure 2. Schematic representation of the *H-2* complex on chromosome 17 in the mouse. Loci encoding class I glycoproteins are shown at either end (H-2^k and H-2^D), while the loci encoding class II glycoproteins are in the center, the I region.

cluster of genes designated as the major histocompatibility complex (MHC). In the mouse this is designated *H-2* and is located on chromosome 17 (Fig. 2). These genes code for proteins that enable T lymphocytes to recognize antigens (Klein *et al.*, 1983). The MHC loci can be divided into two classes, designated class I and class II (Klein, 1979). Class I gene products are characteristically glycopeptides of molecular weight 44,000 with a chain length of approximately 350 amino acids, associated with β_2-microglobulin. A currently accepted representation is shown in Figure 3. Class II gene products are also glycoproteins with two different chains, α-chains (28,000 M_r) and chains (34,000 M_r), as shown in Figure 4. For a more detailed discussion of MHC genetics, the reader is referred to a recent review by Klein *et al.* (1983).

The K and D regions of the mouse *H-2* system constitute the class I antigens or the classic histocompatibility antigens. These are the loci whose products were first demonstrated by alloantisera and are the source of much of the complexity of the *H-2*. Each individual possesses two haplotypes, a paternal and a maternal one. *H-2* specificities limited to a putatively independent haplotype are termed "private," while those that will crossreact with several haploytpes are termed "public." The crossreactivity between *H-2K* and *H-2D*

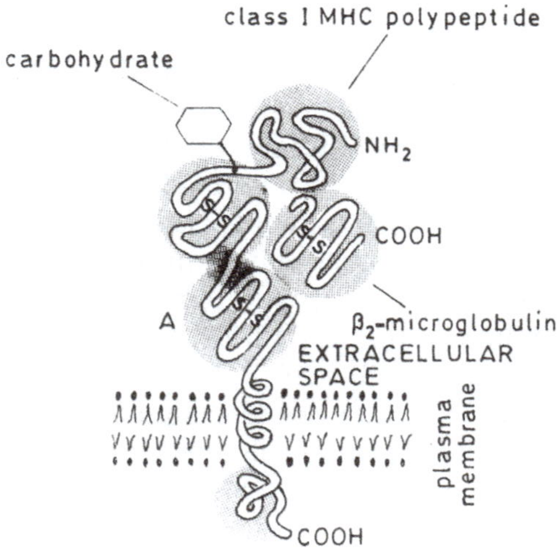

Figure 3. Diagrammatic representation of a class I MHC glycoprotein. The molecule is non-covalently associated with a smaller polypeptide chain, β_2-microglobulin, which is not encoded by the MHC complex.

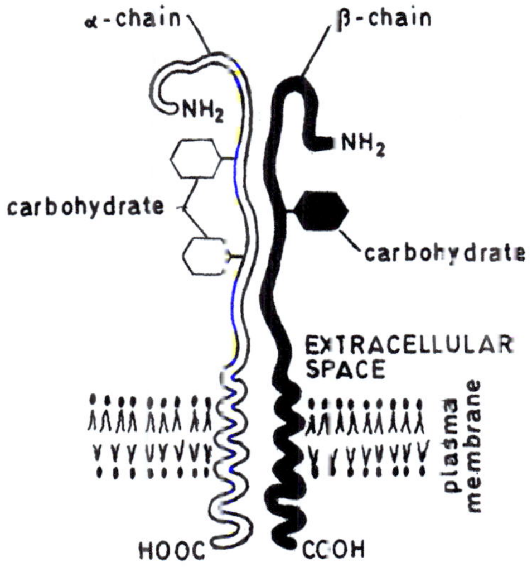

Figure 4. Diagrammatic representation of a class II MHC glycoprotein. The α- and the β-chain are noncovalently associated and are transmembrane.

antigens suggests that they may have arisen from the duplication of an ancestral locus.

Class II antigens, characteristic of the I regions are termed Ia antigens. The function of these genes is not entirely clear, but they are generally recognized to be involved with antigen recognition by T-helper and T-suppressor cells.

In addition to antigens of the MHC, other histocompatibility loci, termed non-*H-2* loci, are found on other chromosomes as well as on chromosome 17. Although their precise number is not known, current estimates put their number in excess of 30. Very little is known of their biological function or their chemical structure, although it has been long recognized by transplantation biologists that they present an obstacle to tissue graft acceptance when the host and donor are compatible at the MHC. Much of the interest in the expression of histocompatibility antigens on mammalian embryos has stemmed from the paradoxical success of the maternal–fetal relationship. In all outbreeding populations, fetal and maternal tissues are in intimate juxtaposition, and there is ample evidence to show that the maternal organism responds immunologically to the paternally derived histocompatibility antigens of the fetus (reviewed by Head and Billingham, 1983). This situation has led a number of biologists to consider the fetus as an allograft and to propose various mechanisms whereby the fetus evades immunological rejection (Medawar, 1953; Beer and Billingham, 1976; Gill and Repetti, 1979).

This analogy has stimulated considerable research effort in immunoreproduction and in analysis of the antigenic status of the embryo; however, it is an oversimplification to compare the surgical exchange of tissues into prepared vascular beds with the development of a fetus separated from the maternal circulation by an extraembryonic organ. It may be that a well-developed immune system is a requirement for establishing a healthy maternal–fetal relationship, for example, to prevent the mutual invasion of cells through the placenta, as suggested by Celada (1974). Recent studies on women who suffer from habitual abortion have shown that a certain disparity between the MHC of mother and father is required for reproductive success.

This review is concerned mainly with the expression of histocompatibility antigens on the surface of eggs and early embryos and discusses what is known of their role in development. The mouse has been the species most utilized for these studies because of the amount of knowledge that has been accumulated concerning murine genetics, and because of the ability to manipulate the murine embryo *in vitro*.

3. Alloantigen Expression on Oocytes

Evidence for the expression of MHC antigens on gametes has been conflicting (Heyner, 1983; Rocklin *et al.*, 1979). One of the problems with assaying MHC components on sperm is the presence of contaminating cells present in the ejaculate. However, a recent report by Anderson *et al.* (1982) appears to have circumvented this problem. The use of a cell-binding radioimmunoassay in conjunction with a panel of highly specific monoclonal antibodies with well-defined reactivities against determinants on HLA class I and class II molecules revealed binding that was similar to that obtained with negative monoclonal controls. In contrast to the numerous studies on sperm, there have been relatively few studies of alloantigen expression on the surface of oocytes.

Studies in the mouse (see Table I) have utilized a number of different techniques; Edidin *et al.* (1974) reported some positive results using complement-mediated cytotoxicity, but Billington *et al.* (1977) were unable to detect MHC antigens on the oocyte membrane using an immunoperoxidase labelling technique. Studies on the mouse oocytes at the dictyate and metaphase II stages have been carried out using multispecific sera, raised in congenic strains of mice and rendered monospecific by absorptions, using indirect immunofluorescence. Serologically specific reactivity to *H-2* antibody was observed on the oocyte membrane. In contrast, zygotes were found to exhibit variable reactivity, and 2-cell stages were found to be negative (Heyner and Hunziker, 1979). These findings suggested that fertilization results in altered expression of membrane

Table I. Serological Detection of Antigens on Oocytes

Antibody directed against	Method of study	Results	Reference
H-2	Complement-mediated cytotoxicity	Some positive Some negative	Edidin *et al.* (1974)
H-2	Immunoperoxidase	Negative	Billington *et al.* (1977)
H-2	Indirect immunofluorescence	Positive	Heyner and Hunziker (1979)
H-1[a] *H-3*[a] *H-7*[a]	Indirect immunofluorescence	Positive	Heyner *et al.* (1980)
H-4[a] *H-9*[a] *H-13*[a]	Indirect immunofluorescence	Positive	Heyner and Giles-Komar (1983)

components, a confirmation of the observations of Johnson and Edidin (1978) that there are marked changes in the lateral mobility of the mouse oocyte membrane following fertilization. Many alloantisera elicited against MHC class I antigens may actually contain antibodies to other cellular antigens, for example, Ly antigens found on lymphocyte membranes, and Ea antigens found on erythrocytes. Caution therefore advised studies on oocytes with monoclonal antibody directed against *H-2K*^{k}, together with polyclonal NIH sera directed against K-region specificities *H-2*-11, 23, and 25. In addition to these class I reagents, a broadly reactive class II, anti-I region alloantiserum was used in the indirect immunofluorescence assay. Oocytes reacted positively with polyclonal reference sera but did not react with the broadly reactive class II antibody. The lack of reactivity to the monoclonal antibody directed against H-2K^k suggested the possibility that *H-2* sites might be sparsely distributed on the oocyte surface, since lymphocytes were positive even at a dilution of 1 : 10,000.

This hypothesis was tested by using a three-step procedure with a goat anti-mouse IGg antibody as a second antibody, followed by incubation with fluorescein-conjugated goat anti-mouse gamma globulin. This amplification still did not yield positive results, so it was repeated, using a cocktail of well-characterized monoclonal antibodies directed against the K region of the MHC. This resulted in a very weak reaction, suggesting that antigen density may be a significant problem in the detection of class I antigens on the oocyte surface (Heyner and Hunziker, 1981).

The finding that class I antigens are distributed at a low density explains previous conflicting data and also explains the reports of inconsistent results with complement-mediated cytotoxicity (Edidin *et al.*, 1974; S. Heyner, unpublished observations). Similarly, S. Hausman (personal communication) was unable to detect a cytotoxic effect with alloantisera directed against the MHC in the rat, although immunofluorescence and mixed hemadsorption yielded positive results.

A question of some significance is whether the class I antigens are actually synthesized by the oocyte, or whether they are carried over from a previous stage. This was tested by eliciting sera with potential reactivities against *H-2*^{k} and *H-2*^{d} antigens as well as against Ia antigens in congenic strains of mice, and testing by direct immunoprecipitation and two-dimensional (2-D) gel electrophoresis. These methods are highly sensitive, and synthesis of individual polypeptides can be distinguished, further, *H-2* and Ia specificities can be distinguished from each other on the basis of differences in their isoelectric points and molecular weights. Analysis of [^{35}S]methionine-labeled immunoprecipitates on 2-D gels revealed that synthesis of class I and Class II antigens was not detectable in unfertilized eggs (J. A. Sawicki, T. Magnusen, and C. J. Epstein, personal communication). By contrast, since expression on this stage was detected with one of the antisera used, it is possible that either *H-2* antigens are absorbed from the maternal milieu, or, more likely, that they are carried over from a previous stage.

If oocytes are grown *in vitro*, they will mature in a few hours, exhibiting breakdown of the germinal vesicle. This occurs whether the zona pellucida is present or not (Heyner and Hunziker, 1979). It is worth noting that the ovum will still go through maturation in the presence of antibody directed against class I antigens, even in the presence of complement (Heyner, unpublished observations). This observation reinforces the idea that class I antigens are very sparsely distributed on the surface of the oocyte, and further, supports the conclusion of Anderson *et al.* (1982) that class I antigens are probably not involved in sperm–egg recognition events.

Table I shows that a number of non-*H-2* antigens can also be detected on the surface of the unfertilized oocyte. These have been recognized using antisera elicited against a number of non-*H-2* specificities, and subsequently rendered monospecific by the appropriate absorption (Heyner *et al.*, 1980; Heyner and Giles-Komar, 1983). Nothing is known of corresponding antigens on the membrane of sperm, so it is not clear whether these molecules may be involved in gamete recognition.

4. Class I and Class II Antigen Expression on Embryos

Early transplantation experiments showed that when mouse tubal embryos were placed under the kidney capsule of specifically sensitized recipients, the graft was destroyed, establishing the presence of alloantigens on the embryo (Simmons and Russell, 1966). Since this experiment, there has been extensive investigation of the expression of class I antigenic determinants on early embryos (reviewed by Edidin, 1976; Solter, 1977; Heyner, 1983). Early studies using indirect immunofluorescence (Palm *et al.*, 1971; Muggleton-Harris and Johnson, 1976) indicated that MHC antigens could not be detected on tubal cleavage stages or blastocysts. The earliest detection of MHC antigens by indirect immunofluorescence on blastocyst outgrowths was around day 6, and that reactivity was restricted to the ICM (Heyner, 1973). These results were in agreement with transplantation studies by Patthey and Edidin (1973).

Most studies are in agreement that MHC antigens are first detected on the ICM of the blastocyst (see Table II); Webb *et al.* (1977) grew mouse embryos in media containing radiolabelled amino acids and used sensitive immunoprecipitation methods to show that *H-2* synthesis could be detected in the cells of the ICM of the late blastocyst. However, in conflict with these data are two experiments of Krco and Goldberg (1977) and Cozad and Warner (1982), in which *H-2* antigens were detected on 8-cell mouse embryos. In both cases, the investigators utilized a complement-dependent cytotoxicity assay in conjunction with well-defined alloantisera to maternal and paternal antigens. Their results are in contrast to earlier studies in which Heyner *et al.* (1969) showed that embryos would develop *in vitro* from the 8-cell to the blastocyst stage in the presence of congenically raised alloantisera to *H-2* antigens and complement. Perhaps the discrepancy between these results can be explained as a consequence of viral antibodies present in mouse serum, which may react with vertically transmitted virus carried by the mice. This explanation would also fit the finding of Cozad and Warner (1982) that their $H\text{-}2^a$ haplotype target em-

Table II. Detection of H-2 Antigens on the Surface of Mouse Embryos[a]

Zygote	2-cell	4–8 cell	Morula	Blastocyst ICM	Blastocyst trophectoderm	6–7-day blastocyst outgrowth embryo	6–7-day blastocyst outgrowth trophoblast	Technique	Reference[c]
NT	NT	– ; ++	–	NT	– ;+	NT	NT	Cytotoxicity	(1, 2, 3, 4)
+	–	–	–	–	–	+	–	Immunofluorescence	(5, 6, 7, 8)
NT	?	NT	NT	NT	NT	NT	–	Mixed hemadsorption	(9)
NT	NT	NT	NT	NT	+	+	–	Immunoperoxidase	(10, 11)
NT	–	NT	–	+	–	NT	NT	Immunoprecipitation	(12)
NT	NT	NT	NT	NT	–[b]	NT	NT	Isotope antiglobulin	(13)

[a]+, positive results; –, inability to detect the antigen; ?, uncertainty; NT, the particular stage has not been examined.
[b]Blastocysts undergoing experimentally induced delay of implantation were variably positive.
[c]References: (1) Heyner *et al.* (1969); (2) Krco and Goldberg (1977); (3) Cozad and Warner (1982); (4) Cozad and Warner (1981); (5) Palm *et al.* (1971); (6) Heyner (1973); (7) Muggleton-Harris and Johnson (1976); (8) Heyner and Hunziker (1979); (9) Sellens (1977); (10) Searle *et al.* (1976); (11) Billington *et al.* (1977); (12) Webb *et al.* (1977); (13) Hakansson *et al.* (1975).

bryos did not apparently express *H-2* antigens; it is possible that this strain of mice was free of virus.

The expression of class I antigens on trophectoderm is controversial. Although most investigators have been unable to detect MHC antigens on these cells, Searle *et al.* (1976) used a sensitive immunoperoxidase assay to demonstrate transient expression of class I antigens on early blastocyst trophectoderm. Trophectoderm of blastocysts undergoing delay of implantation showed variable expression of MHC antigens when examined with a sensitive antiglobulin technique, while those blastocysts released from delay were negative (Hakansson *et al.*, 1975).

From the foregoing it is apparent that the majority of studies are in agreement that MHC antigens can be detected first on the ICM of the blastocyst, and are not present in significant amounts on the trophectoderm. An interesting aspect of many early studies is that antibodies were raised without considering the possiblity that class II antigens might have been involved. Indeed, analysis of most of the strain combinations used to raise the reagents shows that class II antibodies were probably present. However, studies by Delovitch *et al.* (1978) utilizing sensitive immunoprecipitation methods have revealed that Ia antigens appear first at day 11 of gestation and are restricted to the liver until day 16. These results have been confirmed by Jenkinson and Searle (1979), using immunoperoxidase labeling and examination by electron microscopy. These investigators found that preimplantation blastocysts and postimplantation embryos (day 7½) were uniformly negative when tested with anti-Ia antibodies.

In the context of immunoreproduction, the ontogeny of paternal class I antigens is clearly of significant interest, and it is therefore of potential importance that although the early embryo expresses *H-2* by the late blastocyst stage, class II antigens have not been detected until almost mid-gestation. The role of class II antigens in the rejection response has been well-documented (Bach *et al.*, 1976; Thorsby *et al.*, 1978) and the inability to detect such antigens on early developmental stages may indicate a mechanism whereby the embryo evades immunological surveillance. On the other hand, it has been postulated recently that T cells which react to foreign class I antigens and mediate graft rejection differ from conventional cytotoxic T lymphocytes, and are identical to those mediating delayed-type hypersensitivity reactions (Lyt-1^+). Once these cells are stimulated by antigen, they recruit lymphokines and recruit effector cells to destroy the graft (Loveland and McKenzie, 1982). Whatever the mechanism underlying graft rejection, it is clear that class I antigen expression on trophoblast is very important in later postimplantation stages of development.

5. Expression of Non-H-2 Alloantigens on Embryos

It is generally recognized from the experimental data that early mouse embryos are immunogenic, and this property has been ascribed to the expression of non-*H-2* histocompatibility antigens on the surface of the early embryo. Early studies using complement-dependent cytotoxity and antisera directed against nonspecified non-*H-2* alloantigens demonstrated that these alloantibodies were cytotoxic, while congenically elicited *H-2* alloantibody was not cytotoxic and did not interfere with development to the blastocyst stage. Experiments using indirect immunofluorescence (Palm *et al.*, 1971) demonstrated that *H-3* and *Ea-6* (formerly designated *H-6*) could be detected on 2-cell stages. Muggleton-Harris and Johnson (1976) confirmed and extended these results; they utilized broad-spectrum alloantisera directed against minor alloantigens, and showed that while *H-2* antigens could not be detected on tubal embryos and blastocysts, non-*H-2* antigens of maternal origin were detectable throughout preimplantation development, and alloantigens of paternal origin could be detected first around the 6–8-cell stage.

These early studies have been very useful in shedding light on the sources of immunogenicity of early embryos. More importantly, they have established that in the mouse it is unlikely that gene products of the MHC have any role in the early processes of differentiation. On the other hand, these studies have several shortcomings. One is that the non-*H-2* alloantisera raised by conventional means had the potential to contain contaminating antibodies to other cellular antigens. A second criticism is that investigators have studied either specific alloantigens at one or two developmental stages (Palm *et al.*, 1971) or have studied a developmental series, using sera that have not been characterized in terms of their target specificities (Muggleton-Harris and Johnson, 1976; Billington *et al.*, 1977). In an attempt to circumvent these problems, Heyner *et al.* (1980) used alloantibodies elicited in mice compatible at the MHC but differing at multiple non-*H-2* loci to map minor alloantigen expression during different stages of preimplantation development In one series of experiments, detailed serological analysis revealed that one alloantibody was directed against $H\text{-}3^a$, $H\text{-}8^a$, and *Ea-6.1*.

Adsorption with cells from the appropriate mouse strain rendered the antibody monospecific for $H\text{-}3^a$, and indirect immunofluorescence tests were carried out on lymphocytes from strains that had the potential to react with $H\text{-}3^a$ antibody. Similar tests were carried out with alloantiserum containing antibody directed against $H\text{-}1^a$, $H\text{-}7^a$, and *Lyt 1.1*. This antiserum was adsorbed by the appropriate cells to remove all activity except that directed against $H\text{-}7^a$. When mouse developmental stages were examined with the monospecific antibody, differential reactivity was seen in early preimplantation stages. While the unfertilized ovum was weakly positive, antigen expression could not be detected at the 2-cell stage, although by the 8-cell stage both $H\text{-}3^a$ and $H\text{-}7^a$ were detectable, and were also expressed on morulae and blastocysts. Similar results have been obtained recently with alloantisera rendered monospecific for $H\text{-}4^a$, $H\text{-}9^a$, and $H\text{-}13^a$ by Heyner and Giles-Komar (1983) as shown in Table III. These data are in conflict with earlier findings that 2-cell mouse embryos expressed minor alloantigens (Palm *et al.*, 1971; Muggleton-Harris and Johnson, 1976). A reasonable explanation is that the sera used by these investigators had the potential to contain antibody directed against *Ea* or *Lyt* specificities (see Heyner 1983, for discussion) or against closely linked but undefined histocompatibility antigens (Flaherty and Bennett, 1973). Studies in our laboratory

Table III. Reactivity of Preimplantation Mouse Embryos with Antibody Directed against Minor Histocompatibility Antigens[a,b]

Stage	Histocompatibility antigen detected				
	H-3[a]	H-4[a]	H-7[a]	H-9[a]	H-13[a]
Fertilized ovum	+	NT	+	NT	NT
Two-cell	−	−	−	−	−
Eight-cell	+	+	+	+	+
Morula	+	+	+	+	+
Blastocyst	+	+	+	+	+

[a]Two different polyclonal antisera were used, and adsorption with appropriate strains in conjunction with selected target strains allowed discrimination of minor histocompatibility expression. Controls included incubation with fluorescein isothiocyanate alone or following incubation with preimmune serum.
[b]Reactions indicate the results of indirect immunofluorescence tests; +, reactivity; −, inability to detect the antigen; NT, not tested.

have been unable to resolve this difficulty (S. Heyner and R. D. Hunziker, unpublished data).

The blastocyst represents two cell lineages; the inner cell mass and the trophectoderm. Alloantigen expression on trophectoderm has been of particular interest since this tissue is in intimate contact with the maternal tissues during implantation. The detection of specific minor alloantigens on the surface of the trophectoderm (see Table III) confirms the antigenicity of this tissue, as shown by studies utilizing indirect immunofluorescence (Muggleton-Harris and Johnson, 1976) mixed hemadsorption (Billington *et al.*, 1977; Sellens, 1977) and transplantation to specifically preimmunized hosts (Johnson 1975). After implantation, trophectoderm differentiates into a variety of cell types. Therefore studies of trophoblast immunogenicity are often difficult to interpret. This is particularly true of many *in vitro* studies in which the normal architecture of the tissue is lost. Thus Carter (1976, 1978) used multispecific alloantisera to demonstrate that in the mouse, primary trophoblastic giant cells expressed maternal and paternal alloantigens, although secondary trophoblastic giant cells from day 8 of pregnancy did not react in the mixed hemadsorption assay. When cultured for 3–5 days *in vitro* several layers of secondary giant cells were produced from the ectoplacental cone, the outermost of which appeared to be antigen positive. These results are interesting because they demonstrate differential antigen expression can be detected *in vitro*. Although the precise nature of the antigens expressed is not known, these studies do provide evidence that antigen expression differs between cell type and developmental stage. Similarly, caution is necessitated by the observation of Cikes and Klein (1972) that the expression of cell surface antigens may be modulated by growth conditions. Interestingly, antigen expression on trophoblastic stages later in pregnancy is believed to be one mechanism whereby the fetus is able to evade immune rejection by the maternal organism due to the immobilization of maternal antibody (Wegmann *et al.*, 1979), but these studies fall outside the scope of this review.

6. Discussion

It is clear from the studies that have been reviewed that with the exception of class II alloantigens, alloantigens expressed on most adult tissues can also be detected on the oocyte, or on some stage of the preimplantation embryo. Antigen expression on oocytes appears to be very weak; on the basis of studies with MHC antigens, this is due to a lack of synthesis of these antigens (J. A. Sawicki, T. Magnusen, and C. J. Epstein, personal communication). This contrasts with expression of β_2-microglobulin, a 12,000-M_r peptide encoded by the *H-2* complex on chromosome 17 in the mouse (Fig. 3) that has been shown to be synthesized in unfertilized eggs although synthesis was not detectable in fertilized eggs (Sawicki *et al.*, 1981). From these studies it appears reasonable to conclude that class I and minor histocompatibility antigens expressed on oocytes reflect insertion at an earlier stage, and that dilution of the antigens occurs as development proceeds, rendering their detection difficult until synthesis occurs in the early embryo.

An early impetus to the study of alloantigens on the surface of mammalian embryos was the hypothesis that cell recognition may result from specific molecules on the surface of the membrane. Theoretically, if cells have different glycoprotein coats then they would be expected to form separate phases and by merely carrying different glycoprotein coats, cells might be expected to show differential adhesion phenomona (Edwards, 1978). Alloantigens of the MHC or non-MHC alloantigens are excellent candidates for such cell recognition molecules. MHC molecules are glycoproteins conforming the biochemical requirements of such recognition entities (Fig. 4). Similarly, although little is known of the biochemical nature of minor histocompatibility antigens due to the difficulty of eliciting antibody with cytotoxic or hemolytic activity, recent studies have successfully utilized immunization with epidermal cells and mitogen-activated lymphocytes in congenic strains differing only at the minor *H-4* barrier. Sera elicited in this manner have been examined in serological and chemical studies, and these studies have revealed that such antibodies crossreact with products of other minor *H* loci. Further, alloantiserum to the *H-4* gene product precipitates a glycoprotein with a molecular weight of approximately 45,000 from radiolabeled concanavalin A-stimulated lymph node cells (Long *et al.*, 1981).

There have been few studies to examine the effect of incubating preimplantation mouse embryos in alloantisera directed against specified alloantigens. Early studies have demonstrated that alloantibody directed against class I antigens did not interfere with blastocyst formation (Heyner *et al.* 1969) however, studies with alloantibody directed against minor alloantigens have been difficult to accomplish due to the paucity of reagents. Studies in our laboratory indicate that such incubation has no effect on compaction and blastocyst development (S. Heyner and J. Giles-Komar, unpublished data). These findings are particularly interesting in view of the reports that allorecognition is governed by an MHC-like gene system in a protochordate ancestor of the vertebrates, *Botryllus*, a colonial tunicate. Furthermore, the same genetic region apparently acts to prevent fertilization between gametes sharing alleles (Scofield *et al.*, 1982). In humans, at least, histocompatibility gene products do not appear to be involved with fertilization and their involvement in other mammalian species is controversial (Anderson *et al.*, 1982). Although class I antigens are not apparently implicated in the early events of development, there is significant evidence that *H-Y* antigen-expression is involved in sex determination (Wachtel, 1983).

If alloantigens do not have a role in differentiation of the early mouse embryo, we may ask what significance they have. It is well documented that the mouse preimplantation embryo is capable of eliciting a humoral maternal response to MHC alloantigens (Bell and Billington, 1980; Roe and Bell, 1982) as well as to non-*H-2* alloantigens (S. Heyner and J. Giles-Komar, unpublished data). The embryo can be regarded therefore, as a potential target for the maternal immune system from the expression of alloantigens alone, setting aside the questions of embryo-specified antigens such as SSEA-1 (see Chapter 2). During the early cleavage stages of development the embryo is surrounded by an acellular layer composed largely of sialic acid and protein, the zona pellucida. The zona unquestionably provides protection against a cell-mediated maternal response although its role as an effective barrier to humoral immunity is open to question. When the embryo implants into the uterine epithelium, the zona is shed, and the blastocyst attaches to the uterine epithelium. It has been suggested that the inability to detect MHC-specified antigens at this time, as noted by Hakansson *et al.* (1975) and Searle *et al.* (1976) allows the embryo to present an outer surface lacking in antigens and therefore favors establishment of the embryo at a time when it is more vulnerable to immune rejection. However, the expression of non-MHC alloantigens at this stage would argue against the requirement for a nonimmunogenic embryo at implantation. Furthermore, *in vitro* studies may differ significantly from those *in vivo*. For example, Parr (1981) showed that a variety of adult epithelial cells when dissociated from the tissues after prefixation to preserve their native antigen distribution have been shown to express MHC antigens in the laterobasal, but not the apical, membranes. If dissociation was carried out without prefixation, MHC antigens migrated to the apical membranes. These studies suggest that antigen sequestration may be a strategy to avoid rejection during the early phases of implantation.

Although the expression of alloantigens has not been correlated with cell recognition, cell adhesion, or cellular differentiation to date, there is some eivdence that the immunogenicity of alloantigens and the concomitant maternal response may play an important role in reproduction. This is borne out by the well-documented difficulty of developing and maintaining inbred strains of laboratory animals (Gill, 1983). Maintenance of histocompatibility heterozygosity may also depend upon the expression of alloantigens, as suggested by the studies of Hull (1969). When an inbred female was mated to an F1 male, an excess of heterozygotes resulted. Similarly, repeated backcrosses between inbred females and hybrid males resulted in alloantigen heterozygote progeny in excess of those predicted by Mendelian segregation of these genes (Hamilton and Hellstrom, 1978). It is intriguing to speculate how and when selection of such progeny occurs. Conceivably, selection could occur at fertilization, although in light of data discussed above, this is unlikely. Or the selection of such progeny may be mediated via the maternal response at the time of implantation. This question bears directly on human fertility, as shown by Beer et al. (1981), that matings between couples who are compatible over a wide range of histocompatibility antigens lead to an unusually high number of unexplained pregnancy losses. These workers show that when the female is treated by means of blood transfusions from an HLA imcompatible donor, fertility is improved.

During development, the regulation of internal micronutrient concentration is of critical importance. For example, Tasca and his colleagues (Borland and Tasca, 1974; DiZio and Tasca, 1977) have demonstrated that major changes occur in amino acid transport during development, from slow rates in early cleavage to high rates in the blastocyst. Other important molecules such as sugars and nucleosides have not been studied so far. It is quite conceivable that the minor alloantigens may be involved with the transport of specific molecules during development. The study of cell-surface antigen involvement in transport is potentially fruitful, since by the use of appropriate antibodies it may be possible to alter transport characteristics, either by direct binding to cell membrane "carrier" molecules or to a hormone receptor that may influence transport processes.

In conclusion, alloantigens have been recognized to contribute to the immunogenicity of the embryo, and have been extensively analyzed. However, their role in early differentiation events has not been established, and from the experimental data gathered so far, appears unlikely. While class I antigens cannot be detected until around the time of implantation, non-*H-2* alloantigens are expressed earlier, and their role remains obscure. While these glycoproteins are not apparently involved in early differentiation events, it now seems likely that they are implicated in normal reproduction.

ACKNOWLEDGMENTS. Original studies reviewed in this chapter have been supported by the USPHS (HD 11566) and by the Pfeiffer Foundation. The author would like to thank Jill Giles-Komar for critical comments.

References

Anderson, D. J., Bach, D. L., Yunis, E. J., and DeWolf, W. C., 1982, Major histocompatibility antigens are not expressed on human epididymal sperm, *J. Immunol.* **129**:452–454.

Bach, F. H., Bach, M. L., and Sondel, P. M., 1976, Differential function of major histocompatibility complex antigens in T-lymphocyte antigen activation, *Nature (Lond.)* **259**:273–281

Beer, A. E., and Billingham, R. E., 1976, The fetus as an allografy: Consideration of some of the factors that may contribute to its success, in: *The Immunobiology of Mammalian Reproduction* (A. E. Beer, and R. E. Billingham, eds.), Prentice-Hall, Englewood Cliffs, New Jersey.

Beer, A. E., Quebbeman, J. F., Ayers, J. W. T., and Haines, R. F., 1981, Major histocompatibility complex antigens, maternal and paternal immune responses, and chronic habitual abortions in humans, *Am. J. Obstet. Gynecol.* **141**:987–997.

Bell, S. C., and Billington, W. D., 1980, Major anti-paternal alloantibody induced by murine pregnancy is non-complement fixing, Nature *(Lond.)* **288**:387–388.

Billington, W. D., Jenkinson, E. J., Searle, R. F., and Sellens, M. H., 1977, Alloantigen expression during early embryogenesis and placental ontogeny in the mouse. Immunoperoxidase and mixed hemadsorption studies, *Transplant. Proc.* **9**:1371–1377.

Borland, R. M., and Tasca, R. J., 1974, Activation of Na^+-dependent amino acid transport system in preimplantation mouse embryos, *Dev. Biol.* **30**:169–182.

Carter, J., 1976, Expression of maternal and paternal antigens on trophoblast, *Nature (Lond.)* **262**:292–293.

Carter, J., 1978, The expression of surface antigens on three trophoblastic tissues in the mouse, *J. Reprod. Fertil.* **54**:433–439.

Celada, F., 1974, Concepts and methods in immunology applicable to the control of human fertility in: *Immunological Approaches to Fertility Control* (E. Diczfalusy, ed.), pp. 419–443, Karolinska Institutet, Stockholm.

Cikes, M., and Klein, G., 1972, Quantitative studies of antigen expression in cultured murine lymphoma cells, cell surface antigens in "asynchronous" cultures, *J. Natl. Cancer Inst.* **49**:1599–1606.

Cohen, N., and Collins, N. H., 1977, Major and minor histocompatibility systems of ectothermic vertebrates, in: *The Major Histocompatibility Systems in Man and Animals* (D. Gotze, ed.), pp. 313–337, Springer-Verlag, New York.

Cozad, K. M., and Warner, C. M., 1982, Detection of *H-2* antigens on 8-cell mouse embryos, *J. Exp. Zool.* **221**:213–217.

Delovitch, T., Press, J. L., and McDevitt, H. O., 1978, Expression of murine Ia antigens during embryonic development, *J. Immunol.* **120**:818–824.

DiZio, S. M., and Tasca, R. J., 1977, Sodium-dependent amino acid transport in preimplantation mouse embryos, *Dev. Biol.* **59**:198–205.

Edidin, M., 1976, Cell surface antigens in mammalian development, in: *The Cell Surface in Animal Embryogenesis and Development* (G. Poste and G. L. Nicholson, eds.), pp. 127–143, Elsevier/North-Holland, Amsterdam.

Edidin, M., Gooding, L. R., and Johnson, M., 1974, Surface antigens of human normal early embryos and a turnover model system useful for their further study, in: *Immunological Approaches to Fertility Control* (E. Diczfalusy, ed.), pp. 336–356, Karolinska Institutet, Stockholm.

Edwards, P. A. W., 1978, Differential cell adhesion may result from nonspecific interactions between cell surface glycoproteins, *Nature (Lond.)* **271**:248–249.

Flaherty, L., and Bennett, D., 1973, Histocompatibilities found between congenic strains which differ at loci determining differentiation antigens, *Transplantation* 16:505–514.

Gill, T. J., 1983, Immunogenetic aspects of the maternal–fetal interaction, in: *Immunology of Reproduction* (T. G. Wegmann and T. J. Gill, eds.), pp. 55–76, Oxford University Press, New York.

Gill, T. J., and Repetti, C. F., 1979, Immunologic and genetic factors influencing reproduction, *Am. J. Pathol.* **95**:465–470.

Hakansson, S., Heyner, S., Sundqvist, K. G., and Bergstrom, S., 1975, The presence of paternal *H-2* antigens on hybrid mouse blastocysts during experimental delay of implantation and the disappearance of these antigens after onset of implantation, *Int. J. Fertil.* **20**:137–140.

Hamilton, M. S., and Hellstrom, I., 1978, Selection for histoincompatible progeny in mice, *Biol. Reprod.* **19**:267–270.

Handyside, A. H., 1980, Distribution of antibody- and lectin-binding sites on dissociated blastomeres from mouse morulae: Evidence for polarization at compaction, *J. Embryol. Exp. Morphol.* **60**:99–116.

Head, J. R., and Billingham, R. E., 1983, Transplantation biology revisited, in: *Immunology of Reproduction* (T. G. Wegmann and T. J. Gill, eds.), pp. 27–52, Oxford University Press New York.

Heyner, S., 1973, Detection of *H-2* antigens on the surface of the early mouse embryo, *Transplantation* **16**:675–678.

Heyner, S., 1983, Alloantigen expression on oocytes and embryos, in: *Immunology of Reproduction* (T. G. Wegmann and T. Gill, eds.), pp. 79–99, Oxford University Press, New York.

Heyner, S., and Giles-Komar, J., 1983, Differential expression of minor histocompatibility antigens on the surface of the early mouse embryo, *J. Cell Biol.* **97**:3a.

Heyner, S., and Hunziker, R. D., 1979, Differential expression of alloantigens of the major histocompatibility complex on unfertilized and fertilized eggs, *Dev. Genet.* **1**:69–76.

Heyner, S., Palm, J., and Brinster, R. L., 1969, Effect of alloantibody on pre-implantation mouse embryos, *Nature (Lond.)* **222**:783–784.

Heyner, S., and Hunziker, R. D., 1981, Oocytes react with antibodies directed against *H-2* but not Ia antigens, *J. Immunogenet.* **8**:523–528.

Heyner, S., Hunziker, R. D., and Zink, G. L., 1980, Differential expression of minor histocompatibility antigens on the surface of the mouse oocyte and preimplantation developmental stages, *J. Reprod. Immunol.* **2**:269–279.

Hillman, N., Sherman, M. I., and Graham, C., 1972, The effect of spatial arrangement on cell determination during mouse development, *J. Embryol. Exp. Morphol.* **28**:263–278.

Hildemann, W. H., and Dix, T., 1972, Transplantation reactions of tropic Australian echinoderms, *Transplantation* **15**:624–633.

Hull, P., 1969, Maternal–foetal incompatibility associated with the *H-3* locus in the mouse, *Heredity* **24**:203–209.

Izqiuerdo, L., 1977, Cleavage and differentiation, in: *Development in Mammals* (M. J. Johnson, ed.), Vol. 2, pp. 99–118, North Holland, New York.

Izquierdo, L., and Ebensperger, C., 1982, Cell membrane regionalization in early mouse embryos as demonstrated by 5′-nucleotidase activity. *J. Embryol. Exp. Morphol.* **69**:115–126.

Izquierdo, L., Lopez, T., and Martincorena, P., 1980, Cell membrane regions in preimplantation mouse embryos, *J. Embryol. Exp. Morphol.* **59**:89–102.

Jenkinson, E. J., and Searle, R. F., 1979, Ia expression on the developing mouse embryo and placenta, *J. Reprod. Immunol.* **1**:3–10

Johnson, M. H., 1975, Antigens of the peri-implantation trophoblast, in: *Immunobiology of Trophoblast* (R. G. Edwards, C. W. S. Howe, and M. H. Johnson, eds.), pp. 87–100, Cambridge University Press, London.

Johnson, M. H., and Edidin, M., 1978, Lateral diffusion in plasma membrane of mouse egg is restricted after fertilization, *Nature (Lond.)* **272**:448–450.

Klein, J., 1977, Evolution and function of major histocompatibility systems: Facts and speculations, in: *The Major Histocompatibility System in Man and Animals* (D. Gotze, ed.), pp. 339–378, Springer-Verlag, New York.

Klein, J., Figueroa, F., and Nagy, Z. A., 1983, Genetics of the major histocompatibility complex: The final act, *Annu. Rev. Immunol.* **1**:119–142.

Krco, C. J., and Goldberg, E. H., 1977, Major histocompatibility antigens on preimplantation mouse embryos, *Science* **193**:1134–1135.

Long, P. M., Lafuse, W. P., and David, C. S., 1981, Serologic and biochemical identification of minor histocompatibility (*H-4*) antigens, *J. Immunol.* **127**:825–828.

Loveland, B. E., and McKenzie, I. F. C., 1982, Which T cells cause allograft rejection?, *Transplantation* **33**:217–219.

Medawar, P. B., 1953, Some immunological and endocrinological problems raised by the evolution of viviparity in vertebrates, *Symp. Soc. Exp. Biol.* **7**:320–338.

Muggleton-Harris, A. L., and Johnson, M. H., 1976, The nature and distribution of serologically detectable alloantigens on the preimplantation mouse embryo, *J. Embryol. Exp. Morphol.* **35**:59–72.

Palm, J., Heyner, S., and Brinster, R. L., 1971, Differential immunofluorescence of fertilized mouse eggs with *H-2* and non-*H-2* antibody *J. Exp. Med.* **133**:1281–1293.

Parr, E. L., 1981, The self-side expression of *H-2* antigens on epithelial cells and the maternal–fetal relationship, *Transplant. Proc.* **13:**973–978.
Patthey, H. L., and Edidin, M., 1973, Evidence for the time of appearance of *H-2* antigens in mouse development, *Transplantation* **15:**211–214.
Reeve, W. J. D., and Ziomek, C. A., 1981, Distribution of microvilli on dissociated blastomeres from mouse embryos: Evidence for surface polarization at compaction, *J. Embryol. Exp. Morphol.* **62:**339–350.
Reeve, W. J. D., 1981, Cytoplasmic polarity develops at compaction in rat and mouse embryos, *J. Embryol. Exp. Morphol.* **62:**351–367.
Rocklin, R. E., Kitzmiller, J. L., and Kaye, M. D., 1979, Immunobiology of the maternal–fetal relationship, *Annu. Rev. Med.* **30:**375–404.
Roe, R., and Bell, S. C., 1982, Humoral immune responses in murine pregnancy. II. Kinetics and nature of the immune response in females preimmunized against paternal alloantigens, Immunology **46:**23–30.
Sawicki, J. A., Magnusen, T., and Epstein, C. J., 1981, Evidence for expression of the paternal genome in the two-cell mouse embryo, *Nature (Lond.)* **294:**450–451.
Scofield, V. L., Schlumpberger, J. M., West, L. A., and Weissman, I. L., 1982, Protochordate allorecognition is controlled by a MHC-like gene system, *Nature (Lond.)* **295:**499–502.
Searle, R. F., Sellens, M. H., Elson, J., Jenkinson, E. J., and Billington, W. D., 1976, Detection of alloantigens during preimplantation development and early trophoblast differentiation in the mouse by immunoperoxidase labelling, *J. Exp. Med.* **143:**348–359.
Sellens, M. H., 1977, Antigen expression on early mouse trophoblast, *Nature (Lond.)* **269:**60–61.
Simmons, R. L., and Russell, P. S., 1966, The histocompatibility antigens of fertilized mouse eggs and trophoblast, *Ann N.Y. Acad. Sci.* **129:**35–45.
Snell, G., 1971, The histocompatibility systems, *Transplant. Proc.* **3:**1133–1138.
Solter, D., 1977, Primary sex determination: Immunogenetic Aspects, in: *Immunobiology of Gametes* (M. Edidin and M. H. Johnson, eds.), pp. 207–226, Cambridge University Press.
Thorsby, E. D., Albrechtson, B. O., Bergholtz, H., Hirschberg, H., and Solheim, B. G., 1978, Identification and significance of products of the HLA-D region, *Transplant. Proc.* **10:**313–318.
Valembois, P., 1974, Cellular aspects of graft rejection in earthworms and other metazoa, *Contemp. Top. Immunol.* **4:**121–126.
Vorbrodt, A., Konwinski, M., Solter, D., and Koprowski, H., 1977, Ultrastructural cytochemistry of membrane-bound phosphatases in preimplantation mouse embryos, *Dev. Biol.* **55:**117–134.
Wachtel, S. S., 1983, in: *Immunology of Reproduction* (T. G. Wegmann and T. J. Gill, eds.), pp. 101–126, Oxford University Press, New York.
Webb. C. G., Gall, W. E., and Edelman, G. M., 1977, Synthesis and distribution of *H-2* antigens in preimplantation embryos, *J. Exp. Med.* **146:**923–932.
Wegmann, T. G., Singh, B., and Carlson, G. A., 1979, Allogeneic placenta as a paternal strain antigen immunoabsorbant, *J. Immunol.* **122:**270–274.
Ziomek, C. A., and Johnson, M. H., 1980, Cell surface interaction induces polarization of mouse 8-cell blastomeres at compaction, *Cell* **21:**935–942.

Chapter 13

The Immunological Approach to the Study of Preimplantation Mammalian Development

CAROLE L. BANKA and PATRICIA G. CALARCO

1. Introduction

For the developmental biologist interested in the mammalian system, the preimplantation embryo offers the unique opportunity of studying the mechanisms and cellular interactions involved in the transition of a single cell, the zygote, into a complex multicellular organism. The mouse has proved the most useful model for such studies because development of the preimplantation embryo *in vitro* appears to parallel development *in vivo* During this period the embryo undergoes three cleavages from a zygote to an 8-cell embryo. The 8-cell embryo then undergoes a process of compaction, during which the cells become closely adherent and individual cell boundaries become indistinct. Two to three further cell divisions, junction formation, and the first cellular differentiation result in formation of the blastocyst, a fluid-filled sphere of cells bounded by an epithelial trophectoderm with the inner cell mass (destined to become the embryo proper) at one pole. It is the late blastocyst that begins to implant in the uterine wall.

Concurrent with the visible changes during the preimplantation period are a host of biochemical and physiological changes (reviewed in Johnson, 1979; Sherman, 1979). The mouse embryo is capable of undergoing the first cleavage division in the apparent absence of mRNA synthesis (Golbus *et al.*, 1973; Braude, 1979), although there is evidence of low levels of mRNA transcription in the 1-cell embryo (Clegg and Piko, 1982). A sixfold increase in total RNA occurs between the 2-cell stage and early blastocyst (Piko and Clegg, 1982), suggesting that transcription of the embryonic genome is important during early development. This is in contrast to the early development in lower spe-

CAROLE L. BANKA • Department of Reproductive Medicine, University of California–San Diego, La Jolla, California 92093 • PATRICIA G. CALARCO • Department of Anatomy, University of California, San Francisco, [redacted] 94143.

cies that occurs primarily under maternal control (Davidson, 1976). Perhaps the most obvious biochemical changes during the preimplantation period are those seen in the rate and pattern of protein synthesis mediated primarily by stored maternal mRNA during the early cleavages and later by activation of the embryonic genome (Flach *et al.*, 1982). These changes result in stage-specific patterns of total cellular proteins (Van Blerkom and Brockway, 1975; Levinson *et al.*, 1978; Van Blerkom, 1981) as well as stage-specific patterns of proteins expressed on the cell surface (Johnson and Calarco, 1980*a*). There are also marked changes in protein phosphorylation during the preimplantation period in the mouse (Lopo and Calarco, 1982).

Embryonal carcinoma (EC) cells, the stem cells of teratocarcinomas, which have proved a useful model for some aspects of peri- and postimplantation development in the mouse (reviewed in Martin, 1980), also exhibit stage-specific patterns of total cellular proteins (Martin *et al.*, 1978) and cell surface proteins (Howe and Solter, 1981) associated with differentiation. It has been assumed that among the stage-specific proteins of embryos and EC cells are structural proteins and enzymes involved in determination and cellular differentiation. Thus, the appearance and disappearance of these specific proteins may be reflections of the differential gene expression responsible for normal development. An important goal, therefore, has been to isolate individual stage-specific proteins, to characterize them biochemically, to determine whether their appearance is vital for normal development and, if so, to identify what function they serve. One of the most useful methods for such studies has been the immunologic approach and the results of this approach will be the subject of this chapter.

The specificity of antibody–antigen interactions has led to the evolution of techniques for identification and isolation of single molecules from complex mixtures. Researchers studying development in a variety of systems have made the assumption that a temporally expressed molecule of functional importance should have a unique structural configuration conferring antigenicity. In fact, immunological methods have proved useful in studying the cell surface interactions in aggregating slime molds (Rosen *et al.*, 1976), in migrating mesodermal cells during amphibian gastrulation (Boucaut *et al.*, 1984) and in the developing chick neural retina (Brackenbury *et al.*, 1977; Thiery *et al.*, 1977; Buskirk *et al.*, 1980). Developmentally regulated surface molecules of neuronal (Cohen *et al.*, 1981; Lindner *et al.*, 1983; Sadoul *et al.*, 1983) and nonneuronal (Schachner, 1979; Schachner *et al.*, 1981) cells of the mammalian nervous system have been studied immunologically, resulting in the definition of useful markers for a variety of cell types. Both monoclonal and polyclonal antibodies have been effective in describing the temporal expression of membrane antigens during mammalian spermatogenesis (Vernon *et al.*, 1982; O'Brien and Millette, 1984). Antibodies to cytoplasmic antigens have proved useful developmental markers as well. For example, monoclonal antibodies have been generated for studying the developmental restrictions of avian neural crest-derived cells (Ciment and Weston, 1983).

The antibodies used as reagents for immunological studies can be generated in a variety of ways. Syngeneic antisera (made within the same inbred strain) will detect only antigens not present (or not recognized by the immune system) in the adult animal. These antisera have a limited number of specificities; thus, it is often difficult to generate a detectable immune response with syngeneic immunization. Syngeneic sera have been employed to study teratocarcinoma cell surface antigens (Artzt *et al.*, 1973; Stern *et al.*, 1975; Hahnel and Eddy, 1983). Allogeneic antisera (made within the same species) will detect only those antigens that differ from one strain to another within a species. These antisera have been useful in the study of histocompatibility antigens in the mouse (Webb *et al.*, 1977; see also Chapter 12, this volume). Heterogeneic antisera (made in a different species) have multiple specificities; however, they may be rendered very specific by appropriate absorptions (Larraga and Edidin, 1979; Johnson and Calarco, 1980*d*). Heterogeneic sera (heterosera) can also be used to isolate an antigen of interest which can then serve as an immunogen for a more specific antiserum. This has been effective in the study of neural retina adhesion molecules (Brackenbury *et al.*, 1977; Thiery *et al.*, 1977). Finally, monoclonal antibodies (Kohler and Milstein, 1975) generated through any of the immunization regimens outlined above have a single antigenic specificity. In the following discussion of the progress made in studying preimplantation development using immunological tools, it should become obvious that antibodies generated by each of these methods have advantages and disadvantages and that the most productive experimental approaches may involve the use of monoclonal antibodies and other antisera in combination.

Once useful antibodies have been generated, a variety of questions can be addressed. In this chapter, the information gained about preimplantation development through immunological studies is discussed in light of the variety of questions that lend themselves to this experimental approach. Representative studies, particularly recent ones, have been chosen to illustrate each topic. It is not our intention to catalog all antibodies that have been employed to study early mammalian development, as several reviews have done just that (Solter and Knowles, 1979; Wiley, 1979; Johnson and Calarco, 1980*b*). In addition, Chapter 12 (this volume) deals comprehensively with the expression of major and minor histocompatability antigens during early development.

An important aspect of the study of embryonic antigens has been the discovery that certain antigens usually detected only during the embryonic period are expressed again by specific tumors in the adult (e.g., see Coggin and Anderson, 1974). The numerous implications of the expression of these "oncofetal" antigens are not dealt with here. The reader is referred to a comprehensive volume on fetal antigens and cancer edited by Evered and Whelan (1983).

A series of studies employing antibodies to the zona pellucida have had important implications for fertility control in mammals. As this chapter deals only with antigens of the embryo cell surface, the reader is referred elsewhere for information on this aspect of immunological studies (Dunbar, 1983).

2. Developmental Tracking of Antigens

2.1. Cell Surface Antigens

This has been the most widely applied use of antibodies in studies of preimplantation development. A variety of antibody labelling techniques have been utilized to demonstrate clearly delineated periods of cell surface antigen expression. For example, a rabbit antiserum to a purified preparation of intracisternal A virus particles (IAP) detects cell surface antigens that appear first on the mouse zygote, are expressed maximally on 1–8-cell embryos, and are not detectable on morulae (16-cell) or later embryos (Huang and Calarco, 1981*a*). This period of surface expression was confirmed using both fluorescein- and [^{125}I]labeled second antibodies as well as ultrastructural immunocytochemistry. The stage-specific embryonic expression of IAP surface antigens is illustrated in Figure 1. These viral antigens are expressed on the cell surface before and after the 1–8-cell stages when the retroviral IAP particles can be identified within the embryos (Calarco and Szollosi, 1973). Of particular interest is the fact that cell lines that synthesize IAP show no surface expression of the antigen detected on embryos, suggesting that embryos process IAP-associated antigens differently than other cells carrying the retrovirus (Huang and Calarco, 1981*a*).

Perhaps the most extensively studied embryonic surface antigens are the F9 antigens detected first by syngeneic antisera developed against F9 cells, a pseudonullipotential mouse teratocarcinoma stem cell line (Artzt *et al.*, 1973). These antigens are present on sperm, but not on oocytes, and appear on embryos 5–6 hr after fertilization (see Jacob, 1977). The syngeneically defined F9 antigens reach maximum expression at the 8-cell to morula stages and are present on both trophoblast and inner cell mass (ICM) of the blastocyst although in decreasing amounts (Jacob, 1977). The antigens are present on postimplantation embryos through day 8, but the tissue distribution has not been determined (Buc-Caron *et al.*, 1978).

A monoclonal antibody (McAb) has been generated using F9 cells as the immunogen (Solter and Knowles, 1978). The antigen detected by this McAb, stage-specific embryonic antigen-1 (SSEA-1) is present on sperm and on embryos beginning at the late 8-cell stage. The trophoblast of the blastocyst ceases to express SSEA-1 in detectable levels 24 hr after blastocele formation, while antigen expression increases on the ICM and is seen on both ectoderm and endoderm. During the post-implantation period, SSEA-1 is restricted to embryonic ectoderm and visceral endoderm (Fox *et al.*, 1981). This type of tissue segregation of a specific antigen raises intriguing questions concerning function; however, SSEA-1 can also be found later in development in the brain, primordial germ cells, kidney tubules, and the epithelial linings of various organs of the genital tract (Fox *et al.*, 1981). The antigenic determinant for anti-SSEA-1 has been identified and is apparently quite small (see Section 6.2). Therefore, the wide variety of seemingly unrelated tissues that express SSEA-1 in

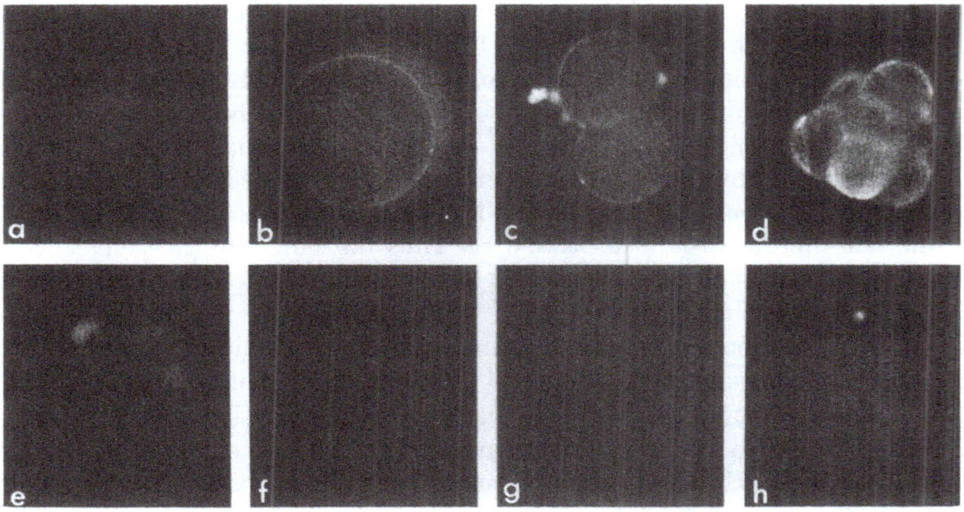

Figure 1. Photomicrographs illustrating cell surface binding of anti-IAP to live mouse oocytes and embryos. Indirect immunofluorescence reveals that the oocyte (a) is negative; the zygote (b), 2-cell (c), and 8-cell (d) are increasingly positive; binding is minimal on the morula (e) and has disappeared on the blastocyst (f). Normal rabbit serum (control) reactivity is illustrated for the 2-cell (g) and 8-cell (h). (From Huang and Calarco, 1981*a*.)

later development may, in fact, bear divergent surface molecules related only in that they share a small antigenic determinant. This drawback in the use of highly specific McAbs must be kept in mind when interpreting results. The different stage and tissue specificities of SSEA-1 and the original F9 antigen(s) (Artzt *et al.*, 1973) indicate that embryonic cell surface antigens of F9 cells are more numerous than originally suggested. In fact, yet another McAb to F9 cells, IIC3, has a pattern of expression distinct from that of the other two reagents (Marticorena *et al.*, 1983).

The Forssman antigen is another that appears on murine preimplantation embryos as well as on cells of adult animals of a variety of species. A rat McAb with Forssman specificity was used to identify the antigen on embryos, where it appears at the late morula stage (Stern *et al.*, 1978; Willison and Stern, 1978). This McAb binds cells within the mouse embryo through at least day 16 of embryogenesis (Stinnakre *et al.*, 1981). Again, the question arises as to whether a small antigenic determinant is borne by different molecules. Milstein and Lennox (1980) use the term "jumping antigens" to discuss this phenomenon. The use of a polyclonal antiserum which recognizes multiple epitopes or the invariant regions (e.g., the protein of a glycoprotein) of an antigen would aid in clarifying questions of tissue distribution of these antigens. In fact, Willison and coworkers (1982) have used a highly specific rabbit anti-Forssman serum on preimplantation embryos and detect the same stage specificity but more intense and uniform binding compared with the McAb. A comparison of post-implantation embryonic tissue distribution of antigen detected by this antiserum would be most interesting. Of further interest in studies of Forssman antigen expression is the observation that the globoside, globotetraosylceramide, appears on embryos at the 2–4-cell stage, 24 hr before the earliest expression of Forssman (Willison *et al.*, 1982). Forssman differs from this globoside only by the addition of one terminal sugar residue, and the authors suggest that the enzyme required for this reaction may be expressed initially at the late morula stage (see Section 6). Thus the immunological detection of antigenic stage specificities may yield information concerning metabolic capabilities of the early embryo.

While antibodies generated against embryonal carcinoma (EC) cells and other antigens have been very useful in the study of embryonic antigens, only antibodies prepared to embryos themselves can be expected to detect antigens specific solely to the preimplantation embryo. Although embryos are believed to be only weakly antigenic, two laboratories have succeeded in generating antibodies by embryo immunization. Immunization of rabbits with mouse blastocysts has produced two antisera, A-BL1 (Wiley and Calarco, 1975) and A-BL2 (Johnson and Calarco, 1980*c*), with similar stage-specific binding patterns, but the antigens defined by A-BL2 have been studied more extensively. A-BL2 detects antigens first expressed on the cell surface at the 4-cell stage with peak expression at the 8-cell stage and a decline by the blastocyst stage. The antibody does not bind to any murine EC cells, nor to any fetal or adult tissues tested, and therefore, appears to be truly embryo specific (Johnson and Calarco, 1980*c*). This reagent has made possible the isolation and biochemical characterization of a set of embryo-specific molecules (see Section 6.1).

A series of 11 embryo-specific monoclonal antibodies with differing stage specificities were generated in our laboratory by immunization of a rat with 8-cell mouse embryos (Banka, 1982; C. L. Banka and P. G. Calarco, unpublished results). Unfortunately, attempts to isolate and identify the antigens detected by these antibodies through immunoprecipitation or electrophoretic gel overlay techniques proved unsuccessful.

A McAb, prepared by immunization of a rat with 4–8-cell embryos, detects an antigenic determinant on mouse oocytes and early embryos that becomes restricted first to the ICM and later to the primitive endoderm (Shevinsky *et al.*, 1982). This antigen, SSEA-3, is present on human teratocarcinoma cells but not those of murine origin.

2.2. Intracellular Antigens

The studies summarized above have all described stage-specific distributions of cell surface antigens which, however, may not be reflective of the stage-specific synthesis of these molecules. Whereas the timing of appearance of an antigen on the surface may be related to function of the surface molecules, the timing of appearance in the cytoplasm is reflective of stage-specific gene transcription and/or mRNA translation. In this regard, studies employing immunoprecipitation have revealed synthesis of the BL antigens at the 2-cell stage, one cleavage division before cell surface expression (Johnson and Calarco, 1980*d*). In addition, BL antigens cease to be synthesized in the expanded blastocyst but remain on the cell surface a few hours longer.

Stage-specific intracellular as well as extracellular expression has been demonstrated for basal lamina components which show an interesting asynchrony in their first developmental appearance. Leivo and *et al.* (1980) demonstrated the appearance of laminin both intra- and intercellularly at the 16-cell stage, while Wu *et al.* (1983) were able to identify laminin as early as the 8-cell stage. In the latter study, the first intracellular appearance of the molecule was found to correlate closely with expression on the blastomere surfaces where it was concentrated at the intercellular contact zones. Another basal lamina component, type IV collagen, appears 2 days later in development. It is first detectable in the blastocyst primarily in the ICM (Leivo *et al.*, 1980). A third basal lamina glycoprotein, entactin, cannot be detected until the time of blastocyst hatching (Wu *et al.*, 1983). Although all three of these molecules will function as basal lamina components later in development, the temporal differences in their early expression may indicate differential functions in early development. In this regard it has been suggested that laminin, the first to appear, may serve not only for intercellular adhesion but also as a factor determining polarization of blastomeres during early differentiation (Wu *et al.*, 1983).

Significant temporal differences have been demonstrated between the first cytoplasmic appearance and the structural organization of several cytoskeletal elements. Although immunoblotting experiments reveal proteins reactive with cytokeratin antibodies in oocytes and all preimplantation stages, a diffuse cytoplasmic staining is seen in these stages, and significant fibrillar staining is not seen until the outgrowth of trophectoderm has begun (Lehtonen *et al.*, 1983). Likewise, Endo A and B, intermediate filament proteins, have been detected by immunoprecipitation as early as the 4–8-cell stage in mouse embryos (Oshima *et al.*, 1983) although intermediate filaments are not detected by electron microscopy until the late morula stage (Jackson *et al.*, 1980). An interesting question remains as to why the embryo synthesizes these molecules well in advance of their known function as cytoskeletal elements.

2.3. Gene Expression

Studies such as those described above chronicle molecular events associated with early embryonic differentiation. Unfortunately, it is unclear whether the stage-specific appearance of any of these antigens reflects activation of the embryonic genome or whether they are synthesized from stored maternal mRNA. This question is vital in determining the underlying genetic control of early mammalian development and has been addressed by seeking the earliest possible evidence of paternal gene expression. Krco and Goldberg (1976) reported the appearance of H-Y (histocompatibility-Y) on approximately half of a population of 8-cell embryos based on a cytotoxicity study, suggesting these were the male embryos. Epstein *et al.* (1980) confirmed this observation by karyotyping the unaffected embryos in a similar cytotoxicity study. As H-Y synthesis is under control of the Y chromosome (Koo *et al.*, 1977), this was clear evidence of embryonic (paternal) gene expression.

Evidence for expression of the paternal genome as early as the 2-cell stage came from an immunological study of electrophoretic variants of β_2-microglobulin (Sawicki *et al.*, 1981). When male and female mice bearing different variants of β_2-microglobulin were bred, the embryos were shown by immunoprecipitation to synthesize both maternal and paternal variants at the 2-cell stage.

Conversely, several lines of evidence suggest that SSEA-3 is not a product of the embryonic genome. Several inbred strains of mice do not express SSEA-3 as a preimplantation antigen. When crosses were made between SSEA-3-positive and SSEA-3-negative strains, the F1 embryos expressed SSEA-3 only when the maternal strain was positive (L. H. Shevinsky, personal communication). In addition, in reciprocal pronuclear transplantations between zygotes of antigen-positive and antigen-negative strains, SSEA-3 expression by the embryo thus derived was determined by the cytoplasm of the zygote and not by the transplanted pronucleus (McGrath and Solter, 1983). These studies strongly suggest that SSEA-3 is synthesized from maternal mRNA contributed by the oocyte.

Thus, we see that the earliest and most frequent use of antibodies, that of determining stage specificity of molecular expression, has provided a wealth of information concerning the controls or manifestations of early differentiation in the mammal. Stage specificity of expression has been interpreted as an indication that antigens serve an important function during that period of expression. This question is discussed in Section 5. Future work should be directed toward resolving whether individual stage-specific antigens are products of stored maternal mRNA or the embryonic genome. In addition to those methods described above, this might be accomplished through study of synthesis of a given antigen in the presence of α-amanitin. This method has proved useful in determining that the enzyme HPRT is translated on maternal mRNA in early embryos until the time of compaction (Harper and Monk, 1983).

3. Cell Marking, Cell Purification, and Separation

Many stage-specific antigens also show tissue- or cell type-specific expression in the later period of preimplantation development and the early postimplantation period and can therefore be used as markers for specific cell types. Anti-SSEA-2 was generated by immunization of BALB/c mice with a BALB/c X human hybrid cell line that carries only one human chromosome and the integrated SV40 genome (Shevinsky *et al.*, 1981). The cell surface antigen detected segregates to the ICM of blastocysts and later to parietal endoderm and can thus be used as a marker for these cell types. Other antibody markers for the ICM include a commercially prepared anti-human chorionic gonadotropin (hCG) (Wiley, 1980) and an antibody to PCC4, a pluripotential EC cell line, which detects antigens only on the ICM of the late blastocyst (Gachelin *et al.*, 1977). SSEA-1 segregates to the ICM of the late blastocyst and subsequently to embryonic ectoderm and visceral endoderm (Fox *et al.*, 1981).

A monoclonal antibody, IIC3, made against F9 TC cells detects antigens first at the morula stage and appears to be a useful marker for early trophoblast cells (Marticorena *et al.*, 1983). The authors suggest that the differential pattern of IIC3 binding might enable discrimination between inside and outside blastomeres of compacted morulae, although it has not been tested.

Markers also exist for later trophoblast and trophectoderm cells. As described above, the fibrillar pattern of cytokeratin immunostaining in the cytoplasm is specific to trophoblast cells at the outgrowth stage (Lehtonen *et al.*, 1983). Monoclonal and polyclonal antibodies prepared to isolated cytoskeletal proteins of trophoblastoma cells (Brulet *et al.*, 1980) or parietal endodermal cells (Oshima, 1981) are specific for intermediate filaments first seen in trophectoderm of the blastocyst and also serve as excellent markers for EC cell differentiation to either parietal (Oshima, 1982) or visceral endoderm (Grover *et al.*, 1983). A cell surface antigen detected by an antiserum produced to ectoplacental cone stage mouse embryos segregates to trophoblastic giant cells and extraembryonic ectoderm following implantation (Searle and Jenkinson, 1978).

A direct application of antibody marking is the technology for separation of cells by virtue of the presence or absence of surface antibody binding. This can be done by affinity binding of whole cells to antibody-coated beads on columns or antibody-coated beads with magnetic cores. Cell sorting on the basis of presence or absence of a fluorescent marker can also be employed once a specific antibody marker has been identified. Pure viable cell populations can be recovered for further study with any of these methods. While it is probable that none of these methods will be employed for separation of cells of the preimplantation embryo due to the limitation of material, they should prove quite useful with EC cells and their derivatives. In fact, a fluorescence-activated cell sorter has been used to quantitate the number of F9 EC cells bearing each of four different glycolipids (Willison *et al.*, 1982). This could easily be adopted as a method for cell purification on the basis of binding of the anti-glycolipid antibodies.

An alternate approach to cell purification is to kill through immune cytolysis all cells bearing a common surface antigen leaving a viable cell population that does not express the antigen. This technique has been applied quite successfully with preimplantation embryos. On the basis of H-Y antigen expression, male embryos have been killed selectively (Krco and Goldberg, 1976;

Epstein *et al.*, 1980). Trophectoderm cells of the blastocyst have also been lysed through the use of cytotoxic antisera. This technique termed "immunosurgery" results in pure populations of ICM cells (Solter and Knowles, 1975).

The use of these antibody markers for identifying and purifying cell populations on the basis of presence or absence of any given antigen has not been fully explored. One obvious potential use is that of identifying and purifying cell types resulting from the differentiation of EC cells in culture.

4. Modulation, Topographical Distribution, and Cellular Localization of Antigens

As a result of crosslinking by the binding of divalent ligands such as IgG antibodies, cell surface molecules are often redistributed within the membrane of the cell. This phenomenon has been most extensively studied on lymphocytes (de Petris and Raff, 1973). Antigens can be passively clustered into multiple discrete patches or swept to one pole of the cell into a single cap in an energy requiring process. Either of these processes can lead to internalization of antibody–antigen complexes, resulting in temporary or permanent disappearance of the antigens from the cell surface. A similar ligand-induced internalization of cell surface molecules, the loss of hormone receptors from a cell in response to hormone binding, termed *downregulation,* has important physiological consequences (e.g., see Gavin *et al.*, 1974).

Although capping of antigens has never been described on cells of the preimplantation embryo, patching has been observed with several antibodies. The BL antigens are visualized as small discrete patches following immunofluorescent staining of all positive stages (Johnson and Calarco, 1980c). An example of this patching on an early 8-cell embryo is shown in Figure 2A. Handyside (1980) demonstrated patching and internalization of surface molecules on 8-cell and 16-cell embryos with a variety of ligands, including a rabbit anti-mouse species antiserum. Antigen patching has also been induced with anti-F9 antibodies (Jacob, 1977) and anti-Forssman as well as anti-globoside sera (Willison *et al.*, 1982).

Several workers using different antibodies have described a sweeping of antigens to the apical pole of the late 8-cell embryo, as illustrated in Figure 2B (Handyside, 1980; Johnson and Calarco, 1980c; Hahnel and Eddy, 1983). This redistribution of antigens is apparently not antibody-induced modulation, but rather a reflection of the reorganization of microvilli to apical regions of the blastomeres associated with compaction (Calarco and Epstein, 1973; Ducibella *et al.*, 1977; Reeve and Ziomek, 1981) and has been termed *polarization* (Handyside, 1980; Johnson *et al.*, 1981). Interesting patterns of antigen redistribution on later embryos were revealed with antiserum from a rat immunized with 8-cell embryos (C. L. Banka and P. G. Carlarco, unpublished results). As illustrated in Figure 3A, the rat antiserum used in a double indirect immunofluores-

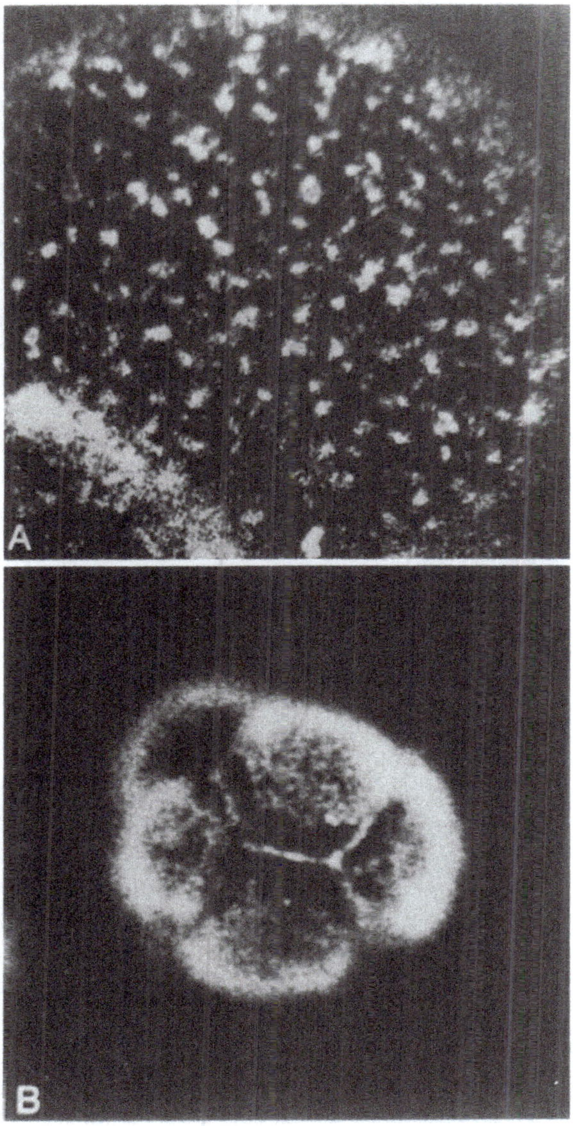

Figure 2. Photomicrographs illustrating cell surface binding of anti-BL2 to live 8-cell mouse embryos (indirect immunofluorescence). (A) Single blastomere of an early 8-cell embryo showing distribution of BL antigens in small, discrete patches. The patching is a consequence of antibody binding. (B) Compacting 8-cell embryo showing polarized redistribution of anti-BL2 binding sites to the apical poles. This redistribution is a reflection of the reorganization of the microvilli to the apical regions during compaction. (From Johnson and Calarco, 1980c.)

cence assay showed markedly different patterns of antigen distribution on individual trophectoderm cells of the blastocyst. Although these patterns might result from antibody-induced redistribution of antigens, it is more likely that they reflect differences in microvillar distribution on the blastomeres, which has been described with scanning electron microscopy (see Figure 3B).

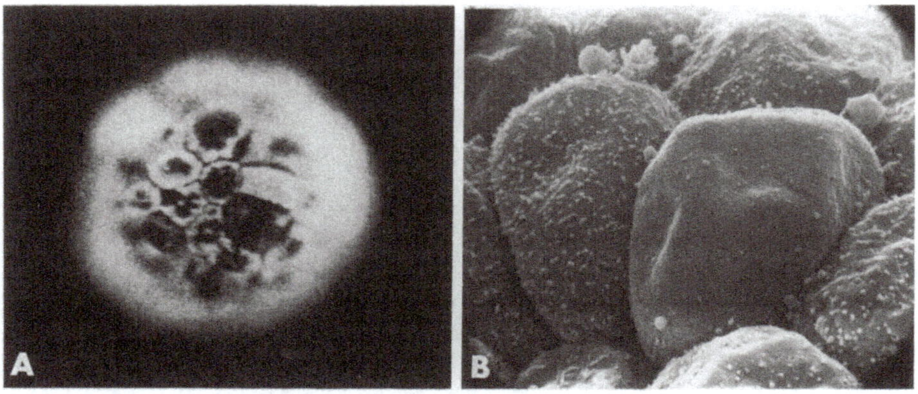

Figure 3. (A) Photomicrograph of live mouse blastocyst following exposure to a rat antiserum to 8-cell mouse embryos (double indirect immunofluorescence). Note differences in antibody binding patterns on individual blastomeres. This may be the result of antibody induced redistribution of antigens or may reflect differences in the number and distribution of microvilli on individual blastomeres. (B) Scanning electron micrograph of a portion of the surface of a mouse blastocyst. Note the differences in number and distribution of microvilli on individual blastomeres.

The implications of the data reported thus far concerning cell surface modulation in the early embryo are unclear. The capacity of a cell for ligand-induced redistribution of surface molecules is probably a reflection of the fluidity of the membrane as well as molecular linkage to the underlying cytoskeleton; this capacity may change during development. It has been shown, for example, that lateral diffusion in the plasma membrane of the mouse oocyte is restricted after fertilization (Johnson and Edidin, 1978), although this restriction may have been specific for the lipid analogue employed in the study. Subsequent experiments revealed no large-scale changes in lateral diffusion rates following fertilization when a variety of lipid analogues were tested but revealed organization of the oocyte membrane into lipid domains (Wolf *et al.*, 1981).

The topographical localization of antigens, a constant feature of embryonic cells rather than a consequence of antibody-mediated redistribution, can be studied by fixing cells prior to antibody exposure. Several intriguing observations concerning antigenic differences associated with free versus apposed cell surfaces have been made using this approach. Using a rabbit anti-mouse species antiserum, Wiley and Eglitis (1981) showed a more intense staining of surface antigens on the free surfaces of the cells of the morula than on the apposed surfaces. The differences in antigen distribution persisted even after dissociation of cells in calcium-free medium, suggesting that it was an intrinsic feature of the cells rather than technical artifact. However, this may be reflective of the presence of microvilli on only the free surfaces of the cells.

Studies of cytoplasmic localization of myosin have revealed a polarized cortical distribution of the cytoskeletal protein that persists throughout the cleavage stages of development (Sobel, 1983*a*). The myosin is restricted to the peripheral region of the blastomeres and is undetectable in regions of cell contact. Disaggregation of embryos resulted in a loss of the polarization of myosin within the blastomeres; the myosin appearing as a continuous cortical layer (Sobel, 1983*b*). When the isolated blastomeres acquired new cell contacts either between two daughter cells as a result of cell division or through aggregation of single blastomeres or whole embryos, polarization of myosin away from the regions of cell contact was reestablished.

Similar observations have been made at the ultrastructural level with A-BL2 (Polak-Charcon *et al.*, 1983*a*,*b*, 1985). At the 2-cell stage, before expression of BL antigens on the surface of the embryo, immunocytochemistry reveals antigens localized only in the cortical region underlying the free surface of blastomeres but undetectable deeper in the cytoplasm or in the area of cell contact. This cortical localization persists through cleavage (Fig. 4) and disappears at the late blastocyst stage concomitant with cessation of BL antigen synthesis. The fact that no rough endoplasmic reticulum is present at this stage, and that free polysomes abound in the cortical region at this stage of embryogenesis (Calarco and Brown, 1969; Hillman and Tasca, 1969) raises the possibility that BL antigens are synthesized in this cortical region before expression on the cell surface. It is probable that continued studies, particularly at the ultrastructural level, of cell surface modulation and topographical localization of a variety of antigens will shed light on cellular processes operational during early development and future studies should be thus directed.

5. Antigen Function

Several rabbit heterosera exhibit a concentration-dependent interference with normal embryonic development in culture when compared with normal rabbit sera. These include A-BL1 (Wiley and Calarco, 1975), A-BL2 (Johnson and Calarco, 1980c), and A-N1, an antiserum prepared against the nullipotential mouse TC stem cell line, Nulli SCC1 (Calarco and Banka, 1979). Unfortunately, the specific function disturbed by antibody binding during embryogenesis has not been identified with any of these sera.

To date, the only specific preimplantation embryonic process successfully blocked immunologically is that of compaction of the 8-cell embryo. Kemler and co-workers (1977) have prevented compaction by culturing embryos in the presence of Fab fragments (univalent antibody fragments that cannot induce crosslinking) made from a rabbit antiserum to F9 cells. This phenomenon is reversible in that embryos washed free of Fab will go on to compact and form blastocysts. Fab added to early compacted embryos causes *decompaction*.

In similar experiments, a rabbit heteroserum generated by immunization

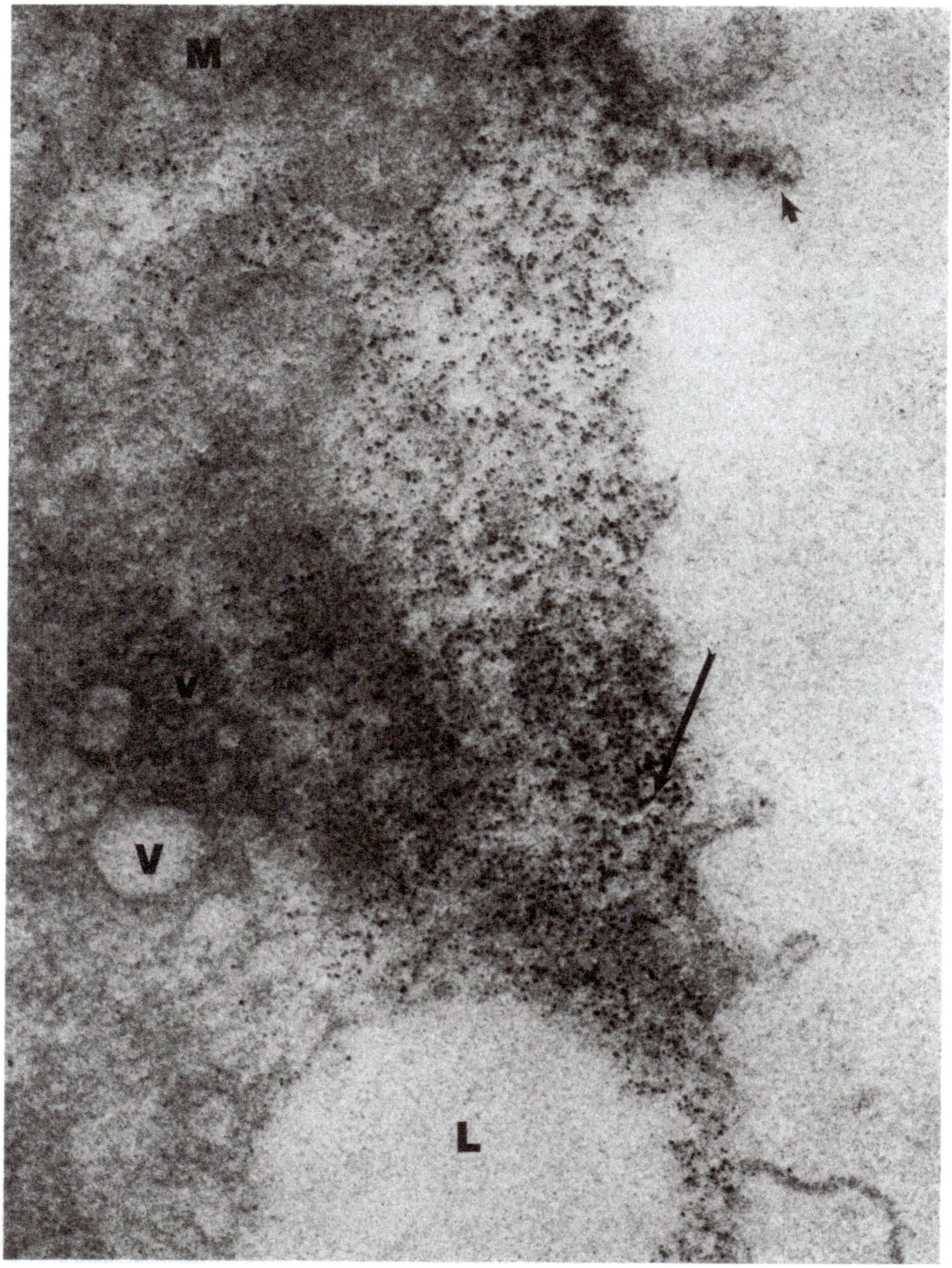

Figure 4. Transmission electron micrograph of an unstained section through the cortical region of an outer cell of a mouse morula permiabilized with saponin and labelled with anti-BL2 serum followed by protein A-5 nm gold. Note the localization of the gold particles in the apical cortical region and microvilli (short arrow). The long arrow denotes a region suggestive of channels or small

with the nullipotential embryonal carcinoma (EC) cell line, LS5770, has been found to prevent compaction of embryos without initially affecting cell division (Johnson *et al.*, 1979). Associated with the prevention of compaction is a general absence of junctional complexes, failure of the establishment of cytoplasmic and cell surface polarization in the trophectoderm cells and reduction or absence of ICM differentiation. Although this study serves as an example of the usefulness of antibodies in dissecting complex cellular interactions during development, the use of an unabsorbed and, therefore, nonspecific heteroserum does not allow for identification of the specific functional cell surface molecules bound by antibody during the prevention of compaction.

Other divalent antibodies have been shown to prevent compaction of the 8-cell embryo (Ducibella, 1980; Damsky *et al.*, 1983; Richa *et al.*, 1985). Interestingly, two of these antibodies, one generated to serum-free medium from cultures of MCF-7 human mammary carcinoma cells (anti-SFMII) and one to a purified glycoprotein from the same source (anti-gp80) block endoderm formation in the early embryo (Richa *et al.*, 1985). In addition, divalent antibodies and Fab fragments have been shown to cause junctional modulation and/or inhibition of a "compactionlike" process in EC cells (Dunia *et al.*, 1979; Hyafil *et al.*, 1980; Kartha *et al.*, 1983). Fab fragments from an antiserum raised against trypsin treated F9 EC cells have been utilized to demonstrate that 1–8-cell embryos share a common calcium-dependent cell adhesion site with teratocarcinoma cells (Ogou *et al.*, 1982). This adhesion system is presumably in effect before compaction begins. Finally, relating to the post-implantation stage, two antisera, goat antiserum to a BHK fibroblast glycoprotein fraction (anti-gp140) and goat antiserum to medium from mouse mammary tumor line (anti-SFMI), have been shown to prevent trophoblast outgrowth in culture (Richa *et al.*, 1985).

The ability to manipulate a specific process such as compaction immunologically during early mammalian embryogenesis was a significant step forward. It was first suggested that the molecule(s) detected by antibodies that block compaction might mediate calcium interactions necessary for intercellular adhesion (Kemler *et al.*, 1977). Alternatively, antibodies may be masking specific molecules involved in the mechanisms of recognition or adhesion required for compaction (Hyafil *et al.*, 1980). However the targets for antibody binding operate in mediating compaction, the development of antibodies that inhibit this process offered the technical means for isolation and characterization of these developmentally important molecules. In fact, such studies have been carried out in several laboratories and serve as examples of the final topic of this chapter, the use of antibodies in biochemical and molecular characterization of specific embryonic antigens.

---

vesicles with gold particles aligned along the walls. The labeling intensity decreases dramatically toward the interior of the cell. Certain organelles are not labeled by the gold: M, mitochondrion; V, vesicles of varying diameters; L, probable lipid droplet. Embryos treated with normal rabbit serum (control) and protein A-5 nm gold do not show this cortical labelling. Magnification: × 65 000.

6. Molecular and Biochemical Characterization of Antigens

One of the most significant advances in the application of immunological methods to the study of early embryogenesis has been the use of mono- and polyclonal antibodies for the biochemical characterization of developmentally important antigens. In some cases this has been done through immunological comparison of unknown and known antigens using polyclonal antibodies. In other cases, comparisons of the epitopes of unknown antigens with those of known antigens have been made using monoclonal antibodies. Yet another approach has been the isolation of molecules by a variety of biochemical separation techniques. Here antibody absorption serves as the assay for antigens following each step in the purification scheme. Finally, antigens have been isolated by immunoprecipitation, permitting further characterization. Although these approaches have been primarily directed toward isolation of embryonal carcinoma (EC) antigens, radiolabeling techniques have allowed for a scaling down of methodology such that a few laboratories have succeeded in immunoprecipitating antigens from small numbers of embryos. A variety of studies have been directed toward characterization of entire protein or glycoprotein antigens, while others have been concerned with characterization of the specific carbohydrate antigenic determinants on glycolipids.

6.1. Protein and Glycoprotein Antigens

The first compaction-related molecule to be purified was an 84,000-M_r glycoprotein detected by rabbit anti-F9, Fab fragments of which are capable of inhibiting compaction of embryos. The glycosylated portion of the gp84 molecule contains galactose, fucose, mannose, and N-acetylglucosamine (Hyafil *et al.*, 1980). The antigen can be released from F9 cell membrane preparations by detergent solubilization or trypsin digestion; calcium protects the antigen from complete proteolysis by the trypsin. Immunization with the tryptic fragments resulted in generation of a rat monoclonal antibody capable of precipitating gp84 (subsequently called *uvomorulin*) in the presence of calcium (Hyafil *et al.*, 1981). Further studies with the mono and polyclonal antibodies have revealed that uvomorulin normally exists as a 120,000-M_r glycoprotein with a transient 135,000-M_r precursor and that the gp84 originally identified is a proteolytic fragment (Peyrieras *et al.*, 1983). In addition, uvomorulin was found to be highly hydrophillic suggesting that it is a nonintegral ("exocellular") molecule associated with the cell surface rather than an integral membrane component (Peyrieras *et al.*, 1983).

A goat antiserum to molecules shed by human mammary carcinoma cells in culture also blocks compaction of the early embryo as well as disrupting cell–cell interactions of epithelial cells (Damsky *et al.*, 1983). This antiserum detects an 80,000-M_r glycoprotein in medium that shares with uvomorulin the property of being protected from proteolysis by calcium. A more specific polyclonal antiserum generated by immunization with purified gp80 also inhibits compaction and recognizes gp80 as well as a 120,000-M_r band from detergent lysates of epithelial cells. The common biological and biochemical properties suggest that the cell adhesion molecule detected in this study is uvomorulin. Thus a molecule of functional significance in both embryonic and adult tissues has been detected and isolated immunologically using very different immunogens. Both studies serve as excellent examples of the use of a complex heterogeneic antiserum for purification of an unknown antigen, which then serves as immunogen for production of a more specific reagent, a monoclonal antibody in one case and a highly specific polyclonal in the other.

A third laboratory has defined a calcium-dependent cell adhesion system in embryos and EC cells using polyclonal and monoclonal antibodies both of which detect a 124,000-M_r protein or glycoprotein with all the properties of uvomorulin (Ogou *et al.*, 1982; Yoshida and Takeichi, 1982; Yoshida-Noro *et al.*, 1984).

Proteins and glycoproteins specific to embryos have also been isolated and characterized biochemically. The first report of stage-specific antigens immunoprecipitated directly from preimplantation embryos was that of Johnson and Calarco (1980*d*) with anti-blastocyst serum (A-BL2). A-BL2 brings down a pair of molecules in the 65,000–70,000-M_r range from 2-cell, 8-cell and morula stage embryos. Two-dimensional electrophoretic analysis of the BL antigens reveals at least three pairs of proteins differing in isoelectric points. This charge heterogeneity may represent varying degrees of glycosylation during intracellular synthesis or following insertion in the membrane (Johnson and Calarco, 1980*d*).

Intracisternal A particle (IAP) -associated antigens have been precipitated from preimplantation embryos with anti-IAP serum (Huang and Calarco, 1981*b*). Three molecular species are detected in the precipitates from 2–8-cell embryos that correlate temporally with the presence of IAP in embryos and have apparent molecular weights of 67,000, 69,000, and 73,000. The 73,000-M_r antigen is thought to be the major core protein, identified as p73 in IAP from neoplastic cells (Patterson *et al.*, 1978). One or more of the precipitated molecules may represent the IAP-associated embryo surface antigens; however, surface labelling studies indicate that the embryonic surface molecules may be composed of two molecules of 25,000 and 50,000 M_r (P. G. Calarco and M. Siebert, unpublished results).

A rabbit antiserum (A-N1) prepared against the nullipotential teratocarcinoma (TC) stem cell line Nulli SSC1, precipitates as the major component from preimplantation embryos a 67,000-M_r glycoprotein (Banka and Calarco, 1986). This molecule appears as a single elliptical spot with a mean pI of 5.3 in two-dimensional gels (Fig. 5). Because gp67 is also synthesized by TC cells, it could be isolated with A-N1 in quantities sufficient to serve as the immunogen for a more specific heteroserum or monoclonal antibody.

Using a rabbit antiserum to the embryoid bodies differentiated from the TC cell line, OTT6050, Webb (1980) characterized a series of antigens from surface labelled embryos. Immunoprecipitation resulted in detection of proteins of 115,000, 70,000, and 48,000 M_r on cleavage stage embryos. The 70,000-M_r

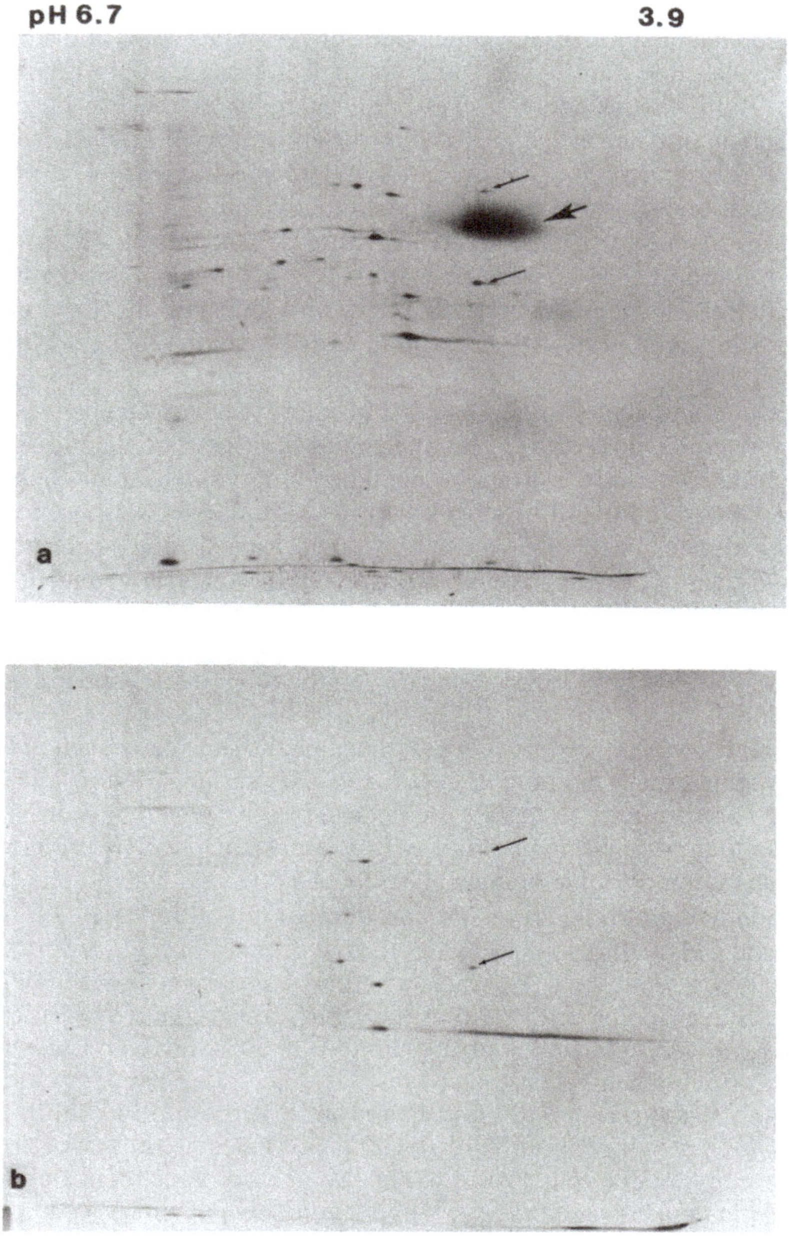

Figure 5. Autoradiographs of two-dimensional electrophoretic gel separation patterns of immunoprecipitates from [^{35}S]methionine-labeled mouse blastocysts with (a) rabbit anti-Nulli SSC1 serum and (b) rabbit anti-Nulli SSC1 serum absorbed with live Nulli SCC1 teratocarcinoma stem cells. Small arrows in (a) and (b) indicate the same two spots for orientation. The large arrow (a) indicates the embryonic antigen, gp67, detected by the antiserum.

protein disappeared at the blastocyst stage while two new proteins of 82,000 and 22,000 M_r appeared.

Other proteins precipitated from preimplantation embryos include the intermediate filament proteins Endo A and Endo B with molecular weights of 55,000 and 50,000, respectively (Oshima *et al.*, 1983). These embryonic proteins also appear to have multiple charged forms on two-dimensional gels due at least in part to differences in phosphorylation (Oshima, 1982).

Muramatsu and co-workers (1979) using a syngeneic anti-F9 serum precipitated a variety of molecules from F9-41 cells labeled with radioactive sugars. The major glycoproteins precipitated ranged from 40,000 to 43,000 M_r. The precipitates contained rather large amounts of fucose, galactose, and glucosamine but only small amounts of mannose. Furthermore, most labeled sugars in the glycoproteins were found to be in large glycopeptides (≥6000) following protein digestion. High-molecular-weight glycopeptides are synthesized in quantity by mouse blastocysts and EC cells including F9-41, but not by the differentiated derivatives of EC cells (Muramatsu *et al.*, 1978). The decrease in large glycopeptides accompanying differentiation correlates with the loss of syngeneically defined F9 antigen(s) (Jacob, 1977) and may be a general characteristic of early embryonic differentiation. Of further interest in the immunoprecipitation studies with F9-41 cells (Muramatsu *et al.*, 1979) was the fact that 20–25% of the galactose-labeled material precipitated by syngeneic anti-F9 was found to be glycolipid. In this respect, Larraga and Edidin (1979) have shown that 88% of the binding activity of an antiserum to the TC cell line, TerC, can be eliminated by absorption with TerC glycolipids. As will be noted in the discussion to follow, most of the early embryonic antigens detected by monoclonal antibodies to date are glycolipids.

6.2. Carbohydrate Antigenic Determinants

A specific antigenic determinant associated with glycolipid has been identified on EC cells and preimplantation embryos beginning at the late morula stage with the anti-Forssman McAb (Stern *et al.*, 1978; Willison and Stern, 1978), while an antiserum to another neutral glycolipid, globoside, first binds embryos at the 2–4-cell stage (Willison *et al.*, 1982) (see Section 2.1). Globoside is structurally identical to the Forssman glycolipid with the exception of its terminal sugar—globoside bears a terminal β-N-acetylgalactosamine, Forssman antigen has an additional α-N-acetylgalactosamine (see Table I). Again, it has been suggested that the enzyme responsible for the linkage of an α-N-acetylgalactosamine may not be synthesized until the morula stage (Willison *et al.*, 1982). An antibody to globotriaosylceramide, which contains one less terminal sugar than globoside (see Table I), binds embryos at only the 2-cell and 4-cell stages.

The monoclonal anti-SSEA-3, generated by immunization with 4–8-cell mouse embryos (Shevinsky *et al.*, 1982), has strongest reactivity with a previously unidentified globoseries glycosphingolipid from human teratocar-

Table I. Terminal Saccharides of Carbohydrate Antigens Detected on Embryos

Globotriaosylceramide	Gal(α1–4)Gal(β1–4)Glc . . .
Globoside	GalNAc(β1–3)Gal(α1–4)Gal(β1–4)Glc . . .
Forssman	GalNAc(α1–3)GalNAc(β1–3)Gal(α1–4)Gal(β1–4)Glc . . .
SSEA-3	Gal(β1–3)GalNAc(β1–3)Gal(α1–4)Gal(β1–4)Glc . . .
i antigen	Gal(β1–4)GlcNAc(β1–3)Gal(β1–4)GlcNAc . . .
I antigen	Gal(β1–4)GlcNAc(β1–6) →Gal(β1–4)GlcNAc . . . Gal(β1–4)GlcNAc(β1–3)
H antigen	Fuc(α1–2)Gal . . .
SSEA-1	Gal(β1–4)GlcNAc . . . ↑ Fuc(α1–3)

cinoma cells (Kannagi *et al.*, 1983). The proposed structure for this glycolipid, GL-5, is shown in Table I. It can be seen that this molecule also differs from globoside by one terminal sugar residue. It remains to be determined whether this glycolipid is responsible for binding of anti-SSEA-3 to oocytes and preimplantation embryos. There is evidence that the SSEA-3 antigenic determinant may be carried by at least one glycoprotein molecule having a molecular weight of 72,000 (Shevinsky *et al.*, 1982).

The concept that simple glycosylation changes might be the basis for stage-specific expression of antigens on preimplantation embryos was first introduced in relationship to studies of the structure of SSEA-1 (Gooi *et al.*, 1981). Anti-SSEA-1 was found to be crossreactive with human erythrocyte membrane glycolipids related to blood group H and Ii components (Nudelman *et al.*, 1980; Hakomori *et al.*, 1981). Unlike the H and Ii antigens, these erythrocyte glycolipids, which crossreact with SSEA-1, were found to share a common terminal trisaccharide, Gal(β1-4)[Fucα1-3]G1cNAc, as shown in Table I (Gooi *et al.*, 1981; Hakomori *et al.*, 1981). A subsequent study has revealed that trace components from group 0 human erythrocyte membranes with the same terminal trisaccharide react with anti-SSEA-1 and that antigenicity of the glycolipids increases with the length of the core carbohydrate structure (Kannagi *et al.*, 1982). Although it is probable that SSEA-1 expressed in the membrane of the preimplantation embryo is characterized by the terminal trisaccharide described above, the core structure of the embryonic antigen remains unknown. However, the presence of related antigens in erythrocyte membranes was a fortuitous finding in that it has allowed for important glycolipid characterizations. In fact, the terminal trisaccharide of SSEA-1 is specifically expressed on at least one tumor antigen (Hakomori *et al.*, 1981), suggesting an association with abnormal differentiation.

It has been proposed that the I or i blood group antigens are precursors to SSEA-1 as their terminal saccharides differ from SSEA-1 only by the single fucosylation of SSEA-1 (Gooi *et al.*, 1981). Human anti-I serum detects antigens on mouse oocytes and all subsequent preimplantation stages (Knowles *et al.*,

1982). Thus appearance of I reactivity on the embryo surface precedes by several days the expression of SSEA-1 that appears at the 8-cell stage (Solter and Knowles, 1978). Embryonal carcinoma cells (PSA1) express I but not i (Kapadia *et al.*, 1981). In postimplantation embryos as well as PSA1 embryoid bodies, primary endoderm expresses both I and i, while embryonic and extraembryonic ectoderm show a decrease in I expression and never contain i (Kapadia *et al.*, 1981). The fact that I and i antigens differ only in that I is branched (see Table I), suggests that stage-specific antigenic differences may also arise through a decrease in branching of a polysaccharide chain. This correlates well with the reported decrease in size of glycopeptides associated with differentiation of early embryonic cells (see above; Muramatsu *et al.*, 1978). Kapadia and co-workers (1981) also determined that epithelia differentiated from the embryonic ectoderm of PSA1 embryoid bodies have lost I antigen but express antigens reactive with antibodies to blood group H antigen. H antigens are oligosaccharides with terminal fucose residues on Ii antigen cores (see Table I), again suggesting, as in the study of globoside and Forssman antigens, that the addition of a single sugar residue can account for stage-specific differences in antigenicity.

Interestingly, two monoclonal antibodies, C6 and A5, with specificities nearly identical to those of I and i antibodies respectively, have been generated by immunization with the nullipotential EC cell line, Nulli-SCC1 (Fenderson *et al.*, 1983). However, C6 and A5 bind preimplantation embryos only after neuraminidase treatment of the embryos, suggesting that antigens are masked during the preimplantation period by terminal sialic acid substitutions. It is possible, therefore, that stage specific antigenic differences may also arise through the removal of terminal sugar residues.

Thus a new concept has arisen through studies with several antibodies in recent years—that stage-specific differences in cell surface antigens can arise through differences in branching or single differences in glycosylation of the oligosaccharide chains of glycolipids or glycoproteins These differences may result from sequential additions of single sugar residues either during synthesis or after insertion into the membrane and may reflect the earliest expression of the enzymes responsible for the linkage of the terminal sugars. The terminal sugars of stage-specific embryonic antigens may be involved with cellular interactions mediated by lectin molecules. The fact that lectinlike molecules have been identified on the surface of teratocarcinoma cells (Grabel *et al.*, 1979) supports this suggestion. In addition to continuing studies on the importance of glycosylation changes during the preimplantation period and possible interactions of the sugar residues with lectins, attention should also be focused on the core structures bearing the oligosaccharide chains.

It is obvious from the work described above that the biochemical analyses of stage-specific embryonic antigens have been quite fruitful in advancing our knowledge of cellular mechanisms active during the early development of the mammalian embryo. It is only through the availability of appropriate immunological reagents that these characterizations have been possible. This immunological approach should continue to provide vital information in the future.

7. Conclusions and Future Directions

We have seen that antibodies with their ability to discern small differences in antigenic determinants have served a unique role in advancing our knowledge of early mammalian development. Antibodies to known antigens (e.g., Forssman, globoside, IAP, H-Y, β_2-microglobulin) and previously unidentified antigens (e.g., BL2, SSEA-1, -2, -3) have enabled investigators to track the developmentally regulated appearance and disappearance of these molecules both at the cell surface and within the cytoplasm. In addition to information gained concerning embryonic synthesis of these antigens, in the case of β_2-microglobulin evidence was obtained for embryonic gene expression as early as the 2-cell stage. In many cases antigens have been shown to segregate to specific cells or tissues as development proceeds. These descriptive studies have led to the potential use of many of these antibodies as cell or tissue markers. Those markers that bind to the cell surface have a further technical application in cell purification or cell sorting. The antibodies can also be utilized for antigen quantitation.

In other sections of this chapter we have discussed the biological information gained through the use of antibodies to describe cell surface modulation of antigens and topographical distribution or localization of antigens. Although specific functions have not been identified for most of the antigens discussed here, exciting experiments from several laboratories have defined a calcium-dependent adhesion molecule involved in compaction of the 8-cell embryo.

Remarkable advances have been made in the biochemical characterizations of protein, glycoprotein, and glycolipid antigens detected by antibodies that bind to preimplantation embryos. Unfortunately, few of these studies have characterized the antigens as they are expressed on embryos themselves, but rather, due to lack of embryonic material, have relied on characterization of crossreactive antigens on other cell types. Caution must always be invoked when extrapolating results from other cell types to the embryo itself. In fact, Willison *et al.* (1982) have shown that the molecule carrying the Forssman determinant on embryos differs from the molecules carrying the same determinant on other cell types. The lack of quantity of embryonic material required for large scale biochemical analyses may be overcome in the future due to a recent advance in the field. Two laboratories (Evans and Kaufman, 1981; Martin, 1981) have succeeded in generating pluripotent cell lines from normal preimplantation mouse embryos. Cell lines generated from embryos with varied genetic backgrounds should facilitate rapid progress in biochemical characterizations of those antigens expressed by the ICM and its developmental derivatives.

Other rapid advances can be expected with the advent of technology for gene cloning. In this respect, Brulet and Jacob (1982) isolated a cDNA clone from a trophoblastoma cell line whose complementary mRNA directs synthesis of a trophoblast-specific intermediate filament protein. The protein appears to be that detected by the monoclonal antibody, TROMA-1 (Brulet *et al.*, 1980) and the polyclonal anti-Endo A (Oshima, 1981). *In situ* hybridization studies of the preimplantation embryo with cDNA probes such as this one should provide exciting results. An experimental approach combining the use of cDNA probes and embryo-specific antibodies for identification of the products of the genes thus defined should prove fruitful. This approach may aid in answering questions concerning maternal versus embryo genome control, gene amplification, and gene regulation during the early embryonic period.

Gene cloning should also allow for the synthesis of quantities of product sufficient to be used as the immunogen for generation of monoclonal or polyclonal antibodies. Alternatively, sequencing of embryonic genes should allow for the prediction of peptide sequences that, produced synthetically, could also serve as immunogens for appropriate immunologic reagents for the study of the preimplantation period.

The generation of antibodies to short synthetic peptides the sequence of which has been predicted by studies of a larger natural protein represents another advance in immunological technology (Sutcliffe *et al.*, 1983); it is hoped that it will be applied to the study of preimplantation embryonic development.

Although the advent of monoclonal antibodies has had perhaps the single greatest impact on immunological studies of the early embryo, the reader should be aware that certain limitations are inherent in their high degree of specificity. Because each antibody detects a single, usually small, epitope, they are not always ideal for studying all aspects of a large antigen. The phenomenon of "jumping antigens" (the detection of small epitopes on different core molecules) (Milstein and Lennox, 1980) has been mentioned above. This has significant implications in studies such as developmental tracking of a single antigen. In addition, monoclonal antibodies are often ineffective in immunoprecipitating antigens. For example, the intermediate filament protein, Endo A, can be precipitated by several polyclonal antibodies (Brulet *et al.*, 1980; Oshima, 1981) but apparently not by the monoclonal TROMA-1. A polyclonal antibody precipitates the calcium dependent adhesion molecule (Yoshida and Takeichi, 1982), while a McAb to the same antigen cannot (Yoshida-Noro *et al.*, 1984). Similar difficulties were encountered with the 11 embryo-specific monoclonals generated in our laboratory (Banka, 1982) (see Section 2). The monoclonal antibody to uvomorulin has not been reported to block compaction, while several polyclonals to the same molecule do (Damsky *et al.*, 1983; Peyrieras *et al.*, 1983). Finally, the ability of polyclonal antibodies to detect multiple regions on a single antigen confers important advantages for certain studies. For example, when preimplantation embryos are cultured for 24 hr in the presence of tunicamycin (which blocks glycosylation of glycoproteins) a molecule of 60,000 M_r (presumably the invarient protein core of the BL2 antigens) can still be precipitated by the A-BL2 polyclonal antibody (Johnson and Calarco, 1980*c*). The embryo-specific monoclonal antibodies, which have predominantly carbohydrate specificities, would no longer detect antigens following such alterations. Certainly the advantages of monospecificity and "immortalization" of monoclonal antibodies outweigh the disadvantages mentioned here, but the advantages of polyclonal antibodies that detect multiple epitopes on a specific antigen should be kept in mind. It is clear that the ideal immunologic approach to the study of preimplantation development will continue to be one employing monoclonal and polyclonal antibodies in combination. The addition of new molecular techniques such as gene cloning to the immunological approach should lead to rapid and exciting advances in our understanding of the controls of preimplantation mammalian development.

ACKNOWLEDGMENTS. The authors wish to thank Dr. L. Johnson for reviewing the manuscript and for helpful comments. We also thank Dr. L. Shevinsky for reading the manuscript.

References

Artzt, K., Dubois, P., Bennett, D., Condamine, H., Babinet, C., and Jacob, F., 1973, Surface antigens common to mouse cleavage stage embryos and primitive teratocarcinoma cells in culture, *Proc. Natl. Acad. Sci. USA* **70:**2988–2992.

Banka, C. L., 1982, Developmentally regulated antigens of early mouse embryos and teratocarcinomas, Doctoral dissertation, University of California, San Francisco.

Banka, C. L., and Calarco, P. G., 1986, A new embryonic antigen, p67, common to preimplantation embryos and pluripotent embryonic cells, *Gamete Res.* (in press).

Boucaut, J. C., Darribere, T., Boulekbache, H., and Thiery, J.-P., 1984, Prevention of gastrulation but not neurulation by antibodies to fibronectin in amphibian embryos, *Nature (Lond.)* **307:**364–367.

Brackenbury, R., Thiery, J.-P., Rutishauser, U., and Edelman, G. M., 1977, Adhesion among neural cells of the chick embryo. I. An immunological assay for molecules involved in cell–cell binding, *J. Biol. Chem.* **252:**6835–6840.

Braude, P. R., 1979, Time-dependent effects of α-amanitin on blastocyst formation in the mouse, *J. Embryol. Exp. Morphol.* **52:**193–202.

Brulet, P., and Jacob, F., 1982, Molecular cloning of a cDNA sequence encoding a trophectoderm-specific marker during mouse blastocyst formation, *Proc. Natl. Acad. Sci. USA* **79:**2328–2332.

Brulet, P., Babinet, C., Kemler, R., and Jacob, F., 1980, Monoclonal antibodies against trophectoderm-specific markers during mouse blastocyst formation, *Proc. Natl. Acad. Sci. USA* **77:**4113–4117.

Buc-Caron, M. H., Condamine, H., and Jacob, F., 1978, The presence of F9 antigen on the surface of mouse embryonic cells until day 8 of embryogenesis, *J. Embryol. Exp. Morph.* **47:**149–160.

Buskirk, D. R., Thiery, J. P., Rutishauser, U., and Edelman, G. M., 1980, Antibodies to a neural cell adhesion molecule disrupt histogenesis in cultured chick retinae, *Nature (Lond.)* **285:**488–490.

Calarco, P. G., and Banka, C. L., 1979, Cell surface antigens of preimplantation mouse embryos detected by an antiserum to an embryonal carcinoma cell line, *Biol. Reprod.* **20:**699–704.

Calarco, P. G., and Brown, E., 1969, An ultrastructural and cytological study of preimplantation development of the mouse, *J. Exp. Zool.* **171:**253–283.

Calarco, P. G., and Epstein, C. J., 1973, Cell surface changes during preimplantation development in the mouse, *Dev. Biol.* **32:**208–213.

Calarco, P. G., and Szollosi, D., 1973, Intracisternal A particles in ova and preimplantation stages of the mouse, *Nature New Biol.* **243:**91–93.

Ciment, G., and Weston, J. A., 1983, Enteric neurogenesis by neural crest-derived branchial arch mesenchymal cells, *Nature (Lond.)* **305:**424–427.

Clegg, K. B., and Piko, L., 1982, RNA synthesis and cytoplasmic polyadenylation in the one-cell mouse embryo, *Nature (Lond.)* **295:**342–345.

Coggin, J. H., and Anderson, N. G., 1974, Cancer, differentiation and embryonic antigens. Some central problems, *Adv. Cancer Res.* **19:**106–166.

Cohen, J., Mirsky, R., Rattray, S., Selvendren, S., and Vulliamy, T., 1981, Monoclonal antibodies which define neuron-specific cell surface molecules in the mammalian central and peripheral nervous system, in: *Monoclonal Antibodies Against Neural Antigens* (R. McKay, M. Raff, and L. Reichardt, eds.), pp. 61–68, Cold Spring Harbor Press, New York.

Damsky, C. H., Richa, J., Solter, D., Knudsen, K., and Buck, C. A., 1983, Identification and purification of a cell surface glycoprotein mediating intercellular adhesion in embryonic and adult tissue, *Cell* **34:**455–466.

Davidson, E. H., 1976, *Gene Activity in Early Development*, Academic Press, New York.

De Petris, S., and Raff, M., 1973, Normal distribution, patching and capping of lymphocyte surface immunoglobulin studied by electronmicroscopy, *Nature New Biol.* **241:**257–259.

Ducibella, T., 1980, Divalent antibodies to mouse embryonal carcinoma cells inhibit compaction in the mouse embryo, *Dev. Biol.* **79:**356–366.

Ducibella, T., Ukena, T., Karnowski, M., and Anderson, E., 1977, Changes in cell surface and cortical cytoplasmic organization during early embryogenesis in the preimplantation mouse embryo, *J. Cell Biol.* **74:**153–167.

Dunbar, B. S., 1983, Morphological, biochemical, and immunochemical characterization of the mammalian zona pellucida, in: *Mechanism and Control of Animal Fertilization* (J. F. Hartmann, ed.), pp. 139–155, Academic Press, New York.

Dunia, I., Nicholas, J. F., Jakob, H., Benedetti, E. L., and Jacob, F., 1979, Junctional modulation in mouse embryonal carcinoma cells by Fab fragments of rabbit anti-embryonal carcinoma cell serum, *Proc. Natl. Acad. Sci. USA* **76:**3387–3391.

Epstein, C. J., Smith, S., and Travis, B., 1980, Expression of H-Y antigen on preimplantation mouse embryos, *Tissue Antigens* **15:**63–67.

Evans, M. J., and Kaufman, M. H., 1981, Establishment in culture of pluripotential cells from mouse embryos, *Nature (Lond.)* **292:**154–156.

Evered, D., and Whelan, J., 1983, *Fetal Antigens and Cancer*, CIBA Foundation Symposium 96, Pitman, London.

Fenderson, B. A., Hahnel, A. C., and Eddy, E. M., 1983, Immunohistochemical localization of two monoclonal antibody-defined carbohydrate antigens during early murine embryogenesis, *Dev. Biol.* **100:**318–327.

Flach, G., Johnson, M. H., Braude, P. R., Taylor, R. A. S., and Bolton, V. N., 1982, The transition from maternal to embryonic control in the 2-cell mouse embryo, *EMBO J.* **1:**681–686.

Fox, N., Damjanov, I., Martinez-Hernandez, A., Knowles, B. B., and Solter, D., 1981, Immunohistochemical localization of the early embryonic antigen (SSEA-1) in postimplantation mouse embryos and fetal and adult tissues, *Dev. Biol.* **83:**391–398.

Gachelin, G., Kemler, R., Kelly, F., and Jacob, F., 1977, PCC4, a new cell surface antigen common to multipotential embryonal carcinoma cells, spermatozoa, and mouse early embryos, *Dev. Biol.* **57:**199–209.

Gavin, J. R., III, Roth, J., Neville, D. M., Jr., DeMeyts, P., and Buell, D. N., 1974, Insulin-dependent regulation of insulin receptor concentrations: A direct demonstration in cell culture, *Proc. Natl. Acad. Sci. USA* **71:**84–88.

Golbus, M. S., Calarco, P. G., and Epstein, C. J., 1973, The effects of RNA synthesis (α-amanitin and actinomycin D) on preimplantation mouse embryogenesis, *J. Exp. Zool.* **186:**207–216.

Gooi, H. C., Feizi, T., Kapadia, A., Knowles, B. B., Solter, D., and Evans, M. J., 1981, Stage specific embryonic antigen involves 1–3 fucosylated type 2 blood group chains, *Nature (Lond.)* **292:**156–158.

Grabel, L. B., Rosen, S. D., and Martin, G. R., 1979, Teratocarcinoma stem cells have a cell surface carbohydrate binding component implicated in cell–cell adhesion, *Cell* **17:**477–484.

Grover, A., Oshima, R. G., and Adamson, E. D., 1983, Epithelial layer formation in differentiating aggregates of F9 embryonal carcinoma cells, *J. Cell Biol.* **96:**1690–1696.

Hahnel, A. C., and Eddy, E. M., 1983, Syngeneic antiserum to Nulli SCC1 embryonal carcinoma cells recognizing surface antigens of embryonic cells, *J. Reprod. Immunol.* **5:**371–382.

Hakomori, S., Nudelman, E., Levery, S., Solter, D., and Knowles, B. B., 1981, The hapten structure of a developmentally regulated glycolipid antigen (SSEA-1) isolated from human erythrocytes and adenocarcinoma: A preliminary note, *Biochem. Biophys. Res. Commun.* **100:**1578–1586.

Handyside, A. H., 1980, Distribution of antibody- and lectin-binding sites on dissociated blastomeres from mouse morulae: Evidence for polarization at compaction, *J. Embryol. Exp. Morphol.* **60:**99–116.

Harper, M. I., and Monk, M., 1983, Evidence for translation of HPRT enzyme on maternal mRNA in early embryos, *J. Embryol. Exp. Morphol.* **74:**15–28.

Hillman, N., and Tasca, R., 1969, Ultrastructural and autoradiographic studies of mouse cleavage stages, *Am. J. Anat.* **126:** 151–173.

Howe, C. C., and Solter, D., 1981, Changes in cell surface proteins during differentiation of mouse embryonal carcinoma cells, *Dev. Biol.* **84:**239–243.

Huang, T. T., and Calarco, P. G., 1981*a*, Evidence for the cell surface expression of intracisternal A particle-associated antigens during early mouse development, *Dev. Biol.* **82:**388–392.

Huang, T. T. F., Jr., and Calarco, P. G., 1981*b*, Immunoprecipitation of intracisternal A particle-associated antigens from preimplantation mouse embryos, *J. Natl. Cancer Inst.* **67:**1129–1134.

Hyafil, F., Morello, D., Babinet, C., and Jacob, F., 1980, A cell surface glycoprotein involved in the compaction of embryonal carcinoma cells and cleavage stage embryos, *Cell* **21:**927–934.

Hyafil, F., Babinet, C., and Jacob, F., 1981, Cell–cell interactions in early embryogenesis: A molecular approach to the role of calcium, *Cell* **26:**447–454.

Jackson, B. W., Grund, C., Schmid, E., Burki, K., Franke, W., and Illmensee, K., 1980, Formation of cytoskeletal elements during mouse embryogenesis, *Differentiation* **17:**161–179.

Jacob, F., 1977, Mouse teratocarcinoma and embryonic antigens, *Immunol. Rev.* **33:**3–32.

Johnson, L. V., and Calarco, P. G., 1980*a*, Electrophoretic analysis of cell surface proteins of preimplantation mouse embryos, *Dev. Biol.* **77:**224–227.

Johnson, L. V., and Calarco, P. G., 1980*b*, Mammalian preimplantation development. The cell surface, *Anat. Rec.* **196:**201–219.

Johnson, L. V., and Calarco, P. G., 1980*c*, Immunologic characterization of embryonic cell surface antigens recognized by antiblastocyst serum, *Dev. Biol.* **79:**208–223.

Johnson, L. V., and Calarco, P. G., 1980*d*, Stage-specific embryonic antigens detected by an antiserum against mouse blastocysts, *Dev. Biol.* **79:**224–231.

Johnson, M. H., 1979, Extrinsic and intrinsic factors in early mammalian development, *J. Reprod. Fertil.* **55:**255–265.

Johnson, M. H., and Edidin, M., 1978, Lateral diffusion in plasma membrane of mouse egg is restricted after fertilization, *Nature (Lond.)* **272:**448–450.

Johnson, M. H., Chakraborty, J., Handyside, A. H., Willison, K., and Stern, P., 1979, The effect of prolonged decompaction on the development of the preimplantation mouse embryo, *J. Embryol. Exp. Morphol.* **54:**241–261.

Johnson, M. R., Pratt, H. P. M., and Handyside, A. H., 1981, The generation and recognition of positional information in the preimplantation mouse embryo, in: *Cellular and Molecular Aspects of Implantation* (S. R. Glasser and D. W. Bullock, eds.), pp. 55–74, Plenum Press, New York.

Kannagi, R., Nudelman, E., Levery, S. B., and Hakomori, S., 1982, A series of human erythrocyte glycosphingolipids reacting to the monoclonal antibody directed to a developmentally regulated antigen, SSEA-1, *J. Biol. Chem.* **257:**14865–14874.

Kannagi, R., Levery, S. B., Ishigami, F., Hakomori, S., Shevinsky, L. H., Knowles, B. B., and Solter, D., 1983, New globoseries glycosphingolipids in human teratocarcinoma reactive with the monoclonal antibody directed to a developmentally regulated antigen, stage-specific embryonic antigen 3, *J. Biol. Chem.* **258:**8934–8942.

Kapadia, A., Feizi, T., and Evans, M. J., 1981, Changes in the expression and polarization of blood group I and i antigens in post-implantation embryos and teratocarcinomas of mouse associated with cell differentiation, *Exp. Cell Res.* **131:**185–195.

Kartha, S., Felix, J. S., and Littlefield, J. W., 1983, Calcium-induced compaction and its inhibition in embryonal carcinoma cell aggregates, *Dev. Biol.* **100:**172–180.

Kemler, R., Babinet, C., Eisen, H., and Jacob, F., 1977, Surface antigen in early differentiation, *Proc. Natl. Acad. Sci. USA* **74**:4449–4452.
Knowles, B. B., Rappaport, J., and Solter, D., 1982, Murine embryonic antigen (SSEA-1) is expressed on human cells and structurally related human blood group antigen I is expressed on mouse embryos, *Dev. Biol.* **93**:54–58.
Kohler, G., and Milstein, C., 1975, Continuous cultures of fused cells secreting antibody of predefined specificity, *Nature (Lond.)* **256**:495–497.
Koo, G. C., Wachtel, S. S., Krupen-Brown, K., Mittl, L. R., Breg, W. R., Genel, M., Rosenthal, I. M., Borgaonkar, D. S., Miller, D. A., Tantravahi, R., Schreck, R. R., Erlanger, B. F., and Miller, O. J., 1977, Mapping the locus of the H-Y gene on the human Y-chromosome, *Science* **198**:940–942.
Krco, C. J., and Goldberg, E. H., 1976, H-Y (male) antigen: Detection on eight-cell mouse embryos, *Science* **193**:1134–1135.
Larraga, V., and Edidin, M., 1979, Immunochemical characterization of surface antigens of TerC, a teratocarcinoma-derived cell line, *Proc. Natl. Acad. Sci. USA* **76**:2912–2916.
Lehtonen, E., Lehto, V-P., Vartio, T., Badley, R. A., and Virtanen, I., 1983, Expression of cytokeratin polypeptides in mouse oocytes and preimplantation embryos, *Dev. Biol.* **100**:158–165.
Leivo, I., Vaheri, A., Timpl, R., and Wartiovaara, J., 1980, Appearance and distribution of collagens and laminin in the early mouse embryo, *Dev. Biol.* **76**:100–114.
Levinson, J., Goodfellow, P., Vadeboncoeur, M., and McDevitt, H., 1978, Identification of stage specific polypeptide synthesis during murine preimplantation development, *Proc. Natl. Acad. Sci. USA* **75**:3332–3336.
Lindner, J., Rathjen, F. G., and Schachner, M., 1983, L1 mono- and polyclonal antibodies modify cell migration in early postnatal mouse cerebellum, *Nature (Lond.)* **305**:427–429.
Lopo, A., and Calarco, P., 1982, Stage-specific changes in protein phosphorylation during preimplantation development in the mouse, *Gamete Res.* **5**:283–290.
Marticorena, P., Hogan, B., DiMeo, A., Artzt, K., and Bennett, D., 1983, Carbohydrate changes in pre- and peri-implantation mouse embryos as detected by a monoclonal antibody, *Cell Diff.* **12**:1–10.
Martin, G. R., 1980, Teratocarcinomas and mammalian development, *Science* **209**:768–776.
Martin, G. R., 1981, Isolation of a pluripotent cell line from early mouse embryos cultured in medium conditioned by teratocarcinoma stem cells, *Proc. Natl. Acad. Sci. USA* **78**:7634–7638.
Martin, G. R., Smith, S., and Epstein, C. J., 1978, Protein synthetic patterns in teratocarcinoma stem cells and mouse embryos at early stages of development, *Dev. Biol.* **66**:8–16.
McGrath, J., and Solter, D., 1983, Nuclear transplantation in mouse embryos, *J. Exp. Zool.* **228**:355–362.
Milstein, C., and Lennox, E., 1980, The use of monoclonal antibody techniques in the study of developing cell surfaces, *Curr. Top. Dev. Biol.* **14**:1–32.
Muramatsu, T., Gachelin, G., Nicolas, J. F., Condamine, H., Jakob, H., and Jacob, F., 1978, Carbohydrate structure and cell differentiation: Unique properties of fucosyl-glycoproteins isolated from embryonal carcinoma cells, *Proc. Natl. Acad. Sci. USA* **75**:2315–2319.
Muramatsu, T., Gachelin, G., Damonneville, M., Delarbre, C., and Jacob, F., 1979, Cell surface carbohydrates of embryonal carcinoma cells: Polysaccharidic side chains of F9 antigens and of receptors to two lectins, FBP and PNA, *Cell* **18**:183–191.
Nudelman, E., Hakomori, S., Knowles, B. B., Solter, D., Nowinski, R., Tam, M. R., and Young, W. W., Jr., 1980, Monoclonal antibody directed to the stage-specific embryonic antigen (SSEA-1) reacts with a branched glycosphingolipid similar in structure to Ii antigen, *Biochem. Biophys. Res. Commun.* **97**:443–451.
O'Brien, D. A., and Millette, C. F., 1984, Identification and immunochemical characterization of spermatogenic cell surface antigens that appear during early meiotic prophase, *Dev. Biol.* **101**:307–317.
Ogou, S.-I., Okada, T. S., and Takeichi, M., 1982, Cleavage stage mouse embryos share a common cell adhesion system with teratocarcinoma cells, *Dev. Biol.* **92**:521–528.
Oshima, R. G., 1981, Identification and immunoprecipitation of cytoskeletal proteins from murine extra-embryonic endodermal cells, *J. Biol. Chem.* **256**:8124–8133.

Oshima, R. G., 1982, Developmental expression of murine extra-embryonic endodermal cytoskeletal proteins, *J. Biol. Chem.* **257**:3414–3421.

Oshima, R. G., Howe, W. E., Klier, F. G., Adamson, E. D., and Shevinsky, L. H., 1983, Intermediate filament protein synthesis in preimplantation murine embryos, *Dev. Biol.* **99**:447–455.

Patterson, B. M., Segal, S., Lueders, K., and Kuff, E. L., 1978, RNA associated with murine intracisternal type A particles codes for the main particle protein, *J. Virol.* **27**:118–126.

Peyrieras, N., Hyafil, F., Louvard, D., Ploegh, H. L., and Jacob, F., 1983, Uvomorulin: A nonintegral membrane protein of early mouse embryo, *Proc. Natl. Acad. Sci. USA* **80**:6274–6277.

Piko, L., and Clegg, K. B., 1982, Quantitative changes in total RNA, total Poly (A), and ribosomes in early mouse embryos, *Dev. Biol.* **89**:362–378.

Polak-Charcon, S., Johnson, L. V., and Calarco, P. G., 1983*a*, Cortical synthesis of a stage-specific surface glycoprotein during early mammalian development, *Proc. 41st E.M. Soc. Am.* 508–509.

Polak-Charcon, S., Johnson, L. V., and Calarco, P. G., 1983*b*, Intracellular localization and surface expression of stage-specific glycoprotein antigens during early mouse development, *J. Cell Biol.* **97**:33a.

Polak-Charcon, S., Johnson, L. V., and Calarco, P. G., 1985, Intracellular localization and surface expression of a stage-specific embryonic glycoprotein, *Gamete Res.* (in press).

Reeve, W. J. D., and Ziomek, C. A., 1981, Distribution of microvilli on dissociated blastomeres from mouse embryos: evidence for surface polarization at compaction, *J. Embryol. Exp. Morphol.* **62**:339–350.

Richa, J., Damsky, C. H., Buck, C. A., Knowles, B. B., and Solter, D., 1985, Cell surface glycoproteins mediate compaction, trophoblast attachment, and endoderm formation during early mouse development, *Dev. Biol.* **108**:513–521.

Rosen, S. D., Haywood, P. L., and Barondes, S. H., 1976, Inhibition of intercellular adhesion in a cellular slime mould by univalent antibody against a cell-surface lectin, *Nature (Lond.)* **263**:425–427.

Sadoul, R., Hirn, M., Deagostini-Bazin, H., Rougon, G., and Goridis, C., 1983, Adult and embryonic mouse neural cell adhesion molecules have different binding properties, *Nature (Lond.)* **304**:347–350.

Sawicki, J. A., Magnuson, T., and Epstein, C. J., 1981, Evidence for expression of the paternal genome in the two-cell mouse embryo, *Nature (Lond.)* **294**:450–451.

Schachner, M., 1979, Cell surface antigens of the nervous system, *Curr. Top. Dev. Biol.* **13**:259–279.

Schachner, M., Kim, S. K., and Zehnle, R., 1981, Developmental expression in central and peripheral nervous system of oligodendrocyte cell surface antigens (O antigens) recognized by monoclonal antibodies, *Dev. Biol.* **83**:328–338.

Searle, R. F., and Jenkinson, E. J., 1978, Localization of trophoblast-defined surface antigens during early mouse embryogenesis, *J. Embryol. Exp. Morphol.* **43**:147–156.

Sherman, M. I., 1979, Developmental biochemistry of preimplantation mammalian embryos, *Ann. Rev. Biochem.* **48**:443–470.

Shevinsky, L. H., Knowles, B. B., Howe, C. H., Aden, D. P., and Solter, D., 1981, A murine stage-specific embryonic antigen (SSEA-2) is expressed on some murine SV40-transformed cells, *J. Immunol.* **217**:632–636.

Shevinsky, L. H., Knowles, B. B., Damjanov, I., and Solter, D., 1982, Monoclonal antibody to murine embryos defines a stage-specific embryonic antigen expressed on mouse embryos and human teratocarcinoma cells, *Cell* **30**:697–705.

Sobel, J. S., 1983*a*, Localization of myosin in the preimplantation mouse embryo, *Dev. Biol.* **95**:227–231.

Sobel, J. S., 1983*b*, Cell–cell contact modulation of myosin organization in the early mouse embryo, *Dev. Biol.* **100**:207–218.

Solter, D., and Knowles, B. B., 1975, Immunosurgery of mouse blastocyst, *Proc. Natl. Acad. Sci. USA* **72**:5099–5102.

Solter, D., and Knowles, B. B., 1978, Monoclonal antibody defining a stage-specific mouse embryonic antigen (SSEA-1), *Proc. Natl. Acad. Sci. USA* **75**:5565–5569.

Solter, D., and Knowles, B. B., 1979, Developmental stage-specific antigens during mouse embryogenesis, *Curr. Top. Dev. Biol.* **13**:139–165.

Stern, P. L., Martin, G. R., and Evans, M. J., 1975, Cell surface antigens of clonal teratocarcinoma cells at various stages of differentiation, *Cell* **6**:455–465.

Stern, P. L., Willison, K., Lennox, E., Galfre, G., Milstein, C., Secher, D., and Ziegler, A., 1978 Monoclonal antibodies as probes for differentiation and tumor associated antigens: A Forssman specificity on teratocarcinoma cells, *Cell* **14**:775–784.

Stinnakre, M., Evans, M. J., Willison, K. R., and Stern, P. L., 1981, Expression of Forssman antigen in the post-implantation mouse embryo, *J. Embryol. Exp. Morphol.* **61**:117–131.

Sutcliffe, J. G., Shinnick, T. M., Green, N., and Lerner, R. A., 1983, Antibodies that react with predetermined sites on proteins, *Science* **219**:660–666.

Thiery, J.-P., Brackenbury, R., Rutishauser, U., and Edelman, G. M., 1977, Adhesion among neural cells of the chick embryo. II. Purification and characterization of a cell adhesion molecule from neural retina, *J. Biol. Chem.* **252**:6841–6845.

Van Blerkom, J., 1981, Intrinsic and extrinsic patterns of molecular differentiation during oogenesis, embryogenesis, and organogenesis in mammals, in: *Cellular and Molecular Aspects of Implantation* (S. R. Glasser and D. W. Bullock, eds.), pp. 155–178, Plenum Press, New York.

Van Blerkom, J., and Brockway, G., 1975, Qualitative patterns of protein synthesis in preimplantation mouse embryos. I. Normal pregnancy, *Dev. Biol.* **44**:148–157.

Vernon, R. B., Muller, C. H., Herr, J. C., Feuchter, F. A., and Eddy, E. M., 1982, Epididymal secretion of a mouse sperm surface component recognized by a monoclonal antibody, *Biol. Reprod.* **26**:523–535.

Webb, C. G., 1980, Characterization of antisera against mouse teratocarcinoma OTT6050: Molecular species recognized on embryoid bodies, preimplantation embryos, and sperm, *Dev. Biol.* **76**:203–214.

Webb, C. G., Gall, W. E., and Edelman, G. M., 1977, Synthesis and distribution of H-2 antigens in preimplantation mouse embryos, *J. Exp. Med.* **146**:923–932.

Wiley, L. M., 1979, Early embryonic cell surface antigens as developmental probes, Curr. Top. *Dev. Biol.* **13**:167–197.

Wiley, L. M., 1980, Early mouse embryonic cell surface antigens that are detectable by antisera to human chorionic gonadotropin (hCG), *Exp. Cell Res.* **129**:47–54.

Wiley, L. M., and Calarco, P. G., 1975, The effects of anti-embryo sera and their localization on the cell surface during mouse preimplantation development, *Dev. Biol.* **47**:407–418.

Wiley, L. M., and Eglitis, M. A., 1981, Cell surface and cytoskeletal elements: Cavitation in the mouse preimplantation embryo, *Dev. Biol.* **86**:493–501.

Willison, K. R., Karol, R. A., Suzuki, A., Kundu, S. K., and Marcus, D. M., 1982, Neutral glycolipid antigens as developmental markers of mouse teratocarcinoma and early embryos: An immunologic and chemical analysis, *J. Immunol.* **129**:603–689.

Wolf, D. E., Edidin, M., and Handyside, A. H., 1981, Changes in the organization of the mouse egg plasma membrane upon fertilization and first cleavage: Indications from the lateral diffusion rates of fluorescent lipid analogs, *Dev. Biol.* **15**:195–198.

Wu, T.-C., Wan, Y.-J., Chung, A. E., and Damjanov, I., 1983, Immunohistochemical localization of entactin and laminin in mouse embryos and fetuses, *Dev. Biol.* **100**:496–505.

Yoshida, C., and Takeichi, M., 1982, Teratocarcinoma cell adhesion Identification of a cell-surface protein involved in calcium-dependent cell aggregation, *Cell* **28**:217–224.

Yoshida-Noro, C., Suzuki, N., and Takeichi, M., 1984, Molecular nature of the calcium-dependent cell–cell adhesion system in mouse teratocarcinoma and embryonic cells studied with a monoclonal antibody, *Dev. Biol.* **101**:19–27.

Index

GPSR Compliance

The European Union's (EU) General Product Safety Regulation (GPSR) is a set of rules that requires consumer products to be safe and our obligations to ensure this.

If you have any concerns about our products, you can contact us on ProductSafety@springernature.com

In case Publisher is established outside the EU, the EU authorized representative is:

Springer Nature Customer Service Center GmbH
Europaplatz 3
69115 Heidelberg, Germany

Batch number: 10692442

Printed by Printforce, the Netherlands